ECO-FRIENDLY WATERBORNE POLYURETHANES

ECO-FRIENDLY WATERBORNE POLYURETHANES

Synthesis, Properties, and Applications

Edited by

Ram K. Gupta and Ajay Kumar Mishra

CRC Press
Taylor & Francis Group
Boca Raton London New York

CRC Press is an imprint of the
Taylor & Francis Group, an **informa** business

First edition published 2022
by CRC Press
6000 Broken Sound Parkway NW, Suite 300, Boca Raton, FL 33487-2742

and by CRC Press
2 Park Square, Milton Park, Abingdon, Oxon, OX14 4RN

CRC Press is an imprint of Taylor & Francis Group, LLC

Library of Congress Cataloging-in-Publication Data
Names: Gupta, Ram K., editor. | Mishra, Ajay Kumar, 1965- editor.
Title: Eco-friendly waterborne polyurethanes : synthesis, properties, and applications / edited by Ram K. Gupta and Ajay Kumar Mishra.
Description: First edition. | Boca Raton, FL : CRC Press, 2022. | Includes bibliographical references and index. | Summary: "Waterborne polyurethanes (WPUs) exhibit many advantages over conventional volatile organic compounds (VOCs) based polyurethanes and have emerged as an environmentally friendly alternative. WPUs offer an opportunity to use sustainable raw materials to produce environmentally sustainable polymers, particularly, polyols derived from vegetable oils. This book provides state-of-the-art knowledge of the synthesis, application, and property enhancement of WPUs. Written for polymer chemists, materials scientists, and other researchers and industry, this book serves as a comprehensive reference for readers interested in the development and application of sustainable polymers"-- Provided by publisher.
Identifiers: LCCN 2021047626 (print) | LCCN 2021047627 (ebook) | ISBN 9781032002866 (hbk) | ISBN 9781032002873 (pbk) | ISBN 9781003173526 (ebk)
Subjects: LCSH: Polyurethanes. | Polymerization–Environmental aspects. | Water-soluble polymers. | Green chemistry.
Classification: LCC TP1180.P8 E35 2022 (print) | LCC TP1180.P8 (ebook) | DDC 668.4/239–dc23/eng/20211122
LC record available at https://lccn.loc.gov/2021047626
LC ebook record available at https://lccn.loc.gov/2021047627

ISBN: 978-1-032-00286-6 (hbk)
ISBN: 978-1-032-00287-3 (pbk)
ISBN: 978-1-003-17352-6 (ebk)

DOI: 10.1201/9781003173526

Typeset in Times
by MPS Limited, Dehradun

Contents

Preface ix
Editors' Biographies xi

Chapter 1 Introduction to Waterborne Polyurethanes 1
Felipe M. de Souza, Prashant Kote, and Ram K. Gupta

Chapter 2 Chemistry and Technology of Waterborne Polyurethanes 17
Naresh A. Rajpurohit, Kaushalya Bhakar, Dinesh Kumar, and Meena Nemiwal

Chapter 3 Green Materials for Waterborne Polyurethanes 31
Felipe M. de Souza, Prashant Kote, and Ram K. Gupta

Chapter 4 Eco-Friendly Synthesis for Waterborne Polyurethanes 47
Pavan M. Paraskar, Vinod M. Hatkar, and Ravindra D. Kulkarni

Chapter 5 Synthesis, Properties, and Applications of Waterborne Polyurethanes 65
Sanam Amiri, Hossein Nazokdast, Gity Mir Mohamad Sadeghi, Shervin Ahmadi, and Sahar Amiri

Chapter 6 Nanocomposites of Waterborne Polyurethanes 83
Samiran Morang and Niranjan Karak

Chapter 7 Waterborne Polyurethanes for Flexible and Rigid Foams 101
Hilal Olcay and E. Dilara Kocak

Chapter 8 Flame-Retardant Waterborne Polyurethanes 119
Giulio Malucelli

Chapter 9 Synthesis, Characterization, and Applications of Smart Waterborne Polyurethanes 137
Ronglan Wu and Wei Wang

Chapter 10 Shape Memory Waterborne Polyurethanes 157

Arunima Reghunadhan, Sabu Thomas, and Jiji Abraham

Chapter 11 Waterborne Polyurethanes for Self-Healing Applications 177

Masoumeh Kianfar, Shiva Mohajer, and Mir Saeed Seyed Dorraji

Chapter 12 Waterborne Polyurethanes for Biomedical Applications 193

Abbas Mohammadi, Mahtab Eslamieh, Negar Salehi, and Saman Abrishamkar

Chapter 13 Waterborne Polyurethanes for Tissue Engineering 213

D.E. Mouzakis and Styliani Papatzani

Chapter 14 Waterborne Polyurethanes for Biodegradable Coatings 231

Sukanya Pradhan, Smita Mohanty, and Sanjay Kumar Nayak

Chapter 15 Recent Developments in Waterborne Polyurethanes for Coating Applications 253

Verónica L. Mucci, Mirta I. Aranguren, Javier I. Amalvy, and María E. V. Hormaiztegui

Chapter 16 Waterborne Polyurethanes for Weather Protective Coatings......... 267

Sonalee Das

Chapter 17 Recent Developments in Waterborne Polyurethanes for Corrosion Protection.. 287

Felipe M. de Souza, Muhammad Rizwan Sulaiman, and Ram K. Gupta

Chapter 18 Waterborne Polyurethane for Electrically Conductive Coating.. 305

Sheraz Iqbal, Tauqir A. Sherazi, Tahir Rasheed, and Muhammad Bilal

Chapter 19 Waterborne Polyurethanes for Electrical Applications 321

V. Dhinakaran, P.M. Bupathi Ram, S. Narain Kumar, M. Tharun Kumar, K.P. Manoj Kumar, and M. Varsha Shree

Chapter 20 Waterborne Polyurethanes for Sensors.......................................333

Charles Oluwaseun Adetunji, Abel Inobeme, Kshitij RB Singh, Olugbemi T. Olaniyan, John Tsado Mathew, Jay Singh, Vanya Nayak, and Ravindra Pratap Singh

Chapter 21 Waterborne Polyurethanes for Sealants.......................................355

Mehrdad Fallah, Amir Ershad Langroudi, and Aida Alavi

Chapter 22 Waterborne Polyurethanes for Packing Industries.......................375

Saima Zulfiqar, Rida Badar, and Muhammad Yar

Chapter 23 Waterborne Polyurethanes for Automobile Industries..................393

Ayesha Kausar

Chapter 24 Non-Isocyanate-Based Waterborne Polyurethanes........................407

Marcin Włoch and Iga Carayon

Chapter 25 Waterborne Polyurethanes: Challenges and Future Outlook.........433

Felipe M. de Souza and Ram K. Gupta

Index...441

Preface

Polyurethane is one of the fastest-growing polymer industries. Polyurethanes are heavily used in consumer as well as industrial products. Based on recent advances in polymer science, polyurethanes find advanced applications in a range of areas, such as automobiles, adhesives, coatings, flame retardants, smart polymers, and biomedical products. Waterborne polyurethanes have exhibited many advantages over conventional volatile organic compound-based polyurethanes and have emerged as one of the environmentally friendly alternatives. With the growing environmental awareness, waterborne polyurethanes offer an opportunity to use sustainable raw materials to produce environmentally sustainable polymers, particularly polyols derived from vegetable oils. Waterborne polyurethanes' manufacturing process is known to be similar to the existing manufacturing process used for conventional polyurethanes; hence, there can be a seamless transfer of the technology. Waterborne polyurethanes are derived from the same reaction used to synthesize polyurethanes; however, eco-friendly solvents are used instead of volatile organic compounds.

The main purpose of this book is to provide current state-of-the-art knowledge on the chemistry involved in the synthesis of waterborne polyurethanes to attract a wider audience. Also, this book will provide a new guideline for the preparation of green polyurethanes for the industry. This book covers the current status and development in the field of waterborne polyurethanes. A wide range of synthetic approaches used to prepare waterborne polyurethanes from biosources, including green internal emulsifiers, neutralizers, and isocyanate-free routes are covered. The approaches discussed in this book for the synthesis and applications of waterborne polyurethanes not only provide eco-friendly and green polymers but also enable the use of bio-renewable materials for a sustainable future. This book provides fundamentals as well as advanced concepts to readers for developing new ideas and approaches to obtain eco-friendly polyurethanes. Many advanced applications of waterborne polyurethanes, such as coatings, corrosion inhibition, adhesives, sealants, antibacterial, self-healing polyurethanes, etc., are covered.

Editors' Biographies

Dr. Ram Gupta is an Associate Professor at Pittsburg State University. Dr. Gupta's research focuses on conducting polymers and composites, green energy production and storage using nanomaterials, optoelectronics and photovoltaics devices, organic-inorganic hetero-junctions for sensors, bio-based polymers, flame-retardant polymers, bio-compatible nanofibers for tissue regeneration, scaffold and antibacterial applications, corrosion-inhibiting coatings, and biodegradable metallic implants. Dr. Gupta has published over 230 peer-reviewed articles; made over 300 national, international, and regional presentations; chaired many sessions at national/international meetings; edited many books; and written several book chapters. He has received over two and a half million dollars for research and educational activities from many funding agencies. He is serving as Editor-in-Chief, Associate Editor, and editorial board member of numerous journals.

Ajay Kumar Mishra (MSc, MPhil, Ph.D., CSci, FRSC) is a Director at Academy of Nanotechnology and Waste Water Innovations. He worked as Professor at University of South Africa until December 2020. He also holds an adjunct professorship at Jiangsu University, China, and also is Visiting Professor at Bashkir State University, Russia, and Robert Gordon University, UK. His research interests include composites, nanocomposites, photocatalysis, multifunctional nanomaterials, polymers, carbon-based nanomaterials, and wastewater research. Prof. Mishra has authored and edited numerous peer-reviewed scientific international journal articles and books in this subject area.

1 Introduction to Waterborne Polyurethanes

Felipe M. de Souza, Prashant Kote, and Ram K. Gupta
Department of Chemistry, Kansas Polymer Research Center, Pittsburg State University, Pittsburg, KS, USA

CONTENTS

1.1 Introduction......1
1.2 Synthesis of Waterborne Polyurethanes......3
  1.2.1 Prepolymer Method......3
  1.2.2 Acetone Process......4
  1.2.3 Melt-Dispersion Process......5
  1.2.4 Ketamine–Ketazine......5
1.3 Classification of Waterborne Polyurethanes......6
  1.3.1 Cationic Waterborne Polyurethanes......6
  1.3.2 Anionic Waterborne Polyurethanes......8
  1.3.3 Zwitterionomers......9
1.4 Applications of Waterborne Polyurethanes......10
1.5 Conclusion......14
References......15

1.1 INTRODUCTION

Eco-friendly materials that can be used in commercial applications are very attractive for a sustainable future compared to fossil-based chemicals. Polyurethanes are widely used in commercial sectors; however, current efforts are to use eco-friendly materials and routes for their synthesis. Waterborne polyurethanes (WPUs) using renewable resources are emerging as a new class of polymers that are eco-friendly and sustainable. WPUs find wide applications in coatings, inks, adhesives, sealants, the biomedical field, etc. The market for WPU is increasing every year and is expected to reach $2.1 billion USD by 2025. The continuous increase in the market is due to the need for eco-friendly materials and many regulations that limit the amount of volatile organic compounds (VOCs) that can be released from commercial products, such as solvent-borne polyurethanes. That also includes other

DOI: 10.1201/9781003173526-1

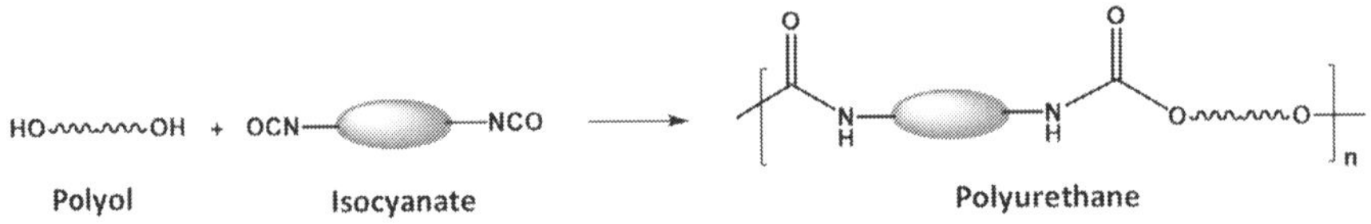

FIGURE 1.1 A chemical reaction between polyol and diisocyanate.

ingredients, such as physical blowing agents used in the synthesis of polyurethane foams like chlorofluorocarbons (CFCs), which were banned due to their harmful effects on the environment. The chemical reaction for the synthesis of polyurethane is very simple, as seen in Figure 1.1. The reaction proceeds through a polyaddition reaction between compounds containing several hydroxyl groups (polyols) and di- or multi-functional isocyanates that form the urethane linkage. The traditional polyols are mostly obtained through the polymerization of ethylene or propylene oxide, which leads to polyether polyols. These components are difunctional hydroxyl-terminated polymers with an aliphatic chain. Due to the flexibility, mobility, and hydrophobicity of the structure, polyols are referred to as the soft domains of polyurethane. On the other hand, isocyanates are composed of rigid groups with low mobility and polarized segments that properly interact with water through hydrogen bonding; hence, they are more dispersible in water or polar solvents, such as acetone.

Thus, to improve the water dispersibility, more polarized, ionic groups or surfactants should be introduced to form a stable polyurethane–water suspension, such as polyester polyols that contain a –C(O)–O– group that interacts with water through hydrogen bonding. High hydrophilicity might be an issue for many applications as these polymers can be more susceptible to react with moisture or degrade under the presence of oxidizing agents, such as acid or alkaline media [1]. Therefore, to counter such issues, other approaches such as the use of emulsifiers (external or internal) can be applied. External emulsifiers are generally surfactants, and their applications are cost-effective; however, sometimes they lead to the formation of unstable big particles. On the other hand, chemical implementation of an ionic group into a polyurethane's backbone as a pendant group that can contain a negative (anionic), positive (cationic), or species that contains both charges (zwitterion) can provide a balance of hydrophilic and hydrophobic nature to obtain stable dispersion in aqueous media.

Di-methylol propionic acid (DMPA) as an effective internal emulsifier is regularly used to introduce a side carboxylic acid group into the polyurethane's chain to synthesize anionic WPU. It can then be neutralized by a quaternary ammonium salt, such as triethylamine (TEA), to further improve its water dispersibility [2]. Cationic WPU is another type that consists mostly of introducing a pending tertiary amine group, such as methylene diethanolamine (MDEA), into the main chain that can be converted into a quaternary amine salt by the addition of an acid. There are many applications of both anionic and cationic WPUs. However, a recent approach used by Zhang et al. demonstrated an interesting example of simultaneous use of both anionic and cationic WPUs to serve as a pesticide delivery system using castor oil-based WPUs [3]. The WPUs were synthesized by reacting castor oil as the bio-based polyol and isophorone diisocyanate (IPDI) with 2,2-bis(hydroxymethyl) butyric acid (DMBA) or *N*-methyl diethanolamine (MDEA). The latter two were responsible for yielding an anionic or cationic WPU, respectively. The use of WPUs enhanced the pesticide's retention time and, thus, its efficiency.

Zwitterionic-based WPUs present lower adhesion of bacteria or biomaterials, such as proteins, blood, fibrinogen, and others on its surface. Additionally, these polymers have effective biocompatibility and chemical stability compared with other polymers, such as poly(ethylene glycol) (PEG). A group of researchers developed an anti-biofouling polyurethane with viable applications for the transport of proteins [4]. The synthesis consisted of a prepolymer method that introduced positive and negative charges into the backbone through a quaternary ammonium salt and sulfate group, respectively. [2-(dimethylamino)ethyl methacrylate dihydroxy terminated ($DMA(OH)_2$)] was synthesized and used as a chain extender. Then 1,3-propane sulton (1,3-PS) was introduced through a ring-opening reaction. The result showed an improvement in the resistance for adhesion of protein to the polyurethane due to the zwitterionic segment, suggesting the importance of such polyurethanes for biomedical equipment to prevent microbial infections and clot formation during the transportation of fluids or prolonged periods exposed to blood or other body fluids.

1.2 SYNTHESIS OF WATERBORNE POLYURETHANES

Most of the synthetic approaches for WPUs consist of two steps. In the first step, a prepolymer is prepared using the desired di or polyols along with the di or polyisocyanate. In the second step, the terminal isocyanate groups from the prepolymer are used as reactive sites to introduce chain extenders that contain hydrophilic groups to disperse the polymer into aqueous media. Several other synthetic routes, such as acetone process, melt-dispersion, and ketamine–ketazine, have been developed. This session briefly describes the main aspects of these procedures.

1.2.1 Prepolymer Method

The main motivation behind developing WPUs was to find alternative synthetic routes that use eco-friendly solvents or reduce the consumption of solvents while providing properties comparable to petrochemical-based polyurethanes. The prepolymer or emulsification method consists of dispersing the prepolymer in water by initially introducing an excess of isocyanate to obtain isocyanate function as an end group to increase the hydrophilicity. Then, a chain extender is introduced in the heterogeneous phase. There are some requirements to successfully perform this method, such as executing the dispersion step in a short period and maintaining a suitable temperature so that isocyanate does not react with water. Cycloaliphatic isocyanates are desired for this approach due to their lower reactivity toward water. For the success of synthesis, careful control of viscosity and proper functionality are also required. Generally, 15% of organic solvents, such as *N*-methyl-2-pyrrolidone, are used to reduce viscosity. The main purpose of adopting this method is to avoid VOCs and decrease overall production costs.

However, some components, such as DMPA, that introduce the ionic segment are derived from non-renewable sources.

To find an alternative, researchers developed an approach that used soybean oil for the synthesis of WPU [5]. The synthesis consisted of performing a ring-opening reaction of epoxidized soybean oil with adipic and pimelic acid, separately, to introduce a carboxylic acid as an ionic segment. The synthesized WPU was used as a pressure-sensitive adhesive. Shear strength of about 1 h to more than 100 h, a tack of 1.78–5.66 N, and peel strength of 1.23–2.77 N/mm were observed for this adhesive. Castor oil was used as a natural polyol source and provided a total bio content of about 77% in the WPU. Also, castor oil demonstrated good compatibility with the bio-derived ionic segments. The results demonstrated that the WPU obtained through the emulsification process had effective thermal stability that presented the max thermal decomposition temperature (~383°C) along with good adhesive characteristics. These properties, accompanied by renewable sources of materials, show potential for large-scale applications.

1.2.2 Acetone Process

The acetone process consists of synthesizing a prepolymer and implementing a chain extender in acetone to obtain a homogenous phase. After completion of the synthesis, a solvent exchange process is performed by adding water and removing acetone through distillation or rotary evaporation. Acetone is used as it is a convenient solvent due to its chemical inertness, proper dispersion of polyurethanes in it, water solubility, and low boiling point, which facilitates its easy removal. Some advantages of this method include the formation of a homogenous system that yields small particles and reduces the reaction rate between chain extenders toward isocyanate, leading to higher WPU reproducibility. Acetone also decreases the viscosity of the system, which eases processing. However, this process requires a high volume of acetone, which increases the cost and makes it less eco-friendly. It also adds an extra distillation step and a requirement for the polyurethane structure to be soluble in acetone to avoid precipitation [2]. Despite these factors, the acetone process is a viable industrial method for the production of adhesives due to its high reproducibility and better control of the process. As an alternative option, methylethylketone was used in place of acetone. Nanda et al. compared the properties of polyurethanes prepared using acetone and prepolymer methods [6]. The study showed that a low concentration of ionic segments may be required to achieve a stable dispersion. A stable dispersion of polyurethanes was achieved by using 2% and 4% of DMPA in acetone and prepolymer methods, respectively.

Sardon et al. synthesized a WPU using the acetone process to understand the influence of DMPA concentration, amount of acetone in dispersion, temperature for solvent exchange, and conditions of solvent evaporation [7]. Use of 60 wt.% of acetone, lower than 30°C solvent exchange temperature, and 0.1 mmol of DMPA per gram of polyol (at PU of 75 wt.%) were some of the optimal experimental conditions to form stable dispersions. The procedure consisted of using poly(propylene glycol) as the diol, DMPA as an internal ionic emulsifier, TEA to neutralize the acidic groups, dibutyltin diacetate (DBTDA) as the catalyst, and acetone to dissolve the system. After stirring the system and reaching the temperature of 60°C, the isocyanate (IPDI) was slowly added, followed by the addition of 1,4-butanediol (BD) after 5 h to react with terminal isocyanate groups. To create dispersion and perform the solvent exchange, water was slowly added under high stirring, and the solvent was removed through rotary evaporation.

Another variation for this procedure was performed by Wang et al. [8]. A polyurethane dispersion was prepared through the acetone process by reacting toluene diisocyanate (TDI) and poly(oxypropylene glycol) at 80°C for 1 h. After that, DMPA was added and the temperature was raised to 85°C for 2–3 h to promote the reaction between the isocyanate with DMPA. The reaction system was cooled down to 50°C, followed by the addition of acetone and TEA to reduce the viscosity and to neutralize the carboxylic acid groups from DMPA, respectively. The acetone was removed through a solvent exchange process with water. This study was performed to analyze the properties of WPU dispersions under two conditions: (1) grafting of casein on polyurethane structure and (2) mixing of casein with WPU. It was observed that the mixing of casein with WPU led to smaller particle size, lower viscosity, and poly dispersibility, compared with the case of grafted casein. The decrease of these parameters led to a more stable polyurethane dispersion. In addition, an enhancement in mechanical property was observed when the WPU was mixed with casein. The improved properties were due to the proper dispersion of casein nanoparticles (30–50 nm) within the WPU sheets.

1.2.3 Melt-Dispersion Process

The melt-dispersion, or hot-melt, process offers an alternative path that does not require organic solvents. The first step of this process consists of the synthesis of a prepolymer with terminal isocyanate groups followed by the addition of urea to yield biuret groups. The biuret capped prepolymer is then dispersed in water followed by a reaction with formaldehyde. The latter has the function to form methylene bridges (act as crosslinker) that go through a condensation reaction in acidic media. A higher temperature of around 130°C is required to perform the reaction between urea and isocyanate. The high temperature also decreases the viscosity, which facilitates the mixing of the reagents. It is a convenient method due to the relatively facile synthetic process and non-requirement of organic solvents. A general example of the hot-melt process to synthesize a WPU is explicated in Figure 1.2.

1.2.4 Ketamine–Ketazine

The ketamine–ketazine method is a deviation of the prepolymer method that consists of mixing a prepolymer with a blocked amine (ketamine) or hydrazine (ketazine) before the addition of water into the system. Ketamines are obtained by reacting a diamine with acetone, while ketazines are formed by the reaction between hydrazine and acetone. Both functions are not reactive toward the isocyanate-terminated prepolymer, which facilitates the proper mixing. Water is

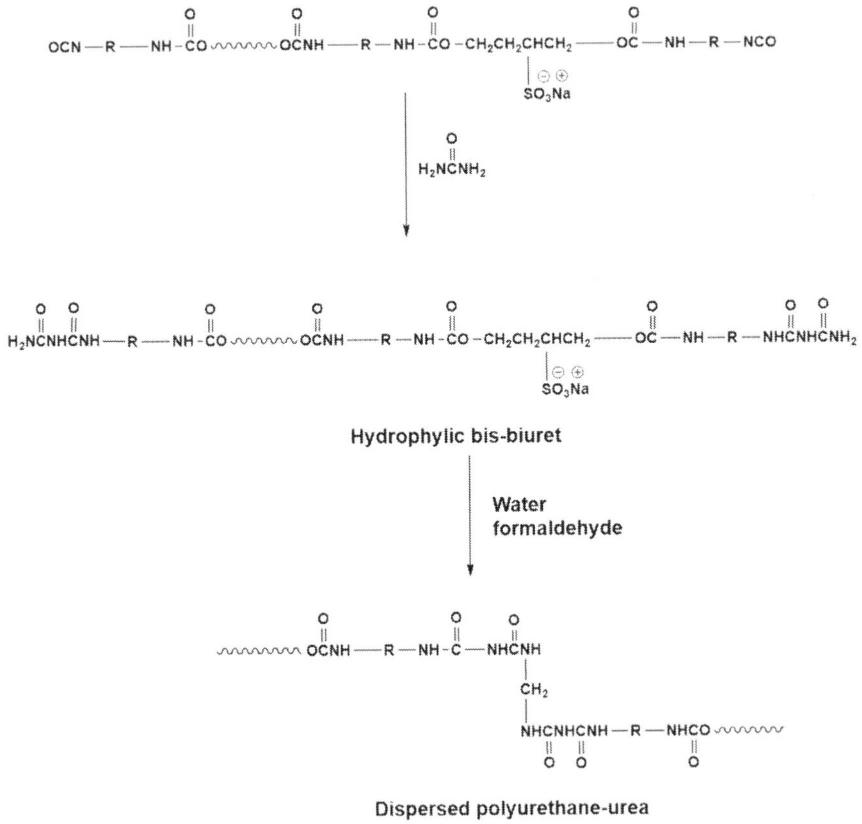

FIGURE 1.2 Synthesis of a WPU through the hot-melt process. Adapted with permission from reference [9]. Copyright 2012, Taylor & Francis.

then added to form a stable dispersion. However, it also reverts the ketamine or ketazine to a free amine or hydrazine, respectively, which can then react with isocyanate and start the chain extension step. Generally, the ketamine chain extension occurs faster than ketazines. It is a viable process as high-performance coatings can be obtained without the need for organic solvents to stabilize the dispersion. The ketamine–ketazine process is shown in Figure 1.3.

1.3 CLASSIFICATION OF WATERBORNE POLYURETHANES

1.3.1 Cationic Waterborne Polyurethanes

In a simple term, cationic polyurethanes contain a cation into their backbone. The cation can be introduced by reacting an isocyanate with a diol having nitrogen or sulfur as a heteroatom. The positive charge on the backbone of polyurethanes can be obtained by performing a quaternization of nitrogen or sulfur. The introduction of a positive charge not only improves the dispersibility of polyurethanes in water but also grants antibacterial properties. Therefore, most of the cationic WPUs are

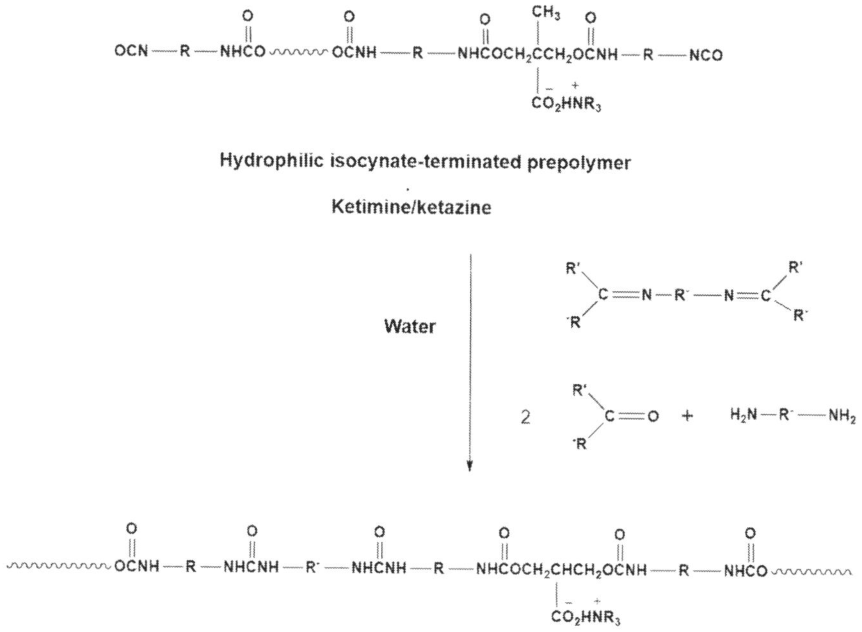

FIGURE 1.3 General ketamine–ketazine process for the synthesis of WPU. Adapted with permission from reference [9]. Copyright 2012, Taylor & Francis.

used for biomedical applications. A simple synthetic approach to obtain an antibacterial cationic WPU consists of a reaction between IPDI and polycarbonate diol along with 3-dimethylamino-1,2-propanediol (DMAPD) as an aminated chain extender [10]. Generally, to obtain a stable emulsion and small particle size, water must be added slowly and dropwise accompanied by high stirring. This yields coatings with an even distribution of solid particles, which avoid peeling, fouling, poor adhesion to a substrate, and heterogeneous distribution of solid particles. The synthesized cationic WPU films are able to effectively kill bacteria, such as *Escherichia coli* and *Staphylococcus epidermis*, Gram-negative and Gram-positive, respectively [10]. The films also display a tensile strength of about 36 MPa and elongation at break around 620%. Figure 1.4 shows the zone of inhibition for the coatings, demonstrating high efficiency in killing bacteria.

One of the challenges of synthesizing cationic WPUs is to overcome their poor resistance to water and many chemicals. Internal emulsifiers are added during the synthesis to improve dispersibility but also decrease the chemical resistance of WPUs. MDEA and 3-(dimethylamino)propane-1,2-diol (DMAD) are common internal emulsifiers for cationic WPUs. The excess incorporation of these emulsifiers into the WPU backbone increases the water intake, which leads to a plasticizer effect, deteriorating mechanical properties, and promotes corrosion through hydrolysis. Neutralization of internal emulsifiers with acid and the introduction of emulsifiers at different positions are used to solve this issue. Gong et al. studied the effect of emulsifier positions on the properties of a cationic WPU

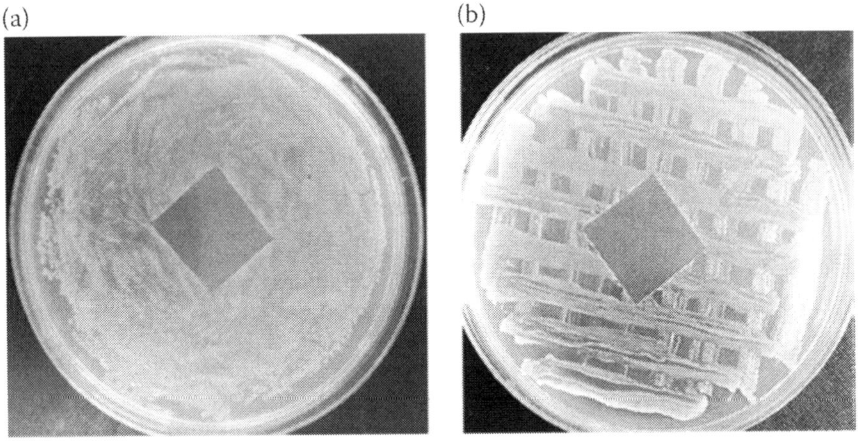

FIGURE 1.4 Test result for the inhibition of bacteria growth in the squared centered area. Adapted with permission from reference [10]. Copyright 2020, American Chemical Society.

synthesized using CO_2-based polyol and IPDI [11]. Two internal emulsifiers were used (MDEA and DMAD) and introduced in different positions in the polymeric structure as terminal groups (t), side chain (s), and backbone (b). The results from this study demonstrated that terminal cationic groups were more effective to disperse the polymer, which was likely attributed to the lesser steric hindrance and mobility of the terminal groups. This facilitated the neutralization of the tertiary ammonium-based internal emulsifier with acid. Besides the consumption of a low concentration of internal emulsifiers (1 wt.%), the use of sustainable carbonate polyol accounted for a decrease in water absorption. The water absorption decreased from 6% to 1.5% when 65 wt.% of CO_2-based polyol was added into the polymeric structure. Hence, demonstrating an effective improvement in water resistance. The synthesis of cationic WPUs requires specific conditions to have satisfactory properties. Factors such as water addition rate, stirring, and concentration of internal emulsifier are some of the core parameters that should be considered for designing new experiments.

1.3.2 Anionic Waterborne Polyurethanes

Anionic WPUs contain a negatively charged group in their structure which is usually provided by carboxylic acid, phosphonate, or sulfonic groups. Conductive properties in anionic WPUs can be introduced by using some metal cations such as Na^+ or Li^+ which bond to anionic sites of WPUs and allowing ionic conductions through the polymeric network of the dispersed PUs. It has been observed that larger cations caused a decrease in the glass transition temperature (T_g) since their charges are more dispersed, leading to weaker Coulombic attractions. Besides the introduction of conductibility, the increase of ionic segments into a polymer also leads to an increase in tensile strength and hardness accompanied by a decrease in crystallinity. These properties can also be controlled by other parameters such higher NCO/OH ratio, which leads to a higher number of rigid segments. Other factors such as particle size and dispersion stability can be influenced by synthetic methods.

DMPA and DMBA are commonly used internal emulsifiers for anionic WPUs since their hydroxyl groups can react with the isocyanate of PU's backbone as well as provide pending carboxylic acid groups (negative charge) to improve the dispersion in water. However, different approaches have been adapted which use polycarbonate diols as a main soft segment, carboxylic diol as an internal emulsifier, and *m*-tetramethyl xylylene di-isocyanate (TMXDI) to avoid the over hydrolysis of the isocyanate groups during dispersing in water [12]. This approach yielded WPU dispersions with lower particle size than those compared with DMPA and DMBA, indicating a more stable dispersion with higher interaction with water, and thus, the coating showed a more homogenous structure. Hsiao et al. designed an anionic WPU that was processed into a composite consisting of graphene nanosheets [13]. Graphite was chemically exfoliated through modified Hummer's process which created oxygenated groups onto the surface such as epoxy, carbonyl, and carboxyl. Then, the thermal reduction was performed to revert to the aromaticity of graphene and increase the interlayer distance. To improve the dispersibility of graphene in water, aminoethyl methacrylate (AEMA) was chemically bonded to the graphene's structure through free radical polymerization. The amine groups were quaternized by decreasing the pH. This led to the formation of positive charges that were chemically grafted into the graphene structure. These positive charges were countered with an anionic WPU containing SO_3^- as dandling groups. This system formed an ionic interaction between the graphene nanosheets and the WPU allowing them to properly disperse in water. The composite coating showed high electrical conduction (43.6 S/m) along with high electromagnetic shielding (38 dB between the frequency range of 8.2–12.4 GHz).

1.3.3 Zwitterionomers

The zwitterionomers are species that contains both negative and positive charges. Zwitterionomers can be synthesized by quaternization of *N*-alkyl diols with sultones or lactones. Some examples of the latter are sulfobetaine, phosphobetaine, and carboxybetaine. One of the challenges for the synthesis of zwitterionic-based polymers is the phase separation that arises from hydrophobic and ionic zwitterionic segments, which often makes it difficult to form a homogenous system. As a result, fouling and uneven distribution can occur in the coatings. The use of interpenetrating polymer networks (IPN) and grafted polymers on the PU's surface can avoid these issues [14,15]. A group of researchers synthesized poly(2-(dimethylamino)ethyl methacrylate) (PDEM$(OH)_2$) as a diol and then reacted with 1,3-PS to yield zwitterions in the side chain (zPDEM) [16]. The ionic nature of zwitterions allowed the WPU to properly interact with water to provide stable dispersions. The most notable property of zwitterionic-based materials is their resistance to protein adsorption and thus their wide application in microbial antifouling. The results demonstrated that the antifouling effect could be controlled based on the number of zwitterionic side chains.

1.4 APPLICATIONS OF WATERBORNE POLYURETHANES

WPUs find their wide applications in adhesives; paint additives; thickeners; textile dyes; pigment paste; drug delivery systems; biomedical equipment; and anti-corrosion, anti-bacterial, and mechanical resistant coatings, to name a few. The diversity in applications of WPUs is due to a range of polyol available for their synthesis. The high reactivity of the reagents and facile synthetic approaches make them commercially important.

The processing of polymeric materials for coating applications could be challenging if the viscosity of the melt/polymer solution is high, particularly for high molecular weight compounds. One of the solutions to this problem is the use of WPUs for coating applications, as the viscosity of an aqueous colloidal dispersion is virtually independent of molecular weight. WPUs prepared using prepolymer and acetone processes are widely used for coating applications. To further describe the application of WPUs, it is important to understand how the coatings are formed through the processes. When water is evaporated from the WPU dispersion, a film is formed through coalescence, and polymeric particles tend to nucleate, which leads to the film's physical entanglement. To enhance the effect, strong intermolecular interactions, such as hydrogen bonding and electrostatic forces, lead to the development of elasticity and tensile strength. Such effects are intrinsically related to the WPU's chemical structure. A few general cases of structure–property correlation can be described, where the presence of aromatic rings improves the resistance to mechanical shock and scratch; aliphatic segments lead to flexible coatings, better mixing with pigments, and color retention; and ionomeric segments increase the resistance to protein adsorption [17–19]. Thus, such versatile properties of WPU coatings allow it to be implemented in a variety of sectors.

A cationic WPU dispersion based on poly(urethane-co-vinyl pyridine) was used as a binder in the base coat for leather finishing [20]. The coating was synthesized through the internal emulsifier method, where the prepolymer was composed of poly(tetramethylene oxide) glycol (PTMG) and TDI. The vinyl pyridine was used as an internal emulsifier with bromobutane as a quaternization agent. The quaternization of pyridine moieties led to the formation of a positive charge that added anti-bacterial properties. Also, cationic polymeric structures presented high adhesion toward anionic substrates, such as leather and glass. The tensile strength of leather was increased after the WPU's application due to the proper interaction between the polymer and the fibrous structure of the leather. Continuing on the applications for anti-bacterial coatings, Xia et al. developed a cationic WPU based on soybean oil and different amino polyols [21]. The coating was obtained through the internal emulsifier method, which consisted of the prepolymer synthesized using methoxylated soybean-oil polyol and IPDI. Triethanolamine and *N*-methyl diethanolamine (MDEA) were used as internal emulsifiers. The coating provided a storage modulus of 641 MPa along with the bactericidal effect toward *Listeria monocytogenes* and *Salmonella minnesota*. It was observed that the crosslink density, which depends on amino polyols' functionality and concentration, has a great influence on the properties of the coatings. Higher crosslink density improved the mechanical properties. Besides, the optimum concentration of cations improved the surface area, facilitating contact with bacteria, enhancing its antibacterial properties.

The search for natural components that provide competitive properties for coatings is perhaps the main challenge faced by both academia and industry. Liang et al. used octahydro-2,5-pentalenediol (OPD), a bio component derived from citric fruits, and castor oil to synthesize a WPU with considerable tensile strength (22.3 MPa), Young's modulus (382.8 MPa), glass transition temperature (50–75°C), transparency, and high corrosion inhibition efficiency (94.7%) [22]. OPD's structure was composed of two condensed cycloaliphatic rings that introduced rigidity to polyurethane and hence increased the tensile strength. On the other hand, due to low mobility, the coatings became more brittle with an increase in the concentration of OPDs, which led to an increase in the glass transition temperature. The transparency was likely related to the saturated structure that does not absorb the light in the visible spectrum. The anti-corrosion properties were obtained due to the formation of a tortuous structure after the addition of OPD that created a shielding effect for corrosive agents. Also, the cycloaliphatic ring's chemical stability played a role in improving these properties.

Polyurethane coatings are used for protection against several external agents, such as UV radiation, heat, mechanical shocks, moisture, corrosion, etc. High hydrophobicity of these coatings is required for many applications, such as self-cleaning, chemical resistance against corrosion, anti-icing, and anti-biofouling. However, the synthesis of hydrophobic coatings often requires the use of organic solvents to form a stable dispersion and sometimes fluorinated compounds to reduce the surface energy to prevent the interaction with water. A large number of organic solvents and costly fluorinated reagents can be challenging for commercial applications, and thus alternative approaches should be developed. In addition, the coatings must also present satisfying mechanical robusticity to retain their properties under a wide range of temperatures and varying environmental conditions to become competitive in the market.

Based on those requirements, Li et al. reported an eco-friendly, self-healing, and superhydrophobic WPU coating, which was able to withstand several cycles of freezing and melting while maintaining its properties [23]. The superhydrophobic components were synthesized through acid catalysis by a reaction between SiO_2, tetraethoxysilane (TEOS), and hexadecyltrimethoxysilane (HDTMS). The synthetic process and microstructure are provided in Figure 1.5. The spray-coating method was used to fabricate the WPU composite as a PU aqueous solution was applied over the substrate followed by the SiO_2@HD-POS. A compact coating was obtained after water evaporation, which strengthened the contact between the PU and SiO_2@HD-POS. Such coating yielded a satisfactory hydrophobicity as the contact angle between a water droplet and the coating's surface was 163.9°. Also, the coating was able to form a layer of air when submerged in water and completely bounce water off and maintain its properties. The coating was exposed to several organic solvents, solutions with varying pH (1–14), and UV radiation to test the chemical stability. The results showed good stability of the coating in all environments. The self-healing properties were studied by exposing the coating to O_2 plasma to destroy the hydrophobic surface. After the plasma treatment for only 5 s,

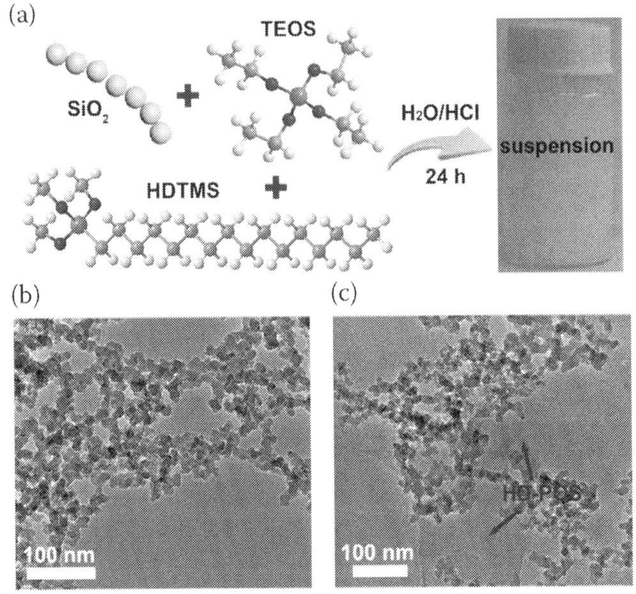

FIGURE 1.5 (a) Synthesis of superhydrophobic suspensions of hexadecyl polysiloxanes-modified SiO_2 (SiO_2@HD-POS) and transmission electron microscopy (TEM) for (b) SiO_2 nanoparticles and (c) SiO_2@HD-POS. Adapted with permission from reference [23]. Copyright 2018, American Chemical Society.

the coating became hydrophilic with a contact angle of 0°. However, after thermal healing at 150°C for 2 h, the coating was able to reestablish its hydrophobicity by displaying a contact angle of 160°. The synthesis and procedure demonstrated in this study provide interesting insight into the development of high-performance coatings through an eco-friendly route.

The biomedical field often requires devices designed to transport and store body fluids, such as whole blood, plasma, and fibrinogen, without adsorbing these biological components into its surface to avoid thrombogenicity, microbial growth, and fouling. Also, biocompatibility, biodegradability, and flexibility are some of the required properties for the device components to be used as a scaffold, or as a vascular graft. Zwitterionic-based WPUs are polymers that can meet these requirements as zwitterionic units, like phosphorylcholine, carboxybetaine, and sulfobetaine are efficient to mitigate protein adsorption. In that regard, Ye et al. developed a biodegradable and elastic WPU that had high resistance to fibrinogen adsorption and thrombogenicity [24]. The synthesis was performed by using polycaprolactone-diol (PCL-diol) and sulfobetaine-diol (SB-diol), in different proportions to react with diisocyanatobutane. During the synthesis, putrescine was used as a chain extender. The obtained polyurethane was processed into fibrous structures to form conduits through the electrospinning process.

Regarding mechanical properties, the study demonstrated that the addition of 50% SB-diol into the polyurethane's structure (PESBUU-50) yielded the optimum results in both dry and wet conditions in comparison to the other proportions of 0%,

25%, 75%, and 100%. The values obtained in dry and wet conditions for tensile strength were 48 and 33 MPa, while strain at break was 851% and 749%, respectively. In addition to high mechanical strengths, these coatings were able to reduce protein adsorption. Hydrophilicity played a major role in protein adsorption capacity as higher hydrophilicity leads to higher water intake (Figure 1.6). A similar effect was observed as the bovine blood adsorption was given from both macroscopic perspectives and through scanning electron microscopy (SEM). The study demonstrated that designing the polyurethane structure with flexible groups allowed it to be a robust material. This effect must be in coordination with the number of ionic segments that are responsible for decreasing protein adsorption. However, under higher concentrations, it may decrease the mechanical properties and swell in contact with water.

Self-healing of coatings is another very interesting property that allows the coatings to heal by changing their external/internal conditions. The self-healing property of WPUs leads to several advanced applications. Shahabadi et al. fabricated a self-healing WPU that consisted of non-covalent interaction between lignin and modified graphene (LMG) [25]. The fabrication of LMG was performed by sonicating pristine graphite in the presence of lignin in aqueous media for 8 h. The LMG was then incorporated into the WPU matrix and allowed the composite to self-heal from a cut after exposure to infrared radiation for 150 s. The mechanism for self-healing was associated with the viscoelasticity recovery of LMG. After the

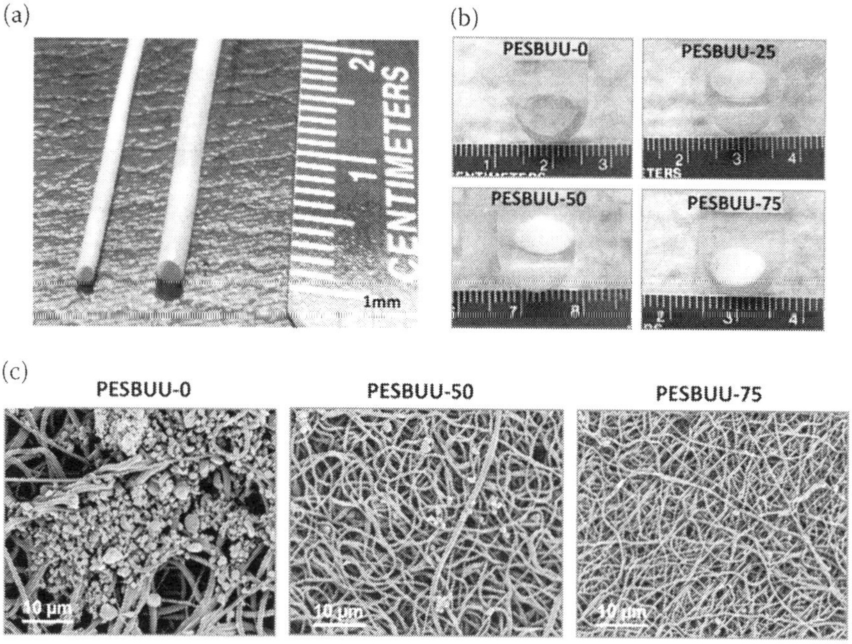

FIGURE 1.6 (a) Electrospun fibrous conduit samples, (b) naked eye, and (c) SEM images for the electrospun PESBUU samples after 3 h of interaction with ovine blood at 37°C. Adapted with permission from reference [24]. Copyright 2014, American Chemical Society.

cuts through the WPU/LMG, the energy accumulated in response to the cut was released, making the cut surfaces heal back. After exposure to infrared radiation, the polymerics' composite diffusion rate increased, which increased the healing rate of the material.

Protection against corrosion is very important as it brings considerable economical savings for industries and society. WPUs contain urethane linkage in their structure, which is chemically stable and provides better adhesion to many surfaces, and thus can be used as corrosion protective coatings. Khatoon et al. used safflower oil-based polyurethane (SFPU) that contains nanocomposite of V_2O_5 encapsulated with polydiphenyl amine (PDPA) for corrosion protection [26]. The polyurethane combined with the nanofillers presented satisfactory results, such as low corrosion current density (7.45×10^{-11} A/cm^2), corrosion potential (−0.04 V), and relatively high hydrophobicity (contact angle of 84°). The synthesis of the composite coating consisted of an in-situ polymerization of safflower-based polyurethane in the presence of V_2O_5–PDPA, at which the safflower polyol was dispersed in xylene with V_2O_5–PDPA and TDI was added dropwise in the presence of catalyst dibutyltin dilaurate (DBTDL). The inherent anti-corrosion properties of SFPU provided a barrier effect due to the uniform adherence to carbon steel. However, the neat SFPU could not prevent the diffusion of ions. Hence, the use of nanoparticles of V_2O_5–PDPA improved the corrosion protective properties as the presence of nanoparticles improved the barrier property. They acted as anion storage preventing the flow of electrons, thus inhibiting the corrosion. The durability of coatings is very important for such applications as ions can penetrate the metal substrate over time to reduce their effectiveness as a protective coating. However, V_2O_5–PDPA creates a tortuous effect that hinders the penetration of the ions, which retards the corrosion process. Eventually, after these ions reach the surface of carbon steel, the Fe^{2+} and Fe^{3+} undergo oxidation, forming a passive layer of Fe_2O_3 and Fe_3O_4 that further protects the metal. Thus, the anti-corrosion mechanism combines three key elements for an efficient process: (1) proper adhesion to the surface, (2) barrier/tortuous effect, and (3) formation of a passive layer [27,28].

1.5 CONCLUSION

WPUs are versatile materials that can have positive, negative, or both charges in their structure. These unique characteristics allows them to be used in many applications, such as adhesives, sealants, protective coatings, anti-microbial coatings, self-healing coatings, and components for biomedical devices. The current challenges are to reduce production cost and make the process more sustainable. The use of renewable resources and the inclusion of ionic or non-ionic hydrophilic segments into the polyurethanes for improved dispersibility in water to avoid the use of organic solvents are some of the green approaches. In this regard, bio-derived materials, such as corn oil, soybean oil, and castor oil, have opened a broad range of possibilities to synthesize WPUs. The use of nanoparticles has shown significant improvement in the properties of WPUs. The approaches discussed in this chapter for the synthesis and applications of WPUs not only provide eco-friendly and green polymers but also enable the use of bio-renewable materials for a sustainable future.

REFERENCES

1. Ionescu M (2005) *Chemistry and Technology of Polyols for Polyurethanes*. Rapra Technology, Shawbury, UK.
2. Honarkar H (2018) Waterborne Polyurethanes: A Review. *J Dispers Sci Technol* 39:507–516.
3. Zhang Y, Liu B, Huang K, Wang S, Quirino RL, Zhang Z, Zhang C (2020) Eco-Friendly Castor Oil-Based Delivery System with Sustained Pesticide Release and Enhanced Retention. *ACS Appl Mater Interfaces* 12:37607–37618.
4. Wang C, Ma C, Mu C, Lin W (2014) A Novel Approach for Synthesis of Zwitterionic Polyurethane Coating with Protein Resistance. *Langmuir* 30:12860–12867.
5. Liu L, Deng H, Zhang W, Madbouly SA, He Z, Wang J, Zhang C (2021) Novel Internal Emulsifiers for High Biocontent Sustainable Pressure Sensitive Adhesives. *ACS Sustain Chem Eng* 9:147–157.
6. Nanda AK, Wicks DA (2006) The Influence of the Ionic Concentration, Concentration of the Polymer, Degree of Neutralization and Chain Extension on Aqueous Polyurethane Dispersions Prepared by the Acetone Process. *Polymer (Guildf)* 47: 1805–1811.
7. Sardon H, Irusta L, Fernández-Berridi MJ, Luna J, Lansalot M, Bourgeat-Lami E (2011) Waterborne Polyurethane Dispersions Obtained by the Acetone Process: A Study of Colloidal Features. *J Appl Polym Sci* 120:2054–2062.
8. Wang N, Zhang L, Lu Y (2004) Effect of the Particle Size in Dispersions on the Properties of Waterborne Polyurethane/Casein Composites. *Ind Eng Chem Res* 43:3336–3342.
9. Szycher M (1999) *Szycher's Handbook of Polyurethanes*, 1st ed. CRC Press, New York.
10. Wang Y, Chen R, Li T, Ma P, Zhang H, Du M, Chen M, Dong W (2020) Antimicrobial Waterborne Polyurethanes Based on Quaternary Ammonium Compounds. *Ind Eng Chem Res* 59:458–463.
11. Gong R, Cao H, Zhang H, Qiao L, Wang F, Wang X (2020) Terminal Hydrophilicity-Induced Dispersion of Cationic Waterborne Polyurethane from CO_2-Based Polyol. *Macromolecules* 53:6322–6330.
12. Lee D-K, Tsai H-B, Yang Z-D, Tsai R-S (2012) Polyurethane Dispersions Derived From Polycarbonatediols by a Solvent-Free Process. *J Appl Polym Sci* 126: E275–E282.
13. Hsiao ST, Ma CCM, Tien HW, Liao WH, Wang YS, Li SM, Yang CY, Lin SC, Yang R Bin (2015) Effect of Covalent Modification of Graphene Nanosheets on the Electrical Property and Electromagnetic Interference Shielding Performance of a Water-Borne Polyurethane Composite. *ACS Appl Mater Interfaces* 7:2817–2826.
14. Chang Y, Chen S, Yu Q, Zhang Z, Bernards M, Jiang S (2007) Development of Biocompatible Interpenetrating Polymer Networks Containing a Sulfobetaine-Based Polymer and a Segmented Polyurethane for Protein Resistance. *Biomacromolecules* 8:122–127.
15. Yuan J, Zhang J, Zang X, Shen J, Lin S (2003) Improvement of Blood Compatibility on Cellulose Membrane Surface by Grafting Betaines. *Colloids Surfaces B Biointerfaces* 30:147–155.
16. Ma C, Zhou H, Wu B, Zhang G (2011) Preparation of Polyurethane with Zwitterionic Side Chains and Their Protein Resistance. *ACS Appl Mater Interfaces* 3:455–461.
17. Fan W, Jin Y, Shi L, Zhou R, Du W (2020) Developing Visible-Light-Induced Dynamic Aromatic Schiff base Bonds for Room-Temperature Self-Healable and Reprocessable Waterborne Polyurethanes With High Mechanical Properties. *J Mater Chem A* 8:6757–6767.

18. Xu Y, Yang Y, Yan D-X, Duan H, Zhao G, Liu Y (2018) Gradient Structure Design of Flexible Waterborne Polyurethane Conductive Films for Ultraefficient Electromagnetic Shielding with Low Reflection Characteristic. *ACS Appl Mater Interfaces* 10:19143–19152.
19. Poussard L, Burel F, Couvercelle J-P, Merhi Y, Tabrizian M, Bunel C (2004) Hemocompatibilty of New Ionic Polyurethanes: Influence of Carboxylic Group Insertion Modes. *Biomaterials* 25:3473–3483.
20. Sundar S, Vijayalakshmi N, Gupta S, Rajaram R, Radhakrishnan G (2006) Aqueous Dispersions of Polyurethane–Polyvinyl Pyridine Cationomers and Their Application as Binder in Base Coat for Leather Finishing. *Prog Org Coatings* 56:178–184.
21. Xia Y, Zhang Z, Kessler MR, Brehm-Stecher B, Larock RC (2012) Antibacterial Soybean-Oil-Based Cationic Polyurethane Coatings Prepared From Different Amino Polyols. *ChemSusChem* 5:2221–2227.
22. Liang H, Li Y, Huang S, Huang K, Zeng X, Dong Q, Liu C, Feng P, Zhang C (2020) Tailoring the Performance of Vegetable Oil-Based Waterborne Polyurethanes through Incorporation of Rigid Cyclic Rings into Soft Polymer Networks. *ACS Sustain Chem Eng* 8:914–925.
23. Li Y, Li B, Zhao X, Tian N, Zhang J (2018) Totally Waterborne, Nonfluorinated, Mechanically Robust, and Self-Healing Superhydrophobic Coatings for Actual Anti-Icing. *ACS Appl Mater Interfaces* 10:39391–39399.
24. Ye S-H, Hong Y, Sakaguchi H, Shankarraman V, Luketich SK, D'Amore A, Wagner WR (2014) Nonthrombogenic, Biodegradable Elastomeric Polyurethanes with Variable Sulfobetaine Content. *ACS Appl Mater Interfaces* 6:22796–22806.
25. Seyed Shahabadi SI, Kong J, Lu X (2017) Aqueous-Only, Green Route to Self-Healable, UV-Resistant, and Electrically Conductive Polyurethane/Graphene/Lignin Nanocomposite Coatings. *ACS Sustain Chem Eng* 5:3148–3157.
26. Khatoon H, Ahmad S (2019) Vanadium Pentoxide-Enwrapped Polydiphenylamine/Polyurethane Nanocomposite: High-Performance Anticorrosive Coating. *ACS Appl Mater Interfaces* 11:2374–2385.
27. Matějovský L, Macák J, Pospíšil M, Baroš P, Staš M, Krausová A (2017) Study of Corrosion of Metallic Materials in Ethanol–Gasoline Blends: Application of Electrochemical Methods. *Energy & Fuels* 31:10880–10889.
28. Sababi M, Pan J, Augustsson P-E, Sundell P-E, Claesson PM (2014) Influence of polyaniline and ceria nanoparticle additives on corrosion protection of a UV-cure coating on carbon steel. *Corros Sci* 84:189–197.

2 Chemistry and Technology of Waterborne Polyurethanes

Naresh A. Rajpurohit, Kaushalya Bhakar, and Dinesh Kumar
School of Chemical Sciences, Central University of Gujarat, Gandhinagar, India

Meena Nemiwal
Department of Chemistry, Malaviya National Institute of Technology, Jaipur, India

CONTENTS

2.1 Introduction 18
2.2 Chemistry of Waterborne Polyurethane 18
2.3 Waterborne Polyurethane-Based Nanocomposites 22
2.4 Application and Technology of Waterborne Polyurethanes 23
  2.4.1 Printing Ink 23
  2.4.2 Surface Sizing Treatment 24
  2.4.3 Coating 24
  2.4.4 Adhesives 25
2.5 Miscellaneous Application of Waterborne Polyurethanes 26
  2.5.1 Appliances 26
  2.5.2 Automotive 26
  2.5.3 Building and Construction 26
  2.5.4 Packaging 27
  2.5.5 Composite Wood 27
  2.5.6 Wastewater Treatment 27
2.6 Conclusion 27
Acknowledgments 27
References 28

DOI: 10.1201/9781003173526-2

2.1 INTRODUCTION

Polyurethanes (PUs) are high-performing polymers which can be prepared in two different steps. In the first step, isocyanate and polyol react and form a thick viscous liquid polymer. In the second step, this viscous liquid compound reacts with a diol or diamine compound and forms high molecular polymers [1]. PUs generally disperse volatile organic compounds (VOCs), which harm the environment. Environmental awareness has been a reason for increasing research efforts and developments of environmentally friendly green synthesis routes of PUs; therefore, in the late 1960, the system of aqueous PU dispersion was introduced, normally termed waterborne polyurethanes (WPUs). It is a binary colloidal system in which PU particles are continuously spread in the aquatic medium. Meanwhile, it produces polymers in which many hydrophilic groups are present for obtaining high water solubility. WPUs are made up of two segments. The first main portion, called the backbone, is hydrophobic and is called a soft segment. The second one, the emulsifier, is attached to many ionic materials. They are also known as rigid segments. The emulsifier can be further divided into two types: internal and external portions. The internal emulsifiers include nonionic centers—polyethylene oxide and ionic centers—anionic, cationic, and zwitterions. Since the aqueous dispersions of PU are nontoxic and nonpolluting for the air, they have been broadly used in coatings and adhesives for fabric, leatherwood, and glass, etc. WPU dispersions are more stable, are easy to handle, and show similar character to solvent-based systems. There are many unique properties, such as high tensile strength, chemical resistance, high elongation, low-temperature flexibility, water resistance, nonflammability, etc. Similarly, many other applications, such as printing ink, surface sizing treatment, coating, adhesive, and more, play an important role in modern technology for green environment resolution. This chapter describes the chemistry of WPU and WPU-based nanocomposites with improving properties and application as coating adhesive and other vital role applications.

2.2 CHEMISTRY OF WATERBORNE POLYURETHANE

In recent years, concern for the environment is one of the reasons to develop materials based on the green chemistry approach. The aqueous-based PUs and polyurethane-ureas (PUU) are one trending example. Aqueous PU dispersion is a binary colloidal system where PU particles have to be continuously spread in a water-based solvent medium. PU dispersion is a highly sticky liquid and is usually independent of the molecular weight of the polymer chain. Particularly, their water-based dispersion, commonly named WPU and waterborne polyurethane urea (WBPUU), is attracting the attention of researchers due to its eco-friendlier synthesis approach [2]. The synthesis of these polymers takes place through the green route, which is a solvent-free method. The water is used as a primary solvent and reduces the generation of organic solvents and other toxicity in the environment compared to the ordinary synthesis process of solvent-borne PU. Aqueous PU dispersions can be conveyed into coatings and adhesives containing little or no cosolvent and have a great application in various fields, including textile, coating, painting, adhesive, and so on [3].

The polyaddition reaction of isocyanate with polyol lead to formation of an intermediate polymer, and further reaction with diol or diamine resulted in the final polymer. Generally, PU-based polymers are prepared by the reaction of two segments. The excess of the isocyanate group of polyisocyanate reacted with the hydroxyl group of polyols to form a urethane linkage (–HNCOO–), named a polyaddition reaction. It is a versatile family of synthetic polymer composed of repeating units of PU linkage in the polymer chain. Various functionalities, including polyols and polyisocyanates in the backbone chain, are responsible for chain extension and crosslinking. These highly efficient polymers are prepared by alternating two segments, known as soft and hard segments. The soft segment contains the polyol units where the hydroxyl group presents as a functional group. The polyol can be a low or high molecular weight chain and may be biodegradable or non-biodegradable [4]. The soft segment also acts a vital role in giving out flexibility to the polymer system. Concerning polyols, different types of linear polyether, polyester, and polycarbonate polyols can be used depending on the desired properties of the final product [5]. However, polyethers are more likely to ameliorate the dispersion of PU in an aqueous medium and give extra flexibility to the system. Polycarbonates generally give highly tensile strength material with better hydrolysis and oil resistance properties [6]. Typically polytetramethylene ether glycol (PTMG), poly(tetramethylene adipate) glycol (PTA), and polycaprolactone glycol (PCL) are often used. The functionality of polyol can play a vital role during the synthesis of PU and PUU. By altering the functionality of polyol, different kinds of PU and PUU can be obtained. The macrodiols difunctional polymer consisting functionality of two are used to get thermoplastic films, commonly used WPU and WPUU [7]. Higher functionality polyols are used to synthesized highly crosslinked WPU and WPU thermoset system. The hard segment consists of polyisocyanate. The isocyanate is a highly reactive functional group that can combine with several functional groups, including carboxylic groups, amines, urea, etc. There are numerous isocyanates, but the most regularly used in the production of WPU and WPUU dispersions are diisocyanates. The isocyanate can have different structures, including aromatic or aliphatic (cyclic or linear) [8], and it impacts the reactivity of the reaction system. Aromatic isocyanate gives high thermal retardant properties to the PU because of the increased stability of aromatic rings. For instance, aromatic diisocyanates, including 2,4- or 2,6-toluene diisocyanate (TDI) or 4,4′-diphenylmethane diisocyanate (MDI), is widely used at the industrial level in PU synthesis but does not go well with the synthesis of WPU and WPUU because of their high reactivity in aqueous medium and tendency to result in extreme viscosities, which make the diffusion process difficult [9]. The aliphatic cycloaliphatic diisocyanates, like 5-isocyanate-1-(isocyanate methyl)-1,3,3-trimethyl cyclohexane (IPDI), 4,4′-dicyclohexyl methane diisocyanate ($H_{12}MDI$), and 1,6-hexamethylene diisocyanate (HDI), are more commonly used in coating-related applications due to their UV-resistant properties. These aliphatic diisocyanates also give protection against thermal and hydrolytic attacks and enhance resistance against the yellowing effect in films [9,10]. IPDI is more often used in the synthesis of the dispersion because of the low reactivity of their NCO group with water, which makes the reaction system controllable, and the highly stable final product is formed [10]. Additionally, its unsymmetrical composition leads to less ordered structures, which accelerates the diffusion of water toward hard segment domains during the dispersion step [9]. The development of PU from using $H_{12}MDI$ in place of IPDI gives a highly dispersed product with improved mechanical properties. Chain extenders are mostly difunctional, low molecular weight compounds [11]. The block length of the hard segment can be increased by using low molecular weight diols or diamines known as chain extenders. These chain extenders are aromatic or aliphatic compounds with a hydroxyl group or amine group terminated, which reacts with a diisocyanate to form PU. The reaction of an amine with diisocyanate is kinetically very fast and exothermic in nature [12]. The type of chain extender used during the synthesis process plays an important role in developing WPU and WPUU dispersion. The diol-based chain extenders, such as ethylene glycol, butanediol, etc., react with the hard segment's NCO group to form urethane linkage [13]. The reaction of diamine-based chain extenders with diisocyanate results in urea linkage. Chain extension with diamines terminates, namely diethylene triamine (DETA) and triethylenetetramine (TETA), leads to the internal crosslinking [14]. In urethane linkage, only one hydrogen contributes to hydrogen bonding, as shown in Figure 2.1. In the case of urea linkage, both hydrogens can simultaneously participate in bidentate hydrogen bonding, resulting in high tensile strength and modulus with decreased elongation product. This stronger interaction of urea linkage in the WPUU system gives the stiffer product. Generally, a diol-

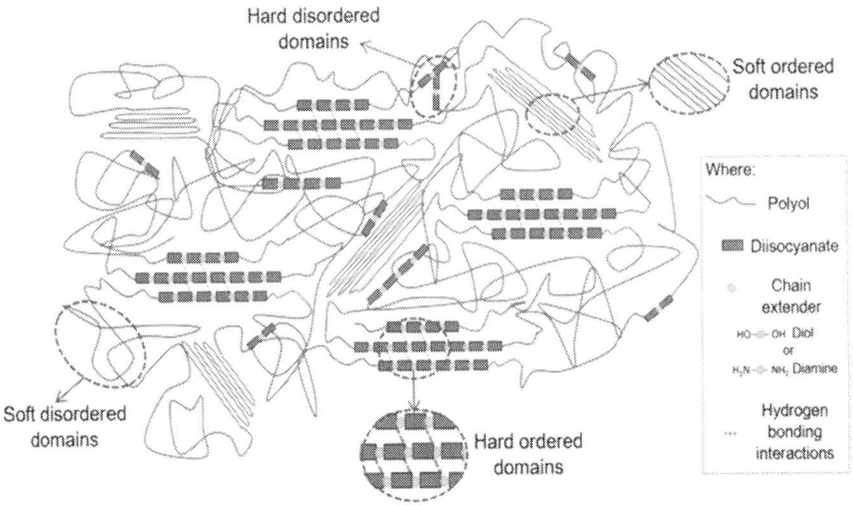

FIGURE 2.1 The interactions—segment conformations and hydrogen bonding of polyurethanes or polyurethane-ureas—and the resulting organized and cluttered microdomains. Adapted with permission from reference [15]. Copyright The Authors, some rights reserved; exclusive licensee [MDPI]. Distributed under a Creative Commons Attribution License 4.0 (CC BY) https://creativecommons.org/licenses/by/4.0/.

based chain extender is used in WPU synthesis, and a diamine-based chain extender is used to synthesize WPUU [15].

Traditional PU, like most polymeric materials, is not capable of forming a stable homogenous solution with water. NCO-terminated hydrophobic PU pre-polymers can be mixed with an appropriate external emulsifier and strong shear forces. However, the dispersions obtained in this way are coarse and storage unstable. Therefore, a certain type of hydrophilic change is essential before dispersion in water is possible, and it usually is done by incorporating the ionic group into the polymer chain. The ionic groups present in the PU chain are known as ionomers. These ionic groups work as internal emulsifiers and are hydrophilic. The emulsifiers are amphiphilic polymer-like grafted polymers, which stabilize the PU dispersion in an aqueous medium. There are various types of emulsifiers used in WPU and WPUU synthesis. The emulsifier is covalently combined with the polymer's backbone and builds part of the polymer chain. Depending on nature, the emulsifier can classify into two categories: internal and external emulsifiers. The internal emulsifier can subdivide into ionic and nonionic emulsifiers. The ionic emulsifier can be further distributed into anionic and cationic emulsifiers, depending on the charge the functional group confers on the stability of the system. The nonionic WPU and WPUU are not extensively used because of their less hydrophilic nature, making them less dispersed in an aqueous medium. The synthesis of WPU and WPUU dispersions containing both internal emulsifiers, i.e. nonionic and ionic types simultaneously, has gained more population to acquire films with superior and stable properties by compiling advantageous properties provided by each emulsifier. Regarding ionic emulsifiers, the presence of an ionic center pendant in the polymer chain differentiates the cationic and anionic WPU and WPUU. In the case of anionic emulsifier, the sulfonated and carboxylic acid is normally used. For cationic emulsifiers, tertiary amines are integrated, which have to be last neutralized or quaternized, respectively, to form salt [16]. In the synthesis of anionic WPU and WPUU, the most frequently used anionic emulsifiers are 2,2-bis(hydroxymethyl) butyric acid (DMBA) or sulfonated agents and 2,2-bis(hydroxymethyl) propionic acid (DMPA). Instead, *N*-methyl diethanolamine (MDEA) is typically employed as a cationic emulsifier [16]. Both cationic and anionic systems have wide-ranging applications in various fields, including adhesives, textiles, and coatings. Generally, the reactivity of WPU and WPUU reactants can be altered by using a catalyst during the reaction between –OH and –NCO groups at a lower temperature. Numerous types of catalysts are employed during the reaction. Normally, tertiary amines such as 1,4-diazabicyclo octane (DABCO), triethylamine (TEA) ,or organotin compounds, especially dibutyltin dilaurate (DBTDL) and stannous octoate (SnOc), are used [17]. The polymerization can take place through one or two steps. In two-step polymerization, the chemical composition of PU can be more precisely specified. The synthesis procedure of WPU and WPUU via two-step polymerization is shown in the Figure 2.2.

After the polymerization process, the dispersion of WPU and WPUU takes place. However, depending upon the nature of the chain extender, homogenous or heterogenous reaction medium is formed. The diol-based chain extender was added before the water addition reaction due to its low reactivity with the –NCO group

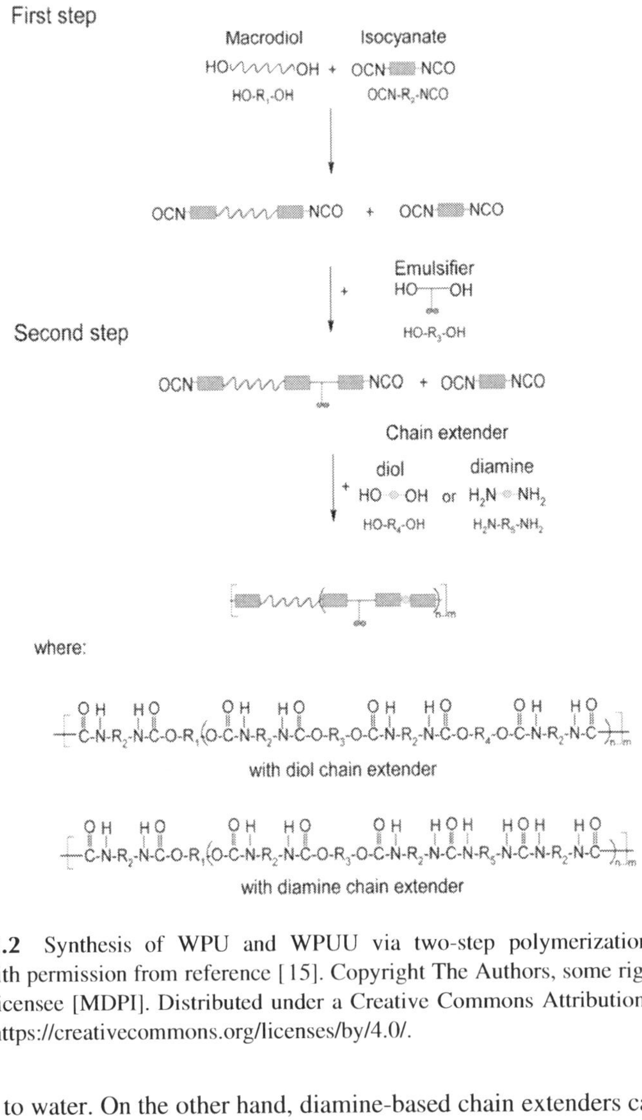

FIGURE 2.2 Synthesis of WPU and WPUU via two-step polymerization procedure. Adapted with permission from reference [15]. Copyright The Authors, some rights reserved; exclusive licensee [MDPI]. Distributed under a Creative Commons Attribution License 4.0 (CC BY) https://creativecommons.org/licenses/by/4.0/.

compared to water. On the other hand, diamine-based chain extenders can be added before the water addition reaction to form a homogenous medium or after the water addition reaction to complete the PU reaction in a heterogeneous medium [18]. The stabilization of dispersed particles also depends upon the emulsifier.

2.3 WATERBORNE POLYURETHANE-BASED NANOCOMPOSITES

PU-based material exhibits great potential for many applications in different fields, but it is also important to point out the various disadvantages of the material, including weak mechanical properties, lower tensile strength, and stability than solvent-borne PU polymer. In order to defect these drawbacks, the polymer is reinforced with different kinds of nanostructure to improve its properties and open a new door for the material application. Researchers are continuously working on developing various types of WPU-based nanocomposites with superior mechanical and thermal properties.

Laraza et al. [19] developed WPU and graphene/graphene oxide-based nanocomposites. Biobased waterborne polyurethane is combined with graphene, graphene oxide (GO), and reduced GO in this work. The combination of graphene and GO with WPU improved mechanical and thermomechanical properties and did not show any electrical conductivity. Coatings of graphene and reduced graphene oxide were studied to give electrical properties to the composites. Electrical conductor materials were obtained after coating the systems. Nike et al. [20] synthesized and characterized bio-nano composite based on chitosan, WPU, and modified nanosilica. The presence of chitosan in the PU matrix can increase the system's tensile strength and elastic modulus due to the urea bond and hydrogen bond formed between the polymer chain. It is also improving the thermal and mechanical properties. The hydrogen bonding between the silanol group of silica and the amine group of chitosan may decrease the interaction between chitosan and pre-polymer to decrease the covalent bond density. Bernard et al. [21] synthesized GO/WPU nanocoatings and studied the effects of GO content on performance properties. The superior mechanical and physical properties of GO with good dispersion ability in water make it a good nanofiller candidate for WPU coating. It was reported that GO is well dispersed in an aqueous medium and has good adhesion to WPU. GO/WPU nanocoating has superior thermal conductivity, stability, degree of dispersion, flammability, and water sorption at a relative humidity compared to nanocoating without GO/WPU.

2.4 APPLICATION AND TECHNOLOGY OF WATERBORNE POLYURETHANES

For the last two decades, WPUs have been used in various fibers as adhesive properties, sealing materials, coatings, paint colors, defoamers, pigment pastes, binders for cement, and dyes and biomaterials for the textile industry, resulting in the increasing demand to design and synthesize novel WPUs with unique properties. The following applications are described below.

2.4.1 Printing Ink

In the printing industry, ink has a significant role. In packing materials, ink demands a lot due to good binding capacity; therefore, in the packaging area, the demand for printing inks is increasing worldwide. According to a survey, more than one million tons of ink are produced in Europe annually [22]. New types of water-based ink have been considered the most popular sustainable printing ink due to its eco-friendly and low viscosity. It is collected for four types of elements: colorant, vehicle or binder, additives, and water. Vehicles or binders have numerous functions in ink. The old-style binders are generally aromatic alcohol resins, which are derivatives of oil resin, which is not environmentally benign. Ink performance should be improved for an eco-friendly environment by applying new resins as binders, using water-based materials. Therefore, water-based inks play an important role in modern times. They are also nontoxic and environmentally friendly. WPU dispersions contain a creditable grouping of physical properties, such as high ductile capacity and tear strength, strong springiness, a long-range of stiffness, worthy abrasion resistance capacity, good resistance to chemicals, and stability at low temperatures; all these factors apply well in the ink, surface coating, and adhesive industries [23].

2.4.2 Surface Sizing Treatment

Surface sizing treatment can improve the performance of paper. It has a tremendous effect and can increase liquid wetting resistance, strength of surface, and tensile strength of sheets [24]. For these changes, many surface sizing agents are used, such as starch [25], chitosan [26], vinyl alcohol and butadiene rubber latex [27], etc. Nowadays, WPUs are also applicable for the surface sizing agent due to their environmentally friendly nature and nontoxic behavior. WPUs have a high tolerance capacity regarding mechanical flexibility, toughness, and resistance capacity [28].

2.4.3 Coating

Coating is an important technique for preventing corrosion on metal and other materials. Therefore, coating should be designed that can provide high durability, increased hardness, and high capacity for bearing chemical resistance of the substrate. Coating of steel is one of the major concerns in the steel industry, to protect steel against corrosion. Coating of steel can be done in a variety of ways. One of them is an organic coating, which is a ubiquitous and effective method against corrosion. Solvent-based coating is one such approach. It acts as a barrier to water, oxygen, and corrosive ions like Cl^- and H^+. However, the disadvantage of solvent-based coating is that it is not eco-friendly and releases many VOCs. So, an alternate approach is a prerequisite to overcoming these disadvantages, as mentioned earlier. An eco-friendly approach that has a less harmful impact on the environment can be an alternative to solvent-based coating [29–31]. One such approach is the green coating of the substrate, which is eco-friendly and releases a minimal amount of VOC.

Green coating by WPU has clear weather ability, is environmentally friendly, and possesses good adhesion to steel substrate [32–34]. Thus, WPU coating of steel structure is a good approach against corrosion and is extensively used in the corrosion protection field. The waterborne coating has the disadvantage of hydrophilic moiety, allowing water to enter the coating, weakening corrosion protection of coatings. So, WPU has limited applications in protecting steel structures as it has low water resistance and possesses a meager capacity of anticorrosion. To overcome this problem, the coating surface is loaded with multifunctional micro nanoparticles, e.g. graphene, GO [35], montmorillonite [36], carbon nanotubes [37], SiO_2 [38], Zn oxide, Ti_{O2} [39,40], and Ce Oxide [41]. Such methods have been used widely in recent years. Nanomaterial graphene has a conducting property and can be used in the electrochemical protection of metal substrates. These nanomaterials are tiny in size and can supply compactness to the coating, fill up the micropores, and crack at the substrate coating surface. Hence, they supply a compact and smooth coating. This compact property of coating does not allow diffusion of corrosive agents and improves their shielding ability for the substrate [42].

2.4.4 Adhesives

Adhesive materials are one of the higher demands lately because adhesives have a wide range of applications and are used in various ways. Good adhesive material possesses unique properties, such as low starting solidity, low viscosity, and low cure time. There are many polymeric adhesives, but PU/acrylic has a unique property that makes it different from other adhesives [43]. PU/acrylic hybrid has good adhesion property, can be easily isolated, possesses color stability, and resists UV degradation. Recently, solvent-free and emulsifier-free WPU/acrylic hybrid adhesive materials have been introduced. These altered synthesized hybrids were used in footwear applications to prepare adhesive formulation. A wide range of waterborne hybrids have been prepared by varying the molar ratio of the substrate. Variation in the molar ratio of crosslinkable monomer trimethyl propane (TMP) or chain extender ethylene diamine (EDA) can result in various waterborne hybrids. These hybrids act as a high-performance adhesive for a synthetic leather or ethylene–vinyl acetate sole at the dry state and the wet state. The adhesive strength of the PU/acrylic hybrid works differently in the dry and wet states. The difference of property in both states depends on the variation in the molar ratio of TMP/EDA. In dry strength, the adhesive strength of formulated adhesives was not dependent on TMP/EDA molar ratio, while in the wet state, the adhesive strength of formulated adhesives increased enormously with an increase in TMP/EDA molar ratio. If the ratio is highest, the adhesive strength was higher than the footwear adhesion requirement.

Gruber and Derby developed a process for PU/ acrylic hybrids. The developed process results in modified PU/acrylic hybrids, which find their use for laminating adhesives in flexible packaging [44]. They used conventional PU/acrylic hybrids in their newly developed process but changed the chain extension step. Their new process at the chain extension step introduced a mixture of monofunctional and difunctional amines. The purpose of mixing these functionalized amines was to alter some of their property at a helpful level. As after introducing monofunctional amines, PU/acrylic hybrid provides good peel strength, even soaking in water. The developed process provides higher peel strength than the solvent-borne system and also makes an excellent laminating adhesive. There are other types of adhesives named pressure-sensitive adhesives. This special type of pressure-sensitive adhesive has a unique property that is to be sticky. When these pressure-sensitive adhesives are exposed to light pressure, they exhibit excellent tackiness and have very low or no residue left when removed from the substrate. This specification in property can be achieved by going through the required process regarding the polymer microstructure. This obtained polymer microstructure provides a balanced adhesive–cohesive nature. To provide better cohesive property, the polymer chain must be crosslinked, but only at a definite level. As the adhesive property cannot be

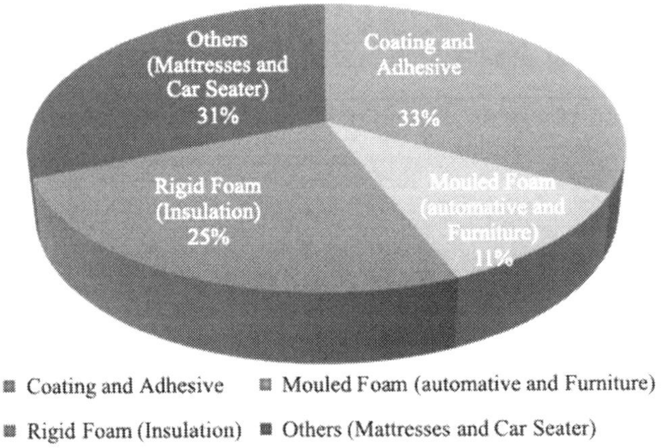

FIGURE 2.3 Different uses of PUs.

compromised, there should be balance in cohesiveness and adhesiveness. Most of the PUs are used in coating, and adhesive properties from the data and survey show almost 33% of users in both applications, as shown in Figure 2.3.

Most PUs of soft foam are used as coating and adhesive technology. The second applies to rigid foam as the major insulation technique.

2.5 MISCELLANEOUS APPLICATION OF WATERBORNE POLYURETHANES

2.5.1 Appliances

Most common PUs used in rigid foams in the freezer and refrigerator are essentially low-cost materials. There is a possibility of generating a WPU system applicable to rigid foam systems.

2.5.2 Automotive

PUs are applicable for car seats. Nowadays, some newly developed environmentally friendly PUs are used in car seats as foam. This type of PU foam makes car seats comfortable. It is also useful in reducing weight due to low weight properties, increasing fuel economy.

2.5.3 Building and Construction

In recent times, home demand is increasing continuously. There is also demand for high-performance materials that are strong and lightweight. A new type of synthesized PU can perform well. These WPUs are used in floors, flexible carpets, and the cooling system that generates through a color wall, etc.

2.5.4 Packaging

WPU packaging foam is cost-effective, easy to handle, and extensively used to protect and carry many things, such as electrical and medical apparatus, glassware, and bulky manufacturing parts.

2.5.5 Composite Wood

WPUs play an essential role in most materials, such as composite wood. WPU-based binders are used in composite wood as permanent glue-like organic PUs if moisture is absent. Therefore, a huge possibility for a bright future of WPUs in composite wood.

2.5.6 Wastewater Treatment

Removing pollutants from wastewater using mixed microbial cultures is a stable process from an ecological point of view. WPUs have good physical, mechanical, and chemical stability, so they can be easily adapted with nanocomposites, resulting in excellent carriers of immobilized cells. For this reason, they can protect bacterial communities. They can provide a promising strategy for wastewater treatment. For example, WPUs were used as carriers for immobilization and decontamination of bacterial strains for wastewater treatment. Thus, they provide a promising strategy for high nitrogen wastewater treatment [45].

2.6 CONCLUSION

This chapter gives an overview of the structure of WPUs and their chemistry in the formulation mechanism of WPU. Dispersion of waterborne systems exhibits some unique properties similar to a solvent-based system, but with no harmful effect on the environment. The increasing price of solvents and increasing awareness of environmental concerns are factors in increasing the use of water as the primary solvent in the synthesis of PUs. The applications of WPUs are increasing in different fields, such as coating, adhesives, etc. In coating, WPUs are mainly used to prevent corrosion. They can also be used as an adhesive for increasing binding capacity, as packaging that is easy to handle, and as building block material for bearing large weight. The poor mechanical and thermal properties of WPUs restrict their technological application in some fields. To overcome these drawbacks, the PU is fused with nanostructure to form a nanocomposite with improved mechanical properties and a more stable structure.

ACKNOWLEDGMENTS

Naresh A. Rajpurohit, Kaushalya Bhakar, and Dinesh Kumar are thankful to the Central University of Gujarat, Gandhinagar, and Meena Nemiwal is thankful to MNIT, Jaipur, for providing support to carry out this work.

REFERENCES

1. Hepburn, C, (2012) *Polyurethane Elastomers*. Springer Science & Business Media.
2. Garrison, T and Kessler, M, (2016) Plant oil-based polyurethanes. *Bio-Based Plant Oil Polymers and Composites*, pp. 37–54. doi:10.1016/B978-0-323-35833-0.00003-7
3. Santamaria-Echart, A, Fernandes, I, Barreiro, F, Corcuera, MA and Eceiza, A, (2021) Advances in waterborne polyurethane and polyurethane-urea dispersions and their eco-friendly derivatives: A review. *Polymers*, *13*(3), pp. 409.
4. Wang, Z, Hou, Z and Wang, Y, (2013) Fluorinated waterborne shape memory polyurethane urea for potential medical implant application. *Journal of Applied Polymer Science*, *127*(1), pp. 710–716.
5. Hao, H, Shao, J, Deng, Y, He, S, Luo, F, Wu, Y, Li, J, Tan, H, Li, J and Fu, Q, (2016) Synthesis and characterization of biodegradable lysine-based waterborne polyurethane for soft tissue engineering applications. *Biomaterials Science*, *4*(11), pp. 1682–1690.
6. Cakic, SM, Spirkova, M, Ristic, IS, B-Simendic, JK, Milena, M and Poręba, R, (2013) The waterborne polyurethane dispersions based on polycarbonate diol: Effect of ionic content. *Materials Chemistry and Physics*, *138*(1), pp. 277–285.
7. Bullermann, J, Friebel, S, Salthammer, T and Spohnholz, R, (2013) Novel polyurethane dispersions based on renewable raw materials—Stability studies by variations of DMPA content and degree of neutralisation. *Progress in Organic Coatings*, *76*(4), pp. 609–615.
8. Yilgor, I, Yilgor, E and Wilkes, GL, (2015) Critical parameters in designing segmented polyurethanes and their effect on morphology and properties: A comprehensive review. *Polymer*, *58*, pp. A1–A36.
9. Cakic, SM, Stamenkovic, JV, Djordjevic, DM and Ristic, IS, (2009) Synthesis and degradation profile of cast films of PPG-DMPA-IPDI aqueous polyurethane dispersions based on selective catalysts. *Polymer Degradation and Stability*, *94*(11), pp. 2015–2022.
10. Wang, K, Peng, Y, Tong, R, Wang, Y and Wu, Z, (2010) The effects of isocyanate index on the properties of aliphatic waterborne polyurethaneureas. *Journal of Applied Polymer Science*, *118*(2), pp. 920–927.
11. Lee, SK and Kim, BK, (2009) High solid and high stability waterborne polyurethanes via ionic groups in soft segments and chain termini. *Journal of Colloid and Interface Science*, *336*(1), pp. 208–214.
12. Padsalgikar, AD, (2017) 3-Speciality plastics in cardiovascular applications. *Plastics in Medical Devices for Cardiovascular Applications*. William Andrew Publishing, pp. 53–82.
13. Chen, TK, Shieh, TS and Chui, JY, (1998) Studies on the first DSC endotherm of polyurethane hard segment based on 4,4′-diphenylmethane diisocyanate and 1,4-butanediol. *Macromolecules*, *31*(4), pp. 1312–1320.
14. Blackwell, J, Nagarajan, MR and Hoitink, TB, (1982) Structure of polyurethane elastomers: Effect of chain extender length on the structure of MDI/diol hard segments. *Polymer*, *23*(7), pp. 950–956.
15. Santamaria-Echart, A, Fernandes, I, Barreiro, F, Corcuera, MA and Eceiza, A, (2021) Advances in waterborne polyurethane and polyurethane-urea dispersions and their eco-friendly derivatives: A review. *Polymers*, *13*(3), p.409.
16. Barikani, M, Valipour Ebrahimi, M and Seyed Mohaghegh, SM, (2007) Preparation and characterization of aqueous polyurethane dispersions containing ionic centers. *Journal of Applied Polymer Science*, *104*(6), pp. 3931–3937.
17. Fernandez-d'Arlas, B, Corcuera, M, Labidi, J, Mondragon, I and Eceiza, A, (2013) Poly (urea) urethanes based on amorphous quaternizable hard segments and a

crystalline polyol derived from castor oil. *Colloid and Polymer Science*, *291*(5), pp. 1247–1254.
18. Kim, BK, (1996) Aqueous polyurethane dispersions. *Colloid and Polymer Science*, *274*(7), pp. 599–611.
19. Larraza, I, Alonso-Lerma, B, Gonzalez, K, Gabilondo, N, Perez-Jimenez, R, Corcuera, MA, Arbelaiz, A and Eceiza, A, (2020) Waterborne polyurethane and graphene/graphene oxide-based nanocomposites: Reinforcement and electrical conductivity. *Express Polymer Letters*, *14*(11).
20. Nikje, MMA and Tehrani, ZM, (2010) Synthesis and characterization of waterborne polyurethane-chitosan nanocomposites. *Polymer-Plastics Technology and Engineering*, *49*(8), pp. 812–817.
21. Dias, G, Prado, M, Le Roux, C, Poirier, M, Micoud, P, Ligabue, R, Martin, F and Einloft, S, (2020) Synthetic talc as catalyst and filler for waterborne polyurethane-based nanocomposite synthesis. *Polymer Bulletin*, *77*(2), pp. 975–987.
22. Noreen, A, Zia, KM, Zuber, M, Tabasum, S and Saif, MJ, (2016) Recent trends in environmentally friendly waterborne polyurethane coatings: A review. *Korean Journal of Chemical Engineering*, *33*(2), pp. 388–400.
23. Zhou, X, Li, Y, Fang, C, Li, S, Cheng, Y, Lei, W and Meng, X, (2015) Recent advances in synthesis of waterborne polyurethane and their application in water-based ink: A review. *Journal of Materials Science & Technology*, *31*(7), pp. 708–722.
24. Zhang, C, Liang, H, Liang, D, Lin, Z, Chen, Q, Feng, P and Wang, Q, (2021) renewable castor-oil-based waterborne polyurethane networks: Simultaneously showing high strength, self-healing, processability and tunable multi shape memory. *Angewandte Chemie International Edition*, *60*(8), pp. 4289–4299.
25. Guo, YH, Guo, JJ, Li, SC, Li, X, Wang, GS and Huang, Z, (2013) Properties and paper sizing application of waterborne polyurethane emulsions synthesized with TDI and IPDI. *Colloids and Surfaces A: Physicochemical and Engineering Aspects*, *427*, pp. 53–61.
26. Lipponen, J and Gron, J, (2005) The effect of press draw and basis weight on woodfree paper properties during high-solids surface sizing. *Tappi Journal*, *4*(1), pp. 15–20.
27. Kjellgren, H, Gällstedt, M, Engström, G and Järnström, L, (2006) Barrier and surface properties of chitosan-coated greaseproof paper. *Carbohydrate Polymers*, *65*(4), pp. 453–460.
28. Pelton, R, (2009) Bioactive paper provides a low-cost platform for diagnostics. *TrAC Trends in Analytical Chemistry*, *28*(8), pp. 925–942.
29. Yoon, SH, Park, JH, Kim, EY and Kim, BK, (2011) Preparations and properties of waterborne polyurethane/allyl isocyanate-modified graphene oxide nanocomposites. *Colloid and Polymer Science*, *289*(17), pp. 1809–1814.
30. Mo, Q, Li, W, Yang, H, Gu, F, Chen, Q and Yang, R, (2019) Water resistance and corrosion protection properties of waterborne polyurethane coating enhanced by montmorillonite modified with Ce3+. *Progress in Organic Coatings*, *136*, p.105213.
31. Liu, M, Mao, X, Zhu, H, Lin, A and Wang, D, (2013) Water and corrosion resistance of epoxy–acrylic–amine waterborne coatings: Effects of resin molecular weight, polar group and hydrophobic segment. *Corrosion Science*, *75*, pp. 106–113.
32. Noreen, A, Zia, KM, Zuber, M, Tabasum, S and Saif, MJ, (2016) Recent trends in environmentally friendly waterborne polyurethane coatings: A review. *Korean Journal of Chemical Engineering*, *33*(2), pp. 388–400.
33. Gurunathan, T and Chung, JS, (2017) Synthesis of aminosilane crosslinked cationomeric waterborne polyurethane nanocomposites and its physicochemical properties. *Colloids and Surfaces A: Physicochemical and Engineering Aspects*, *522*, pp. 124–132.

34. Wu, J and Chen, D, (2016) Synthesis and characterization of waterborne polyurethane-based on covalently bound dimethylol propionic acid to e-caprolactone based polyester polyol. *Progress in Organic Coatings*, *97*, pp. 203–209.
35. Li, J, Cui, J, Yang, J, Li, Y, Qiu, H and Yang, J, (2016) Reinforcement of graphene and its derivatives on the anticorrosive properties of waterborne polyurethane coatings. *Composites Science and Technology*, *129*, pp. 30–37.
36. Zheng, H, Shao, Y, Wang, Y, Meng, G and Liu, B, (2017) Reinforcing the corrosion protection property of epoxy coating by using graphene oxide–poly (urea-formaldehyde) composites. *Corrosion Science*, *123*, pp. 267–277.
37. Al-Maharma, AY and Sendur, P, (2018) Review of the main factors controlling the fracture toughness and impact strength properties of natural composites. *Materials Research Express*, *6*(2), p.022001.
38. Meng, F, Wang, H, Huang, F, Guo, Y, Wang, Z, Hui, D and Zhou, Z, (2018) Graphene-based microwave absorbing composites: A review and perspective. *Composites Part B: Engineering*, *137*, pp. 260–277.
39. Ramezanzadeh, B, Haeri, Z and Ramezanzadeh, M, (2016) A facile route of making silica nanoparticles-covered graphene oxide nanohybrids (SiO_2-GO); fabrication of SiO_2-GO/epoxy composite coating with superior barrier and corrosion protection performance. *Chemical Engineering Journal*, *303*, pp. 511–528.
40. Rashvand, M and Ranjbar, Z, (2013) Effect of nano-ZnO particles on the corrosion resistance of polyurethane-based waterborne coatings immersed in sodium chloride solution via EIS technique. *Progress in Organic Coatings*, *76*(10), pp. 1413–1417.
41. Sabzi, M, Mirabedini, SM, Zohuriaan-Mehr, J and Atai, M, (2009) Surface modification of TiO2 nanoparticles with silane coupling agent and investigation of its effect on the properties of polyurethane composite coating. *Progress in Organic Coatings*, *65*(2), pp. 222–228.
42. Zhou, S, Zhu, X, Ma, L, Yan, Q, and Wang, S, (2018) Outstanding superhydrophobicity and corrosion resistance on carbon-based film surfaces coupled with multi-walled carbon nanotubes and nickel nano-particles. *Surface Science*, *677*, pp. 193–202.
43. Parhizkar, N, Ramezanzadeh, B and Shahrabi, T, (2018) Corrosion protection and adhesion properties of the epoxy coating applied on the steel substrate pre-treated by a sol-gel based silane coating filled with amino and isocyanate silane functionalized graphene oxide nanosheets. *Applied Surface Science*, *439*, pp. 45–59.
44. Mehravar, S, Ballard, N, Tomovska, R and Asua, JM, (2019) Polyurethane/acrylic hybrid waterborne dispersions: Synthesis, properties and applications. *Industrial & Engineering Chemistry Research*, *58*(46), pp. 20902–20922.
45. Dong, H, Wang, W, Song, Z, Dong, H, Wang, J, Sun, S, Zhang, Z, Ke, M, Zhang, Z, Wu, WM and Zhang, G, (2017) A high-efficiency denitrification bioreactor for the treatment of acrylonitrile wastewater using waterborne polyurethane immobilized activated sludge. *Bioresource Technology*, 239, pp. 472–481.

3 Green Materials for Waterborne Polyurethanes

Felipe M. de Souza, Prashant Kote, and Ram K. Gupta
Department of Chemistry, Kansas Polymer Research Center, Pittsburg State University, Pittsburg, KS, USA

CONTENTS

3.1 Introduction.....31
3.2 Materials for Polyurethanes.....32
3.3 Green Materials and Chemistries for WPU.....33
3.3.1 Thiol-Ene Coupling.....33
3.3.2 Epoxidation/Ring-Opening Reaction.....34
3.3.3 Hydroformylation and Hydrogenation Reactions.....35
3.4 Current Developments and Prospectives of WPUs.....36
3.5 Conclusion.....42
References.....43

3.1 INTRODUCTION

Polyurethanes are used in many fields, such as in bedding and furniture, automotive, footwear, thermal insulation, packaging, medical devices, paint, coatings, adhesives, sealants, elastomeric fibers, etc. The vast applications of polyurethanes are possible due to their unique tunable characteristics and the availability of a wide range of raw materials for their synthesis [1]. Polyurethanes are conventionally synthesized by a polyaddition reaction between multifunctional polyol (–OH) and polyisocyanates (–N=C=O). Although the polyaddition reaction is a simple process, for commercial applications of polyurethanes, many other chemicals, including volatile organic compounds (VOCs), are added in different steps of processing. The VOCs utilized during the manufacturing process have implied health, safety, and environmental concerns. Due to these issues, many restrictions were imposed on these industries, encouraging them to reduce the use of VOCs and move towards water-based polyurethanes. These restrictions led to the use of renewable resources for the production of VOC-free polyurethanes.

DOI: 10.1201/9781003173526-3

Waterborne polyurethanes (WPUs) are polymeric materials capable of overthrowing the problem of toxicity, flammability, along with low or no content of organic solvents. In addition, many bio-resources such as vegetable oils and fruit oils can be used as the starting materials for the synthesis of WPU, bringing more sustainable aspects to the industries. Based on a report, the use of WPU can reduce about 70% of VOCs compared to solvent-based polyurethanes [2,3]. A pivotal component for WPU synthesis is the use of vegetable oils, which are renewable, environmentally friendly, and inherently biodegradable. Despite their higher price compared to petrochemicals, bio-renewable materials can achieve a sustainable and perhaps cheaper production cost once they establish a solid base in the market. The plethora of renewable starting materials available also allow the synthesis of new polyurethanes with a range of properties.

3.2 MATERIALS FOR POLYURETHANES

The properties of polyurethanes depend on the type of polyol and isocyanate used for their synthesis. Aromatic and aliphatic are the two main classes of isocyanates for polyurethanes. Aromatic isocyanates lead to more rigid chemical structures due to the inherent low mobility of the aromatic ring, whereas aliphatic ones tend to make them flexible. Isocyanates are toxic, and researchers are finding ways to find safer alternatives for isocyanates. Isocyanate-free polyurethanes have been developed recently, but more research is needed to improve their properties for commercial applications.

Apart from isocyanates, polyols also play an important role in polyurethanes. Functionality, pending groups, molecular weight, and chain flexibility of polyols can affect the properties of polyurethanes. Different types of polyols, such as polyether, polyester, polycarbonates, polybutadienes, etc., are some of the common polyols for the synthesis of polyurethanes. Polyether and polyester polyols are the main polyols dominating the polyurethane industries. Polyether polyols are synthesized by the ring-opening reaction of the oxirane ring, which leads to bifunctional polyols. The most common starting materials for polyether polyols are ethylene oxide, propylene oxide, and butylene oxide. Among those, propylene oxide is largely used because of its established market and relatively good chemical stability. Polyether polyols based on ethylene oxide present a primary hydroxyl group, which is more reactive toward isocyanate than a secondary and tertiary group. However, due to the hydrophilic character of ethylene oxide-based polyol, the polyurethane's stability against moisture is lower than polyurethanes synthesized from propylene oxide and butylene oxide-based polyols. Polyester polyols are another major group that is used for the synthesis of polyurethanes. They are synthesized through a polycondensation reaction of multifunctional carboxylic acids and polyhydroxy compounds. Compared to polyether polyols-based polyurethanes, polyester ones tend to have higher thermal stability and chemical resistance.

Another major component for the synthesis of polyurethanes is catalysts. Tertiary amines and metal complex compounds based on zinc, bismuth, tin, or lead are common catalysts used in polyurethanes. Due to the presence of heavy metals in organometallic catalysts, there has been a growing interest to find environmentally friendly and safer materials that could serve as an alternative for these catalysts. Tin- and lead-based catalysts are the most harmful to the environment and health.

Surfactants are another important component used during the formulation of polyurethanes to improve the mixability of polyols with isocyanates and allow a good foaming process. Surfactants can be divided into four different types: anionic, cationic, nonionic, and amphoteric. Cationic and anionic surfactants are commonly used in polyurethanes. The first one is used for better emulsification action, while the second provides stability against corrosion [1]. For polyurethane foams, surfactants play an important role in forming and stabilizing the cellular structure, air permeability, and prevention of coalescence, while for WPU, surfactants improve dispersibility. Surfactants can be introduced externally or internally during the synthesis of WPU. External surfactants are blended into the reaction mixture to create the emulsion. Internal emulsifiers are chemically bonded to the starting chemicals to improve the dispersion of the polyurethanes in water. The second approach is a preferred method as it leads to a more stable WPU dispersion. With the basics of polyurethanes discussed above, the second part of this chapter addresses the main points in terms of materials and chemistries for WPU, showing the vast number of materials that can be employed as well as the chemical process for their synthesis. The third session covers many green approaches used for WPU, along with other important characteristics.

3.3 GREEN MATERIALS AND CHEMISTRIES FOR WPU

One of the major challenges in the synthesis of WPU is obtaining stable dispersions using mild reaction conditions and green solvents. This process is inherently demanding as the polyurethane chains do not properly interact with water as most of the polymer chains are composed of aliphatic and aromatic segments originated from polyol and isocyanate, respectively. This session discusses the materials used to overcome these issues as well as the chemistry involved. Developing greener procedures accompanied by a lower processing cost are two important steps toward sustainable production of WPU for commercial applications. Several plant-derived oils are suitable for the synthesis of WPU through well-known chemical modifications by taking advantage of unsaturations in their structure. Among many vegetable oils, castor oil is a convenient starting material as it contains carbon–carbon double bonds as well as hydroxyl groups, which allow its use in WPU without any chemical modifications. However, many bio-renewable compounds with carbon–carbon double bonds can be transformed into polyols using facile approaches, such as thiol-ene coupling (TEC), epoxidation/ring-opening, hydroformylation/reduction, ozonolysis/reduction, transesterification, etc. [4–6].

3.3.1 THIOL-ENE COUPLING

Thiol-ene coupling is a radical addition reaction between a double bond (C=C) and a mercaptan (H–S–R) in the presence of UV radiation and a photocatalyst.

The mechanism for this reaction consists of the formation of a thyil radical (•S–R) that reacts to the double bond, leading to a secondary or tertiary carbon radical, which bonds with the hydrogen radical (H•) derived from the mercaptan bond cleavage [7,8]. This is a very facile and versatile approach as the R group in mercaptan can be virtually any organic group. For example, Fu et al. demonstrated that castor oil can be converted into a polyol and an isocyanate to make a fully bio-based WPU by performing thiol-ene coupling [9]. A dicarboxylic acid-based compound was first synthesized through thiol-ene reaction between undecylenic acid and 3-mercaptopropionic acid, followed by Curtius rearrangement to obtain a bio-diisocyanate. Castor oil chain extender was synthesized using 3-mercaptopropionic acid via thiol-ene reaction, which also acted as an internal emulsifier. The isocyanate-terminated polyurethane reacted with a castor oil chain extender to introduce carboxylic acids into the polymeric chain, hence increasing its dispersibility in water. Finally, the carboxylic acid was neutralized with trimethylamine. It is worth noting that the isocyanate was synthesized using undecylenic acid that can be obtained by the pyrolysis of ricinoleic acid found in castor oil. Also, the thiol-ene coupling can be performed using low quantity or no solvent.

3.3.2 Epoxidation/Ring-Opening Reaction

Many bio-oils, such as soybean, olive, linseed, canola, grape seed, and castor, contain unsaturation that can act as a reactive site for many chemical transformations. One of the most performed chemical modifications in these oils is epoxidation followed by ring-opening reaction. This consists of breaking the double bond and forming an oxirane ring, traditionally by forming a peracid in-situ. For that, hydrogen peroxide and acetic or formic acid can be used. Then, a ring-opening reaction can be performed virtually by using any nucleophile, for example, methanol, butanol, ethanol, HCl, and many others. A general chemical reaction for this method is shown in Figure 3.1. Alcohols or water in the presence of acid catalysts, organic acids, inorganic acids, and hydrogenation can be used to open the epoxy rings [10]. Epoxidation can be done in four ways: (i) peracids including peracetic acid or perbenzoic acid in the presence of an acid catalyst; (ii) organic and inorganic peroxides, including transition metal catalysts; (iii) halohydrins using hypohalous acids and their salts; and (iv) molecular oxygen. Epoxidation with molecular oxygen does not require a catalyst and is considered the cleanest method. However, it is not applied to the industry because it demands too specific equipment and extensive safety measures due to the risk of explosions [11,12]. While formic and acetic acids are the most commonly used acids because they are readily available in liquid form at room temperature and are relatively inexpensive. Zhang et al. explored a greener and viable path utilizing castor oil fatty acid to perform the alkoxylation reaction (ring-opening) for polyurethanes [13]. The synthesis was performed without solvent or catalyst and introduced an extra hydroxyl group in the structure to increase the functionality of the polyols (Figure 3.2).

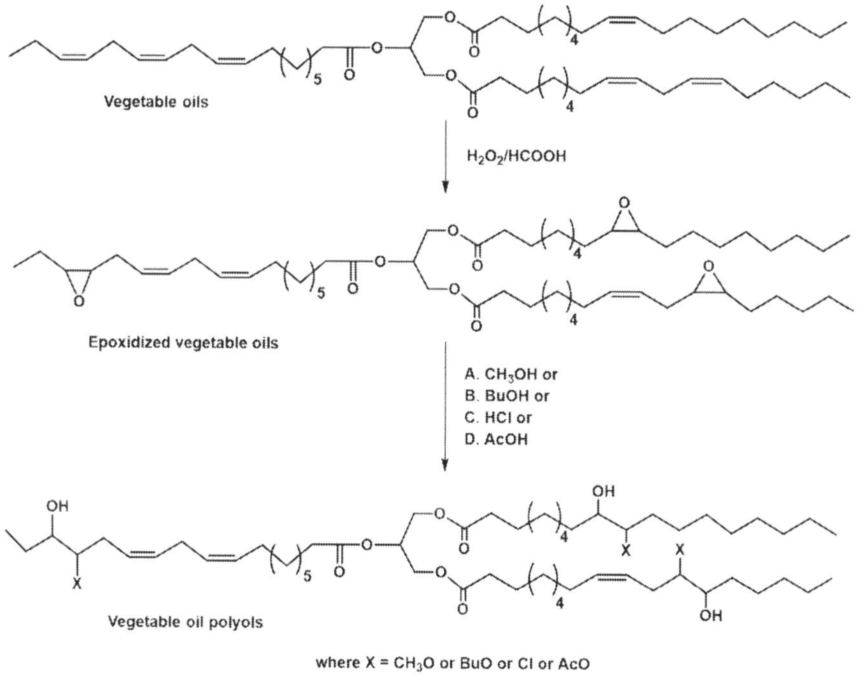

FIGURE 3.1 General epoxidation followed by ring-opening reaction for unsaturated oils. Adapted with permission from reference [14]. Copyright 2014, Elsevier.

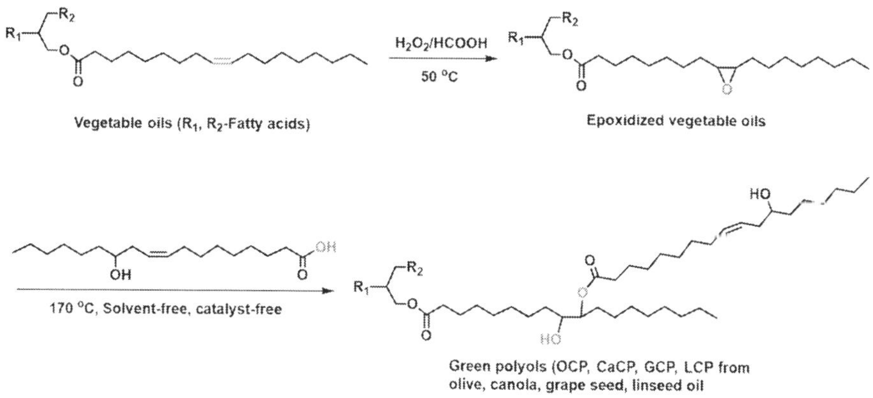

FIGURE 3.2 Epoxidation followed by ring-opening reaction of castor-oil fatty acid. Adapted with permission from reference [13]. Copyright 2015, American Chemical Society.

3.3.3 Hydroformylation and Hydrogenation Reactions

Hydroformylation involves the addition of CO and H_2 to an alkene to form an aldehyde. The aldehydes produced by hydroformylation are normally reduced to alcohol via hydrogenation, which is a reaction with hydrogen under pressure and catalysts such as Pt or Raney Ni. Hydroformylation is usually carried out in the presence of Rh or Co carbonyls as catalysts. Rh catalysts are very efficient, avoid oligomerization, and deliver almost 100% conversions; however, they are not cost-effective. On the other hand, hydroformylation with Co allows some oligomerization, likely due to transesterification at higher temperatures. Petrovic et al. made an interesting study on the property of several polyols based on soybean that was synthesized through the hydroformylation/reduction method [15]. This approach presents some advantages, such as the introduction of primary hydroxyl groups, which are more reactive toward isocyanate and the high conversion of aldehydes into hydroxyl groups, leading to a higher functionality of the polyols. When targeting for WPU, a high functionality may not be desired in the early stages of WPU synthesis, since it may lead to crosslinking yielding polymers that swell in solution instead of dispersing. It is possible to control the functionality by partially blocking the hydroxyl groups by performing esterification.

3.4 CURRENT DEVELOPMENTS AND PROSPECTIVES OF WPUs

The unique properties and eco-friendly nature of WPUs make them suitable for many applications, such as coating, adhesive, sealant, and ink industries. Because of their vast applications, they share almost 20% of the global polyurethane market [16]. This session discusses some of the techniques and approaches used to synthesize green WPU as well as the structure–property correlation. Nonionic, cationic, and anionic are the three main types of WPU. Nonionic WPUs are generally composed of an ethylene oxide group in its chain that acts as an internal emulsifier to provides stable dispersion. This type of WPU is used for finishing, cosmetics, and dyes; however, nonionic WPUs tend to become unstable at high temperatures. On the other hand, ionic WPUs are more stable at higher temperatures, and they tend to be mechanically stronger due to intermolecular interactions [17]. Cationic WPUs have good adhesion on several substrates, such as textiles, glass, and leather. Due to the positive charge on the surface of cationic WPUs, they can interact with biomolecules with an opposite charge to provide antimicrobial and antioxidant properties [18]. Anionic WPUs typically present higher mechanical strength compared to cationic WPUs [19,20].

Despite these differences, several factors can influence the properties of WPUs. Some of them are (i) polyol's chemical structure that is mostly aliphatic chains considered as soft segments in the WPU's structure. Longer aliphatic chains may lead to a more compact film; nonetheless, it is likely to decrease the dispersibility of the derived polyurethane dispersion in water, (ii) isocyanate's chemical structure (rigid segments) leads to an increase in crystallinity, mechanical, and thermal properties. For example, the use of 4,4′-dicyclohexylmethane diisocyanate leads to the formation of polyurethanes with higher tensile strength due to its structural rigidity and proper packing of aromatic rings that induce more crystallinity, whereas IPDI (more flexible structure) leads to a lower tensile strength [21], (iii) post-crosslinking allows control of the reaction system and yields WPU with relatively higher resistance to moisture and temperature. This process can be performed in several ways, such as blending method, interpenetration network, polymerization, and graft emulsion [22], (iv) introducing nanoparticles [23,24] and (v) hydrophilic chain extenders which is an important step of the synthetic process of WPUs as it increases the dispersibility of the polymer in water as well as the molecular weight. Traditionally, dimethylol propionic acid (DMPA) and derivatives are used as anionic chain extenders while diethanolamine (DEA) and derivatives can be used as cationic chain extenders.

Most of these materials, specially DMPA, are derived from nonrenewable sources; hence, researchers are looking for alternatives. For example, Gaddam et al. synthesized a DMPA-free anionic waterborne polyurethane-imide (WPUI) by incorporating maleated cottonseed oil polyol (MAHCSO) as the internal emulsifier and different dianhydrides as chain extenders, which provided two valuable aspects for green chemistry [25]. First, the introduction of a renewable material (MAHCSO) as an internal emulsifier provided a green alternative for DMPA. Also, since MAHCSO is a relatively large molecule for traditional internal emulsifiers, it aided in decreasing the microphase separation between the soft and rigid segments in the WPUI, which resulted in a smaller particle size that made the dispersion stable up to six months. The decrease in microphase separation was analyzed by performing dynamic mechanical thermal analysis (DMTA), which demonstrated a single glass transition temperature (T_g), suggesting both soft and rigid segments of the polymer were properly dispersed in each other. Second, the addition of aromatic dianhydrides moieties introduced rigid groups and increased the molecular weight, which enhanced their hydrophobicity, mechanical, and thermal properties. Some of the values obtained were around 10.3 MPa for tensile strength, 7.1 MPa for Young's modulus, 144% elongation at break, $T_g \sim 62°C$, and 100° of water contact angle.

Aside from the internal emulsifier as a core component for the synthesis of WPU, neutralizing agents are also very important to provide a stable dispersion. Neutralizing agents contain a charge opposite to that of the internal emulsifier in the polymeric dispersion. Zhang et al. used several bio-based phenolic acids as efficient neutralizers [26]. The chemical structure of the neutralizer is presented in Figure 3.3. A neat cationic WPU composed of castor oil, IPDI, and *N*-methyl diethanolamine was synthesized. The phenols were separately incorporated into it and compared with acetic acid (AA) and HCl. The presence of phenols not only neutralized the positive charges in the polymer surface but also promoted higher intermolecular interactions between the chains through hydrogen bonding, leading to a bridged crosslinked structure. The phenolic groups also acted as anchor groups to reduce the mobility of soft segments, thus avoiding agglomeration. The mechanism of the improvement of phase mixability can be seen in Figure 3.4. The higher number of hydrogen bonds provided higher T_g. It also increased the mechanical properties as one phenolic hydrogen in the compound led to tougher coatings, and as the number of phenolic hydrogens increased to three, the coatings behaved like a fiber, presenting high tensile strength but lower elasticity. Lastly, the presence of phenolic groups endowed high UV absorption for the coating as well as antibacterial properties for both Gram-positive and Gram-negative bacteria.

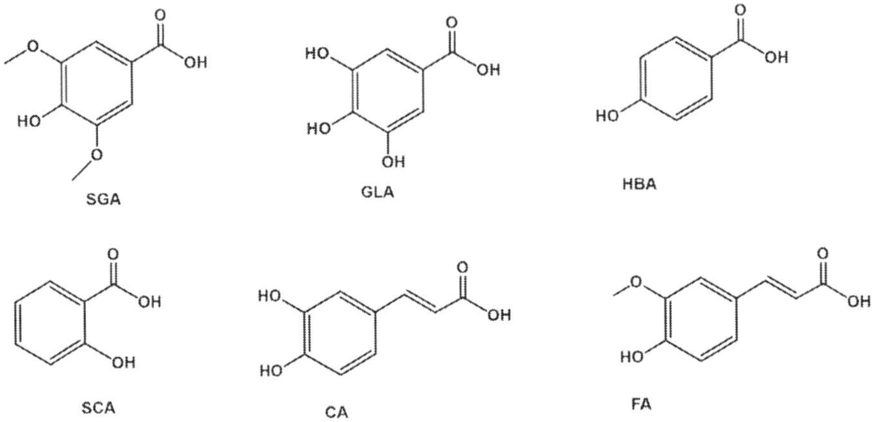

FIGURE 3.3 Phenolic acids used as neutralizers in castor oil-based WPU expressed as syringic acid (SGA), gallic acid (GLA), 4-hydroxybenzoic acid (HBA), salicylic acid (SCA), caffeic acid (CA), and ferulic acid (FA). Adapted with permission from reference [26]. Copyright 2015, American Chemical Society.

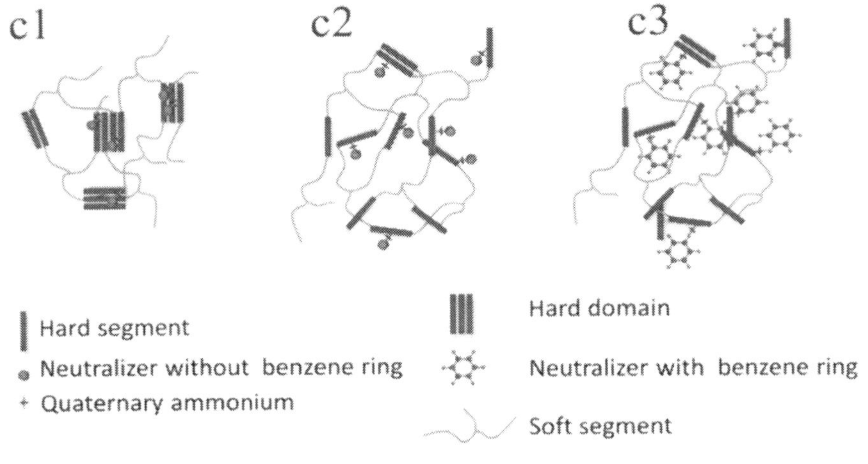

FIGURE 3.4 Phase separation for the castor oil-based WPU with (c1) high phase separation, (c2) low phase separation, and (c3) neutralization with phenolic acids. Adapted with permission from reference [26]. Copyright 2015, American Chemical Society.

Carbon dioxide plays a crucial role in vital plants to maintain life on the earth, but excess of this gas is harmful as it enhances the greenhouse effect. The scientific community finds ways to diminish the concentration of excess CO_2, such as using it for synthesizing polymers. CO_2 can be chemically structured in WPU to form carbonate ether bonds, which introduce new properties such as anticorrosive, hydrophobicity, shape-memory, etc. [27,28]. For example, Gong et al. synthesized a CO_2-based WPU using CO_2-based polyol, IPDI, and 3-(dimethyl-amino) propane-1,2 diol (DMAD) [29]. The researchers performed a study regarding the position of the internal emulsifier in the polymeric chain. For that, DMAD was chemically bonded as a terminal group (t-DMAD) or as a side group (s-DMAD). It was observed that the terminal groups of tertiary amines (t-DMAD) provided the highest dispersibility with the use of only 1 wt.% of this internal emulsifier. High dispersibility occurred because of a less steric hindrance at the terminal group, which facilitated the neutralization of the tertiary amines with acid. Such an approach provided an important aspect to decrease cost and optimized the synthetic route to decrease the use of internal emulsifiers to achieve satisfactory dispersion.

One may consider that higher dispersibility with water correlates with higher hydrophilicity and therefore less resistance to moisture, leading to an increase in water intake for the polymers. However, it was observed that due to the incorporation of carbonate groups, relatively rigid and chemically stable group, there was lower water absorption. The lowest value of water intake was obtained when 65% of the WPU was composed of carbonate groups derived from CO_2, reaching about 2% of water absorption while immersed in a solution for 70 h. On the other hand, when a traditional WPU dispersion was submitted to the same test, the absorption of water went up to almost 7% in 70 h. Thus, this type of approach revealed an important aspect for the emulsifying process, indicating that the positioning of the polar groups has more influence than the quantity, which is incorporated into the polymeric backbone.

As demonstrated so far, the use of bio-renewable sources to obtain new materials heavily relies on the chemistry of double bonds, and scientists look for eco-friendly new materials suitable for such transformations. Shao et al. focused their work on using a bio-based compound named rosin, which is known for its biodegradability, biocompatibility, and abundance [30]. Rosin was then chemically modified and used to prepare a stable dispersion of hydrophilic aromatic PU (HAPU) microspheres with potential applications in the medical field. Rosin was first chemically bonded with acrylic acid (AA) through a Diels–Alder reaction to make a dicarboxylic acid named rosin acid (RA), followed by an esterification reaction between RA and glycidyl methacrylate (GMA) to introduce double bond and hydroxyl groups to be used as a polyol (RAG). RAG reacted with TDI to make the PU backbone, and DMPA was used as a chain extender as well as an internal emulsifier that was further neutralized with TEA. The remaining double bonds in the polymer were used as a reactive site for radical polymerization of styrene to form stable microsphere nanostructures. The synthetic process is described in Figure 3.5. The rosin-based nanospheres had a size of 120 nm and were able to adsorb golden orange II dye around 17–9 mg/g in the pH range of 5.5–7. This approach showed a possible path for the use of rosin or other chemically similar bio-based compounds for a drug-delivery system.

The development of biocompatible and biodegradable polymeric materials carries great importance in the biomedical area, particularly using renewable materials. Liu et al. offer a greener approach to synthesize WPU for biomedical applications [31]. They used PEG and polycaprolactone (PCL) diols along with methylene diisocyanate (MDI) to form polyurethane that was crosslinked with cellulose nanocrystals (CNC) to improve mechanical properties. The WPU presented memory-shape properties in contact with water around 37°C, along with

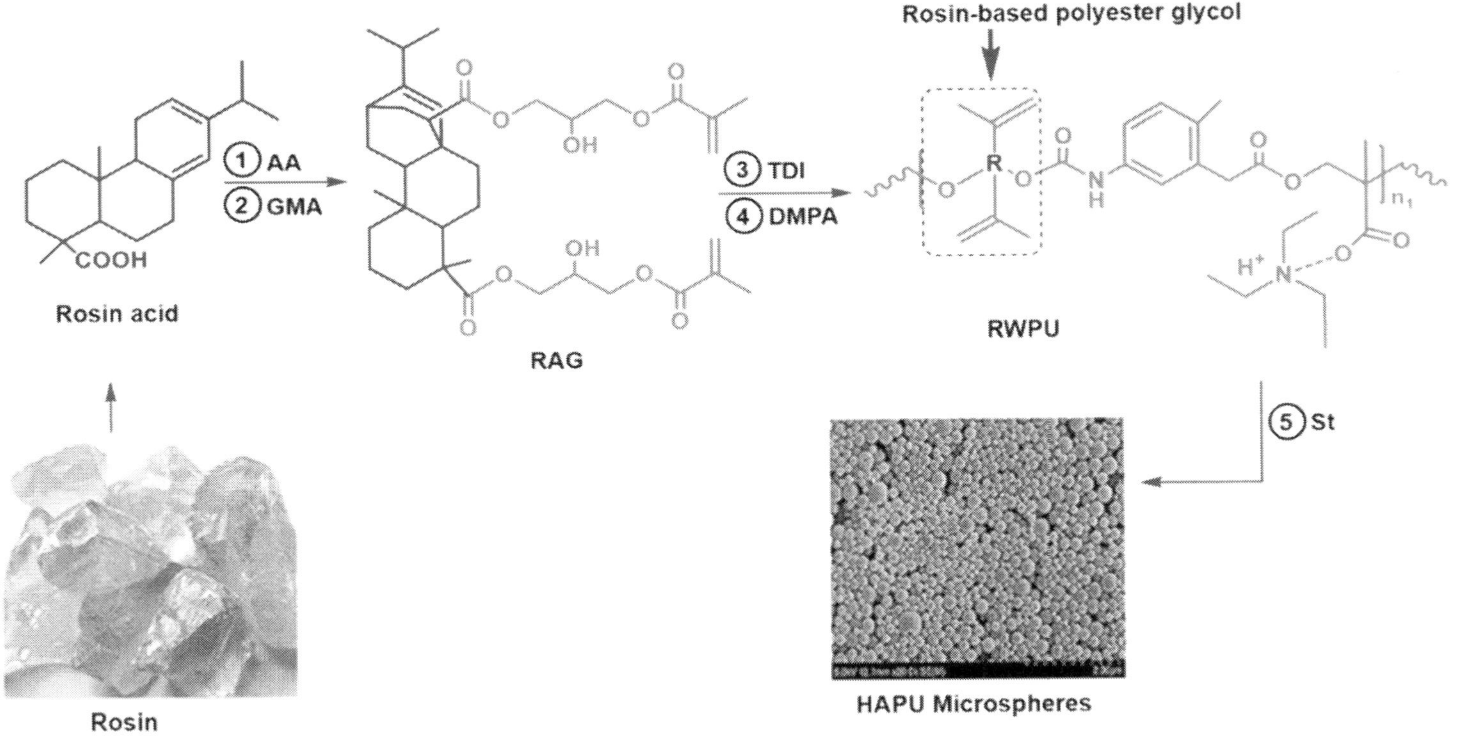

FIGURE 3.5 Synthetic process of rosin-based polyol through reaction with (1) AA and (2) GMA. Followed-up with a traditional WPU synthetic process with (3) TDI and (4) DMPA to obtain a rosin-based WPU (RWPU). Finally, it reacted with (5) styrene to obtain HAPU microspheres. Adapted with permission from reference [30]. Copyright 2019, American Chemical Society.

good interaction with osteoblasts (bone cells with a high metabolism to induce bone growth). The compounds utilized to make the polyol were highly hydrophilic, which caused the water absorption to occur relatively fast. This characteristic is desired for biomedical materials as they are required to interact well with water to induce cell growth. The crosslinked structure promoted swelling, which allowed the WPU composite to return to its original form after bending and exposure to 37°C in a water solution, which was mostly attributed due to the presence of CNC. In addition, PEG is a hydrophilic polymer that eases water absorption. Hence, these factors, such as shape memory properties close to the body's temperature, good interaction with water, biodegradability, and low cytotoxicity, are good indications that this WPU is a viable candidate for a cell-growing tissue.

The isocyanate moieties are key components for the synthesis of PU due to their high reactivity toward hydroxyls, carboxylic acids, and amines. In the case of WPU, the use of isocyanate demands special care during synthesis as it can easily react with water instead of polyol. Therefore, parameters such as temperature and water addition need to be controlled to obtain a stable WPU dispersion. This is one of the reasons why the acetone process is mostly used as it gives better control of the reaction due to its chemical inertness and volatility, which permits a proper solvent exchange with water to create dispersion with low particle size.

The high toxicity of isocyanate is a concern, and finding a green alternative is challenging. Some researchers have synthesized non-isocyanate-based polyurethanes, particularly for biomedical applications. In the biomedical field, decreasing the amount of isocyanate in polyurethane can facilitate some processes. For example, WPU hydrogels can function as drug delivery systems or enzyme immobilizer, which requires chemical bonding with bioactive materials. These biological components are often sensitive to temperature, solvents, and can also react with isocyanate. Hence, a common approach to fabricate PU gels is by synthesizing the polymer and then impregnating it with the drug, enzyme, or bioactive material [32]. Yet, envisioning a process performed in aqueous media to incorporate biomolecules in WPU hydrogels is challenging. A new organic moiety, poly(hydroxyl urethane) (PHU), has gained ground as a promising isocyanate-less polymer synthesized via a polyaddition reaction between di or polycyclic carbonates along with di or polyamines. Polycyclic carbonates are vital components of this process. They are formed through the reaction between epoxies and CO_2 [33,34]. Bourguignon et al. proposed a novel synthetic process for an isocyanate-free route that was performed in water and at room temperature by chemically bonding PEG dicyclic carbonate with polyethyleneimine (PEI) [35]. The synthetic process for this reaction is shown in Figure 3.6.

Quienne et al. synthesized an isocyanate-free polyester-urethane polymer in a two-step process [36]. In the first step, hydroxyurethane was synthesized using a glycerol carbonate and a diamine. In the second step, the primary and secondary hydroxyl groups were reacted with 2,5-furan dicarboxylate to synthesize a polyester–urethane. Thermal stability was studied and observed that the onset degradation varied between 170 up to 220°C, which is comparable to traditional PU.

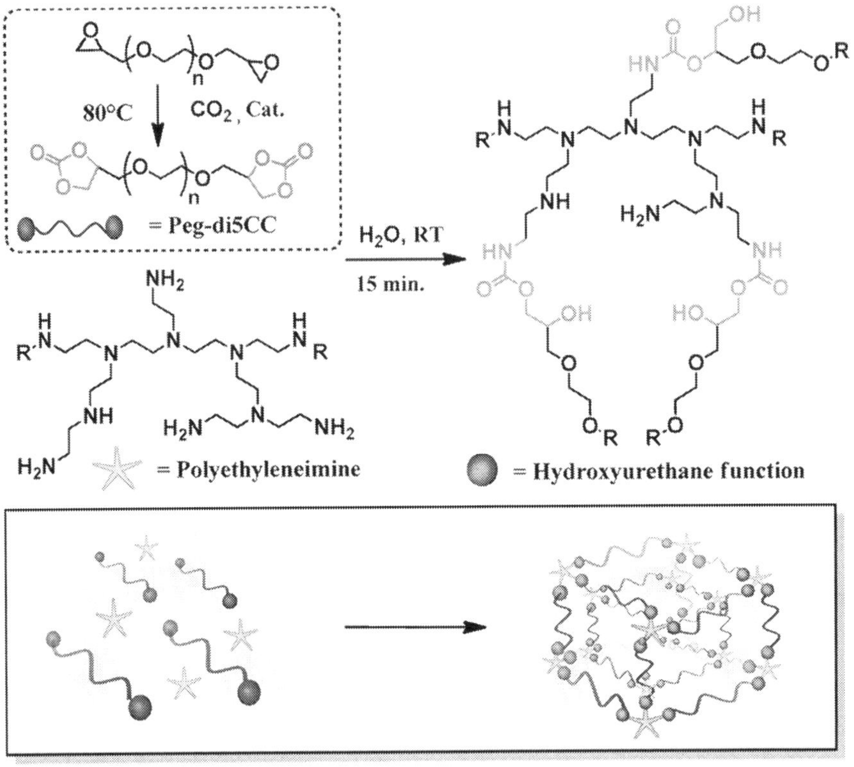

FIGURE 3.6 Synthetical approach to obtain an isocyanate-free WPU through the facile and eco-friendly methodology. Adapted with permission from reference [35]. Copyright 2019, American Chemical Society.

Despite the similarities in thermal degradation with PU, the PHU can undergo a process of dynamic bonding exchange that occurs by trans-carbamoylation reactions, which allows PHU to be reprocessed or recycled [37]. It can occur through thermal or mechanical activation [38–40]. Hence, this type of technology introduces novel materials that can be obtained through sustainable sources, for example, glycerol carbonate is synthesized through the reaction between glycerol and CO_2. On top of that, polymers that can be reprocessed or recycled are more sustainable and imply on lower cost in the long term as it does not require virgin reagents to be used for every synthesis. Thus, exploring different paths to incorporate bio-renewable materials, reduce the use of toxic chemicals, and perform the synthesis in aqueous media are practical approaches that the scientific community can act to contest the status quo.

3.5 CONCLUSION

By analyzing the synthesis processes and techniques developed over the years, it was notable that the PU industry thrived due to several factors, such as facile synthesis and processing, a vast number of starting materials, and a wide range of properties that covered many industrial sectors. As the concern with the environment increases, the industries are required to adopt green processes and materials without compromising the properties of polyurethanes. Even though this change is extremely challenging, it is possible to see through this chapter that the scientific community has developed several sustainable routes. As an example, techniques such as thiol-ene, hydroformylation, and epoxidation/ring-opening, among others used to synthesize polyols derived from bio-renewable resources, are viable and applicable for industrial applications. The use of aqueous media to obtain high-performance WPUs makes the processing safer and has decreased overall cost. Also, the biomedical field can greatly benefit from this type of approach since it facilitates the incorporation of bioactive materials for drug delivery, enzyme immobilization, scaffold, and more. Another important new addition to the green WPU synthesis is the development of isocyanate-free routes. This approach decreases the dependence on only one type of chemical group that is derived from petrochemical sources and provides a safer alternative to obtain urethane linkages. Despite the relative infant stage for this technology at the moment, several researchers have obtained promising materials that can potentially compose the market shortly.

REFERENCES

1. Akindoyo JO, Beg MDH, Ghazali S, Islam MR, Jeyaratnam N, Yuvaraj AR (2016) Polyurethane types, synthesis and applications-a review. *RSC Adv* 6:114453–114482.
2. Sukhawipat N, Raksanak W, Kalkornsurapranee E, Saetung A, Saetung N (2020) A new hybrid waterborne polyurethane coating synthesized from natural rubber and rubber seed oil with grafted acrylate. *Prog Org Coatings* 141:105554.
3. Galbis JA (2017) In: de García-Martín M (ed). *Bio-Based Polyurethanes from Carbohydrate Monomers*. IntechOpen, Rijeka.
4. Kong X, Narine SS (2007) Physical properties of polyurethane plastic sheets produced from polyols from canola oil. *Biomacromolecules* 8:2203–2209.
5. Palaskar DV, Boyer A, Cloutet E, Le Meins J-F, Gadenne B, Alfos C, Farcet C, Cramail H (2012) Original diols from sunflower and ricin oils: Synthesis, characterization, and use as polyurethane building blocks. *J Polym Sci Part A Polym Chem* 50:1766–1782.
6. Stemmelen M, Pessel F, Lapinte V, Caillol S, Habas J-P, Robin J-J (2011) A fully biobased epoxy resin from vegetable oils: From the synthesis of the precursors by thiol-ene reaction to the study of the final material. *J Polym Sci Part A Polym Chem* 49:2434–2444.
7. Hoyle CE, Lee TY, Roper T (2004) Thiol-enes: Chemistry of the past with promise for the future. *J Polym Sci, Part A Polym Chem* 42:5301–5338.
8. Desroches M, Caillol S, Lapinte V, Auvergne RM, Boutevin B (2011) Synthesis of biobased polyols by thiol–ene coupling from vegetable oils. *Macromolecules* 44:2489
9. Fu C, Zheng Z, Yang Z, Chen Y, Shen L (2014) A fully bio-based waterborne polyurethane dispersion from vegetable oils: From synthesis of precursors by thiol-ene reaction to study of final material. *Prog Org Coatings* 77:53–60.
10. Petrović ZS, Cvetković I, Hong DP, Wan X, Zhang W, Abraham T, Malsam J (2008) Polyester polyols and polyurethanes from ricinoleic acid. *J Appl Polym Sci* 108:1184–1190.

11. Bohnet M (2003) *Ullmann's Encyclopedia of Industrial Chemistry*, Third Edition. SpringerBerlin.
12. Cai C, Dai H, Chen R, Su C, Xu X, Zhang S, Yang L (2008) Studies on the kinetics of in situ epoxidation of vegetable oils. *Eur J Lipid Sci Technol* 110:341–346.
13. Zhang C, Madbouly SA, Kessler MR (2015) Biobased polyurethanes prepared from different vegetable oils. *ACS Appl Mater Interfaces* 7:1226–1233.
14. Garrison TF, Kessler MR, Larock RC (2014) Effects of unsaturation and different ring-opening methods on the properties of vegetable oil-based polyurethane coatings. *Polymer (Guildf)* 55:1004–1011.
15. Petrović ZS, Guo A, Javni I, Cvetković I, Hong DP (2008) Polyurethane networks from polyols obtained by hydroformylation of soybean oil. *Polym Int* 57:275–281.
16. Gama N V, Ferreira A, Barros-Timmons A (2018) Polyurethane foams: Past, present, and future. *Materials (Basel)* 11:1841
17. Yen M-S, Tsai H-C, Hong P-D (2006) The physical properties of aqueous cationic–nonionic polyurethane with poly(ethylene glycol methyl ether) side chain and its blend with aqueous cationic polyurethane. *J Appl Polym Sci* 100: 2963–2974.
18. Li J, Zhang X, Gooch J, Sun W, Wang H, Wang K (2015) Photo- and pH-sensitive azo-containing cationic waterborne polyurethane. *Polym Bull* 72:881–895.
19. Liang H, Liu L, Lu J, Chen M, Zhang C (2018) Castor oil-based cationic waterborne polyurethane dispersions: Storage stability, thermo-physical properties and anti-bacterial properties. *Ind Crops Prod* 117:169–178.
20. Liang H, Wang S, He H, Wang M, Liu L, Lu J, Zhang Y, Zhang C (2018) Aqueous anionic polyurethane dispersions from castor oil. *Ind Crops Prod* 122:182–189.
21. Lu Y, Larock RC (2010) Aqueous cationic polyurethane dispersions from vegetable oils. *Chem Sus Chem* 3:329–333.
22. Deng Y, Zhou C, Zhang Q, Zhang M, Zhang H (2020) Structure and performance of waterborne polyurethane-acrylate composite emulsions for industrial coatings: Effect of preparation methods. *Colloid Polym Sci* 298:139–149.
23. Tamayo L, Acuña D, Riveros AL, Kogan MJ, Azócar MI, Páez M, Leal M, Urzúa M, Cerda E (2018) Porous nanogold/polyurethane scaffolds with improved antibiofilm, mechanical, and thermal properties and with reduced effects on cell viability: A suitable material for soft tissue applications. *ACS Appl Mater Interfaces* 10:13361–13372.
24. Zhao H, She W, Shi D, Wu W, Zhang Q, Li RKY (2019) Polyurethane/POSS nanocomposites for superior hydrophobicity and high ductility. *Compos Part B Eng* 177:107441.
25. Gaddam S kumar, Palanisamy A (2017) Anionic waterborne polyurethane-imide dispersions from cottonseed oil based ionic polyol. *Ind Crops Prod* 96:132–139.
26. Zhang Y, Zhang W, Deng H, Zhang W, Kang J, Zhang C (2020) Enhanced mechanical properties and functional performances of cationic waterborne polyurethanes enabled by different natural phenolic acids. *ACS Sustain Chem Eng* 8:17447–17457.
27. Alagi P, Ghorpade R, Choi YJ, Patil U, Kim I, Baik JH, Hong SC (2017) Carbon dioxide-based polyols as sustainable feedstock of thermoplastic polyurethane for corrosion-resistant metal coating. *ACS Sustain Chem Eng* 5:3871–3881.
28. Subhani MA, Köhler B, Gürtler C, Leitner W, Müller TE (2016) Transparent films from CO_2-based polyunsaturated poly(ether carbonate)s: A novel synthesis strategy and fast curing. *Angew Chemie Int Ed* 55:5591–5596.
29. Gong R, Cao H, Zhang H, Qiao L, Wang F, Wang X (2020) Terminal hydrophilicity-induced dispersion of cationic waterborne polyurethane from CO_2-based polyol. *Macromolecules* 53:6322–6330.

30. Shao J, Yu C, Bian F, Zeng Y, Zhang F (2019) Preparation and properties of hydrophilic rosin-based aromatic polyurethane microspheres. *ACS Omega* 4:2493–2499.
31. Liu Y, Li Y, Yang G, Zheng X, Zhou S (2015) Multi-stimulus-responsive shape-memory polymer nanocomposite network cross-linked by cellulose nanocrystals. *ACS Appl Mater Interfaces* 7:4118–4126.
32. Shekunov BY, Chattopadhyay P, Tong HHY, Chow AHL, Grossmann JG (2007) Structure and drug release in a crosslinked poly(ethylene oxide) hydrogel. *J Pharm Sci* 96:1320–1330.
33. Alves M, Grignard B, Mereau R, Jerome C, Tassaing T, Detrembleur C (2017) Organocatalyzed coupling of carbon dioxide with epoxides for the synthesis of cyclic carbonates: Catalyst design and mechanistic studies. *Catal Sci Technol* 7:2651–2684.
34. Büttner H, Longwitz L, Steinbauer J, Wulf C, Werner T (2017) Recent developments in the synthesis of cyclic carbonates from epoxides and CO_2. *Top Curr Chem* 375:50.
35. Bourguignon M, Thomassin J-M, Grignard B, Jerome C, Detrembleur C (2019) Fast and facile one-pot one-step preparation of nonisocyanate polyurethane hydrogels in water at room temperature. *ACS Sustain Chem Eng* 7:12601–12610.
36. Quienne B, Kasmi N, Dieden R, Caillol S, Habibi Y (2020) Isocyanate-free fully biobased star polyester-urethanes: Synthesis and thermal properties. *Biomacromolecules* 21:1943–1951.
37. Li L, Chen X, Torkelson JM (2019) Reprocessable polymer networks via thiourethane dynamic chemistry: Recovery of cross-link density after recycling and proof-of-principle solvolysis leading to monomer recovery. *Macromolecules* 52:8207–8216.
38. Fang Y, Du X, Jiang Y, Du Z, Pan P, Cheng X, Wang H (2018) Thermal-driven self-healing and recyclable waterborne polyurethane films based on reversible covalent interaction. *ACS Sustain Chem Eng* 6:14490–14500.
39. Hu S, Chen X, Torkelson JM (2019) Biobased reprocessable polyhydroxyurethane networks: Full recovery of crosslink density with three concurrent dynamic chemistries. *ACS Sustain Chem Eng* 7:10025–10034.
40. Fortman DJ, Brutman JP, Cramer CJ, Hillmyer MA, Dichtel WR (2015) Mechanically activated, catalyst-free polyhydroxyurethane vitrimers. *J Am Chem Soc* 137:14019–14022.

4 Eco-Friendly Synthesis of Waterborne Polyurethanes

Pavan M. Paraskar, Vinod M. Hatkar, and Ravindra D. Kulkarni
Department of Oils, Oleochemicals and Surfactants Technology, Institute of Chemical Technology, Matunga (E), Mumbai, MS, India

CONTENTS

4.1 Introduction........48
4.2 Synthesis of Bio-Based Polyols by Various Routes and Materials........49
4.2.1 Epoxidation/Oxirane Ring-Opening........52
4.2.2 Transesterification........52
4.2.3 Transamidation........53
4.2.4 Thiol-Ene Coupling........53
4.2.5 Hydroformylation/Reduction........53
4.2.6 Ozonolysis/Reduction........54
4.3 Utilization of Bio-Based Isocyanate in WPUs........54
4.3.1 Sugar-Based Pentamethylene Diisocyanate (PDI)........56
4.3.2 Amino Acid-Based Diisocyanates........56
4.3.3 Furan-Based Diisocyanates........56
4.3.4 Lignin-Based Diisocyanates........56
4.3.5 CNSL-Based Diisocyanates........57
4.3.6 Vegetable Oil/Fatty Acid-Based Diisocyanates........57
4.4 Synthesis of Internal Emulsifier From Bio-Based Feedstock........57
4.4.1 Anionic Internal Emulsifiers........58
4.4.2 Cationic Internal Emulsifiers........58
4.4.3 Non-Ionic and General Internal Emulsifiers........58
4.5 Bio-Based Chain Extender........58
4.6 Water-Based Non-Isocyanate Polyurethane Dispersion From Vegetable Oil........60
4.7 Conclusion........61
References........62

DOI: 10.1201/9781003173526-4

4.1 INTRODUCTION

Polyurethanes (PUs) have been enormously cherished since their discovery because their adaptability makes them suitable for many diverse applications, such as foams, paints, coatings and varnishes, adhesives, elastomers, packaging, thermal and acoustic insulators, sealants, furniture, biomedical devices, and construction materials [1,2]. Consequently, the worldwide market for PU was worth $95.13 billion in 2019. By 2023, it is predicted to grow to $149.91 billion and compound an annual growth rate (CAGR) of 12% [3]. PUs are traditionally synthesized by reacting polyols containing hydroxyl groups and diisocyanate to form a PU linkage. Due to the growing global desire for environmental conservation, the development of waterborne polyurethane dispersions (WPUDs) has received significant interest as a substitution for petroleum-based solvents (volatile organic compounds, or VOCs) via water, an eco-friendly solvent [4]. These polymers' advantages must overtake the shortcomings of using a non-eco-friendly component in the manufacture, for example, isocyanate.

Moreover, attention to incorporating materials that might have lesser ecological impact inspired scientists and technocrats to substitute petroleum-based reactants. There are few shortcomings of the conventional PU system (Figure 4.1). Comparatively, modern scientific endeavors are lined to the utilization of non-isocyanate polyurethanes (NIPUs), and it is only likely that this route will lead to the financial effectiveness of these alternatives in years to come [5,6]. Besides this, the development and manufacturing of green polyols have been more straightforward and have had swifter growth. Vegetable oils utilized as a basis for diverse and abundant chemicals have been pursued, and there was an enormous achievement of manufacturing polyols with various chemical edifices that can be utilized to synthesize PUs [7].

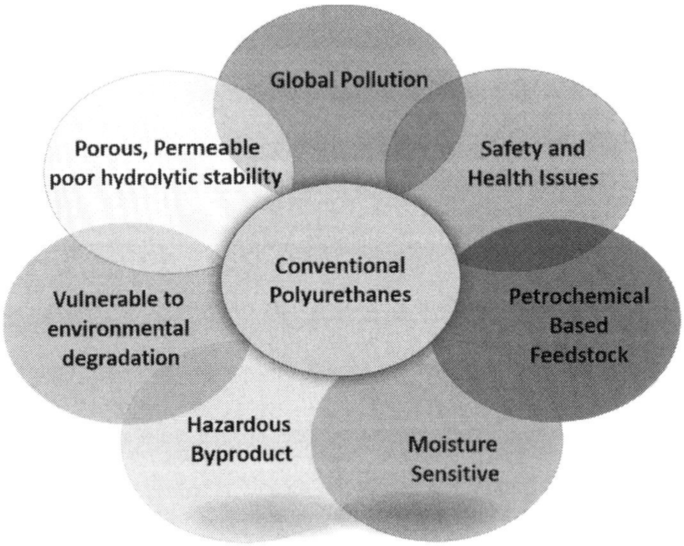

FIGURE 4.1 The drawbacks of the conventional PU system.

WPUDs could be utilized to fabricate sealants, paint additives, fibers, textile dyes, metal primers, adhesives, biomaterials, etc., which describes the rising fascination in the constant advancement of these materials. WPUDs are materials whose cost significance owes to rising ecological apprehensions about hazardous air pollutants and volatile organic chemicals [8]. Waterborne PUs presents various advantages contrasted with traditional solvent-borne PUs: good spreading ability during application, high molecular weight with low viscosity, and low VOC emission. An emulsifier should be introduced to avoid coagulation of the organic solvable PU in the aqueous media during the manufacturing of WPUDs [9]. Furthermore, the utilization of an internal emulsifier (also called a co-reactant) has eliminated transudation of the material because the solution medium disappears and the PU piece is established. The diols with a carboxylic or a carboxylate group (dimethylol propionic acid (DMPA)) are used to synthesize anionic WPUDs, whereas a diol usually contains an amine group that is used to produce cationic WPUDs. Nonionic internal emulsifiers (which are segmental polyether moieties) must be considered, resulting in less stable dispersions than the ionic emulsifiers used [1]. The production of these PUs and their additional distribution in the aqueous phase results in water-based suspensions of nanoparticles that are steady within the 20–200 nm particle size range. Lately, vegetable oil is earning interest in the WPUDs market because of active functional sites such as unsaturation, carboxyl groups, reactive phenolic structures, and hydroxyl groups. The utilization of bio-based chain extenders in WPUDs was also reported [10].

4.2 SYNTHESIS OF BIO-BASED POLYOLS BY VARIOUS ROUTES AND MATERIALS

Numerous renewable feedstocks have been exploited to synthesize bio-based polyols, including natural rubber, vegetable oil, fatty acids, sugar, glucan, chitosan, heparin, and chitin [11]. Vegetable oils, such as castor and Lesquerella oils, inherently contain hydroxyl groups, which can be directly used as polyol to synthesize bio-based PU. A comprehensive study on the synthesis of polyols using a biomass exclusively of vegetable oils, including sunflower oil, castor oil, jatropha oil, soybean oil, palm oil, and linseed oil, for waterborne PU was carried out [1]. The key functionalities, including epoxide groups, carbon–carbon double bonds, and ester groups, are available in vegetable oil, which can be further utilized for chemical modification to introduce hydroxyl groups into vegetable oil structures. Numerous synthetic channels recently have been explored to manufacture polyols to establish –OH groups in their structure. Various bio-based polyols used for waterborne PUs are shown in Table 4.1.

Vegetable oil-based polyol has been utilized with DMPA, diisocyanates, and chain extenders to synthesize eco-friendly anionic waterborne PUs [12]. Vegetable oil triglyceride with integral hydrophobic chains provides some challenges for the synthesis of waterborne polyols. Besides this, the hydrophobic nature of triglycerides improves the physicochemical properties of waterborne polyols with improved hydrolytic and thermal stability. Gogoi and co-workers reported that the dimer technology-based fatty acid was used to synthesize waterborne polyols. They observed the hydrolytic and storage stability improvement in PU dispersions with superior mechanical strength, adhesion, hydrophobicity, gloss, and chemical

TABLE 4.1
Bio-based polyols

Sr. No.	Polyol	Properties
1	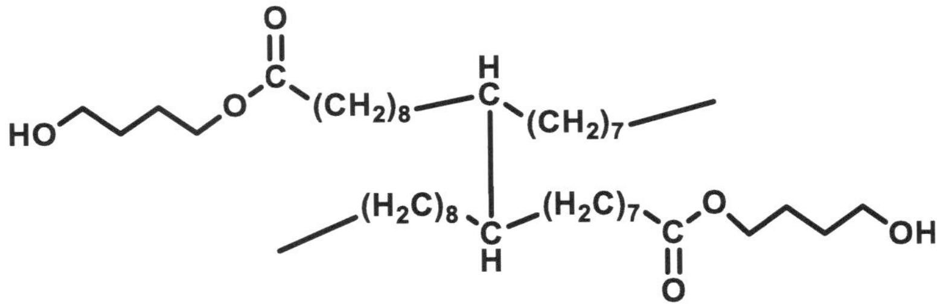 Dimer acid based polyol	i. High flexibility ii. Low volumetric shrinkage iii. Biocompatibility iv. Hydrophobicity
2.	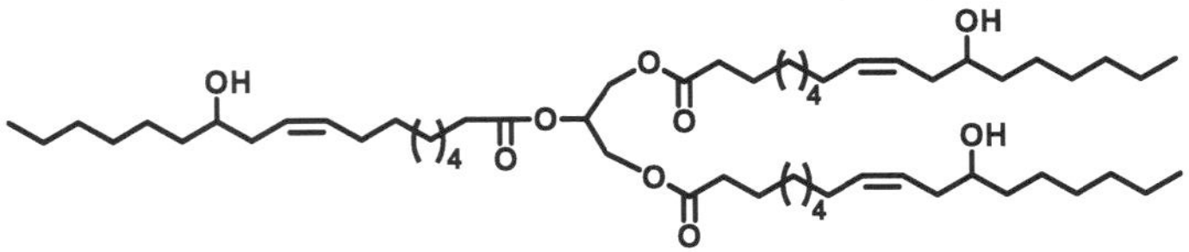 Castor oil based Polyol	i. Biocompatibility ii. Hydrophobicity iii. Biodegradability iv. High thermal stability v. Solvent resistance
3.	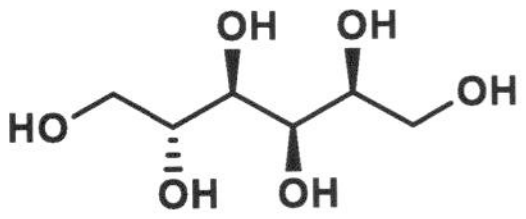 Sorbitol 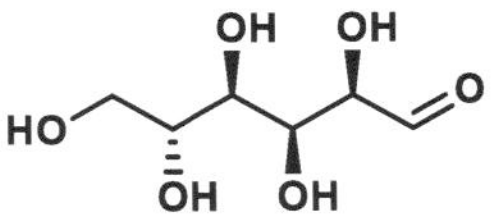 Glucose	i. Rigidity ii. Biodegradability iii. Biocompatibility

4.	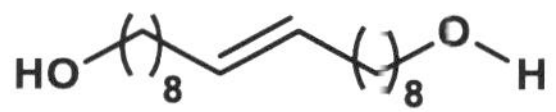 Fatty acid based diol	i. High flexibility ii. Biocompatibility
5.	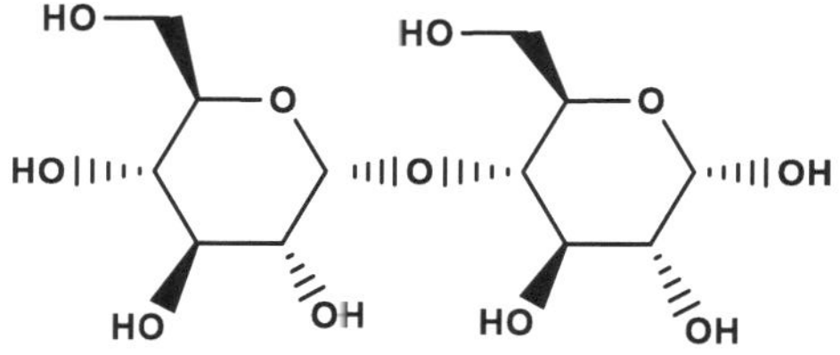 Starch	i. Rigidity ii. Biodegradability iii. Biocompatibility

resistance properties [13]. Lu and Larock prepared the WPUDs by modifying soy-based polyol with MDI and NMDA (*N*-methyldiethanolamine), followed by neutralization using stoichiometric amounts of acetic acid and mixing with excess water with vigorous mixing to form the prepolymer [14]. Kim and Lee synthesized a series of castor oil-based WPUs with an improved dispersion procedure, which exhibits higher tensile property and molecular weight [15]. The common synthetic routes for modifications of vegetable oil to polyols were recently cited by numerous studies, such as hydroformylation followed by hydrogenation, epoxidation of double bond followed by oxirane ring-opening, ozonolysis followed by reduction, transesterification, thiol-ene coupling, and transamidation. They are briefly explained as follows.

4.2.1 Epoxidation/Oxirane Ring-Opening

The epoxidation of carbon–carbon double bonds and subsequent oxirane ring-opening by proton donors, such as alcohols, amines, and carboxylic acids, is the largely exploited synthetic path for manufacturing bio-based polyols. Generally, in situ peracids from hydrogen peroxide and formic or acetic acid are the most used methods for the epoxidation of vegetable oil [7]. Moreover, numerous factors influence the performance of final bio-based polyols, such as the count and position of the hydroxyl moieties, variety of ring-opening agents, and residual degree of unsaturation. The hydroxyl number influences the PU coatings' mechanical performance, glass transition temperature, and cross-linking density. Furthermore, higher tensile strength, tensile modulus, and toughness will be achieved by a higher residue of unsaturation in PU coatings. Li et al. (2021) utilized pentaerythritol triacrylate (PETA) and acrylated epoxidized soybean oil (AESO) for a UV-curable waterborne polyurethane acrylate (PUA) resin [16]. Liao et al. (2016) prepared jatropha oil-based WPUDs by epoxidation and ring-opening using dimethylolbutanoic acid (DMBA) as an internal emulsifier [17].

4.2.2 Transesterification

Transesterification of vegetable oils/triglyceride structures is an important route to utilize the ester functionality to prepare bio-based polyols. Glycerol is a bio-based polyol commonly used for transesterification vegetable oil; however, other petro-based polyols like trimethylolpropane and pentaerythritol have also been researched. Numerous scientists have reported synthesizing vegetable oils based on monoglycerides, such as neem, castor, soybean, Nahar, and linseed, for manufacturings bio-based PUs [1]. The transesterification of glycerol and vegetable oils catalyzed by metal oxide provides a blend of polyols, such as mono-, di-, and triglycerides, as well as residual glycerol. A series of castor oil-based WUPDs was recently prepared using an oligoester's transesterification achieved by glycolysis of recycled PET. Prepared WPUDs reacted with APTES, which established nano-silica particles (solid fraction of 0.5–2 wt % silicon content in the composite). The hybrid WPUD composites indicated growth in thermal permanence with growing silica strength because of the robust network between nano-filler and PU [18].

4.2.3 Transamidation

Transamidation is one of the prominent technologies used to synthesized WPUDs. The primary advantages of amide functionality include cross-linking density, curing time thermal stability, adhesion, and flexibility. Therefore, the initiation of amine functionality in polyol structure is a significant intention in bio-based WPUD manufacturing. Diethanolamide is the product of a catalyzed amidation reaction with a triglyceride molecule and diethanolamine [1]. Polyesteramide polyols are produced through a condensation reaction between diethanolamine and polybasic acid. The variation in final polyol occurs because of the utilization of the different types of polybasic acids. More et al. prepared polyesteramide polyols from mustard oil as the initial feedstock and utilized them for PU water dispersion. The prepared polyesteramide polyol was cross-linked with isophorone diisocyanate (IPDI) to produce the prepolymer, and after reaction with DMPA, to produce dispersion in water, hydroxyl ethyl methacrylate to initiate unsaturation in the PU network [19].

4.2.4 Thiol-Ene Coupling

Due to the versatility and benign reaction conditions, thiol-ene coupling/thiol-ene chemistry is frequently used for diverse bio-based polyol synthesis. Over the typical acrylate system, incorporating thiols into carbon–carbon double bonds provides various advantages, including oxygen insensitivity, reduced shrinkage stress, and delayed gelation. Thiol-ene coupling is a single-stage reaction involving a free radical mechanism with unsaturated fatty acid double bonds combined with thiols (2-mercaptoethanol, glyceryl dimercaptoacetate, and 3-mercapto propionate) to form primary hydroxyl moieties. Thiol-ene polymerization is an important technique due to the involvement of photoreaction with high conversion and a fast reaction rate [1]. Fu et al. reported that castor oil-based undecylenic acid was modified into a diisocyanate. WUPDs were prepared with the synthesized diisocyanate. Furthermore, the fatty acid chains found in vegetable oils enhanced the flexibility of coating while exhibiting poor chemical resistance, tensile modulus, and strength [20]. A series of bio-based polyols were synthesized from rice bran, corn, olive, linseed, canola, castor, and grape seed oil by using the thiol-ene photoreaction and investigated the relationship between WBPU properties and hydroxyl functionality [21].

4.2.5 Hydroformylation/Reduction

Hydroformylation followed by reduction is another major route for bio-based polyols synthesis. The hydroformylation of vegetable oils indicates converting fatty acid chains with carbon-carbon double bonds to aldehydes with cobalt or rhodium catalysts and syngas (a mixture of CO and H_2) at 70–130°C, followed by Ni catalyzed hydrogenation of aldehyde to manage primary hydroxyl groups [22]. Soybean and linseed oils are more commonly used as initial feedstock for the hydroformylation process than other oils. However, because of the wide range of hydroxyl functionality distribution in hydroformylated polyester polyols, it has a negative impact on WPUD properties. Smoothing coatings with a low abrasion resistance were produced by polyols with a wide functionality distribution, whereas coatings with a narrower functionality distribution developed coatings with an admirable balance of flexibility, hardness, and resistance to abrasion. Dow Chemical Co. commercially manufactured and supplied bio-based polyols prepared from soybean oil under the trade name RENUVATM^{TM} for coatings, adhesives, sealants, flexible foams, and elastomers [1]. Petrović et al. synthesized crude alga oil-derived polyols by hydroformylation using rhodium Rh $(CO)_2$ as a catalyst and compared them to alga polyols made by ozonolysis and epoxidation [23].

4.2.6 Ozonolysis/Reduction

Ozonolysis/reduction is one of the main routes for bio-based polyol synthesis of WUPDs. Bio-based polyol is obtained from vegetable oil by a double-stage ozonolysis process that involves ozonide at the carbon–carbon double bonds of vegetable oil in the first stage, continued by reducing an aldehyde toward polyols in the second stage with the Raney nickel catalyst [24]. Narine et al. effectively manufactured polyols through unsaturated canola oil, continued by ozonolysis and hydrogenation [25]. The fatty acid content of vegetable oils obtained from ozonolysis accumulates different mono, diol, and triol components with saturated triglycerides and triglyceride structures. The ozonolysis method removed half of each fatty acid chain and prevented the dangling chain insertion into the PU. As a result, ozonolysis-based polyols cure at a faster rate than isocyanates. A one-step reaction also prepared hydrogenation-free ozonolysis-based polyols. In general, polyols such as ethylene glycol and glycerol have also been combined in the presence of a catalyst (e.g., sodium hydroxide, calcium carbonate) with vegetable oil to form a combination of primary and secondary alcohols [26].

4.3 UTILIZATION OF BIO-BASED ISOCYANATE IN WPUs

Diisocyanates are one of the essential constituents of PU chemistry. The –N=C=O functional group-containing diisocyanate is a highly reactive compound that plays a vital role in forming the urethane linkage of PU. Usually, based on their chemical constitution, diisocyanates are divided into two types: aliphatic and aromatic. Isophorone diisocyanate (IPDI) and hexamethylene diisocyanate (HDI) are the most commonly used aliphatic diisocyanates. In contrast, methylene diphenyl diisocyanate (MDI) and toluene diisocyanate (TDI) are the most commonly used aromatic diisocyanates. High molecular weight di/polyisocyanate adducts like HDI-biuret and HDI isocyanurate have been used to prepare special PUs. This type of isocyanate is beneficial for low volatile applications, such as paints, coatings, adhesives, etc. Thus, as per the application, both types of diisocyanates are employed to modify PUs with required properties [27]. Due to the problems of toxicity and hazards at work arising from the use of these existing diisocyanates, there have been alternative pathways to synthesize isocyanates from renewable raw material, e.g., sugar-based pentamethylene diisocyanate

(PDI) [28], 1,7-heptamethylene diisocyanate (HPMDI), DDI (dimer acid-based diisocyanate), ethyl or methyl ester of L-LDI (L-lysine diisocyanate), and cashew nut shell liquid-based diisocyanates [1], etc. Various bio-based isocyanates used for waterborne PUs are shown in Table 4.2.

TABLE 4.2
Bio-based isocyanate

Sr. No.	Name	Structure
1	Sugar-based pentamethylene diisocyanate	 **Pentamethylene diisocyanate (PDI)** 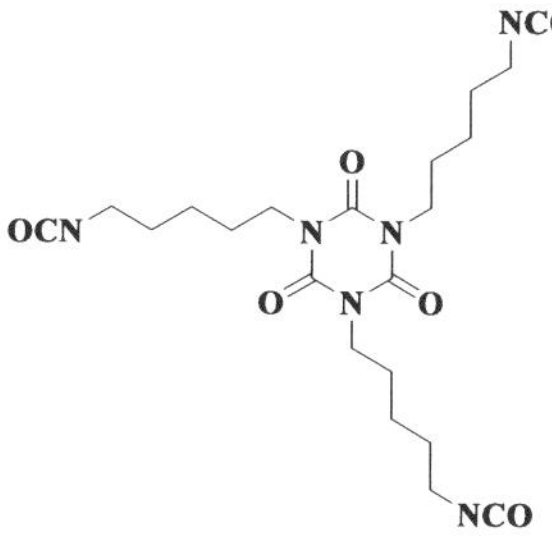 **(DESMODUR® eco N 7300)**
2	Amino acid-based diisocyanates	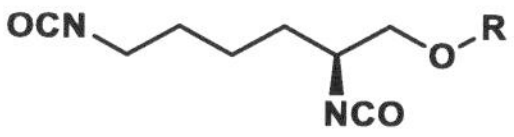 **L-Lysine Diisocyanate**
3	Furan-based diisocyanates	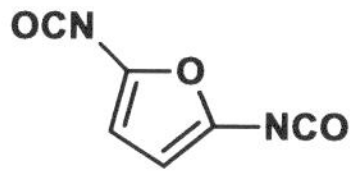 **2, 5 Diisocyanatofuran**
4	Lignin-based diisocyanates	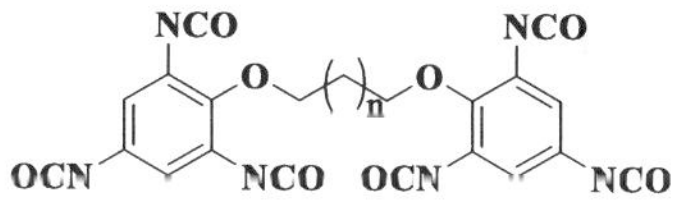 **bis(4-isocyanato-2-methoxyphenoxy)alkane**
5	CNSL-based diisocyanates	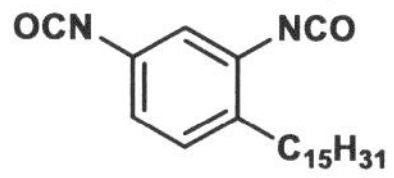 **2,4 Diisocyanato-1-pentadecylbenzene**
6	Vegetable oil/fatty acid-based diisocyanates	 **1,7-Heptamethylene Diisocyanate (HPMDI)**
7	Vegetable oil/fatty acid-based diisocyanates	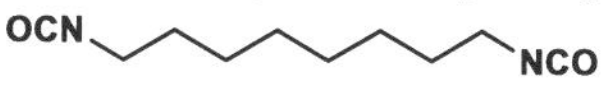 **1,8-Diisocyanatooctane**
8	Vegetable oil/fatty acid-based diisocyanates	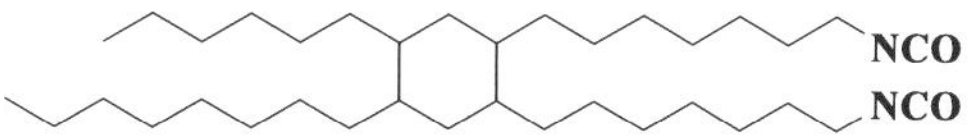 **Dimer acid diisocyanate (DDI)**

4.3.1 Sugar-Based Pentamethylene Diisocyanate (PDI)

The first commercialized bio-based diisocyanate is sugar-based pentamethylene diisocyanate (PDI), the main building block for PUs with considerable bio-content (71% renewable carbon). It is highly effectively manufactured from biomass utilizing biotechnological and chemical methods, such as energy-efficient gas phase technology. Covestro has marketed the trimeric PDI as a hardener under the trade name DESMODUR® eco N 7300. Zenner et al. successfully prepared stereochemically pure diisocyanates on a multigram scale from succinic anhydride and isosorbide or isomannide [28].

4.3.2 Amino Acid-Based Diisocyanates

L-lysine is a well-known biomass-derived amino acid utilized effectively as a starting material for producing ethyl or methyl esters of L-lysine diisocyanate (EELDI or MELDI, respectively). The asymmetric aliphatic structure of L-lysine diisocyanates leads to the amorphous nature of PUs. Similarly, partly bio-based diisocyanate (PBDI) was synthesized from hexamethylene diisocyanate (HMDI) and L-lysine and polymerized using bio-based diols to synthesize PUs. Sahoo et al. successfully prepared low viscous bio aliphatic isocyanate from bio-based resources [29].

4.3.3 Furan-Based Diisocyanates

Furan-based diisocyanate is the first bio-based diisocyanate, reported by Garber in 1962, for the synthesis of PUs. Tawade et al. review the protocol for synthesizing a reactive diisocyanate monomer based on hydroxymethylfurfural by catalyst-free synthesis of linear-chain thermoplastic PUs. Additionally, cross-linked PU foams were prepared from diisocyanate utilizing in situ generated nitrogen as a blowing agent while converting the intermediate dicarboxylic acyl azide into diisocyanate [27]. Tawade et al. used a variety of furan-based diisocyanates and investigated their reaction kinetics toward PU synthesis as well as the effects of furan integration on PU properties.

4.3.4 Lignin-Based Diisocyanates

Ferulic acid, vanillic acid, guaiacol, syringic acid, eugenol, and syringol are the major chemical constituents available in lignin that attracted polymer scientists' attention to use lignin as a starting ingredient for the preparation of difunctional monomers suitable in the synthesis of step-growth polymers. Bis(4-isocyanato2, 6-dimethoxyphenoxy)alkane, aromatic diisocyanates, and bis(4-isocyanato-2-methoxyphenoxy)alkane were produced using syringic acid and vanillic acid and polycondensed with bio-based diols 1,12-dodecanediol and 1,10-decanediol to produce poly(ether urethane)s [27]. Chauhan et al. synthesized lignin-based diisocyanates by functionalizing lignin using 4,4′-diphenyl methane diisocyanate (MDI) at 90°C for 60 min and evaluated their usage in PU formulation by determining phase miscibility, wettability, rheology, and mechanical properties [30].

4.3.5 CNSL-Based Diisocyanates

The cashew nut shell liquid (CNSL) is non-edible oil; its usage as an industrial raw material does not affect the food supply chain. The synthesis of aromatic difunctional monomers from CNSL is attractive because of its three reactive sites (phenolic –OH, unsaturation in a side chain, an aromatic ring). A variety of difunctional monomers, including diisocyante, containing pendent pentadactyl chain, were synthesized from CSNL. A TDI "look-alike" diisocyanate, namely 2,4-diisocyanato-1-pentadecylbenzene, was synthesized from CNSL [27].

4.3.6 Vegetable Oil/Fatty Acid-Based Diisocyanates

Vegetable oils are interesting renewable raw materials for long-chain diacids, which are then used to make diisocyanates. Aliphatic dicarboxylic acids are transformed into dicarboxylic acyl azides, thermally Curtius rearrangement in hydrocarbon solvents to produce diisocyanates [31,32]. This approach's industrial-scale production of diisocyanates is restricted due to the strict safety precautions required due to the recognized risks of handling azido compounds on a bigger scale. Henkel Corporation commercially manufactured the DDI (dimer acid diisocyanate), prepared from a dimer of the fatty acid. The chain of DDI has 36 carbon atoms, and the presence of dangling chains in the structure limits stress transduction causes and steric hindrance for cross-linking [1]. Cramail and colleagues reported the preparation of aliphatic diisocyanate, namely 1-isocyanato-10-[(isocyanatomethyl)thio] decane from fatty acid. Dimer acid diisocyanate (DDI), castor oil (CO), and alkoxysilane-modified castor oil (MCO) were used to prepare PU dispersions, and the metal coatings showed excellent corrosion resistance, mechanical properties, and high thermal stability [31].

4.4 SYNTHESIS OF INTERNAL EMULSIFIER FROM BIO-BASED FEEDSTOCK

WUPDs are prepared from two central portions: the emulsifier and the main structure or backbone. The emulsifiers stabilize the dispersion of PU in water due to their hydrophilicity. Internal emulsifiers and exterior emulsifiers are the two types of emulsifiers. Internal emulsifiers include ionic centers, such as cationic, anionic, and zwitterions, and nonionic centers, such as polyethylene oxide. As a result, the hydrophilic soft segment pendant group, like polyethylene oxide, is present in nonionic PU dispersions [33,34]. Internal emulsifiers are favored since they prevent any component relocation upon film amalgamation. The emulsifying agents will define the type of WBPU, i.e., nonionic, cationic, or anionic. As an emulsifier, ionic group (quaternary ammonium salt, sulfonate, or carboxylic) based diols are often used. DMPA (di-methylol propionic acid) is a standard internal emulsifier regularly used in a polymer as dual hydroxyl functionality acts as a reactive chain site, and carboxylic acid functionality acts as a neutralized site, which will neutralize with a primary group (e.g., TEA, or tri-ethyl amine) [35,36]. Non-bio-renewability of DMPA reduced the bio-based content of the WPUD system, so researchers focused on DMPA-free anionic/cationic WPUDs by incorporating vegetable oil [37].

4.4.1 Anionic Internal Emulsifiers

Anionic internal emulsifiers are mostly used in WPUD formulation. In an anionic internal emulsifier, the ionic centers in this acid are situated in the hard segments, expanding its affinity to water. Subsequently, these ionic groups are accountable for distribution in an aqueous medium of micelles. Researchers developed DMPA-free WPUDs using an ionic soft segment like phosphorylated polyol. Using ring-opening/hydrolysis with ortho-phosphoric acid, phosphorylated polyol (both hydroxyl and ionizable phosphoryl groups) were synthesized from epoxidized cottonseed oil [37]. Using ring-opening epoxidized linseed oil, Chen et al. synthesized anionic WUPDs with hydrochloric acid and glycol [38].

4.4.2 Cationic Internal Emulsifiers

Although an anionic internal emulsifier is used most often, a few investigations have also been offered to produce cationic-WPUDs (c-WPUDs). They provide distinct benefits of stability in lower pH conditions, show good wear resistance, show excellent use with softeners and fixing agents, and are mostly utilized for applications including adhesives, leather, coating, dyeing, textiles, paper, etc. The cationic emulsifiers utilized in the preparation of c-WPUDs are amine salts, with *N*-methyl diethanolamine (N-MDEA) and quaternary ammonium salts being used most widely [39]. A series of bio-based polyols were prepared from various vegetable oils through the thiol-ene photo-click reaction [21]. Liu et al. report improvement in the flexibility and water-resistance of cationic WPUDs. In addition, the feature of the organosilicon compound, the polysiloxane-containing hydroxyalkyl-bifunctional group, and trifunctional group modifier PPS were utilized to alter the cationic WPU [40].

4.4.3 Non-Ionic and General Internal Emulsifiers

Comparatively low viscosity at considerably high solid contents is a distinct characteristic of nonionic PU dispersion with stable dispersions against electrolytes, frost, mechanical influence, fluctuations in the pH level, and solvents. These synthetic groups are PU ionomers, and they are often introduced into the PU chains by polymerization PU with an ionic reagent or reaction agents (ionic diisocyanate or ionic diols) [41]. Liu et al. synthesized a fully bio-based emulsifier using glutaric acid and epoxidized soybean oil via a catalyst-free solvent-free protocol. Citric acid was employed as a bio-based internal emulsifier by Gogai et al., using phthalic anhydride as the dibasic moiety and a unique citric acid-based polyamide polyol as the branching unit [13]. Garrison et al. prepared antibacterial soybean oil-based PUDs, and thin films were synthesized with different molar ratios of an amine diol internal emulsifier [42].

4.5 BIO-BASED CHAIN EXTENDER

Chain extenders are low molecular weight (molecular weight is less than 400 g mol^{-1}) compounds with terminal hydroxyl or amine groups. Chain extenders are reacting with the prepolymer or diisocyanate during PU synthesis. Chain extenders enlarge the chain length and increase the molecular weight of PU. Chain extenders contribute to the hard segment, along with the diisocyanate group, and help to control the hydrolytic stability, thermal, and mechanical properties of the PU [43]. Similar to diisocyanates, chain extenders can be either aliphatic or aromatic. PU chain extenders are mainly divided into linear and branched classes. The most commonly used bio-based chain extenders are listed in Table 4.3 [44].

In the last two decades, the feasibility of renewable resources as chain extenders has seen a considerable rise due to economic and environmental factors. For example, vegetable oils and their derivatives have been innovated as chain extenders in PU and HPU synthesis [45]. The bio-based chain extenders are synthesized by esterifying citric acid and glycerol and combining them with (poly)-caprolactone diol and macroglycol (dimer acid and polyethylene glycol) to form biodegradable water dispersion thermoplastic PUs. The resulting coatings have superior scratch resistance, high impact and tensile strength, and self-healing

TABLE 4.3
Most commonly used biobased chain extenders in PU synthesis

Sr. No.	Chain Extender	Sr. No.	Chain Extender
1.	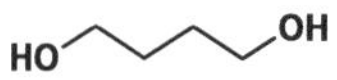 1,4-butanediol	2.	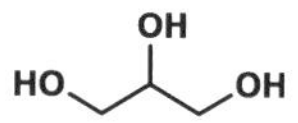 Glycerol
3.	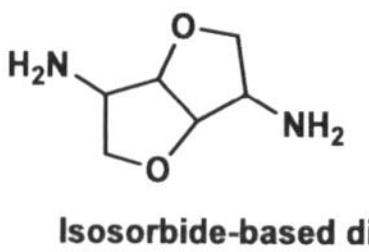 Isosorbide-based diamine	4.	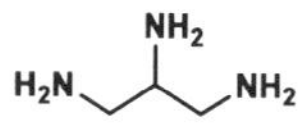 Propane-1,2,3- triamine
5.	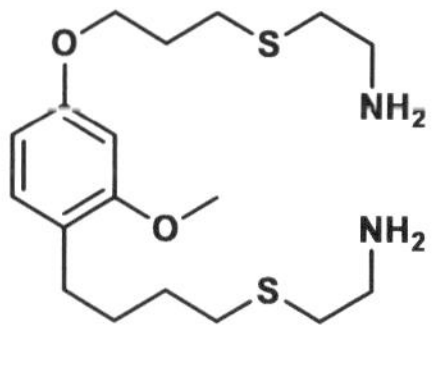 Vanillin based diamine	6.	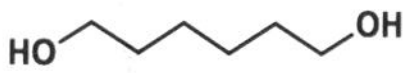 1,4-hexanediol
7.	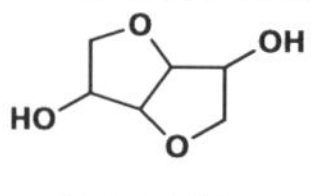 Isosorbide	8.	 Trimethylol propane
9.	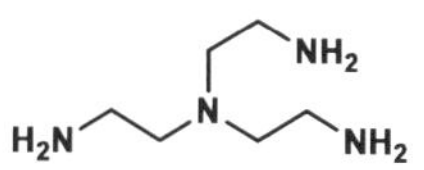 Tris (2-aminoethyl) amine	10.	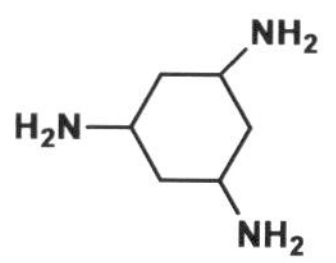 Cyclohexane-1,3,5-triamine

capabilities [10]. Wang et al. (2021) reported that phosphorus–nitrogen chain extender [bis (2-hydroxyethyl) amino]-methyl-phosphonic acid dimethyl ester effectively modifies novel bio-based WPU dispersions produced from soy polyol and castor oil [46].

4.6 WATER-BASED NON-ISOCYANATE POLYURETHANE DISPERSION FROM VEGETABLE OIL

In January 2009, European regulators enforced new regulations, which strictly monitor and regulate free isocyanate content in PU synthesis; therefore, polymer scientists have been extensively searching for alternatives for the synthesis of non-isocyanate-based PU precursors. Moreover, NIPU-based material is eco-friendly, such as a two-pack system, which is an easy-to-control and less volatile organic compound. It is also moisture insensitive and can be used in damp surfaces with high storage stability [47]. Various approaches have also been examined to manufacture NIPUs, including rearrangement reactions, polycondensation, step-growth polyaddition, and ring-opening polymerization (ROP). Preparation of cyclic carbonates from renewable feedstocks with diamines using a polyaddition approach seems to be the most exciting pathway to manufacture the waterborne NIPUs, among the various routes reported in the literature [1]. In the synthesis of traditional waterborne PUs, the waterborne NIPU is formulated by dissolving NIPU prepolymer into the water by counteracting the internal dispersion monomer NIPU prepolymer chain. NIPU prepolymers are generally produced in water-miscible with acetone as a low boiling point solvent. Dimethylolpropionic acid (DMPA) is an internal dispersing agent typically introduced into the design and permits dispersal in the aqueous phase upon counteraction. A short diol or a diamine were used to extend the chain, followed by acetone removal to get NIPU dispersion [48]. The most recent instances of isocyanate-free PU or NIPU dispersion employ the acetone methodology; specifically, the copolymerization of three monomers, including an internal dispersing agent, is done in a low boiling point solvent or bulk and then dissolved in a water-based organic solvent. Ma et al. reported a sequence of experiments investigating NIPUs for the fabrication of WUPDs for coatings [49]. The synthetic reaction scheme shown in Figure 4.2 demonstrates one of the reaction pathways for the synthesis of waterborne NIPU.

Bizet et al. reviewed several cyclic carbonate synthesis methodologies, such as aminolysis and transurethanization, typically performed in bulk and solvents [50]. Zhang et al. prepared waterborne amine-terminated NIPUs, which were obtained from fatty acid diamine (FDA), 3,3′-diamino-*N*-methyldipropylamine, and diglycerol decarbonate, for waterborne NIPU epoxy hybrid coatings from waterborne epoxy chain extender and amine-terminated NIPUs [51]. Ochiai et al. selected the right choice of monomers and the polymerization temperatures to synthesize NIPUs by polyaddition of bifunctional diamines and cyclic carbonates in water without any surfactants and organic solvents [52]. Meng et al. reported the use of urethane methacrylate (UMA) in waterborne systems as a reactive diluent [53].

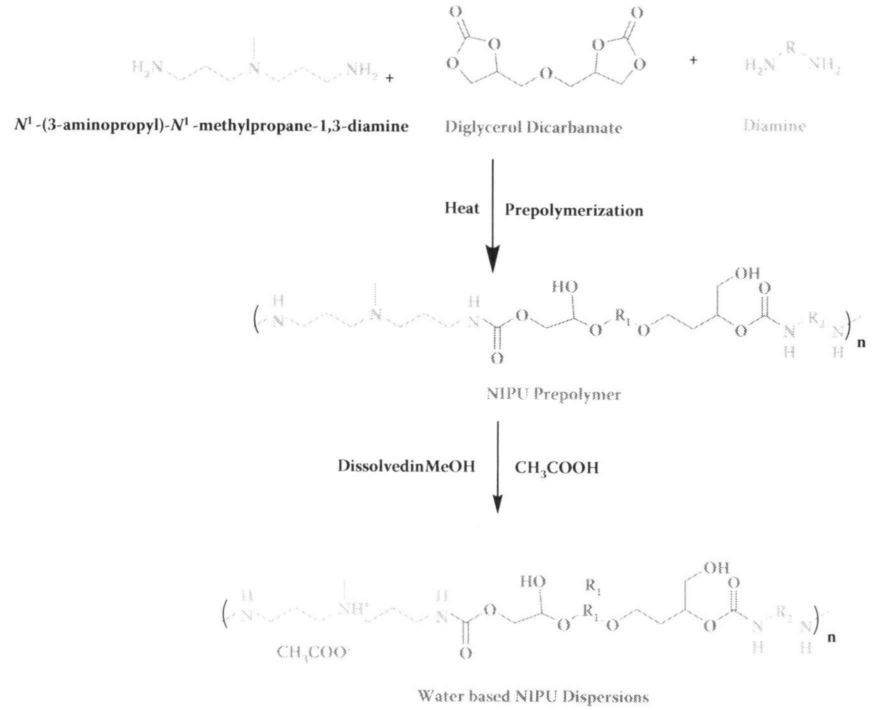

FIGURE 4.2 Synthesis of waterborne NIPU dispersions.

4.7 CONCLUSION

This chapter proposed an overview of the synthesis of bio-based polyols, isocyanate and chain extenders, various emulsifiers, different synthesis methods, and WPUs. The distinctive structural property of bio-based materials, such as vegetable oil, fatty acids, carbohydrates, and cellulose, prove to be established raw feedstock for WPU systems because of their reactive functionality. Undoubtedly, bio-based polyols, isocyanates, and chain extenders and their structural transformation efficiently produce waterborne PU dispersion with excellent final high-end applications. To avoid the harmful impact of phosgene, vegetable oil and its derivatives are interesting renewable raw materials to synthesize bio-based diisocyanate, which yields WPUs. Vegetable oil-derived polyol with emulsifiers is considered to encourage feedstock to produce a new class of waterborne PUDs with a range of industrial applications. With the growing industrial and academic concerns about the utilization of isocyanates, a new era of PU systems that are isocyanate-free has been established. These NIPUs for the fabrication of waterborne PU systems could reach the current industry demand, which is a possible substitution to conventional PU systems. WPU development has yet to attain its peak due to shortcomings in conventional drying agents, modifiers, curing agents, and diluents, which could be corrected by assimilation through eco-friendly sustainable raw material alternatives.

REFERENCES

1. Paraskar P M, Prabhudesai M S, Hatkar V M, Kulkarni R D (2021) Vegetable oil based polyurethane coatings—A sustainable approach: A review. *Prog Org Coatings* 156:106267.
2. Paraskar P M, Prabhudesai M S, Kulkarni R D (2020) Synthesis and characterizations of air-cured polyurethane coatings from vegetable oils and itaconic acid. *React Funct Polym* 156:104734.
3. Global Polyurethane Market (2019 to 2023) Identify growth segments for investment—ResearchAndMarkets.com, published on March 20, 2020, retrieved on April 25, 2021, https://apnews.com/press-release/pr-businesswire/ab0e7c1d524a45719279 1a5a448c76c4.
4. Mucci V L, Hormaiztegui M E V, Aranguren M I (2020) Plant oil-based waterborne polyurethanes: A brief review. *J Renew Mater* 8(6):579–601.
5. Doley S, Dolui, S K (2018) Solvent and catalyst-free synthesis of sunflower oil based polyurethane through non-isocyanate route and its coatings properties. *Eur Polym J* 102:161–168.
6. Błażek K, Datta J (2019) Technology renewable natural resources as green alternative substrates to obtain bio-based non-isocyanate. *Crit Rev Environ Sci Technol* 49:1–39.
7. Paraskar P M, Prabhudesai M S, Deshpande P S, Kulkarni R D (2020) Utilization of oleic acid in synthesis of epoxidized soybean oil based green polyurethane coating and its comparative study with petrochemical based polyurethane. *J Polym Res* 27:1–10.
8. Zhou X, Li Y, Fang C, Li S, Cheng Y, Lei W, Meng X (2015) Recent advances in synthesis of waterborne polyurethane and their application in water-based ink: A review. *J Mater Scien & Techno* 31:708–722.
9. Madbouly S A, Xia Y, Kessler M R (2013) Rheological behavior of environmentally friendly castor oil-based waterborne polyurethane dispersions. *Macromolecules* 46(11):4606–4616.
10. Chandra S, Karak N (2018) Environmentally friendly polyurethane dispersion derived from dimer acid and citric acid. *ACS Sustain Chem Eng* 6:16412–16423.
11. Paraskar P M, Kulkarni R D (2020) Synthesis of isostearic acid / dimer fatty acid-based polyesteramide polyol for the development of green polyurethane coatings. *J. Polym. Environ* 29:57–70.
12. Lokhande G P, Chambhare S U, Jagtap R N (2017) Anionic water-based polyurethane dispersions for antimicrobial coating application. *Polym Bull* 74:4781–4798.
13. Gogoi G, Gogoi S, Karak N (2017) Dimer acid based waterborne hyperbranched poly (ester amide) thermoset as a sustainable coating material. *Prog Org Coat* 112: 57–65.
14. Lu Y, Larock R C (2010) Aqueous cationic polyurethane dispersions from vegetable oils. *ChemSusChem* 3(3):329–333.
15. Kim B K, Leez Y M (1994) Aqueous dispersion of polyurethanes containing ionic and nonionic hydrophilic segments. *J Appl Polym Sci* 54:1809–1815.
16. Li X, Wang D, Zhao L, Hou X, Liu L, Feng B, Li M, Zheng P, Zhao X, Wei S (2021) UV LED curable epoxy soybean-oil-based waterborne PUA resin for wood coatings. *Prog Org Coat* 151:105942.
17. Liao L, Li X, Wang Y, Fu H, Li Y (2016) Effects of surface structure and morphology of nanoclays on the properties of jatropha curcas oil-based waterborne polyurethane/ clay nanocomposites. *Ind Eng Chem Res* 55(45):11689–11699.
18. Cakić S M, Ristić I S, Stojiljković D T, Nikolić N N, Todorović B Z (2018) Effect of the silica nanofiller on the properties of castor oil-based waterborne polyurethane hybrid dispersions based on recycled PET waste. *Polym Bull* 76:1217–1238.

19. More A P, Mhaske S T (2018) Synthesis of polyurethane dispersion from polyesteramide polyol, *Pigment Resin Technol* 47:54–163.
20. Fu C, Hu X, Yang Z, Shen L, Zheng Z (2015) Preparation and properties of waterborne bio-based polyurethane / siloxane cross-linked films by an in situ sol–gel process. *Prog Org Coatings* 84:8–27.
21. Liang H, Feng Y, Lu J, Liu L, Yang Z, Luo Y, Zhang Y, Zhang C (2018) Bio-based cationic waterborne polyurethanes dispersions prepared from different vegetable oils. *Ind Crops Prod* 122:448–455.
22. Ghasemlou M, Daver F, Ivanova E P, Adhikari B (2019) Polyurethanes from seed oil-based polyols: A review of synthesis, mechanical and thermal properties. *Ind Crops Prod* 142:111841.
23. Petrović Z S, Bilić O, Zlatanić A (2013) Polyols and polyurethanes from crude algal oil, *J. Am Oil Chem Soc* 90(7):073–1078.
24. Sun J, Aly K I, Kuckling D (2017) Synthesis of hyperbranched polymers from vegetable oil based monomers via ozonolysis pathway. *J Polym Sci Part A Polym Chem* 55:2104–2114.
25. Narine S S, Yue J, Kong X (2007) Production of polyols from canola oil and their chemical identification and physical properties. *J Am Oil Chem Soc* 84:173–179.
26. Pfister D P, Xia Y, Larock R C (2011) Recent advances in vegetable oil-based polyurethanes, *ChemSusChem* 4:703–717.
27. Tawade B V, Shingte R D, Kuhire S S, Sadavarte N V, Garg K, Maher D M, Ichake A B, More A S, Wadgaonkar P P (2017) Bio-based di-/poly-isocyanates for polyurethanes: An overview. *PU Today*, 41–46.
28. Zenner M D, Xia Y, Chen J S, Kessler M R (2013) Polyurethanes from isosorbide-based diisocyanates. *ChemSusChem* 6:1182–1185.
29. Sahoo S, Kalita H, Mohanty S ,Nayak S K (2017) Synthesis and characterization of vegetable oil based polyurethane derived from low viscous bio aliphatic isocyanate: Adhesion strength to wood-wood substrate bonding. *Macromol Res* 25:772–778.
30. Chauhan M, Gupta M, Singh B, Singh A K, Gupta V K (2014) Effect of functionalized lignin on the properties of lignin—isocyanate prepolymer blends and composites. *Eur Polym J* 52:32–43.
31. Maisonneuve L, Chollet G, Grau E, Cramail H (2016) Vegetable oils: A source of polyols for polyurethane materials. *OCL* 23(5):1–10.
32. Hojabri L, Kong X, Narine S S (2010) Novel long chain unsaturated diisocyanate from fatty acid: synthesis, characterization, and application in bio-based polyurethane. *J Polym Sci Part A Polym Chem* 48:3302 3310.
33. Panda S S, Panda B P, Nayak S K, Mohanty S (2017) A review on waterborne thermosetting polyurethane coatings based on castor oil: synthesis, characterization, and application. *Polym Plast Technol Eng* 57:500–522.
34. Malkappa K, Jana T (2015) Hydrophobic, water dispersible polyurethane: Role of polybutadiene diol structure. *Ind Eng Chem Res* 54(30):7423–7435.
35. Zhu Z, Li R, Zhang C, Gong S, (2018) Preparation and properties of high solid content and low viscosity waterborne polyurethane—acrylate emulsion with a reactive emulsifier, *Polymers* 10(2):1–17.
36. Liu L, Lu J, Zhang Y, Liang H, Liang D, Jiang J, Lu Q, Quirino R L, Zhang C (2019) Thermosetting polyurethanes prepared with the aid of a fully bio-based emulsifier with high bio-content, high solid content, and superior mechanical properties. *Green Chem* 21:526–537.
37. Gaddam S K, Kutcherlapati S N R, Palanisamy A (2017) Self-cross-linkable anionic waterborne polyurethane–Silanol dispersions from cottonseed-oil-based phosphorylated polyol as ionic soft segment. *ACS Sustain Chem Eng* 5:6447–6455.

38. Chen R, C. Zhang C, Kessler M R (2014) Anionic waterborne polyurethane dispersion from a bio-based ionic segment. *RSC Adv* 4:35476–35483.
39. Llorente I E O, Martín J A L, Irusta A G L (2019) Dispersion characteristics and curing behaviour of waterborne UV crosslinkable polyurethanes based on renewable dimer fatty acid polyesters. *J Polym Environ* 27:189–197.
40. Liu T, Liu D, Zhou W, Ni L, Quan H, Yang Z (2019) Effects of soft segment and microstructure of polysiloxane modified cationic waterborne polyurethane on the properties of its emulsion and membrane. *Mater Res Express* 6:025304.
41. Szycher M (2013) *Handbook of Polyurethane*, 2nd ed. Boca Raton, FL: CRC Press; pp 417–447.
42. Garrison T F, Zhang Z, Kim H J, Mitra D, Xia Y, Pfister D P, Brehm-Stecher B F, Larock R C, Kessler M R (2014), Thermomechanical and antibacterial properties of soybean oil-based cationic polyurethane coatings: Effects of amine ratio and degree of crosslinking. *Macromol Mater Eng* 299:1042–1051.
43. You G, Wang J, Wang C, Zhou X, Zhou X, Liu L, (2019) Effect of side methyl from mixed diamine chain extenders on microphase separation and morphology of polyurethane fiber Effect of side methyl from mixed diamine chain extenders on microphase separation and morphology of polyurethane fiber. *Mater Res Express* 6:095301.
44. J Zhao J, Zhou T, Zhang J, Chen H, Yuan C, Zhang W, Zhang A (2014) Synthesis of a waterborne polyurethane-fluorinated emulsion and its hydrophobic properties of coating films. *Ind Eng Chem Res* 53:19257–19264.
45. Ghosh T and Karak N (2020) Mechanically robust hydrophobic interpenetrating polymer network-based nanocomposite of hyperbranched polyurethane and polystyrene as an effective anticorrosive coating. *New J Chem* 44:5980–5994.
46. Wang C H, Zhang J, Wang H, He M, Ding L, Zhao W (2021) Simultaneously improving the fracture toughness and flame retardancy of soybean oil-based waterborne polyurethane coatings by phosphorus-nitrogen chain extender. *Ind Crops Prod* 163:113328.
47. Schmidt S, Ritter B S, Kratzert D, Bruchmann B, Mülhaupt R (2016) Isocyanate-free route to poly(carbohydrate–urethane) thermosets and 100% bio-based coatings derived from glycerol feedstock. *Macromolecules* 49:7268–7276.
48. Zhang C, Wang H, Zeng W, Zhou O (2019) High biobased carbon content polyurethane dispersions synthesized from fatty acid-based isocyanate. *Ind Eng Chem Res* 58(13):5195–5201.
49. Ma S, Chen C, Rafaël J. Sablong C, Koning E, Benthem R A T M (2018) Non-isocyanate strategy for anionically stabilized water-borne polyurea dispersions and coatings. *J Polym Sci A Polym Chem* 56(10):1078–1090.
50. Bizet B, Grau E, Cramail H, Asua J A, (2020) Water-based non-isocyanate polyurethane-ureas (NIPUUs). *Polym Chem* 11 (23):3786–3799.
51. Zhang C, Wang H, Zhou O (2020) Waterborne isocyanate-free polyurethane epoxy hybrid coatings synthesized from sustainable fatty acid diamine. *Green Chem* 22(4): 1329–1337.
52. Ochiai B, Satoh Y and Endo T (2005) Nucleophilic polyaddition in water based on chemo-selective reaction of cyclic carbonate with amine. *Green Chem* 7:765–767.
53. Meng L, Wang X, Ocepek M, Soucek M D (2017) A new class of non-isocyanate urethane methacrylates for the urethane latexes. *Polymer* 109:146–159.

5 Synthesis, Properties, and Applications of Waterborne Polyurethanes

Sanam Amiri, Hossein Nazockdast, and City Mir Sadeghi
Department of Polymer Engineering and Color Technology, Amirkabir University of Technology, P.O. Box 159163-4311, Tehran, Iran

Shervin Ahmadi
Iran Polymer and Petrochemical Institute, P.O. Box 14965-115, Tehran, Iran

Sahar Amiri
Department of Polymer Engineering, Science and Research Branch, Islamic Azad University, P.O. Box 147789-3855, Tehran, Iran

CONTENTS

5.1 Introduction 66
5.2 Structure of Polyurethane 67
5.3 Introduction to Waterborne Polyurethane 67
5.4 WPU Synthesis 67
5.5 Synthesis of Waterborne Polyurethanes 67
5.6 Cationic Waterborne Polyurethanes (CAWPUs) 67
5.7 Anionic Waterborne Polyurethanes (AWPUs) 68
5.8 Synthesis Methods of Polyurethane Dispersions 68
5.9 Acetone Process 69
5.10 Structure of Waterborne Polyurethanes 70
5.11 Dispersion Formation 71
5.12 Applications of Polyurethane Dispersions 71
5.13 Nanoparticle Incorporation Into Polyurethane Dispersions 72
5.14 WPU/Cellulose 72

DOI: 10.1201/9781003173526-5

5.15 WPU Fluorescent Composite........72
5.16 Cationic Waterborne Polyurethane Based on Waste Frying Oil With Antibacterial Activity........73
5.17 WPUs Based on Castor Oil-Based and Sodium Alginate........74
5.18 WPU/Acrylic Hybrids With Industrial Applications........74
5.19 WPU/Acrylic-Based Coatings........75
5.20 WPU/ Acrylic Adhesives........75
5.21 WPU Latex-Based on Castor Oil........75
5.22 WPUs/ Poly(Butylene Itaconate) Ester as Adhesion........76
5.23 Coating Based on WPU and Poly(Vinyl Chloride)........76
5.24 Antimicrobial WPU Containing Cellulose Crystalline and Silver Nanoparticles........76
5.25 WPU as Electrical and Conductive Polymers........76
5.26 Concluding Remarks........78
References........79

5.1 INTRODUCTION

Polyurethanes (PUs), which are composed of multicomponent systems of soft and hard segments, became important in the 1970s; aliphatic or aromatic isocyanates act as hard segments, and diols or polyols as soft segments (Figure 5.1). The ratio of hard to soft segments in the PU structure, chain extenders, catalysts, blowing agents, emulsifiers, stabilizers, and surfactants changes the properties of the obtained polymer for specific applications. PU formation is based on the formation of hydrogen bonding between the polymer chains of soft and hard segments, which form a polymer network with significant physical/mechanical properties, such as flexible or rigid foams, elastomers, and thermoplastic sand fibers [1,2].

One of the drawbacks of PUs is the use of a high amount of volatile organic compounds (VOCs) with a definite amount of free isocyanate [3,4]. Generally, PUs are incompatible with water, and for increased water solubility of these components, special modification and functional groups with various structures, such as hydrophilic monomers, ionic groups such as quaternary ammonium, carboxylate, or sulfonate groups, are needed [5]. Waterborne polyurethanes (WPUs) are commercially interested because of the lower price of water compared to solvent, and fewer environmental problems and good quality has made them highly suitable for a wide range of applications [6]. Due to the evaporation of water in this process, this method is safe, nontoxic, and nonflammable for the environment, so WPUs can be used for various industrial applications, such as smart coatings and adhesives [7,8].

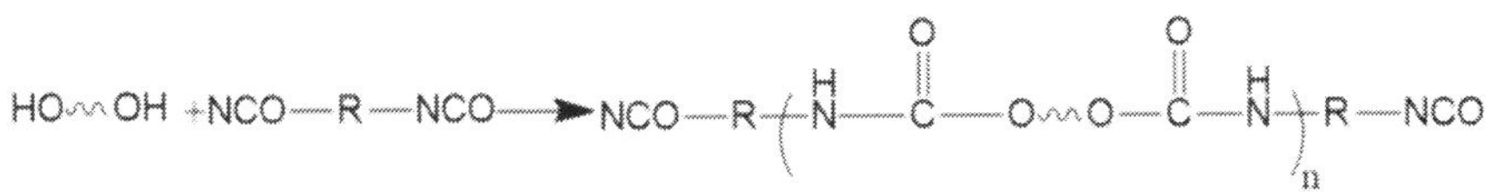

FIGURE 5.1 Schematic diagram of polyurethane synthesis.

5.2 STRUCTURE OF POLYURETHANE

PUs are formed based on polymerization routes such as condensation or the addition polymerization, but addition polymerization is acceptable due to no by-product in the final product. Isocyanate as a precursor is used to form PU to react with active hydrogens of functional groups and CO_2 release as a by-product used in the case of PU foam and the secondary reactions of isocyanate as PU [9,10].

5.3 INTRODUCTION TO WATERBORNE POLYURETHANE

Isocyanate's presence in the structure of PU is a volatile organic solvent, which is a toxic compound for the environment and may cause skin allergy and cancer. One method to overcome this problem is using WPU, which is the stable dispersion of PU in aqueous media in the absence of solvent and divided into anionic, cationic, or nonionic WPU [9,10]. WPUs with various morphologies and physical properties are synthesized with a wide range of molecular weight of PU containing hydrophobic or hydrophilic characteristics of other parts in the PU chain and polar solubilizing group [9,10].

5.4 WPU SYNTHESIS

High molecular weight WPU synthesis is based on four steps of the acetone process, the melt-dispersion process, the pre-polymer mixing process, and the ketimine–ketazine process [10]. The first step is similar to common PU synthesis and starts with a reaction of a soft segment with a hard segment that is then dispersed in water and forms a low viscosity PU. Using a chain extender leads to converting the low molecular PU to a high molecular PU. After formation, the obtained PU is dissolved in a solvent. After aqueous dispersion formation, the samples are dried (Figure 5.2) [10].

WPUs consist of fully reacted chains that are then dispersed in an aqueous medium containing environmentally friendly, nontoxic, nonflammable latex particles or dispersions with various applications. In WPUs, molecular weight does not affect viscosity and flow properties of obtained dispersion. To obtain the melt dispersion process, PU is heated to facilitate dispersion in water, and cationic PUs are obtained [10].

5.5 SYNTHESIS OF WATERBORNE POLYURETHANES

Various methods can be used for the synthesis of WPUs, which include emulsification or prepolymer, acetone process, hot-melt, and ketimine–ketazine. Later in this chapter, each of these methods is explained.

5.6 CATIONIC WATERBORNE POLYURETHANES (CAWPUS)

Cationic WPUs are a solution in which cationic groups can be inserted into the polymer chains to obtain water-dispersible chains with a wide range of areas. The ratio of hard and soft segments, chain extenders, additives, and dispersing centers is

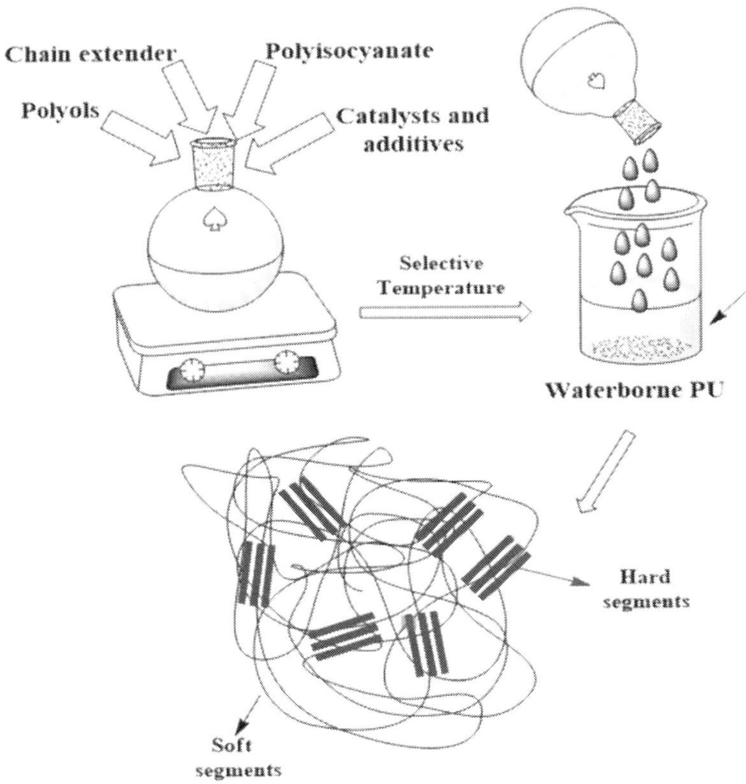

FIGURE 5.2 Schematic diagram of the formation of waterborne PU.

important to determine the physical properties of the CAPU. Elasticity, flexibility, and weather ability depend on the soft segment, whereas mechanical properties of obtained CAPUs depend on the hard segment [11,12].

5.7 ANIONIC WATERBORNE POLYURETHANES (AWPUS)

Anionic waterborne polyurethane (AWPU) is a class of WPUs that is an aqueous dispersion of anionic PUs. For example, in the prepolymer synthesis step, dimethylol-propionic acid (DMPA) is used, and the neutralization step is done with a tertiary amine compound, such as triethylamine; AWPU salt is formed in acetone and water in the presence of a chain extender and a polyol. These AWPUs are used as adhesion to the wood, which in the pre-polymer formation step chain extension occurs immediately before the polymer is separated [13].

5.8 SYNTHESIS METHODS OF POLYURETHANE DISPERSIONS

Due to the formation of WPUs using little or no solvent, in the first step pre-polymers are terminated by the formation of isocyanate groups, and then ionic groups are grafted into the polymer structure. In the aqueous phase, chain extension occurs, and the hydrophilic prepolymer may react with water, so diamine-based chain extenders are used [14,15].

5.9 ACETONE PROCESS

PU dispersions are stable binary colloidal systems with the PU phase as the discontinuous phase and water as the continuous phase [16]. By using acetone, aqueous high molecular weight ionic PU chains with zero-VOC content were obtained by prepolymer formation. Then, acetone was added to the solution, and chain extension occurred in acetone. Latex particles formed with transfer of the ionic PU/acetone into water under high shear, and latex particles formed. Then, VOC-free PU dispersions were obtained by removing acetone from the system via vacuum distillation. Polar solvent such as THF or MEK can also be used to form low viscosity prepolymer, and the molecular weight and viscosity increase via chain extension; conversion also increases (Figure 5.3) [16].

When ionic centers link to each other, molecular weight increases, and when water is added to the system, rearrangement of the ionic centers reduces. By incorporation of hydrophobic segments aligned in between each other, viscosity

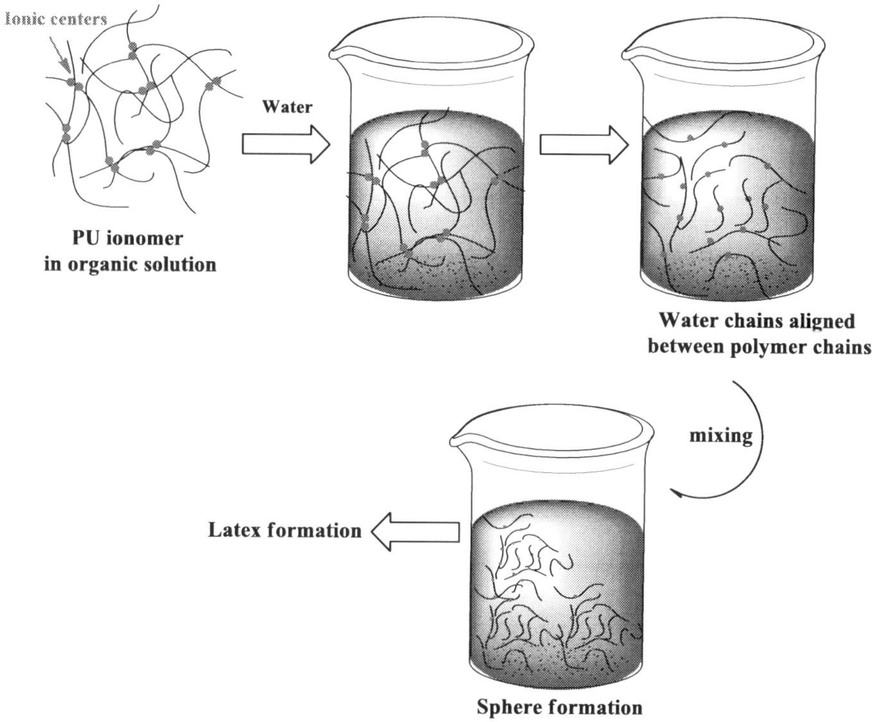

FIGURE 5.3 Schematic diagram of latex formation from PU ionomer in water.

increases, and dispersion forms. The mixture becomes turbid, and the dispersed phase is seen, so the clusters of PU chains are rearranged and form into microspheres in which acetone acts as the core and water is the outer surface [16].

5.10 STRUCTURE OF WATERBORNE POLYURETHANES

For the synthesis of WPUs, hydrophilic and amphiphilic polymers are used as an emulsifier that acts as a stabilizer for PU in water. WPUs consist of soft and hard segments, which are hydrophilic and hydrophobic parts, respectively. There are two kinds of emulsifiers, internal emulsifiers include nonionic centers or ionic centers. Nonionic PU dispersions based on a hydrophilic soft segment pendant group are stable in a wide range of pH levels, electrolytes, mechanical influences, and solvents [17,18].

Similar to WPUs, waterborne polyurethane-ureas (WPUUs) are eco-friendly green PUs that are synthesized without chemical solvents and lead to environmentally friendly products and nontoxic PUs and PUUs [19]. The functional group of PUs and PUUs is urethane groups and urethane/urea groups, respectively (Figure 5.4).

PUU showed a more stable structure compared to PU due to the formation of dual hydrogen bonding interactions, also PU is based on single intermolecular hydrogen bonds [20]. In PUU synthesis, usually a diamine chain extender is used, and phase separation results in the thermodynamical incompatibility of hard and soft segments [21–23]. WPU and WPUU dispersions show specific properties, such as high solid content and high molecular weight, low viscosity and nonflammability, and processability at low temperature [24,25]. Polyols as soft segments are medium to high molecular weight macromolecule terminated hydroxyl groups and make the structure flexible; the molecular weight affects the viscosities of obtained PU or PUU [26–28]. Chain extenders that are di-functional with low molecular weight compounds, such as diamines, can be used for the formation of crosslinked WPU or WPUU dispersions, which form crosslinks between urea groups or

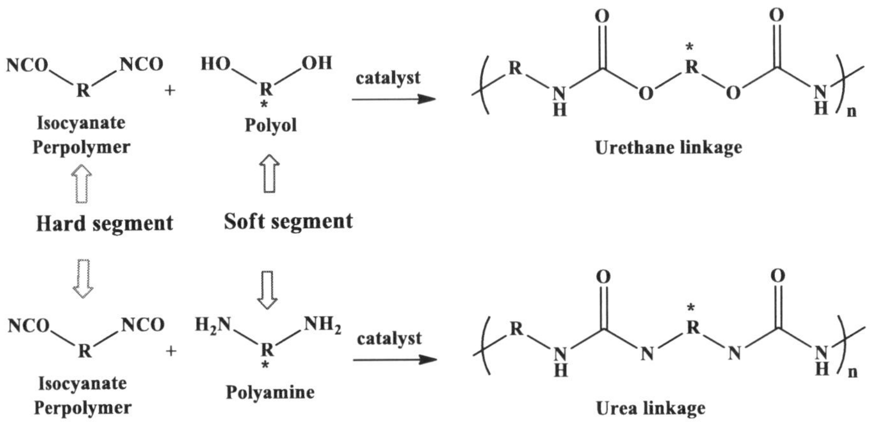

FIGURE 5.4 Scheme of urethane and urea group reactions.

urethane linkages, leading to systems with better stability and strength [27–29]. By incorporating hydrophilic soft segments into nonionic WPU and WPUU, dispersion in water is obtained with a high agitation rate due to the weak hydrophilic character of these types of emulsifiers [29]. In the case of ionic emulsifiers, salts form by neutralizing acid or tertiary nitrogen groups [30–33].

5.11 DISPERSION FORMATION

The dispersion step is an important part of the synthesis of WPUs or WPUUs, which are affected by the structure of the chain extender and reaction rate, change the stabilization mechanism of particles varies such as entropic repulsions. By insertion of hydrophilic segments in the surface of the particles and dispersed into the water, particles stretch restricted segment mobility and entropy, so dispersion becomes stable due to spontaneous repulsion between particles [34]. Scheme 5.1. shows a schematic diagram for the formation of ionic WPU and WPUU dispersions. Dispersion into water causes moving ionic groups into particle surfaces and counter ions surrounded surfaces, so the electrical double layer acts as a shell of the particle, and hydrophobic domains act as a core agent [35].

5.12 APPLICATIONS OF POLYURETHANE DISPERSIONS

Due to the elastomeric and polar properties of PUs (related to soft segments and urethane groups, respectively), PUs have a wide range of applications in the

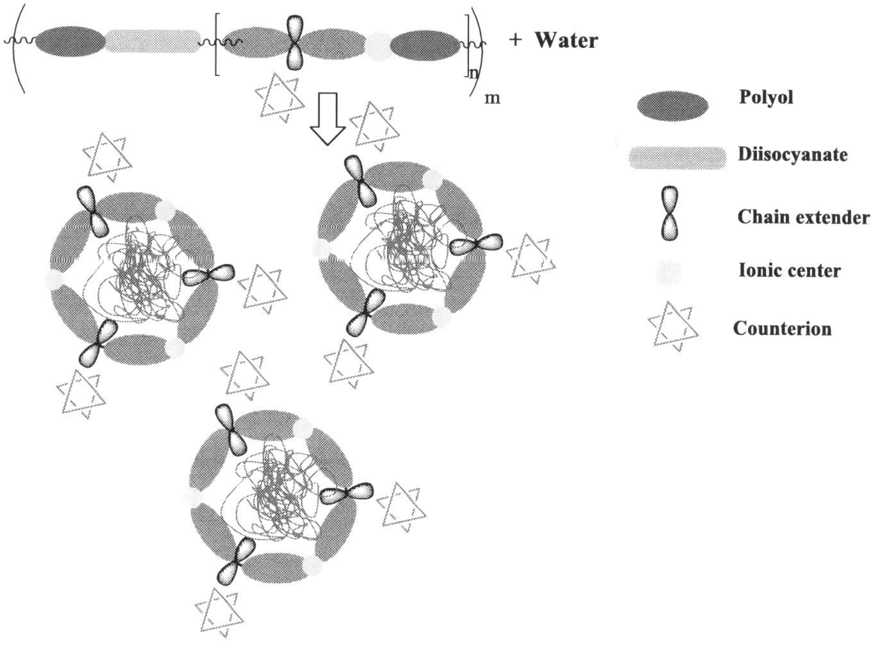

SCHEME 5.1 Schematic diagram of ionic WPU and WPUU particle formation.

adhesive area [16]. By changing the ratio of hard to soft segments, properties of PU dispersion and particle size can be changed such as decreasing NCO/OH molar ratio, decreasing mean particle size, and increasing pre-polymer viscosity, so a decrease in the molecular weight of PU will increase PU crystallinity [34,35].

5.13 NANOPARTICLE INCORPORATION INTO POLYURETHANE DISPERSIONS

To create new applications and new properties, various kinds of nanosized materials may be incorporated into PU dispersion, which may change the properties of nanoparticles, such as shape, size, hydrophobicity, hydrophilicity, or surface roughness. Carbon nanotubes (CNTs) change nanocomposite properties due to high aspcct ratios [36]. Λ self-healing coating can be obtained by adding an encapsulation healing agent into HNTs and then dispersing it in an aqueous solution of acrylic paint to prevent scratch formation on the surface; the encapsulated active agent is released from the HNTs and prevents crack propagation [37,38].

5.14 WPU/CELLULOSE

Due to the increasing preference to use environmentally friendly materials, many researchers are interested in introducing WPU and WPUU nanocomposites containing biomaterials, such as cellulose derivatives, and improving hydrophilic dispersing of obtained WPUs. Cellulose, a renewable biopolymer, is composed of D-glucose units linked by β(1,4) glycosidic bonds with low cost, nontoxicity, biocompatibility, and biodegradability. Cellulose nanocrystals (CNCs) show high specific mechanical properties and length/diameter (L/D) aspect ratio, so they are an interesting prospect in the nanocomposites field [39].

To obtain a good organic dispersion medium of nonpolar polymers, using an appropriated surfactant or surface chemical modification of nano-cellulose is critical. On the other hand, to increase compatibility between the polymer and nano-cellulose in water and the formation of a homogeneous nanocomposite, good aqueous polymer dispersions are needed [40]. Electrospinning of WPUs containing CNC led to the formation of mats with high porosity and large surface area to volume ratio with tailored fiber diameters, which can be used as tissue engineering scaffolds, protective clothing, filtration, cosmetics, and membranes, among others [41].

5.15 WPU FLUORESCENT COMPOSITE

WPUs based on isophorone diisocyanate (IPDI), polyethyleneglycol (PEG), dimethylolpropionic acid (DMPA), and trimethylolpropane (TMP) have been synthesized. After the formation of WPUs, the modification was done by acrylate derivatives and polyurethane/polyacrylate (PU/PA), and crosslinked polyurethane-acrylate (LPUA), respectively [42]. Modified WPU combined with the fluorescent dyestuff was combined and coated cotton fabric, preserving the adhesive property of the composite. Obtained nanocomposites were characterized with various methods; then, the effect of acrylate used to create heat-resistant cotton fabrics was discussed [43-45].

5.16 CATIONIC WATERBORNE POLYURETHANE BASED ON WASTE FRYING OIL WITH ANTIBACTERIAL ACTIVITY

Due to the biodegradability and low toxicity of plant-based oils, these compounds are used in polymers for the formation of new composites with specific properties. The incorporation of waste oil into PU resins is widely used in many industries because of their excellent mechanical properties and flexible processing methods [46].

Antibacterial coatings based on environmentally friendly CWPUs were obtained using methylene diphenyl 4,4′-diisocyanate (MDI) as hard segments, waste frying oil as soft segments, and bis(2-hydroxyethyl) dimethyl ammonium chloride (BHMAC) as an internal emulsifier with good antibacterial activity and adhesion for covering various surfaces [47,48]. By increasing the NCO/MDI ratio (hard segment/soft segment), the antibacterial activity of obtained CWPU increased (Tables 5.1 and 5.2).

Reductions of *S. aureus* bacterial growth of all samples were greater than 99% after 3 h of incubation. By increasing internal emulsifier, the efficiency of

TABLE 5.1
Chemical composition of CWPUs with various internal emulsifier contents [49] Copyright (2019) Hindawi International Journal of Polymer Science

Formula	NCO/MDI	Mole Ratio		
		Monoglyceride	BHMAC	EG
CWPU-1	0.84	1.0	0.1	0.1
CWPU-2	0.91	0.84	0.2	0.1
CWPU-3	0.98	0.84	0.3	0.1

TABLE 5.2
Antibacterial activity of CWPU for 0, 3, 6, and 24 h for *S. aureus* and *E. coli* bacteria [49] Copyright (2019) Hindawi International Journal of Polymer Science

Sample	Bacterial Reduction (%)							
	Gram-Positive Bacterial *S.aureus*				Gram-Negative Bacterial *E. coli*			
	0 h	3 h	6 h	24 h	0 h	3 h	6 h	24 h
Control	0	0.0	0.0	0.0	0.0	0.0	0.0	0.0
CWPU-1	0	99.7 ± 0.6	99.8 ± 0.5	99.9 ± 0.2	0.0	94.7 ± 0.2	99.5 ± 0.1	99.9 ± 0.5
CWPU-2	0	99.9 ± 0.1	99.9 ± 0.5	99.9 ± 0.4	0.0	96.3 ± 0.5	99.9 ± 0.4	99.8 ± 0.5
CWPU-3	0	99.9 ± 0.1	99.9 ± 0.2	100.00.1	0.0	97.6 ± 0.9	99.9 ± 06	100.0 ± 0.1

bacterial reduction increased, which related to greater quaternary ammonium content incorporated in the polymer chains [47].

5.17 WPUS BASED ON CASTOR OIL-BASED AND SODIUM ALGINATE

Nanofillers may be added to PU matrices to form new interactions and improve the mechanical properties of WPUs. The addition of graphene oxide (GO) into WPUs caused the interaction between PUs and GO, so it improved the hydrophobicity of coating, tensile modulus, hardness, and T_g of obtained hybrid films [50–52]. Bio-based composite dispersions were formed by mixing sodium alginate (SA) and WPU dispersion, and results showed the presence of SA led to the formation of hydrogen bonds, which improved tensile strength and storage moduli of obtained WPU nanocomposites. They also had an effect on the tensile properties of WPU, decreased elongation at break, and increased the SA content [53]. Table 5.3 shows the mechanical and tensile properties of pure SA, WPU, and WPU/SA. The lowest tensile strength, 10.8 MPa, and the highest elongation at break of 256% related to pure WPU. When SA content increased in the PU network, due to the brittle manner of SA compound, brittleness of the composite films increased and led to an increase in tensile strength and Young's modulus, but also to a decrease in elasticity and toughness of WPU/SA. For pure WPU, the contact angle was 47.25°, and for WPU/SA, the increase was from 52° to 62°, with the increase of SA content, but all $\theta < 90°$, indicating that all the surfaces of the composite films were hydrophilic [53].

5.18 WPU/ACRYLIC HYBRIDS WITH INDUSTRIAL APPLICATIONS

Due to the high price of the raw materials of WPUs, various additives were used to adjust product price and find the best ratio. For this reason, a combination of PU and acrylic may be used. By changing the ratio of hard and soft segments, emulsifier or additive, the ratio between the two components, and grafting agent, WPU's properties may improve and be adjustable for specific applications, such as modification of chemical structure and improvement of mechanical properties [49].

TABLE 5.3
Electrical conductivity of WPU coating with different MWCNT content [54] Copyright (2020) Express Polymer Letters

Coating	Properties	1 wt.%	0.3 wt.%	0.6 wt.%
Electrostatic spraying	Thickness (μm)	81 ± 4	82 ± 4	83 ± 4
	Square resistance (MΩ)	0	156.2 ± 5	2.6 ± 0.2
	Resistivity (Ωm)	0	12808.4	215.8
Brushing coating	Thickness (μm)	82 ± 4	82 ± 4	83 ± 4
	Square resistance (MΩ)	0	0	155.7 ± 0.5
	Resistivity (Ωm)	0	0	12923.1

5.19 WPU/ACRYLIC-BASED COATINGS

WPU/Acrylic hybrid films showed good chemical structure, water resistance, abrasion resistance, hardness-to-flexibility balance, and substrate adhesion with suitable price. WPU/acrylic coatings are synthesized as solvent-free hybrids in which the acrylic phase adds to WPU dispersion and leads to a grafted PU/acrylic, which is an environmentally friendly coating [55].

5.20 WPU/ ACRYLIC ADHESIVES

PU/acrylic hybrids showed excellent adhesion, flexibility, color stability, and UV degradation resistance and were environmentally friendly. Their polymer chains acted as active sites to form cohesive crosslinks, but not to such an extent as to compromise the adhesiveness. By varying PU and/or acrylic chains or crosslinking, the structure changed and improved shear resistance and tack adhesion [56].

5.21 WPU LATEX-BASED ON CASTOR OIL

Due to environmental problems caused by VOCs, WPUs have become interesting, with good tensile strength, abrasion resistance, chemical resistance, and toughness. To produce bio-based WPU dispersions, vegetable oils or fatty acids can be used as polyols that form new covalent bonds to inorganic materials or polymers [57,58].

In another work, castor oil was grafted to WPU dispersions in which 2-hydroxyethyl acrylate (HEA) was used as a crosslink agent. Hydroxyl groups are applicable crosslinking sites in acrylic latexes. The storage modulus (E′) of both the vinyl-containing WPUs and the grafted hybrid latex are shown in Figure 5.5, respectively, which decreased as the ratio of HEA increased. By increasing the HEA amount, chain termination increased, leading to more dangling chains, which decreased crosslink density [59,60].

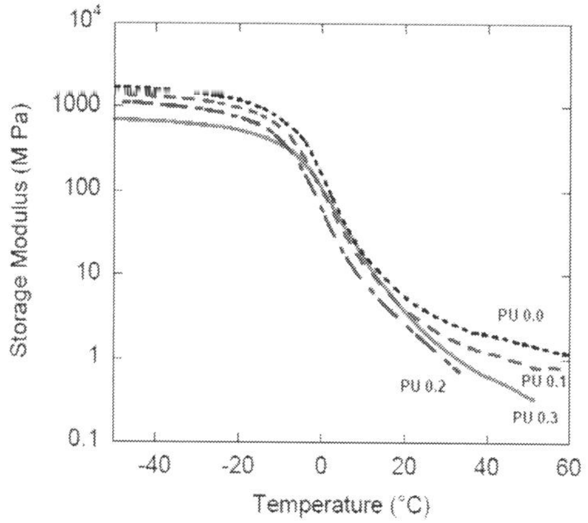

FIGURE 5.5 Storage modulus as a function of temperature for polyurethane films with different amounts of HEA [59] Copyright (2017) Elsevier.

5.22 WPUS/ POLY(BUTYLENE ITACONATE) ESTER AS ADHESION

WPU macromolecules usually possess alternating hard and soft segments in the presence of chain extension reaction and neutralization. Polyether and polyesters, such as soybean oil, modified soybean oil, or modified nature rubber, can be used as polyols and soft segments [61]. Polyester-based PUs showed better mechanical strength and adhesion force because of polar ester groups [62,63].

5.23 COATING BASED ON WPU AND POLY(VINYL CHLORIDE)

WPU-based coating is a promising method to decrease environmental problems and introduce a new version of solvent-free coating. WPU-based coatings are produced via the dispersion of PUs in water in the presence of an emulsifier, which the hydrophilic group inserts into the PU backbone chain. Hydrophilic ionic groups and their structures improve elasticity and mechanical strength, durability, and abrasion resistance [64,65]. Polycarbonate diol has carbonyl groups, which improve tensile strength of obtained films and also increase the reactive silicone amount; they also increase active sites for the formation of new interactions so tensile strength increases, but they decrease elongation due to brittleness of the obtained structure [66]. Due to the presence of hydroxyl groups in the polyol chain, active sites are applicable for the formation of new crosslinks with a carboxylic acid group. Bio-based polyols form dispersions with good compatibility, excellent storage stability, and high solid content (up to 45%); higher toughness; and better thermal stability than traditional solvent-based PUs [67,68].

5.24 ANTIMICROBIAL WPU CONTAINING CELLULOSE CRYSTALLINE AND SILVER NANOPARTICLES

Antimicrobial surfaces and coatings became attractive, with broad developmental prospects that inhibit most harmful bacteria and prevent their growth and proliferation. Applying antibacterial coatings with smart properties is a simple way to obtain safe and clean surfaces from any bacteria for medical applications, especially for wood surfaces. Incorporation of silver nanoparticles (AgNPs) into WPUs as eco-friendly coatings with antimicrobial properties are a good candidate for antimicrobial applications, such as for food and other industrial products [69,70]. WPU/NCC/AgNP coating was synthesized, and the adhesion and antimicrobial activity determined which showed improvement of adhesion of the NCC-WPU composites with the addition of NCC due to interactions formed between the wood board and matrix. Incorporating NCC into the WPU matrix led to the formation of new, strong interactions between the wood board and matrix. The mechanical properties of the WPU will increase, but when NCC is added to a certain extent, its mechanical properties will decrease [71].

5.25 WPU AS ELECTRICAL AND CONDUCTIVE POLYMERS

WPU is a solvent-free compound with a wide range of applications due to its low temperature flexibility, absence of VOCs, water or superior solvent resistance, weathering resistance, and chemical and mechanical properties. Due to the presence of oxygen groups, graphene oxide or graphene extract (GE) is a hydrophilic compound, and oxygen groups are removed to obtain graphene. WPU/G nanocomposites showed shape memory efficiency, oil adsorbency, gas barrier, good mechanical properties, and electrical conductivity, which make this compound a good candidate for the electrical industry. WPU/GO, WPU/GE, and WPU/rGO containing 5 wt.% of GE, GO, and rGO were coated, and conductivity and electrical properties of the obtained coating were determined [72]. Coated WPU showed higher electrical resistance, (155.0 ± 76.0)·106 Ω for WPU/ G and (142.1 ± 20.5)·106 Ω for WPU/rGO, whereas composites had significantly lower resistance values, (6.6 ± 1.7)·106 Ω for 5GE/G and (2.7 ± 1.3)·106 Ω for 5GO/rGO. The remaining oxygen groups in rGO led to improved interaction and formed a thicker coating with better affinity and higher conductivity [72]. The incorporation of metal nanoparticles, such Fe_3O_4, TiO_2, SiO_2, and multiwalled carbon nanotubes (MWCNTs), into WPU led to nanocomposites with good thermal stability. The novel WPU/ Fe_3O_4 nanocomposites were prepared, and oleic acid was used to increase the compatibility of monomers, so Fe_3O_4 dispersed into the WPU network [73]. Fe_3O_4 nanoparticles improved the conductivity of obtained WPU nanocomposites. WPU/ Fe_3O_4 (1.5 wt.%) showed volume higher resistivity and surface resistance, 5.42×10^5 Ω cm and 9.67×10^{12} Ω/cm^{-2}, respectively), and pure WPU showed volume resistivity of 1.04×10^{13} Ω cm and surface resistance of 7.49×10^5 Ω/cm^{-2}. When Fe_3O_4 content increased, crosslinking formation between nanoparticles and WPU increased, Fe_3O_4 mobility decreased, ielectrical properties increased, and there was a slight reduction of resistivity [73].

WPUs have poor mechanical strength, and low electrical conduction is a drawback and restricts their application in this area. This can be modified using nanoparticles to strengthen the physical and mechanical properties of polymers [54,74]. MWCNTs incorporated into WPU networks significantly increase chemical corrosive resistance and conduction of electricity. Figure 5.6 shows the morphology of electrostatic spraying and brushing coatings containing 0.6 wt.% MWCNTs, which shows an agglomeration in the case of the brushing process with a size of

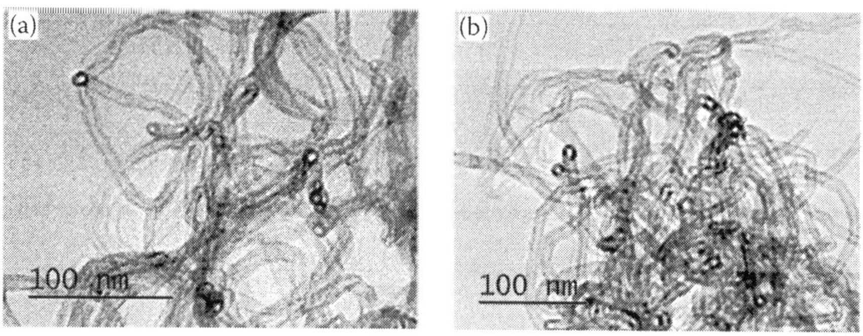

FIGURE 5.6 Morphologies of MWCNTs in uncured (A) electrostatic spraying with 0.6 wt.% MWCNT and (B) brushing with 0.6 wt.% MWCNT on the steel substrates [54] Copyright (2020) Express Polymer Letters.

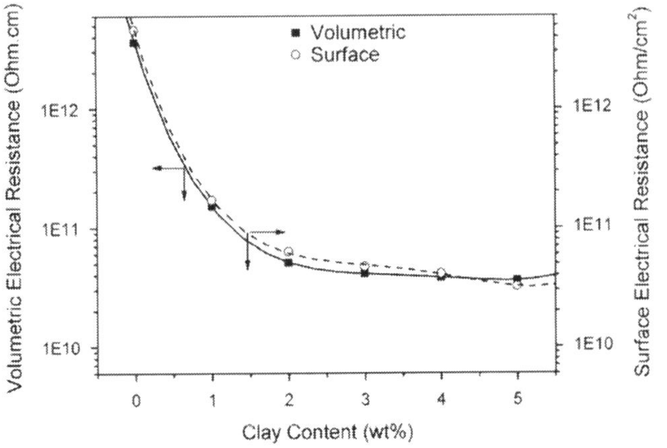

FIGURE 5.7 Volumetric and surface electrical resistances of the WPU/ clay films with different clay contents [75] Copyright (2006) American Chemical Society.

about 150–200 nm. The electrical conductivity of these two coatings indicated the presence of 0.3 wt.% MWCNT in the electrostatic spraying method led to conducting electricity, but this percent is not sufficient in the brushing coatings and showed the effect of good dispersion MWCNTs.

Nanosized layered silicates also can be used to improve the performance and mechanical properties of WPU networks, which is related to the layer-by-layer structure of MMT with good strength, good stiffness, and high aspect ratio [87]. The presence of magnesium or aluminum hydroxide in the structure of MMT led to a reaction between Na^+ and Ca^{2+}, with organic cations rendering the hydrophilic-layered silicate organophilic. Anionic layers were balanced by mobile interlayer cations and formed ionic interactions between the clay and WPU. They modified the morphology, thermal resistance, tensile properties, and ionic conductivities of the obtained nanocomposites. Figure.5.7 shows the effect of clay contents on volumetric and surface electrical resistances of the WPU/clay films, which shows the dependency of ionic conductivity of coatings to cations and anions of the exfoliated clay platelets. It increases with a higher amount of clay content, increases strong ionic interactions between the clay and WPU, and enhances the thermal resistance and tensile properties of WPUs substantially [75].

5.26 CONCLUDING REMARKS

In this chapter, synthesis of WPUs, structure–property relationships, their hybrid dispersions, and applications have been summarized. Soft and hard segments, emulsifiers, chain extenders, and active functional groups changed WPUs. Due to the chemical structure of active functional groups or emulsifiers, cationic and anionic WPUs can be obtained, with a wide range of applications. Incorporation of reactive groups into WPU may change hydrophilicity or hydrophobicity (contact angle), tensile strength, surface energy, elongation, and other properties of WPU as a solvent-free compound that is environmentally attractive. Suprahydrophobic, suprahydrophilic, antibacterial film coatings, abrasion resistance, and anti-corrosion coating can be produced with this kind of WPU. The incorporation of bio-based compounds into the PU network led to the formation of biodegradable WPUs with acceptable and tunable properties that are related to the formation of intermolecular hydrogen bonds and can be used as green coatings and inks. Using long-chain oil as a polyol led to the formation of grafted hybrid latexes with improved thermal and mechanical properties. Due to the low electrical conductivity of WPUs, incorporation of nanoparticles into these networks is a good method to generate thermal properties, magnetic properties, and electrical properties of obtained nanocomposites with potential applications in electrical industrials.

REFERENCES

1. Chattopadhyay DK, Raju KVSN Structural engineering of polyurethane coatings for high performance applications. *Prog Polym Sci.* 2007, 32:352–418.
2. Lei L, Zhong L Synthesis and characterization of waterborne polyurethane dispersions with different chain extenders for potential application in waterborne ink. *Chem Eng J.* 2014, 253:518–525.
3. Wang XR, Shen YD, Lai XJ Micromorphology and mechanism of polyurethane/ polyacrylate membranes modified with epoxide group. *Prog Org Coat.* 2014, 77: 268–276.
4. Guo YH, Li SC, Wang GS Waterborne polyurethane/ poly(n-butyl acrylate-styrene) hybrid emulsions: particle formation, film properties, and application. *Prog Org Coat.* 2012, 74:248–256.
5. Canak TC, Serhatli IE Synthesis of fluorinated urethane acrylate based UV-curable composites. *Prog Org Coat.* 2013, 76:388–399.
6. Can T,Junjie B,Qin C Preparation of waterborne polyurethane adhesives based on macromolecular-diols containing different diisocyanate. *J Adhesion.* 2019, 95:814–833.
7. Fuensanta M, Jofre-Reche JA Structural characterization of polyurethane ureas and waterborne polyurethane urea dispersions made with mixtures of polyester polyol and polycarbonate diol. *Prog Org Coat.* 2017, 112:141–152.
8. Zimmerman LR *Catalysts.* In:Randall, D; Lee, S (Eds). *The Polyurethanes Book.* J. Wiley. United Kingdom, 2002.141–146.
9. Li M, Qiang X, Zhang H Synthesis and characterization of cationic waterborne polyurethane with high solid content. *Gaofenzi Cailiao Kexue Yu Gongcheng.* 2014, 30:37–47.
10. Szycher M *Szycher's Handbook of Polyurethanes*, 1st Edition. CRC press LLC: Florida, 1999.
11. Lu Y, Larock RC Synthesis and properties of grafted latices from a soybean oil-based waterborne polyurethane and acrylics. *J Appl Polym Sci.* 2011, 119:3305–3314.
12. Liu K, Su Z, Miao S, Ma G, Zhang S UV-curable enzymatic antibacterial waterborne polyurethane coating. *Biochem Eng J.* 2016, 113:107–113.
13. Zhou Z, Lin W, Wu XF Electrospinning ultrathin continuous cellulose acetate fibers for high-flux water filtration. *Colloid Surface A.* 2016, 494:21–29.
14. Lee SYL, Kim BK Preparation and properties of water-borne polyurethanes. *Polym Int.* 1997, 42:67–76.
15. Nanda AK, Wicks DA The influence of the ionic concentration, concentration of the polymer, degree of neutralization and chain extension on aqueous polyurethane dispersions prepared by the acetone process. *Polymer.* 2006, 47:1805–1811

16. Dieterich D Aqueous emulsions polyurethanes; dispersions synthesis and solutions and properties. *Prog Org Coat.* 1981, 9:281–340.
17. Naghash HJ, Mohammadidehcheshmeh I, Mehrnia M Synthesis and characterization of a novel hydroxy terminated polydimethylsiloxane and its application in the waterborne polysiloxane-urethane dispersion for potential marine coatings. *Polym Adv Technol.* 2012, 24:307–317.
18. Molla ME Use of the newly synthesized aqueous polyurethane acrylate binders for printing cotton and polyester fabrics. *Adv Chem Eng Sci.* 2012, 02:228–237.
19. Curgul S Effect of chemical composition on large deformation mechanooptical properties of high strength thermoplastic poly(urethane urea)s. *Macromolecules.* 2004, 37:8676–8685.
20. Malay O Polyurethaneurea–silica nanocomposites: Preparation and investigation of the structure–property behavior. *Polymer.* 2013, 54:5310–5320.
21. Saralegi A, Rueda L TPU from renewable resources: Effect of soft segment chemical structure and molecular weight on morphology and final properties. *Polym Int.* 2012, 62:106–115.
22. Guillame SM, Khalil H, Misra M Green and sustainable PUs for advanced applications. *J Appl Polym Sci.* 2017, 134:1–2.
23. Molina GA, Elizalde-Mata A Synthesis and characterization of inulin-based responsive polyurethanes for breast cancer applications. *Polymers.* 2020, 12:865–888.
24. Ospina AC, Orozco VH Study of waterborne polyurethane materials under aging treatments. Effect of the soft segment length. Prog. *Org. Coatings.* 2020, 138:105357.
25. Li Y, Noordover BA Property profile of poly(urethane urea) dispersions containing dimer fatty acid-, sugar- and amino acid-based building blocks. *Eur Polym J.* 2014, 59:8–18.
26. Jiang X, Li J, Ding M Synthesis and degradation of nontoxic biodegrada-ble waterborne polyurethanes elastomer with poly (e-caprolactone) and poly(ethylene glycol) as soft segment. *Eur Polym J.* 2007, 43:1838–1846.
27. Rahman MM, Kim HD Synthesis and characterization of waterborne polyurethane adhesives containing different amount of ionic groups (I) J. *Appl Polym Sci.* 2006, 102:5684–5691.
28. Wang H, Zhou Y, He M, Dai Z Effects of soft segments on the waterproof of anionic waterborne polyurethane. *Colloid Polym Sci.* 2014, 293:875–881.
29. Lee S, Kim BK High solid and high stability waterborne polyurethanes via ionic groups in soft segments and chain termini. *J. Colloid Interface Sci.* 2009, 336: 208–214.
30. Liu X, Hong W, Chen X Continuous production of water-borne polyurethanes: A review. *Polymers.* 2020, 12:2875–2893.
31. Chattopadhyay D, Raju KVSN Structural engineering of polyurethane coatings for high performance applications. *Prog Polym Sci.* 2007, 32:352–418.
32. Rinaudo M Chitin and chitosan: Properties and applications. *Prog Polym Sci.* 2006, 31:603–632.
33. El-Sayed AA, Gabry LE Application of prepared waterborne polyurethane extended with chitosan to impart antibacterial properties to acrylic fabrics. *J Mater Sci Mater Electron.* 2009, 21:507–514.
34. García-Pacios V, Colera M, Iwata Y Incidence of the polyol nature in waterborne polyurethane dispersions on their performance as coatings on stainless steel. *Prog Org Coatings.* 2013, 76:1726–1729.
35. Barikani M, Ebrahimi MV Preparation and characterization of aqueous polyurethane dispersions containing ionic centers. *J Appl Polym Sci.* 2007, 104:3931–3937
36. Lvov Y, Abdullayev E Functional polymer–clay nanotube composites with sustained release of chemical agents. *Prog Polym Sci.* 2013, 38:1690–1719.

37. Shchukin, DG ,Möhwald H Surface-engineered nanocontainers for entrapment of corrosion inhibitors. *Adv Funct Mater*. 2007, 17:1451–1458.
38. Mauser T, Dejugnat C, Möhwald H Microcapsules made of weak polyelectrolytes: Templating and stimuli-responsive properties. *Langmuir*. 2006, 22:5888–5893.
39. Samir MASA, Alloin F Review of recent research into cellulosic whiskers, their properties and their application in nanocomposite field. *Biomacromolecules*. 2005, 6:612–626.
40. Mariano M, Kissi NE Cellulose nanocrystals and related nanocomposites: Review of some properties and challenges. *J Polym Sci Part B Polym Phys*. 2014, 52:791–806.
41. Torres-Giner S, Pérez-Masiá R A review on electrospun polymer nanostructure as advanced bioactive platforms. *Polym Eng Sci*. 2016, 56:500–527.
42. Mokhena TC, Jacobs V A review on electrospun bio-based polymers for water treatment. *Express Polym Lett*. 2015, 9:839–880.
43. Bhardwaj N, Kundu SC Electrospinning: A fascinating fiber fabrication technique. *Biotechnol Adv*. 2010, 28:325–347.
44. Ming L, Can C, Wen-Xin L Synthesis and application of waterborne polyurethane fluorescent composite. *e-Polymers*. 2017, 17:31–37.
45. Guo Y, Li SC, Wang GS Waterborne polyurethane/poly(n-butyl acrylate-styrene) hybrid emulsions: Particle formation, film properties, and application. *Prog Org Coat*. 2012, 74:248–256.
46. Choe E, Min DB Chemistry of deep-fat frying oils. *J Food Sci*. 2007, 72:R77–R86.
47. Phunphoem S, Saravari O Synthesis of cationic waterborne polyurethanes from waste frying oil as antibacterial film coatings. *Hindawi Int J Polym Sci*. 2019. doi:10.1155/2019/2903158
48. Wanga X, Zhanga Y Synthesis and properties of castor oil-based waterborne polyurethane/sodium alginate composites with tunable properties. *Carbohydr Polym*. 2019, 208:391–397.
49. Martin F, Schick William C, HickmanGary T, ClarkRebecca R Aqueous dispersion blends of polyesters and polyurethane materials and printing inks therefrom. Patent US4847316, 1989.
50. Sweetman M, May S, Mebberson N Activated carbon, carbon nanotubes and graphene: Materials and composites for advanced water purification. 2017. doi:10.3390/c3020018
51. Tonnesen HH, Karlsen, J Alginate in drug delivery systems. *Drug Develop Indus Pharmacy*. 2002, 28:621–630.
52. Travinskaya T, Savelyev Y Aqueous polyurethane dispersions – sodium alginate based blends and hydrogels. *J Heterocycl Chem*. 2016, 2:20–25.
53. Son SJ, Kim KB, Lee YH Effect of acrylic monomer content on the properties of waterborne poly(urethane-urea)/acrylic hybrid materials. *J Appl Polym Sci*. 2012, 124:5113–5121.
54. Wang F, Feng L, Lu M Mechanical properties of multi-walled carbon nanotube/waterborne polyurethane conductive coatings prepared by electrostatic spraying. *Polymers*. 2019, 11:714–715.
55. Mehravar S, Ballard N Polyurethane/acrylic hybrid waterborne dispersions: Synthesis, properties and applications. *Ind Eng Chem Res*. 2019, 58:20902–22092
56. Degrandi-Contraires E, Udagama R Influence of composition on the morphology of polyurethane/acrylic latexparticles and adhesive films. *Int J Adhes Adhes*. 2014, 50: 176–182.
57. Xia Y, Larock RC Vegetable oil-based polymeric materials: Synthesis, properties, and applications. *Green Chem*. 2010, 12:1893–1909.
58. Lu Y, Larock RC Synthesis and properties of grafted latices from a soybean oil-based waterborne polyurethane and acrylics. *J Appl Polym Sci*. 2011, 119:3305–3314.

59. Xia Y, Larock RC Castor-oil-based waterborne polyurethane dispersions cured with an aziridine-based crosslinker. *Macromol Mater Eng.* 2011, 296:703–709.
60. Zhang C, Madbouly TFG Recent advances in vegetable oil-based polymers and their composites. *Prog Polym Sci.* 2017. doi:10.1016/j.progpolymsci.2016.12.009.
61. Saetung A, Kaenhin L Synthesis, chara cteristic, and properties of waterborne polyurethane based on natural rubber. *J Appl Polym Sci.* 2012, 124:2742–2752.
62. Wu YM, Xie QW, Gao CH Synthesis and characterization of a novel aliphatic polyester based on itaconic acid. *Polym Eng Sci.* 2014, 54:2515–2521.
63. Jun X, Tong L Synthesis and characterization of waterborne polyurethane emulsions based on poly(butylene itaconate) ester. *Desig Monom Polymers.* 2016, 19:309–318.
64. Zhang W, Xu HZ, Wang MY Properties of aqueous polyurethane prepared from different soft segment. *China Leath.* 2012, 41:9–12.
65. Dong Y, Zhang Z Nitrification characteristics of nitrobacteria immobilized in waterborne Polyurethane in wastewater of corn-based ethanol fuel production. *J Environ Sci China.* 2012, 24:999–1005
66. Sam CS, Han HC, Kyung-Kyu P Synthesis and properties of waterborne polyurethane dispersions for surface coatings on poly(vinyl chloride). *Mole Cryst Liq Cryst.* 2020, 706:101–107.
67. Lingxiao L, Jingyi L, Zhang Y, Haiyan L Thermosetting polyurethanes prepared with the aid of a fully bio-based emulsifier with high bio-content, high solid content, and superior mechanical properties, 2019. *Green Chem.* 2019, 21:526–537.
68. Wang L, Ge S, Liu Z, Zhou Y Properties of antibacterial bioboard from bamboo macromolecule by hot press. *Saudi J Biol Sci.* 2018, 25:465–468.
69. Fei L Synthesis and properties of acrylate modified waterborne polyurethane emulsion. *Adv Mater Res.* 2010, 535:1386–1392.
70. Li J, Liu X, Lu J, Wang Y, Zhao F Anti-bacterial properties of ultrafiltration membrane modified by graphene oxide with nano-silver particles. *J Colloid Interface Sci.* 2016, 484:107–115.
71. Liangsong C Application of eco- friendlywaterborne polyurethanecomposite coating incorporated with nano cellulose crystalline and silver nano particles on wood antibacterial board. *Polymers* 2020, 12, 407–418.
72. Larraza BA, Gonzalez K, Gabilond N Waterborne polyurethane and graphene/graphene oxide based nanocomposites: Reinforcement and electrical conductivity. *eXPRESS Polymer Lett.* 2020, 14:1018–1033
73. Zhang S, Li Y, Peng L Synthesis and characterization of novel waterborne polyurethane nanocomposites with magnetic and electrical properties. *Compos A.* 2013, 55:94–101.
74. Deng J, He CL Magnetic and conducting Fe3O4 -cross-linked polyaniline nanoparticles with core–shell structure. *Synthetic Met.* 2003, 139:295–301.
75. Lee HT, Lin LH Waterborne polyurethane/clay nanocomposites: Novel effects of the clay and its interlayer ions on the morphology and physical and electrical properties. *Macromolecules.* 2006, 39:6133–6141.

6 Nanocomposites of Waterborne Polyurethanes

Samiran Morang, and Niranjan Karak
Advanced Polymer and Nanomaterial Laboratory,
Department of Chemical Sciences, Tezpur University,
Napaam, Assam 784028, India

CONTENTS

6.1 Introduction....84
6.2 Materials and Methods....85
6.2.1 Raw Materials....85
6.2.1.1 Diisocyanate....85
6.2.1.2 Macroglycol....86
6.2.1.3 Chain Extender....86
6.2.1.4 Catalyst....86
6.3 Nanomaterials Suitable for Waterborne Polyurethane Nanocomposites....86
6.3.1 Metal Nanoparticles....87
6.3.2 Silicon-Based Nanomaterials....87
6.3.3 Carbon-Based Nanomaterials....88
6.3.4 Organic Nanomaterials....88
6.4 Preparative Methods for WPU....88
6.4.1 Prepolymer Mixing Process or Emulsification....88
6.4.2 Acetone Process....88
6.4.3 Hot-Melt Process....89
6.4.4 Ketamine–Ketazine Process....89
6.5 Preparative Methods for Nanomaterials....89
6.6 Waterborne Polyurethane Nanocomposite Fabrication....90
6.7 Characterization....91
6.7.1 Spectroscopic Studies....91
6.7.1.1 UV–Visible....91
6.7.1.2 FTIR....91
6.7.1.3 NMR....92
6.7.1.4 Raman....92
6.7.2 Diffraction or Scattering Studies....92
6.7.2.1 X-Ray Diffraction....92

DOI: 10.1201/9781003173526-6

6.7.3 Microscopic Studies 92
6.7.3.1 Scanning Electron Microscope 93
6.7.3.2 Transmission Electron Microscope 93
6.7.3.3 Atomic Force Microscope 93
6.8 Properties of Waterborne Polyurethane Nanocomposite 93
6.8.1 Physical Property 94
6.8.2 Mechanical Property 94
6.8.3 Thermal Property 94
6.8.4 Optical Property 95
6.8.5 Electrical Property 95
6.8.6 Biological Property 95
6.9 Applications 95
6.9.1 Surface Coating 96
6.9.2 Adhesive and Sealant 96
6.9.3 Biomedical Application 96
6.9.4 Smart Material 96
6.9.5 Miscellaneous 97
6.10 Conclusion and Future Aspects 98
References 98

6.1 INTRODUCTION

Polymers are comparable to shadows of human beings, which cannot be detached from their daily activities. Life has become easier, comfortable, and colorful with the development of different forms of polymers. In fact, it is difficult to imagine modern society without synthetic polymers and life without natural polymers. Among the synthetic polymers, polyurethanes (PUs) are renowned, versatile, and most researched polymeric materials. German professor, Otto Bayer, and his coworkers invented PU in 1937.[1] PU contains a hard segment and a soft segment bonded through urethane or carbamate linkages (–NH–(C=O)–O–) in consort with other groups, such as urea, esters, ethers, allophanate, hydrocarbons with aromatic, and aliphatic moieties. The physico-chemical properties of PU can be amended by a change in the composition and structure of its three basic precursors, namely, macroglycol, diisocyanate, and chain extender. PU has drawn substantial interest among researchers and industrialists due to its wide spectra of applications, which is because of its high ability for chemical modifications, as well as high mechanical strength, thermal stability, excellent corrosion resistance, etc. The most common types of PUs are thermoplastic PU, flexible PU, rigid PU, PU ionomer (PUI), and waterborne PU (WPU).[2] Nowadays, diminishing the emissions of VOCs and HAPs to the atmosphere is a very important understanding in the PU industries, which is also an enthusiastic topic for research. One such affinity is the replacement of solvent-borne PU with WPU. In this content, Table 6.1 represents the merits and demerits of WPUs and solvent-borne PUs. WPUs substitute their solvent-borne counterparts in terms of nontoxicity, nonflammability, and low or no content of organic solvents.[3,4] Various synthetic approaches have been developed for preparing WPU, including acetone process, prepolymer emulsification, hot-melt, and ketimine–ketazine processes. All these methods are discussed later in this chapter.

TABLE 6.1
Merits and demerits of solvent-borne PU and WPU

Types	Merits	Demerits
Solvent-borne PU	• High corrosion and abrasion resistance • Excellent mechanical properties • Exhibit smart properties • Less expensive fabrication	• Release VOCs and HAPs to the environment • Nonbiocompatible • Use of toxic solvent(s) • Poor alkali resistance
WPU	• Less or no VOCs • Highly biocompatible and biodegradable, in general • Exhibit comparable properties as solvent-borne PU by modification • Nonflammable	• Very poor water and alkali resistance • Less durability • Low corrosion resistance

Additionally, new methods are also available viz. homogeneous solution polymerization (HSP), miniemulsion polymerization process (MEPP), reversible addition-fragmentation chain transfer (RAFT) polymerization, and atom transfer radical polymerization (ATRP).[5]

The most mysterious and productive modification of WPU is the fabrication of nanocomposites by incorporation of suitable nanomaterials. The most commonly used nanomaterials include cellulose nanofibers (CNF), carbon nanotubes (CNT), nanoclay, functionalized graphene oxide (GO) and reduced GO (rGO), modified GO/zinc oxide (GO/ZnO), polyaniline nanofiber (PAni) and nano-silver, etc. These nanomaterials are prepared by two basic methods, namely, top–down and bottom–up. Again, various techniques are available for fabrications of WPU nanocomposites (WPUNC) viz. solution technique, in-situ polymerization, melt-mixing technique, latex stage fabrication, template synthesis, coagulation spinning, sol–gel process, etc. This chapter discusses different aspects of WPUNC.

6.2 MATERIALS AND METHODS

6.2.1 Raw Materials

The basic raw materials for WPU include macroglycols, diisocyanates, chain extenders containing ionic or ionogenic groups, deionized water, and catalyst. All these raw materials are commercially available and mainly originated from petroleum resources. However, in recent years, renewable resource-based raw materials have received more priority than conventional petroleum-based materials due to environmental concerns.

6.2.1.1 Diisocyanate

The nature of the diisocyanates highly influences the properties of the resultant WPU. Hence, the selection of the appropriate diisocyanate is compulsory to achieve the desired properties. The most commonly used diisocyanates are toluene diisocyanate (TDI), diphenylmethane diisocyanate (MDI), hexamethylene diisocyanate (HMDI), and isophorone diisocyanate (IPDI), etc. Notably, the aliphatic (like HMDI) and cycloaliphatic (like IPDI) diisocyanates are preferred for WPUs because of their low reactivity with water. Also, they provide good color stability and transparency in the films.[6,7] Literature reports various renewable resource-based diisocyanate utilized for synthesis of PU.[8–11]

6.2.1.2 Macroglycol

Macroglycols are aliphatic organic di/polyol compounds with molecular weight in the range of 500–5000 g mol^{-1}. The macroglycols with long chains and low functionality ($f = 2$) are used in flexible PUs, but the macroglycols with a relatively short chain and high functionality ($f > 2$) are used in rigid PUs. Nowadays, different classes of macroglycols are used for the synthesis of WPUs, including the diols of polyesters, polyethers, polycarbonates, polybutadiene, poly(dimethylsiloxane), etc. However, polyether and polyester-based macroglycols are used extensively in the industries as compared to other macroglycols.[2] Further polyols can be obtained from renewable resources such lignin, vegetable oils, starch, sorbitol, etc.

6.2.1.3 Chain Extender

Chain extender is used for increasing the chain length and hence the molecular weight of the WPUs. They are low molecular weight (lower than 500 g mol^{-1}) diols, diamines, and hydroxylamines. Different functionalities (f) are used for different types of WPUs. For examples, glycerine ($f = 3$), dipropylene glycol ($f = 2$), an aqueous solution of sorbitol ($f = 2.75$), sorbitol ($f = 6$), Mannich bases ($f = 4$), sucrose ($f = 8$), toluene-diamine ($f = 4$), etc. The proper selection of chain extender is essential to acquire desire properties, including chemical resistance, heat, and flexural strength, etc.[11]

6.2.1.4 Catalyst

The catalysts applied during WPU synthesis can be grouped into two main categories: organic amino compounds and organometallic compounds. Generally, the amine-based catalysts consist of tertiary amine, e.g. triethylenediamine (TEDA), dimethylethanolamine (DMEA), DABCO, etc. Again, the organometallic catalysts used in WPU synthesis are based on selective metals such as bismuth, lead, tin, zinc, and mercury. Some novel catalysts, based on metal-carboxylates and metal-oxides, e.g., DBTDL, dioctyl tin mercaptide, stannous octoate, dibutyltin oxide, etc., are also used for this purpose.

6.3 NANOMATERIALS SUITABLE FOR WATERBORNE POLYURETHANE NANOCOMPOSITES

The most commonly used nanomaterials include metals, metal oxide, nanoclay, carbon dots, carbon nanotubes, GO, rGO, nano-silica, polyhedral oligomeric

TABLE 6.2
Nanomaterials and their important properties

Nanomaterial (Dimension and Shape)	Salient Features
Ag (0-D, spherical)	Antibacterial, Antifungal, Anti-inflammatory, Catalytic activity
HAp (0-D, spherical)	Biocompatibility, Osteogenic activity
CD (0-D, spherical)	Biocompatibility, Photosensitizing, Photo-catalytic activity, Fluorescence, Phosphorescence, Chemluminoscence
TiO_2 (0-D, spherical)	Antibacterial, Photo-catalytic activity, UV-resistant, Opacity
MWCNT (1-D, cylindrical)	High tensile strength, High thermo-stability, High electrical conductivity
CNF (1-D, Ribbon)	High tensile strength, Biocompatibility
rGO (2-D, sheet)	High Young's modulus, Good electrical conductivity, Energy storage capacity
Clay (2-D, sheet)	High mechanical strength, Excellent absorption, Biocompatibility
PAni (1-D, fiber)	High conducting, Excellent anticorrosive, Good antibacterial

silsesquioxane (POSS), cellulose nanofibers (CNF), PANi nanofibers, etc. The salient features of some nanomaterials are given in Table 6.2.

6.3.1 Metal Nanoparticles

Metal nanoparticles exhbit advanced physico-chemical properties and are useful for various applications. Among the metal nanoparticles, silver nanoparticles (AgNPs) are one of the most interesting metal nanoparticles, showing excellent antimicrobial properties. AgNPs were used in burn treatments, preparation of antimicrobial stainless steel, textiles, sunscreen lotions, and biosensors. For example, Ren et al. prepared WPU/Ag nanocomposites via in situ formations of AgNPs in WPU matrices and concluded that the resultant nanocomposite shows remarkable antibacterial activities by releasing silver ion to suppress proliferation of different bacteria effectively.[12]

6.3.2 Silicon-Based Nanomaterials

The silicon-based nanomaterials exhibit certain properties, such as high porosity, transparency, nontoxicity, and biocompatibility, etc., which make them attractive.[13] Among the silicon-based nanomaterials, montmorillonite (MMT) nanoclay has been used extensively for polymer/layered silicates nanocomposites due to its availability and remarkable intercalation ability.[14] The MMT can be incorporated with pristine WPU to achieve enhanced thermal stability, elasticity, mechanical strength, molecular barrier, and electrochemical properties.[15,16]

6.3.3 Carbon-Based Nanomaterials

The well-known carbon-based nanomaterials (CBN) include fullerenes (C_{60}), carbon nanotubes, graphene oxide (GO), reduced GO (rGO), and carbon-based quantum dots (CDs).[17] The CNT consist of rolled graphene layers with diameter in the nanometer range.[18,19] The CNT can be classified into two types depending on the number of graphene layer(s) rolled into tube(s), namely (1) single-walled CNT (SWCNT) and (2) multi-walled CNT (MWCNT). Due to their unique physical, chemical, and mechanical characteristics, the CBN is used alone as well as in combination with other nanomaterials for various applications, such as anticorrosive materials,[20] pressure sensors,[21] data storage devices,[22] electronics, and semiconductor industries,[23,24] etc.

6.3.4 Organic Nanomaterials

Among the group of conducting polymers, polyaniline (PANi) has achieved great attention because of its intriguing redox and electronic properties. One-dimensional PANi can be incorporated with the polymer matrix by electrospinning, sonication, mechanical stretching, and microwave irradiation. Literature reveals that the electrical conductivity of PANi primarily depends on the degree of oxidation and protonation.[25] CNF is another organic nanomaterial that can be extracted from plants, nonpathogenic bacteria, algae, and fungi found in fruits and vegetables. The CNF-based nanocomposites can be employed in coatings,[26] 3D-paints,[27] etc.

6.4 PREPARATIVE METHODS FOR WPU

WPUs can be prepared by various routes. The basic principles of all these methods consist of two steps. First, preparation of low molecular weight isocyanate-terminated prepolymer via reactions of di/polyols and di/polyisocyanates. Second, the prepolymer chain is prolonged with a suitable chain extender and dispersed in water by introducing hydrophilic solubilizing groups (internal or external emulsifier). The commonly used preparative methods are illustrated herein.

6.4.1 Prepolymer Mixing Process or Emulsification

In this process, a hydrophobically modified isocyanate prepolymer is dispersed in water, and then the chain extension step is performed with diamines in the aqueous phase. Cycloaliphatic diisocyanates are preferred for prepolymer preparation due to their low reactivity with water. In this process, the ionic group is incorporated/in-situ generated with the polyurethane backbone as an internal emulsifier. Further, the viscosity of the reaction mixture is reduced by using the minimum amount of organic solvents like *N*-methyl-2-pyrrolidone (NMP), tetrahydrofuran, acetone, etc.

6.4.2 Acetone Process

In the acetone process, the chain extension step is performed with diamines in acetone after the hydrophilic prepolymer is formed. There are some advantages for the usage of acetone viz. it is unreactive toward isocyanates, is water-miscible, and can be removed from resultant WPU by distillation as it has a low boiling point. Additionally, it diminishes the high reactivity of the amine-based chain extenders toward isocyanates through reversible ketimine formation and hence controls the viscosity.[28] Solvent-free pure WPU can be obtained from the acetone process by removal of acetone.

6.4.3 Hot-Melt Process

Hot-melt process is a solvent-free route for the synthesis of WPUs in which an ionically/nonionically adapted isocyanate-terminated prepolymer having low viscosity is reacted with urea or ammonia to obtain a capped oligomer with terminal biuret groups. Then, this prepolymer is dispersed in water, and chain extension is accomplished via the methylation of the biuret groups with formaldehyde. Finally, polycondensation reaction is performed by lowering the pH, which leads to ultimate product formation of poly(urea-urethane).

6.4.4 Ketamine–Ketazine Process

In the ketamine–ketazine process, the hydrophilic prepolymer is blended with a blocked amine (ketamine) or blocked hydrazine (ketazine) before it comes in contact with water. Subsequently, the mixture is again mixed with water to synthesize dispersion. Afterward, the ketamine and ketazine hydrolyse into an amine and hydrazine, respectively, which leads to the chain extension, and hence WPU is formed. It is important to note that the hydrolysis of ketazines happens more slowly than that of the ketamines. Therefore, the ketazine route is more compatible for isocyanate terminal prepolymer obtained from aromatic isocyanate (e.g., TDI) with high reactivity.

6.5 PREPARATIVE METHODS FOR NANOMATERIALS

The nanomaterials can be prepared by various methods, and these methods can be categorized into two major approaches: (a) top–down and (b) bottom–up, as shown in Figure 6.1.[29] In the top–down approach, nanomaterials are prepared by breaking down the bulk precursor materials into the nano-sized structure. On the other hand, the bottom–up approach involves the controlled aggregation of molecular-level precursors (atoms or molecules). The most commonly used physical methods are ball milling, arc discharge, laser ablation, physical vapor deposition, electro-deposition, aerosol, and inert gas condensation, whereas the chemical methods include solvothermal, hydrothermal, sol–gel, microemulsion, microwave, and co-precipitation.[11] Literature disclosed that the bottom-up approach is more commonly adopted than the top–down approach. It offers some advantages, including it is less expensive, less time consuming, has more purity, and, most importantly, has better control over the morphology and size distribution of the nanomaterial.[29]

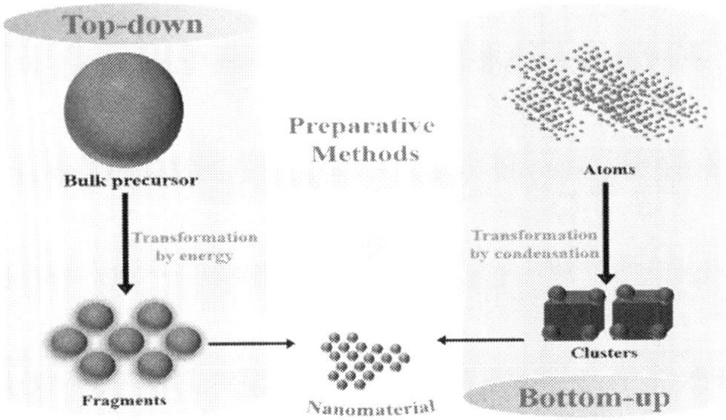

FIGURE 6.1 Preparative methods of nanomaterials.

6.6 WATERBORNE POLYURETHANE NANOCOMPOSITE FABRICATION

The nanomaterials can be uniformly dispersed in the polymer matrix by using various methods, namely, solution technique, in situ technique, and melt mixing method.[30] A brief explanation of these methods is given below.

a. Solution Technique
   This technique involves two primary steps, namely, swelling and dispersion of the selected nanomaterial in a suitable solvent or mixture of solvents. The dispersed nanomaterial is incorporated into the polymer solution with high shearing force, followed by ultra-sonication. Finally, evaporation is performed to remove the solvent and to obtain WPUNC. It is pertinent to mention that the nanomaterial dispersion in the polymer matrix is influenced by interactions between nanomaterial–solvent, polymer–solvent, and nanomaterial–polymer. The last interaction must be greater than the first two for achieving homogeneously dispersed nanocomposites. Subsequently, the nanocomposite developed would be either an exfoliated or an intercalated nanocomposite.
b. In-situ Polymerization
   As the name suggests, in this technique, the selected nanomaterial is swollen and dispersed into the prepolymer or reactants, and then a polymerization reaction is performed to obtain the desired polyurethane nanocomposite. The viscosity of the prepolymer is less than the respective preformed polyurethane, which causes remarkable interaction between polymer and nanomaterial within the system. Generally, the in-situ polymerization technique results in exfoliated nanocomposite.
c. Melt Mixing Technique
   In the melt mixing technique, fabrication is accomplished via direct blending nanomaterials with the polymer in the molten state by using suitable tools such as twin-screw mixers, rollers, injection molding, etc. Intercalation

nanocomposite formation is difficult due to high viscosity of the molten polymer. Therefore, only partially exfoliated nanocomposite can be obtained by using this technique.

Furthermore, other techniques used for the same purpose include the twin-screw pulverization process, latex stage fabrication, template synthesis, coagulation spinning, sol–gel process, plasma treatment, and thermal decompositions.

6.7 CHARACTERIZATION

The hydroxyl value and the isocyanate value are two important parameters for reactant and product. The hydroxyl value can be determined by different methods, namely, acetic anhydride pressure bottle, imidazole-catalyzed phthalic anhydride pressure bottle, and imidazole-catalyzed pyromellitic dianhydride esterification methods. Again, the isocyanate value can be evaluated by using the *n*-butyl amine test or quantitive FTIR spectral analysis.

Further, the WPUs and nanomaterials are characterized by the conventional techniques, such as analytical, diffraction, spectroscopic, and microscopic methods, including elemental, UV–visible, FTIR, and NMR spectroscopies, Raman spectroscopy, X-ray diffraction (XRD) analysis, scanning electron microscope (SEM), atomic force microscope (AFM), transmission electron microscope (TEM), etc.

6.7.1 Spectroscopic Studies

6.7.1.1 UV–Visible

This spectroscopic technique follows Beer–Lambert's law, which provides information allied to absorbance or optical density. The electronic transitions corresponding to various functional groups are evaluated in this technique, e.g., in the UV spectrum of the fluorescent agent 7-amino-4-(trifluoromethyl) coumarin (AFC); the absorption maximum λ_{max} = 366 nm corresponds to benzopyrone units. However, WPU based on AFC showed a bathochromic shift of λ_{max} from 366 to 371 nm because of the interactions among the dipole molecules and ions of WPU-AFC composite.[31] Concurrently, this technique is also useful for the determination of the size and distribution of nanomaterials correlated with the stability or aging of the material in the dispersed medium. The typical surface plasmon resonance (SPR) peaks in the UV–visible region are characteristic of nanomaterials, which can be utilized to confirm the formation of nanomaterials.

6.7.1.2 FTIR

This technique gives clear-cut information about the existence of various infra-red active functional groups and chemical linkages. The completion of WPU synthesis can be confirmed by FTIR spectral analysis, namely, disappearance of the –NCO band at around 2270 cm^{-1}. The presence of hydrogen bonding can perceive from FTIR data because it causes change in the band positions, width of the spectral lines, and intensity. Equally, this technique is also convenient for the identification of various functional groups present in functionalized nanomaterials and their bonding as well as nonbonding interactions with WPU matrices.

6.7.1.3 NMR

The NMR spectroscopy gives information about the numbers of magnetically different types of NMR active nuclei as well as the nature of the immediate environment of each type. The combination of NMR and IR data pledges sufficient structural information of an unknown WPU. The degree of branching (DB) of hyperbranched WPU can be obtained from the NMR spectral analysis. Furthermore, some advanced NMR techniques, such as multi-nuclei solid-state NMR, are applied for characterization. For example, ^{13}C solid NMR is used for the characterization of carbon-based nanomaterials, especially carbon dots, functionalized CNTs, etc.

6.7.1.4 Raman

Raman spectroscopy utilizes the inelastically scattered light originated from a monochromatic light, including laser in the near ultraviolet, visible, or near-infrared range. This technique is based on the "Raman Effect" correlated with the polarizability of electrons in the molecules. This is an efficient tool for investigating defects as well as the electronic and phonon properties of the nanomaterials. Besides, it is useful for the identification of functional groups and specific bonds present in such materials. For example, the reduction of GO to RGO can be confirmed by a decrease in the intensity ratio of the D and G bands in the Raman spectra.

6.7.2 Diffraction or Scattering Studies

6.7.2.1 X-Ray Diffraction

This technique affords vital structural information that is complementary to various microscopic and spectroscopic methods, such as phase identification, degree of crystallinity, purity, size of the crystal, and in some cases, morphology. Generally, two types of XRD techniques are used for these purposes, namely, wide-angle X-ray scattering (WAXS) and small-angle X-ray scattering (SAXS). The WAXS use the techniques to determine the degree of crystallinity, orientations of crystal in PUs, etc., whereas SAXS use the techniques to determine voids and their size, and orientation of small crystalline regions in WPU.

6.7.3 Microscopic Studies

The electron microscopic techniques can transfer small and complex objects into high-resolution images with the help of accelerated electron beams having high energy. These techniques are used for investigating the morphology of nanomaterials, including their shape, size, microstructural entities, lattice defects, and strain, etc. The most frequently used microscopic techniques include the SEM, field emission scanning electron microscope (FESEM), AFM, and TEM.

6.7.3.1 Scanning Electron Microscope

The characteristic surface informations derived from the SEM micrographs are the phase domain size, number of phases, and dispersion patterns of different structural domains in the targeted object. The inhomogeneity present in WPUNC surfaces can be visualized from SEM images. In addition, the chemical composition can be evaluated with an energy dispersive X-ray, an attachment of SEM.

6.7.3.2 Transmission Electron Microscope

Like SEM, the interaction between high-energy electron beam and specimen is the primary phenomenon for TEM. However, the electron beam passes through the sample in TEM rather than scattering from its surface. TEM image analyses provide data, including size, shape, and particle distributions, internal and external diameter (in the case of nanotubes), and uniformity in structure, to characterize a sample. Although TEM provides high-resolution images compared to SEM, it has some disadvantages, such as sample preparation (which should be ultrathin), staining the sample, and localized heating.

6.7.3.3 Atomic Force Microscope

An AFM is operated in three basic modes, namely, contact, noncontact, and tapping modes. The interactions between the tip and surface are dissimilar for each of these modes. In contact mode, the tip is placed in contact with the sample surface continuously. In contrast, the tip does not directly contact the surface in non-contact mode, but the tip position is carefully controlled to hover just above the sample surface. Again, in tapping mode, the tip oscillates above the surface of the sample, but it approaches the surface much more closely such that it intermittently perceives the repulsive or contact regions of the tip/surface interactions. Thus, the AFM is a dominant technique that facilitates the imaging of nanocomposites remarkably.

6.8 PROPERTIES OF WATERBORNE POLYURETHANE NANOCOMPOSITE

Two main factors that differentiate the properties of nanomaterials and bulk materials are the surface and quantum effects.[32] The first factor arises due to the considerable fraction of atoms at the surface, while the second is due to quantum confinement effects in materials with delocalized electrons. Eventually, the properties of nanomaterials such as surface area as well as surface energy, size, shape, and the number of atoms present in the surface significantly vary from that of bulk materials. Successively, the nanocomposites derived from these nanomaterials also exhibited enhanced and unique physico-chemical properties than the pristine system. Thus the section briefly described the general properties of WPUNC.

6.8.1 Physical Property

Physical properties of WPUs are similar to the conventional solvent-borne PUs viz. color, solubility, viscosity, and molecular weight, etc. The molecular weight of the polymer highly affects the solubility, viscosity, and other thermo-mechanical properties.[33] For example, high molecular weight with long-chain WPUs are flexible but low molecular weight WPUs are brittle. Besides, low molecular weight WPUs exhibit higher solubility than that of high molecular WPUs.[34] Generally, the inorporation of nanomaterials does not interfere with the molecular weight of the virgin WPU, but solution viscosity increases.

6.8.2 Mechanical Property

The mechanical properties of WPUNC include tensile strength, elongation, scratch hardness, and impact resistance. State-of-art literature evident the significant improvement of these properties via fabrication of nanocomposite with a small amount of nanomaterial (>5 wt.%) into a WPU matrix. Generally, tensile strength increases exponentially,[35] while flexibility and elongation at break values tend to decrease[36] simultaneously after incorporation of nanomaterials. However, reports are also available where both tensile strength and elongation at break increased with the increase of nanomaterials content. As an example, WPU/clay nanocomposites enhanced both these properties upon the incorporation of 0–3 wt.% nanoclay due to exfoliated structural formation, but decreased at higher loading, maybe due to aggregation of nanomaterial (Table 6.3).[14]

6.8.3 Thermal Property

The thermal stability and transitions of WPUNC depend on several factors such as inherent rigidity, degree of crystallinity, intermolecular forces, cross-linking density

TABLE 6.3
Variation of tensile strength and elongation at break of WPU/clay nanocomposites with loading of clay[a]

Clay Content (wt.%)	Tensile Strength (MPa)	Elongation at Break (%)
0	2.12	189
1	2.86	252
2	3.24	343
3	6.93	606
4	4.94	546
5	3.28	398

Notes
a Reproduced with permission from reference (14). Copyright 2006 American Chemical Society.

and nanomaterial.[11] Metal, clay, and metal oxide-based WPUNC are highly thermostable in which nanomaterials create an obstacle against volatile counterparts from escaping and thereby slowing down the rate of the degradation process.[37] Again CBN such as GO, rGO, MWCNT, etc. also enhance thermal stability by forming char which acts as a thermal insulator.

6.8.4 Optical Property

The most important optical properties of WPUNC include color, gloss, transparency, fluorescence, and phosphorescence. The changes in the optical properties of pristine WPU depend on the nature of the nanomaterials used.[38] For example, CBN such as GO, rGO, and MWCNT decreased transparency and original color of the WPU, while F-SiO_2 nanomaterials don't.[39] Again, some nanomaterials e.g. quantum dots, exhibit special optical properties e.g. luminescence which is useful in the fields of functional surface coating, fluorescent agent for special leather fibers and anti-counterfeiting.

6.8.5 Electrical Property

WPU cannot conduct electricity unlike specially designed conductive polymers or intrinsically conducting polymers (ICPs) e.g. PAni. However, electrical conductivity can be introduced in the WPU by the formation of nanocomposite with conducting nanomaterials. These nanomaterials include rGO, CNT, PAni, etc.[14,35,40] They possess excellent electrical conductivity. For example, Kim et al. reported that fabrication of nanocomposite with only 1 wt.% of graphene-based nanomaterials can make a PU electrically conductive.[35]

6.8.6 Biological Property

WPU exhibits two significant biological properties viz. biocompatibility and biodegradability. Henc, it can be used for biomedical applications. For example, Hsu et al. synthesized a biodegradable PU based on poly(3-hydroxybutyrate) (PHB) which can be used as a potential biomaterial for cardiovascular and other medical applications.[41] Again, nanomaterials like hydroxyapatite (HAp), GO, carbon dots (CD), MWCNT, Fe_3O_4 can be fabricated for various applications including bone tissue engineering, wound healing, etc. The incorporation of Ag, TiO_2 offers antibacterial activity to the WPU.[42,43] Further, some literature reveals decrement in the rate of biodegradability upon formation of nanocomposites.[44]

6.9 APPLICATIONS

WPUs have been originated with versatile properties and flexibility in their modifications, which make them suitable for diversified applications as shown in the Figure 6.2. The modifications with suitable nanomaterials widen their applications exponentially. Some of their important applications are highlighted here.

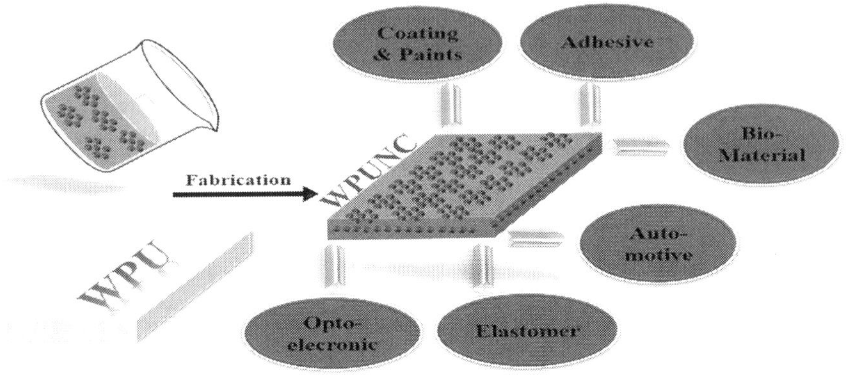

FIGURE 6.2 Applications of WPU nanocomposites (WPUNC).

6.9.1 Surface Coating

Nowadays, WPUNC-based coatings are enormously used for the protection of metal and wood surfaces, building materials, instruments, interior and exterior parts of automobiles from corrosion, scratch, and different deteriorations. This is possibly due to their superior chemical and physical resistance, high mechanical strength with optimum flexibility, transparency along with other advanced qualities such as self-healing, self-cleaning and UV-resistant, anti-bacterial activity, etc.

6.9.2 Adhesive and Sealant

WPUNC can be chosen as adhesive to bind surfaces like metal, wood, glass, etc. The adhesive properties are developed due to its effective wetting and bond (covalent/non-covalent) forming ability with most of the substrates.[45] For example, Zhao et al. prepared a WPU/amine functionalized-GO nanocomposite that possesses excellent adhesive property.[46] On the other hand, WPUNC is also used to seal moisture or water leakages.[47] These nanocomposites create an impermeable barrier against cracking and physical force to join the materials.

6.9.3 Biomedical Application

WPUNC with suitable nanomaterials possess acceptable biocompatibility and hence these materials can be employed for biomedical applications. Mostly, these nanocomposites are used in bone tissue engineering, wound healing, drug delivery, and the preparation of artificial biomaterials.[48,49] Furthermore, few nanocomposites e.g. WPU/Ag exhibit profound antibacterial and often used as infection resistance coating material for different medical equipment and dressing materials.[43]

6.9.4 Smart Material

Smart WPUNC can defend and heal themselves against external stimuli including stress, temperature, moisture, pH, electro-magnetic wave, etc. These materials also possess other advanced features such as shape memory(SM), self-cleaning, superhydrophobicity, scratch resistance, etc. SM WPUNC are extensively used in the aerospace and textile industries. SH WPUNC are also used in aerospace and military equipment, self-healable superhydrophobic coatings, etc.[45] A self-healable elastomeric WPU/acetylene carbon black nanocomposite with double-network structure showed great promise as a healable flexible sensor for next-generation wearable electronic devices, strain sensors, and soft robots, because of its excellent healing efficiency with retaining of high mechanical properties (Figure 6.3).[50]

6.9.5 Miscellaneous

In addition to the above applications, WPUNC find many applications like packaging, elastomer, electronic devices, etc. The major fields that demand the use of packaging films are the food industry, service packaging (e.g. carrier bags),

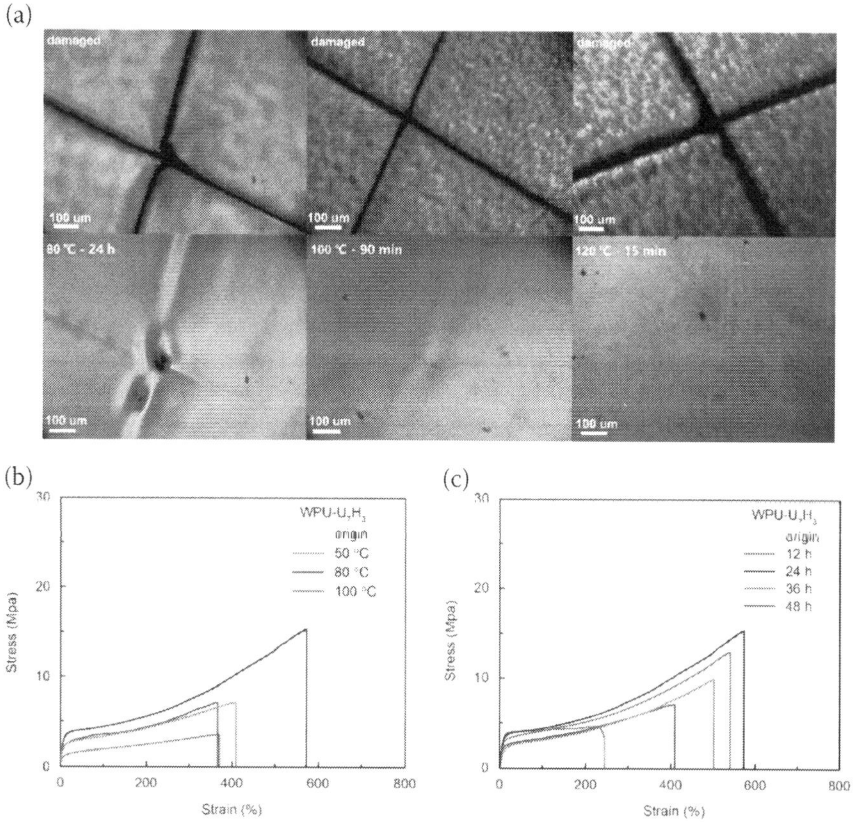

FIGURE 6.3 WPU nanocomposite films (a) optical microscopy images of the scratched specimens before and after healing, a tensile profile of the films after healing (b) at different temperatures for 24 h and (c) at 100°C for different times. Reproduced with permission from reference (50). Copyright 2020 The Royal Society of Chemistry.

packaging for the medical and pharmaceutical industry, etc.[11] Furthermore, WPU can be engineered to build an electromagnetic shield, an organic light-emitting device (OLED), and Infra-red actuator.

6.10 CONCLUSION AND FUTURE ASPECTS

From this chapter it can be concluded that, by fabricating appropriate WPUNC using appropriate nanomaterials and processing techniques, it is possible to develop advanced multifaceted materials to fulfill many modern demands. It is conversant that WPU is a versatile polymer that is further amplified by nanomaterials. The incorporation of nanomaterials not only improves general performances but also adds smart properties to the WPUs. Due to the extraordinary performances of WPUNC, their applications pervade almost all fields of polymer science with great importance. Since it is prepared using low or solvent-free methods and it exhibited acceptable biodegradability, therefore it is favorable for the environment as compared to the conventional solvent-borne PU nanocomposites. However, the replacement of petroleum-based raw material is highly recommended from an environmental point of view. Thus, to explore renewable raw materials and their potential applications to a maximum level, more studies and scientific experiments are very much essential through the collaborative work of material chemists, scientists, engineers, and industrialists.

REFERENCES

1. Delebecq E, Pascault JP, Boutevin B, Ganachaud F (2013) On the versatility of urethane/urea bonds: Reversibility, blocked. *Chem Rev* 113:80–118.
2. Akindoyo JO, Beg MDH, Ghazali S, Islam MR, Jeyaratnam N, Yuvaraj AR (2016) Polyurethane types, synthesis and applications-a review. *RSC Adv* 6:114453–114482.
3. Madbouly SA, Xia Y, Kessler MR (2013) Rheological behavior of environmentally friendly castor oil-based waterborne polyurethane dispersions. *Macromolecules* 46: 4606–4616.
4. Gong R, Cao H, Zhang H, Qiao L, Wang F, Wang X (2020) Terminal hydrophilicity-induced dispersion of cationic waterborne polyurethane from CO_2—based polyol. *Macromolecules* 53:6322–6330.
5. Zhou X, Li Y, Fang C, Li S, Cheng Y, Lei W, Meng X, Li Y, Fang C, Li S, Cheng Y, Lei W, Meng X (2015) Recent advances in synthesis of waterborne polyurethane and their application in water-based ink: A review. *J Mater Sci Technol* 31:708–722.
6. Grepinet B, Pla F, Hobbes P, Monge T, Swaels P (2001) Modeling and simulation of urethane acrylates synthesis. II. Kinetics of uncatalyzed reaction of toluene diisocyanate with a polyether diol. *J Appl Polym Sci* 81:3149–3160.
7. Ulrich H (1976) Chemistry and technology of soybeans. *Adv Cereal Sci Technol* 1:325–377.
8. Holfinger MS, Conner AH, Lorenz LF, Hill CG (1993) Difurfuryl diisocyanates: New adhesives derived from renewable resources. *J Appl Polym Sci* 49:337–344.
9. Marwan R, Kamal D, Kuder, RC (1972) U.S. Patent 3,691,225.
10. Çaylı G, Küsefoğlu S (2008) Biobased polyisocyanates from plant oil triglycerides: Synthesis, polymerization, and characterization. *J Appl Polym Sci* 109:2948–2955.
11. Karak N (2017) *Biobased Smart Polyurethane Nanocomposites From Synthesis to Applications*. Royal Society of Chemistry, London, UK.

12. Zhong Z, Luo S, Yang K, Wu X, Ren T (2017) High-performance anionic waterborne polyurethane/Ag nanocomposites with excellent antibacterial property: Via *in situ* synthesis of Ag nanoparticles. *RSC Adv* 7:42296–42304.
13. Kim BK, Seo JW, Jeong HM (2003) Morphology and properties of waterborne polyurethane/clay nanocomposites. *Eur Polym J* 39:85–91.
14. Lee HT, Lin LH (2006) Waterborne polyurethane/clay nanocomposites: novel effects of the clay and its interlayer ions on the morphology and physical and electrical properties. *Macromolecules* 39:6133–6141.
15. Osman MA, Mittal V, Morbidelli M, Suter UW (2003) Polyurethane adhesive nanocomposites as gas permeation barrier. *Macromolecules* 36:9851–9858.
16. Xu R, Manias E (2001) New biomedical poly(urethane urea)—layered silicate nanocomposites. *Macromolecules* 34: 337–339.
17. Patel KD, Singh RK, Kim HW (2019) Carbon-based nanomaterials as an emerging platform for theranostics. *Mater Horizons* 6:434–469.
18. Zettl A, Collins PG, Bradley K, Ishigami M (2000) Extreme oxygen sensitivity of electronic properties of carbon nanotubes. *Science* 287:1801–1804.
19. Ruoff RS, Lorents DC (1996) Mechanical and thermal properties of carbon nanotubes. *Carbon N Y* :143–148.
20. Wei H, Ding D, Wei S, Guo Z (2013) Anticorrosive conductive polyurethane multiwalled carbon nanotube nanocomposites. *J Mater Chem A* 1:10805–10813.
21. Huang W, Dai K, Zhai Y, Liu H, Zhan P, Gao J, Zheng G, Liu C, Shen C (2017) Flexible and lightweight pressure sensor based on carbon nanotube/thermoplastic polyurethane-aligned conductive foam with superior compressibility and stability. *ACS Appl Mater Interfaces* 9:42266–42277.
22. Qiao ZA, Wang Y, Gao Y, Li H, Dai T, Liu Y, Huo Q (2010) Commercially activated carbon as the source for producing multicolor photoluminescent carbon dots by chemical oxidation. *Chem Commun* 46:8812–8814.
23. Yin Z, Wu S, Zhou X, Huang X, Zhang Q, Boey F, Zhang H (2010) Electrochemical deposition of ZnO nanorods on transparent reduced graphene oxide electrodes for hybrid solar cells. *Small* 6:307–312.
24. Kim KS, Zhao Y, Jang H, Lee SY, Kim JM, Kim KS, Ahn JH, Kim P, Choi JY, Hong BH (2009) Large-scale pattern growth of graphene films for stretchable transparent electrodes. *Nature* 457:706–710.
25. Stejskal J, Sapurina I, Trchová M (2010) Polyaniline nanostructures and the role of aniline oligomers in their formation. *Prog Polym Sci* 35:1420–1481.
26. Cheng D, Wen Y, An X, Zhu X, Ni Y (2016) TEMPO-oxidized cellulose nanofibers (TOCNs) as a green reinforcement for waterborne polyurethane coating (WPU) on wood. *Carbohydr Polym* 151:326–334.
27. Chen R De, Huang CF, Hsu S hui (2019) Composites of waterborne polyurethane and cellulose nanofibers for 3D printing and bioapplications. *Carbohydr Polym* 212: 75–88.
28. Sardon H, Irusta L, Fernández-Berridi MJ, Luna J, Lansalot M, Bourgeat-Lami E (2010). Waterborne polyurethane dispersions obtained by the acetone process: A study of colloidal features. *J App Polym Sci* 120:2054–2062.
29. Biswas A, Bayer IS, Biris AS, Wang T, Dervishi E, Faupel F (2012) Advances in top–down and bottom–up surface nanofabrication: Techniques, applications & future prospects. *Adv Colloid Interface Sci* 170:2–27.
30. Koo JH (2006) *Polymer Nanocomposites: Processing, Characterization, and Applications*. Mcgraw-Hill Nanoscience and Technology Series.
31. Li M, Qiang X, Xu W, Zhang H (2015) Synthesis, characterization and application of AFC-based waterborne polyurethane. *Prog Org Coatings* 84:35–41.

32. Roduner E (2006) Size matters: Why nanomaterials are different. *Chem Soc Rev* 35:583–592.
33. Barrère M, Landfester K (2003) High molecular weight polyurethane and polymer hybrid particles in aqueous miniemulsion. *Macromolecules* 36:5119–5125.
34. Caminade AM, Yan D, Smith DK (2015) Dendrimers and hyperbranched polymers. *Chem Soc Rev* 44:3870–3873.
35. Kim H, Miura Y, MacOsko CW (2010) Graphene/polyurethane nanocomposites for improved gas barrier and electrical conductivity. *Chem Mater* 22:3441–3450.
36. Yadav SK, Cho JW (2013) Functionalized graphene nanoplatelets for enhanced mechanical and thermal properties of polyurethane nanocomposites. *Appl Surf Sci* 266:360–367.
37. Deka H, Karak N, Kalita RD, Buragohain AK (2010) Bio-based thermostable, biodegradable and biocompatible hyperbranched polyurethane/Ag nanocomposites with antimicrobial activity. *Polym Degrad Stab* 95:1509–1517.
38. Kale MB, Luo Z, Zhang X, Dhamodharan D, Divakaran N, Mubarak S, Wu L, Xu Y (2019) Waterborne polyurethane/graphene oxide-silica nanocomposites with improved mechanical and thermal properties for leather coatings using screen printing. *Polymer* 170:43–53.
39. Zheng H, Pan M, Wen J, Yuan J, Zhu L, Yu H (2019) Robust, transparent, and superhydrophobic coating fabricated with waterborne polyurethane and inorganic nanoparticle composites. *Ind Eng Chem Res* 58:8050–8060.
40. Li BQ, Li Y, Zhang X, Chikkannanavar SB, Zhao Y, Dangelewicz AM, Zheng L, Doorn SK, Jia Q, Peterson DE, Arendt PN, Zhu Y (2007) Structure-dependent electrical properties of carbon nanotube fibers. *Adv Mater* 19:3358–3363.
41. Hsu S, Hsieh CT, Sun YM (2015) Synthesis and characterization of waterborne polyurethane containing poly(3-hydroxybutyrate) as new biodegradable elastomers. *J Mater Chem B* 3:9089–9097.
42. Charpentier PA, Burgess K, Wang L, Chowdhury RR, Lotus AF, Moula G (2012) Nano-TiO_2/polyurethane composites for antibacterial and self-cleaning coatings. *Nanotechnol* 23:425606–425615.
43. Hsu S, Tseng H-J, Lin Y (2010) The biocompatibility and antibacterial properties of waterborne polyurethane-silver nanocomposites. *Biometrials* 31:6667–6900.
44. Mishra A, Singh SK, Dash D, Aswal VK, Maiti B, Misra M, Maiti P (2014) Self-assembled aliphatic chain extended polyurethane nanobiohybrids: Emerging hemocompatible biomaterials for sustained drug delivery. *Acta Biomater* 10:2133–2146.
45. Szycher M (2013) *Szycher's Handbook of Polyurethanes*, 2nd Edn. CRC press, Boca Raton
46. Zhao Z, Guo L, Feng L, Lu H, Xu Y, Wang J, Xiang B, Zou X (2019) Polydopamine functionalized graphene oxide nanocomposites reinforced the corrosion protection and adhesion properties of waterborne polyurethane coatings. *Eur Polym J* 120:109249
47. Yousefi N, Gudarzi MM, Zheng Q, Lin X, Shen X, Jia J, Sharif F, Kim JK (2013) Highly aligned, ultralarge-size reduced graphene oxide/polyurethane nanocomposites: Mechanical properties and moisture permeability. *Compos Part A Appl Sci Manuf* 49:42–50.
48. Shaabani A, Sedghi R (2021) Preparation of chitosan biguanidine/PANI-containing self-healing semi-conductive waterborne scaffolds for bone tissue engineering. *Carbohydr Polym* 264:118045.
49. Wang YJ, Jeng US, Hsu SH (2018) Biodegradable water-based polyurethane shape memory elastomers for bone tissue engineering. *ACS Biomater Sci Eng* 4:1397–1406.
50. Yang Y, Ye Z, Liu X, Su J, (2020) A healable waterborne polyurethane synergistically cross-linked by hydrogen bonds and covalent bonds for composite conductors. *J Mater Chem C* 8:5280–5292.

7 Waterborne Polyurethanes for Flexible and Rigid Foams

Hilal Olcay
Department of Textile Engineering, Institute of Science and Technology, Marmara University, Istanbul, Turkey

E. Dilara Kocak
Department of Textile Engineering, Faculty of Technology, Marmara University, Istanbul, Turkey

CONTENTS

7.1 Introduction ... 101
7.2 Polyurethane Foams ... 102
7.3 Waterborne Polyurethanes ... 104
7.4 Structures of Waterborne Polyurethanes ... 105
7.5 Synthesis of Waterborne Polyurethanes ... 106
7.5.1 Emulsification or Prepolymer Method ... 107
7.5.2 Acetone Process ... 107
7.5.3 Hot-Melt Process ... 108
7.5.4 Ketamine–Ketazine Process ... 108
7.6 Bio-based Materials for Waterborne Polyurethane Synthesis ... 108
7.7 Waterborne Polyurethane Composites ... 111
7.8 Conclusion ... 114
References ... 115

7.1 INTRODUCTION

Polyurethanes (PUs) are a high-performance organic polymer material that consists of flexible and rigid sections [1,2]. Due to their versatility in both processing methods and mechanical properties, their use has become widespread in many fields, such as binders, foams, coatings, adhesives, sealants, elastomers, etc. [3–5]. Varieties in the nature of polyols, diisocyanates, catalysts, and other additives allow the synthesis of various final polymers [2]. Global consumption rates in 2016 for different PU applications are given in Figure 7.1. In addition, it is also estimated that the global marketplace for PUs will continue to grow from USD ~65.5 billion in 2018 to USD 91 billion in 2026 [6].

DOI: 10.1201/9781003173526-7

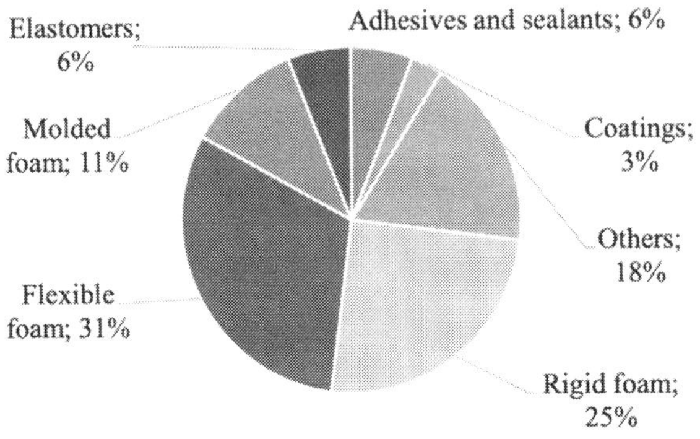

FIGURE 7.1 Global consumption of PU in 2016. Adapted with permission from reference [7]. Copyright (2018) MDPI.

Conventional PU products contain high volatile organic compound (VOC) emissions, which have toxicity and carcinogenicity to the human body and are harmful to the environment and, in some cases, have a definite quantity of free isocyanates [1,2,4]. For this reason, many methods have been tried to reduce VOC emissions and hazardous air pollutants, and it has been determined that one of the most environmentally friendly alternatives is waterborne polyurethane (WPU) [1,4,6]. In the late 1960s, processes were developed that made it possible to synthesize low-solvent or solvent-free aqueous PU dispersions (PUDs) [2,8]. WPUs offer unnumberable advantages, such as good chemical stability and abrasion resistance, low viscosity at high molecular weight, high flexibility, and perfect applicability in regard to conventional solvent-based PUs, and they are one of the fastest developing branches of PU chemistry today [4,9].

7.2 POLYURETHANE FOAMS

PU foam (PUF) represents one of the most significant commercial products of PU and accounts for more than 50% of all PUs. It was developed in the 1960s and contains a large number of cells in a polymer matrix. PUFs are synthesized by the exothermic addition of polymerization between isocyanate groups and polyhydroxy compounds with surfactants, catalysts, and blowing agents to form urethane linkage at room temperature. Foam synthesis includes two essential reactions: blowing, which consists of the reaction between an isocyanate group and water, and gelation, which consists of the reaction between hydroxyl and isocyanate groups (as seen in Figure 7.2).

In the manufacture of PUFs, Reactant A (polyols, blowing agent, catalyst, surfactants, etc.) and Reactant B (isocyanates) are mixed at high speed, and then the blended resin is poured into a mold. Two chemical reactions, blowing and gelation reactions, occur concurrently. The reaction equilibrium between them is affected by chemical compounds and processing conditions [10]. The cellular structure of the

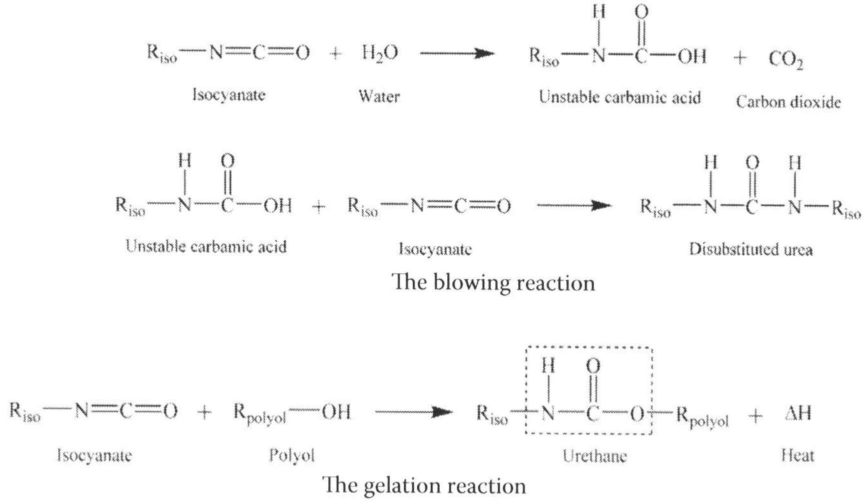

FIGURE 7.2 Two main reactions of PUF synthesis. Adapted with permission from reference [7]. Copyright (2018) MDPI.

foam is generated by stabilizing the bubble formation in the matrix. Therefore, blowing agents and silicone surfactants are fundamentally needed. In the synthesis of PUFs, chemical blowing agent (water), physical blowing agent (*n*-pentane), or both in combination can be used, relying on the desired density of the foam. Surfactants, on the other hand, change the properties of the polymer matrix during the foaming process. They are used to emulsify liquid components, prevent collapse, arrange cell sizes and surface defects, and balance cell structure. The most widely used surfactants are silicone surfactants. Crosslinking agents are used to strengthen the cross bonding of the polymer to provide better dimensional stability to foams [11–13].

Relying on the physical structure and chemical composition of the foams, a very high mechanical endurance, and damping properties, extreme endurance to weathering and organic solvents or oils can be achieved [14]. They also offer advantages such as lightness, low cost, and good adhesion to other types of materials, especially wood [15]. Despite these excellent performances, they have two major drawbacks in practice. First, their raw materials are nonrenewable resources and depend on the price fluctuation of crude oil and depletion of the crude oil stock. On the other hand, they have high flammability when there is a heat source. During combustion, a significant amount of toxic fumes liberates. These toxic combustion products are carbon monoxide (CO), carbon dioxide (CO_2), hydrogen cyanide (HCN), and nitrogen dioxide (NO_2). CO and HCN compounds are particularly hazardous. This combustion behavior needs to be enhanced to meet the necessities of the current regulations, relying on the final application of the material. These days, it has been observed in some studies that replacing fossil-based polyols with renewable bases can contribute to the reduction of oil dependence as well as the reduction of greenhouse gas emissions and even the recycling of industrial waste [16].

PUFs are a class of lightly porous materials that are of great interest because of their specific properties and potential applications in many fields. Foams are commercial products with various product groups that are generally categorized as flexible, semi-rigid, or rigid relying on the mechanical performance, core density, and properties of the reagents, such as functionality and molecular weight of polyols [17–21]. In addition, some flexible foams can be classified as viscoelastic when they have a delayed recovery [20]. Rigid PUF, a highly crosslinked polymer, mainly has unique characteristic properties, such as splendid thermal insulation, durableness, wide-range apparent density (30–200 kg/m^3), relatively improved specific mechanical properties (strength/weight ratio), and a closed-cell structure. These properties make rigid PUF a globally dominant material and a source of great interest for a variety of applications [22,23]. Flexible PUFs, representing the most expanded group among PUs, are block copolymers whose elasticity properties hinge on the phase separation between soft and hard sections [21,24,25]. Rigid blocks are physically crosslinked structures that give the polymer its tightness and consist of substituted urea and urethanes that can interact with hydrogen bonds. Soft blocks are tensible chains that provide flexibility to the polymer and consist of a polyether or polyester chain. The covalent bonding is constituted between these rigid segments and the polyol domains by urethane linkages. Depending on the ratio and composition of these blocks, PUFs can be customized [19,24]. Flexible PUFs have several advantages, such as great variety of flexibility and rigidity, great cushioning, excellent dimensional stability, high endurance during usage, good thermal and acoustic insulation, low density, chemical endurance, low cost, and splendid strength/weight ratio performance [13,26,27]. Nevertheless, they may also have some undesired properties, such as low thermal durability and low mechanical strength [26].

7.3 WATERBORNE POLYURETHANES

PU is obtained by the reaction between a polyol and diisocyanate or polyisocyanate [6]. These resin systems usually contain highly volatile organic solvents (~40–60% by weight) [8,9]. Replacing conventional solvent with water in PU formulations is a more environmentally friendly approach that benefits workers during application as it is nontoxic, easy to use, and decreases the fire risk during storage and transportation [9]. Due to growing global concerns about environmental protection, the improvement of WPUs has attracted ever-increasing attention because of the substitution of VOC with water, an environmentally friendly solvent [6]. The advantages of using these polymers have so far outperformed the disadvantages of using a non-environmentally friendly components such as isocyanate in the synthesis. In the process of preparing WPU dispersions, three reactions take place. They occur between isocyanate/polyol, isocyanate/amine, and isocyanate/water, respectively. Thus, water removes the probability of any unreactive isocyanate groups in the end products, and the toxicity that may occur due to the remaining isocyanate can be eliminated [28]. Especially, WPUs can be used in fibers, textile dyes, adhesives, sealants, paint additives, biomaterials, metal primers, etc., and this explains the continuous improvement of these materials [6].

WPUs are dual colloid aqueous polymers produced by inserting hydrophilic groups, such as carboxylate, sulfonate, quaternary ammonium salt, or hydrophilic segments, into PU molecular chains to make them a hydrophilic characteristic. Since PU is completely incompatible with water, hydrophilic monomers with ionic functionality must be included. These ionic particles are the internal emulsifier and PU ionomer. Compared with a conventional PU system with an organic solvent, WPUs are friendly to the earth, safe, and exhibit splendid compatibility and interchangeability [2,29]. They are nontoxic and nonflammable and do not pollute the air or generate wastewater. These systems are not environmentally hazardous as only water vaporizes during the process. They also have many peerless advantages such as good adhesion, fast dry time, high tensile strength, nonflammability, excellent wear resistance, high flexibility and impact resistance, good low-temperature endurance, low cost, and low viscosity at high molecular weight [1,4,8,29]. Thus, WPU has become a significant component in dyes, sealants, printing inks, adhesives, coatings, synthetic leathers, films, fibers, and foams [1,29].

7.4 STRUCTURES OF WATERBORNE POLYURETHANES

Polyol and diisocyanate are starting materials in the synthesis of WPU. It can be synthesized rapidly with petroleum-based polyols, such as polyesters, polyethers, polycaprolactones, and polycarbonate polyols. Different structures of PUDs can be obtained with different polyols. In a study investigating the effect of polyols on WPU dispersions, TEM micrograms of the obtained PUDs are given in Figure 7.3. It was determined that PUDs prepared with polycarbonate diol had a higher particle size than others. On the other hand, PUDs synthesized with polyether and polyester had small and homogeneous particle sizes [30].

Polyisocyanates can be grouped as aliphatic, aromatic, and cycloaliphatic [29]. There are many studies on WPU synthesis with aliphatic diisocyanates, such as isophorone diisocyanate (IPDI), hexamethylene diisocyanate (HDI), and bis (4-isocyanatocyclohexyl) methane ($H_{12}MDI$) or aromatic diisocyanates, such as di-*p*-phenylmethane diisocyanate (MDI) and toluene diisocyanate (TDI). The former, aliphatic diisocyanates, is preferred to the latter, namely aromatic diisocyanates, because aromatic diisocyanates show higher reactivity to water. If a bio-based polyol is used, the hydroxyl groups are less reactive, and a catalyzer such as

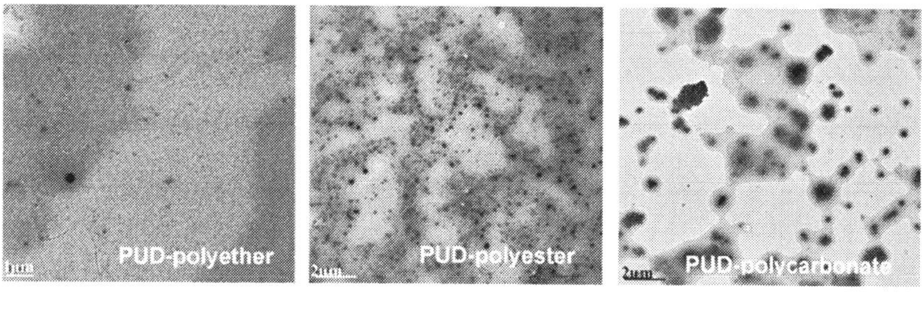

FIGURE 7.3 TEM micrographs of the PUDs prepared with different polyols. Adapted with permission from reference [30]. Copyright (2013) Elsevier.

DBTDL is used to speed up the reaction. In conventional PU synthesis, chain extenders, low molecular weight hydroxyl, and amine-terminated compounds are often used depending on the desired pre-polymer. One of the most widely used extenders is 1.4-butanediol, but other hydroxyl-containing extenders are also used. However, WPU synthesis with or without chain extenders also necessitates the usage of emulsifying agents. It is preferable to use internal emulsifiers as they prevent any probable displacement of this component after film consolidation. The emulsifying agents used will describe whether the WPU is cationic, anionic, or nonionic [6].

Anionic WPUs (a-WPUs) are the most commonly used type and are generally comonomers functionalized with either carboxylate or sulfonate [8]. The most commonly used diacid in this formulation is DMPA. Ionic centers in the hard sections of PU provide this acid and increase hydrophilicity. Since these ionic centers are in charge of the dispersion of micelles in the watery environment, hydrophilic hard areas are favorably located on the particle surface. In some studies, acid addition was avoided to form ionic centers, and double bonds in triacylglycerides were reacted with maleic anhydride. Thereby, the carboxylic acid group that supplies the ionic center is settled in the beginning polyol [6].

Cationic WPU (c-WPU) is prepared with diols, including tertiary nitrogen atoms, which are transformed into quaternary ammonium ions by the adjunct of an appropriate alkylating agent or acid. Commonly used cationic emulsifiers are *N*-methyl diethanolamine (*N*-MDEA), quaternary ammonium salts, or amine salts [6]. c-WPU, which provides the advantage of being determined under acidic conditions, is compatible with softeners and fixing agents and shows good abrasion endurance [31]. In the synthesis of both a-WPUs and c-WPUs, the higher the concentration of the internal emulsifier, the smaller the nanoparticles of the material to be obtained [6]. Nonionic WPUs (n-WPUs) can be prepared with diol or diisocyanate comonomers carrying pendant polyethylene oxide chains. They are colloidally stable in a wide range of pH; however, a high concentration of polyethylene oxide-based comonomer is needed to obtain a dispersion with low particle size. Therefore, combinations of anionic and nonionic stabilization are sometimes used to acquire a combination of low particle size and good steric stability, without the requirement to use an extremely high concentration of nonionic stabilizing comonomer [8].

7.5 SYNTHESIS OF WATERBORNE POLYURETHANES

The use of polymers in an aqueous environment can be achieved by the ability of certain polar functional groups to impart water dispersibility or water solubility to the polymer. The concentration of functional groups, such as carboxylic acid groups, sulfonic acid groups, and tertiary amine groups, in the polymer is an important factor in this regard. The polymer may be water-soluble at high concentrations, and water-dispersible provided that its molecular weight/viscosity is not extreme at lower concentrations. At even lower concentrations, the polar group can provide a charge or steric stabilization to a dispersion of the polymer in water [8].

Generally, the synthesis of WPU can occur in two steps. The first is to obtain the isocyanate-terminated prepolymer using a solvent and elevated temperature to control the viscosity. Then, addition and co-reaction with internal emulsifying agents are actualized. These agents include ionic centers to the chemical structure of the polymer. After that, dispersion of the PU in water is obtained using high shear equipment at room temperature [6]. WPUs consist of two main parts: the main structure or backbone and the emulsifier. Emulsifiers that stabilize the dispersion of PU in water are hydrophilic materials and amphiphilic polymers, such as grafted polymers. Soft segments play a part in hydrophobic, while hard segments with ionic groups are hydrophilic. Emulsifiers can be classified into two groups: internal and external. Internal emulsifiers contain nonionic centers, such as polyethylene oxide, and ionic centers, such as cationic, anionic, and zwitterions. Nonionic PU dispersions include a hydrophilic soft segmented pendent group and are colloidal stable in a wide pH range [2]. WPUs can be synthesized using several methods, such as emulsification or prepolymer, acetone treatment, hot-melt, and ketamine–ketazine [2]. These methods are briefly described below.

7.5.1 Emulsification or Prepolymer Method

In the prepolymer method, the hydrophilically modified NCO-terminated PU prepolymer is chain extended in the heterogeneous phase with diamine or polyamine in the watery dispersion stage. This chain extension can be due to the preferred reactivity of the isocyanate groups with amine rather than water. For the successful implementation of this method, the dispersion stage needs to be carried out at a temperature below the critical point, where the isocyanate groups begin to react with water in a short time. The isocyanate selection is significant in this regard. For example, prepolymers obtained from cycloaliphatic diisocyanates are widely used due to the low reactivity to water. Also, the viscosity and functionality of the mixture must be carefully controlled [2,8]. A prepolymer with medium molecular weight is obtained by the reaction of appropriate amounts of diols with diisocyanate. Then, an internal emulsifier with ionic centers is put into the mixture. After the chain extension, the water dispersion step is carried out. Approximately 12–15% organic solvent (usually *N*-methyl-2-pyrrolidone) is used to decrease the viscosity [32].

7.5.2 Acetone Process

The most prevalently used process after the prepolymer method to obtain PUDs is the acetone process. In this process, the polymer formation is completed in a homogeneous solution, and thus good reproducibility is ensured. Acetone is inert to PU formation reactions and not reactive to isocyanates and miscibility with water. In addition, due to its low boiling point, it can be effortlessly removed during the distillation phase. It also provides the opportunity to control viscosity during the chain extension step [2]. Acetone is one of the most appropriate and widely used solvents for this process. However, it is delimitated by the requirement that the polymer be uncrosslinked and soluble in acetone, and therefore the final products are expected not to resist a solvent. Other disadvantages are the need to use large amounts of acetone and remove the solvent, which necessitates an additional step [2,8]. In the acetone process, the NCO-terminated PU pre-polymer is chain extended to inhibit an extreme viscosity. The chain extender or the water dispersing/solubilizing group is added in the chain extension step. The chain extended polymer is defined as PU urea. Then, water is put in the polymer solution, and after phase reversal, a dispersion is obtained. By removing the solvent by distillation, the desirable watery polymer dispersion is obtained [8].

7.5.3 Hot-Melt Process

In the hot-melt process that provides solvent-free WPU synthesis, urea and an NCO-terminated PU prepolymer react to form a hydrophilic bis-biurea. The capped PU is dispersed at ~100°C in water and reacts with formaldehyde to reduce the viscosity. Then, condensation reaction and polyurethane-urea formation take place [2,8].

7.5.4 Ketamine–Ketazine Process

The ketamine–ketazine process is similar to the pre-polymer mixing process. In this process, an NCO-terminated PU pre-polymer is mixed with ketamine or ketazine as a chain extender. Then, ketamine or ketazine are hydrolyzed to form free diamine or hydrazine, respectively, in the aqueous dispersion step, and thereby chain extension happens [2,8].

7.6 BIO-BASED MATERIALS FOR WATERBORNE POLYURETHANE SYNTHESIS

The main components of WPU are derived from petroleum feedstocks. However, in recent years, due to the factors such as global warming, environmental sustainability, fluctuations in crude oil prices, and decreasing fossil fuel reserves, bio-based materials for industrial production of WPU have received great attention [1,3]. Therefore, the interest in environmentally friendly materials has driven researchers and manufacturers to substitute petroleum-based materials. Studies are carried out on non-isocyanate polyurethanes, and these materials are expected to lead to commercial competitiveness in the future. In addition, the synthesis of bio-based polyols has been simpler and faster [1,6]. These materials include many renewable biomaterials, such as starch, natural oil, cellulose, protein, and sugar [1]. Among these, vegetable oils are one of the most promising alternatives owing to relatively low cost, environmental sustainability, low ecotoxicity, and ready availability [1,6,9,34]. Vegetable oils, mostly triglycerides, have reactive sites such as ester groups and double bonds [35]. Various vegetable oils, such as sunflower, castor, jatropha, palm, cottonseed, soybean, canola, corn, rapeseed, and tung oils, have been used to synthesize bio-based polyols [9]. Thus, the usage of organic solvent and petroleum-based chemicals can be reduced [36]. In addition, several methodologies, such as epoxidation-ring opening [37,38], transesterification [39,40], transforming double bonds by hydroformylation-reduction [41,42], thiol-ene click reactions [43], ozonolysis-reduction [44], and microbial transformation, can be performed to synthesize

WPU from vegetable oils [1,34,45]. The methodology used is another factor that affects the hydroxyl number as well as the iodine value of the oil. For example, polyol synthesized by epoxidation-ring opening may result in the inclusion of one hydroxyl group per double bond, polyol obtained by hydroformylation-reduction of two hydroxyl groups per double bond [29].

In a study, a jatropha oil-based polyol (JOL) was obtained to synthesize an anionic water-based polyurethane dispersion by acetone process. The synthesis scheme of jatropha oil-based WPU (JPU) dispersions is given in Figure 7.4. First of all, JOL and dimethylol propionic acid (DMPA, chain extender) were taken into the four-necked flask, heated up to 78°C, and mixed for 30 min. Dropwise isophorone diisocyanate (IPDI) and a few drops of dibutyltin dilaurate (DBTDL, catalyst) were added. Ethyl methyl ketone (MEK) was added to reduce viscosity, and after 3 h of reaction, hydroxyethyl methacrylate (HEMA) was added as a chain terminator. After all diisocyanate in the reaction was consumed, the system was cooled to 40°C,

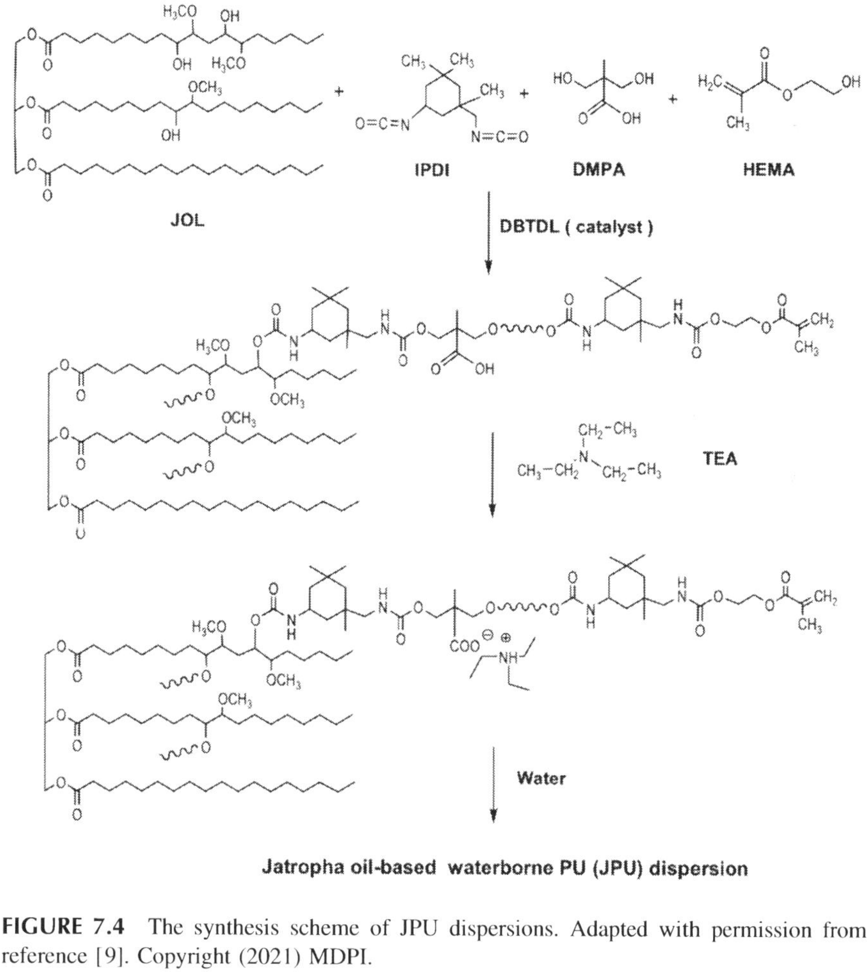

FIGURE 7.4 The synthesis scheme of JPU dispersions. Adapted with permission from reference [9]. Copyright (2021) MDPI.

neutralized with triethylamine (TEA), and dispersion was formed with distilled water at high speed. After removal of MEK, the JPU dispersions with 25% solids content were obtained [9].

In a study, WPU dispersions were obtained from soybean oil-based polyol. Methoxylated soybean oil polyols (MSOLs), DMPA, and IPDI were taken into a four-necked flask, and the reaction took place at 78°C for 1 h in a nitrogen atmosphere. MEK was put in to lower the viscosity of the system. After 2 h of reaction, the reactants were cooled to 40°C and neutralized by the addition of TEA. After MEK was removed under vacuum, PU dispersions with 20% solid content were formed with distilled water at high speed. The resulting SPU dispersions exhibited a uniform particle size increasing from about 12 to 130 nm in diameter with an increment in the OH functionality of MSOL from 2.4 to 4.0 and decreasing with the raising content of hard segments. It has also been found that the functionality and rigid segment content of MSOLs act a major part in controlling the SPU morphology and mechanical and thermal characteristics [4]. In a study about the synthesis of high-performance bio-based WPU, the castor oil–glycerol-based WPU was synthesized. IPDI and DBTDL were put into a flask and stirred for 10 min at 78°C. The mixture of glycerol, castor oil, and DMBA was added dropwise. Then, MEK was added for decreasing the viscosity, and after the reaction for 2 h, the mixture was cooled to ambient temperature. TEA was employed for neutralizing the –COOH groups. Then, water was added with stirring for 2 h and MEK removed from WPU. Consequently, tensile strength, Young modulus, toughness, T_g, and pencil hardness were improved with the addition of glycerol. Moreover, it was found that the addition of glycerol not only increased the biomaterial content but also improved the performance of the WPU [33].

In another study, castor oil-based anionic WPU dispersion was synthesized and blended with various quantities of sodium alginate aqueous solution to prepare bio-based composite dispersions and films. As a result of the tests, it was determined that as the sodium alginate content increased, the elongation at the break of composite films decreased, while the tensile strength and storage modulus increased. Moreover, the contact angle and thermal stability of composite films increased with raising sodium alginate content. These developments have been associated with increased crosslink density and the formation of interlocking networks with the addition of sodium alginate macromolecules [34]. In a study about a fully bio-based WPU, a diisocyanate was obtained from castor oil-based undecylenic acid via Curtius rearrangement, and a hydrophilic chain extender was obtained from castor oil and 3-mercaptopropionic acid by thiol-ene coupling. In other words, all raw materials of WPU were based on vegetable oil. The synthesis scheme of WPU is shown in Figure 7.5. Castor oil, the obtained diisocyanate, DBTDL, and acetone were put into a three-necked flask, and the blend was stirred at 80°C for 3 h. After the addition of carboxyl-functionalized castor oil, the reaction was maintained for 3 h. Following the reaction to completion, the blend was cooled to 50°C, and triethylamine was put into the mixture with stirring for 30 min. The prepolymer was dispersed in water with stirring, and acetone was removed from PUDs. The synthesized WPU dispersion was tested, and it was determined that the WPU with good thermal properties could be substituted the petroleum-based materials [35].

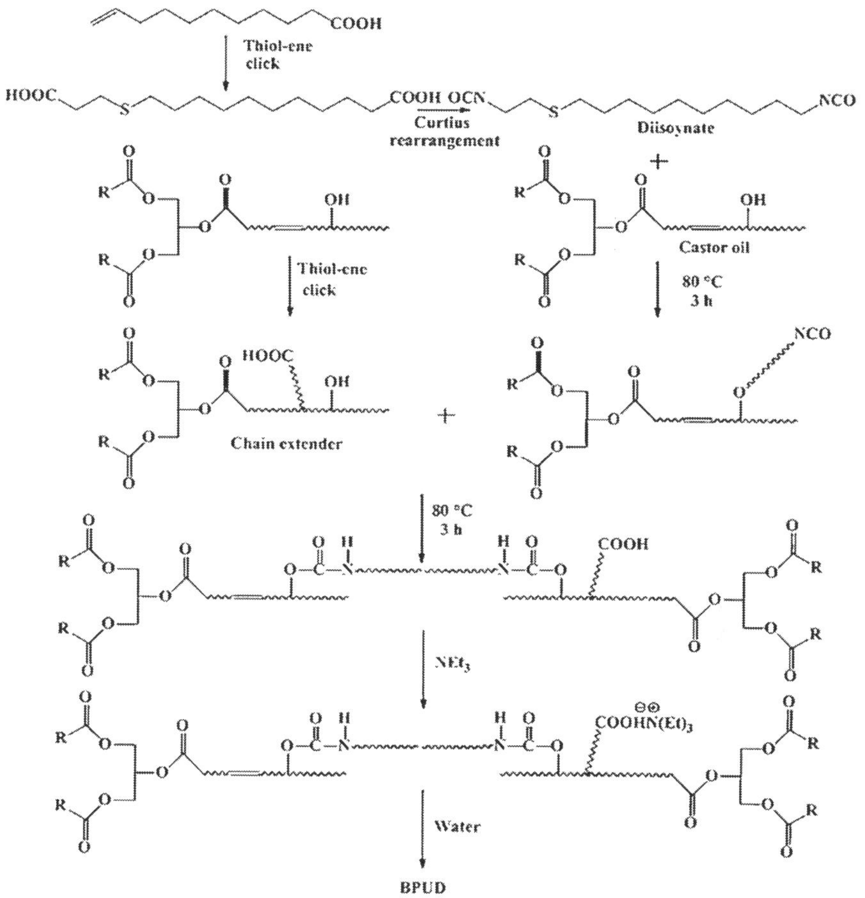

FIGURE 7.5 A synthesis scheme of the fully bio-based WPU. Adapted with permission from reference [35]. Copyright (2014) Elsevier.

7.7 WATERBORNE POLYURETHANE COMPOSITES

In recent times, the research and application of polymer nanocomposite foams have drawn great attention. Incorporating nanoparticles into polymeric foams is an efficient way to improve many features of the foams by changing the morphological structures [46]. Various nanoparticles, such as clay [47], carbon nanofibers (CNFs) [48], and multiwalled carbon nanotubes [49], are used in polymer nanocomposite foams. Thus, significant improvements can be achieved in mechanical, thermal, and other characteristics of foam materials [36,46]. In a study, active carbon WPU (ACWPU) foam was obtained by using WPU instead of conventional binders. It was synthesized with two consecutive steps. Firstly, a solution of WPU is obtained, which expands within the pores and passages of the foam to form an adhesive layer on the surface. Thereafter, AC is inserted into the expanded pores. Isophorone diisocyanate (IPDI, 5.57 g), polypropylene glycol 2000 (PPG2000, 20 g), and stannous octoate (SnOct, used as catalyst, 0.5 g) were put into a flask and heated up to 80°C with stirring, under nitrogen atmosphere. Then, the dihydromethyl propionic acid (DMPA, 1.34 g) was added, and the reaction was maintained for 2 h. Immediately afterward, 1.4-butanediol (BDO, 0.45 g) was added, and the reaction occurred for 3 h at 60–70°C. After neutralization with triethanolamine (1.40 mL) for 30 min, the obtained WPU was dispersed with distilled water under forceful stirring. The morphological structures and dispersion sizes of the WPU emulsion were examined with TEM micrograms, and the nanoscale distribution of the emulsion was proven. When the FTIR spectrum is examined, the observed peaks supported the formation of the WPU structure [50].

Silver nanoparticles (*n*-Ag), one of the most important nanoparticles, have recently drawn great attention in virtue of their naturally quantum effects and particularly their high antibacterial characteristics. Studies have been carried out on the use of *n*-Ag to enhance the properties of WPUs [46]. In a study, WPU/Ag nanocomposite foams with different *n*-Ag contents were synthesized by the mechanical foaming method. First, 92% WPU emulsion and 8% ammonium stearate (foam stabilizing agent) were taken into a beaker and stirred at room temperature for 15 min. Then, 0–4% *n*-Ag was put into the beaker with stirring at a foaming rate of 250% and sonicated for 30 min to prepare a well-dispersed solution. Lastly, the blend was poured on a flat glass using a 0.3 mm thick casting blade at ambient temperature and dried at 80°C [46]. The distribution and morphological structure of *n*-Ag in WPU/Ag nanocomposite foams were analyzed by SEM and AFM. Figure 7.6 shows the images of the foams with 0%, 1%, 2%, 3%, and 4% *n*-Ag contents. It is seen that all samples have an open cell structure. Compared to nanocomposite foams, pure WPU foam has a smaller and uniform cell size. This is because *n*-Ag can be used as a nuclear agent that speeds up cell growth, wall rupture, and cell fusion [46].

Average surface roughness data based on the AFM images in Figure 7.6 are 8.95, 11.0, 13.4, 19.3, and 25.8 nm for WPU/Ag foams with 0%, 1%, 2%, 3%, and 4% *n*-Ag contents, respectively. The pure WPU foam was found to have a smooth surface. With the increase of *n*-Ag content, large *n*-Ag aggregations were observed in the foams, and much rougher surfaces were obtained. When the 4% *n*-Ag content was reached, the unevenness of the surface increased to 25.8 nm. The thermal and mechanical features of nanocomposite foams are significantly affected by the distribution of reinforcement material in the matrix. When the *n*-Ag content increased from 0% to 4% by weight, the average pore sizes of the foams raised from 13.74 to 34.24 μm. Consequently, as the *n*-Ag content increased, cell aggregation and interconnection became more pronounced [46].

The thermal properties of WPU/Ag nanocomposite foams were investigated by TGA. Pure WPU and nanocomposite foams were found to have a similar thermal decomposition tendency. All foams have two degradation stages, between 330–420°C and after 420°C. This situation may be correlated with the rigid part and soft part of the WPU, respectively. WPU foam has better thermal properties in comparison with PU foam. While the *n*-Ag content increased from 0% to 4% by weight, the initial decomposition temperature fluctuated slightly in the range of 292.49–299.27°C and the peak slightly increased from 376.31 to 380.24°C [46].

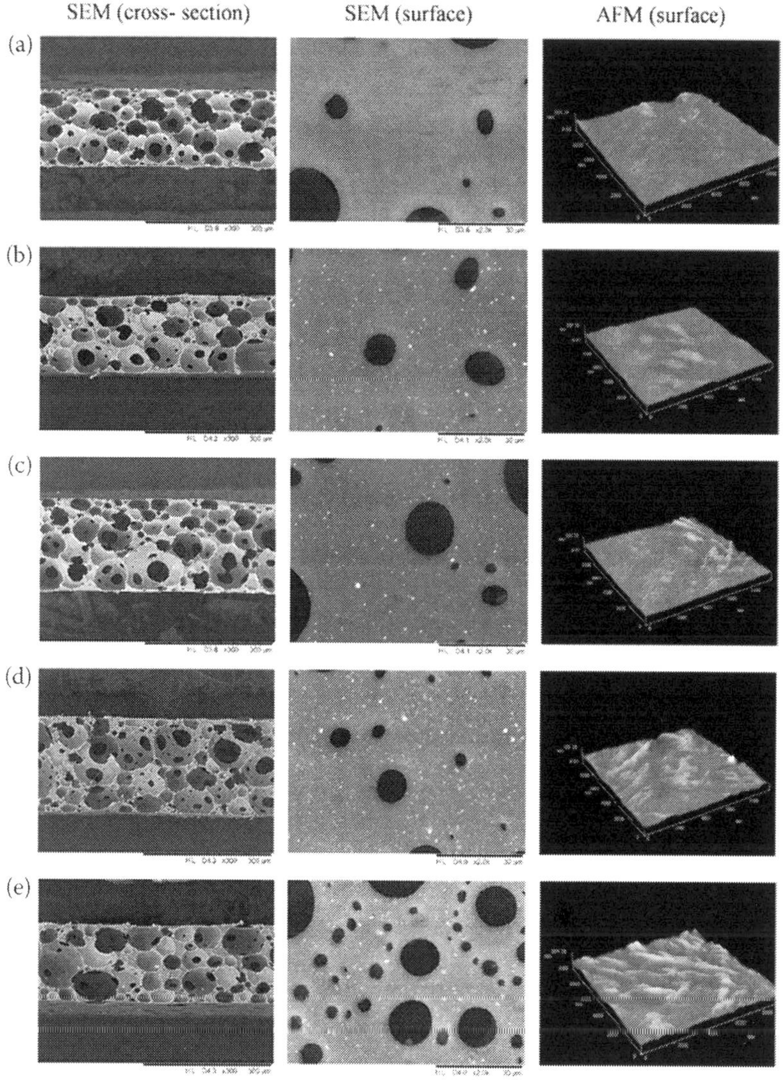

FIGURE 7.6 SEM and AFM images of WPU/Ag nanocomposite foams with 0% (a), 1% (b), 2% (c), 3% (d) and 4% (e) of *n*-Ag. Adapted with permission from reference [46]. Copyright (2018) Wiley.

When looking at the 450–550°CC test temperature range, it was determined that the thermal stability of nanocomposite foams increased with increasing *n*-Ag content from 0% to 4% by weight. Especially when the *n*-Ag contents were 3% and 4% by weight, a significant increase was observed and the degradation rate of the foams decreased. In addition, the char yield of WPU/Ag nanocomposite foams at 700°CC increased with the increase of *n*-Ag content. As a result, the thermal properties of the foams were enhanced with the reinforcement of *n*-Ag [46].

One of the disadvantages of WPU foams is that they have low mechanical strength. To increase their strength, fibers, particulate fillers, and nanoparticles are used as reinforcements. In a study, a tensile test was applied to WPU/Ag nanocomposite foams with different *n*-Ag content and tensile strength and elongation values were determined. Due to its amorphous structure, the tensile strength of pure WPU foam was 0.61 MPa, and the elongation at break was 412.39%. The tensile strength reached 1.26 MPa with a rise of *n*-Ag loading to 2% and decreased to 1.17 MPa with an increase in *n*-Ag loading from 2% to 4% by weight. When the reinforcement ratio is more than 2% by weight, *n*-Ag aggregation led to a reduction in tensile strength [20]. In addition, the elongation was diminished with *n*-Ag reinforcement since harder nanocomposite foams with less deformation were obtained with the addition of *n*-Ag. However, the elongation values for all foams were greater than 340% [46].

Nano-cellulose, one of the most prevalent natural polymers, has many advantages, such as renewability, low weight, biodegradability, high aspect ratio, and low cost. Therefore, nano-cellulose with two morphologies, nano-cellulose fibers, and nano-cellulose crystals, is preferred as reinforcement for the composites [36]. WPU's mechanical and thermal properties can also be enhanced with the reinforcement of cellulose nanocrystals (CNCs). In a study, CNCs obtained with ancestral endoglucanase (EnCNCs) were the reinforcement of WPU and compared with CNCs synthesized by sulfuric acid hydrolysis (AcCNCs). The reinforcement of EnCNCs improved mechanical properties and thermomechanical stability of WPUs, better than AcCNCs. Moreover, EnCNCs enabled greener material to be obtained compared to AcCNCs [51]. In another study, octa(aminopropyl) polyhedral oligomeric silsesquioxane (OPOSS)/2.2.6.6-tetramethylpiperidine-1-oxyl (TEMPO)-oxidized cellulose microfibers (TOCMF) nanocomposite material (named PC) was obtained by peptide-forming reactions. The vegetable oil-based WPU was synthesized by combining the remaining amino groups on the POSS with the –NCO group. Thus, the bulk structure of cellulose and the hydrogen bonding between cellulose and PU were preserved by the grafting POSS. The mechanical and thermal properties and water resistance were enhanced with the synergistic modification [36]. Nano-Fe_3O_4 particles can be used to provide magnetic and electrical properties to composites in several fields such as rechargeable batteries, electromagnetic interface shielding, electrodes, and sensors. In a study, modified nano-Fe_3O_4 with oleic acid were used for preparing WPU nanocomposites. Thus, better thermal, electrical, and magnetic properties were reached compared to pristine WPU [52].

7.8 CONCLUSION

PU is a high-performance organic polymer material, and PUF represents a significant commercial product of PU. PUFs are a class of lightly porous materials that are of great interest because of their specific properties and potential applications in many fields. The performance of PUFs can be optimized for a wide variety of applications with a well-designed formulation of polyols, isocyanates, catalysts, surfactants, and blowing agents. PUs contain high VOC emissions, which have toxicity and carcinogenicity to the human body and are harmful to the environment and, in some cases, have a definite quantity of free isocyanates. So many methods have been tried to reduce VOC emissions and hazardous air pollutants, and it has been determined that one of the most environmentally friendly alternatives is WPU. WPU is nontoxic and nonflammable, and it does not pollute the air or generate wastewater. It also has many advantages, such as good adhesion, fast drying time, high tensile strength, nonflammability, excellent wear resistance, high flexibility and impact resistance, good low-temperature endurance, low cost, and low viscosity at the high molecular weight. WPUs can be synthesized using various methods, such as emulsification or prepolymer, acetone treatment, hot-melt, and ketamine–ketazine.

In recent years, due to the factors such as global warming, environmental sustainability, fluctuations in crude oil prices, and decreasing fossil fuel reserves, bio-based materials for industrial production of WPU have received great attention. Studies have been carried out on non-isocyanate polyurethanes and the synthesis of bio-based polyols. In addition, to change the morphology and improve the properties of foams, nanoparticles can be incorporated into polymeric foams.

REFERENCES

1. Cheng, Z., Li, Q., Yan, Z., Liao, G., Zhang, B., Yu, Y., Yi, C., Xu, Z. (2019). Design and synthesis of novel aminosiloxane crosslinked linseed oil-based waterborne polyurethane composites and its physicochemical properties. *Progress in Organic Coatings*, 127, 194–201.
2. Honarkar, H. (2018). Waterborne polyurethanes: A review. *Journal of Dispersion Science and Technology*, 39, 4, 507–516, doi:10.1080/01932691.2017.1327818.
3. Kattiyaboot, T. and Thongpin, C. (2016). Effect of natural oil based polyols on the properties of flexible polyurethane foams blown by distilled water. *Energy Procedia*, 89, 177–185, doi:10.1016/j.egypro.2016.05.024.
4. Lu, Y. and Larock, R.C. (2008). Soybean-oil-based waterborne polyurethane dispersions: effects of polyol functionality and hard segment content on properties. *Biomacromolecules*, 9, 3332–3340.
5. Santiago-Calvo, M., Tirado-Mediavilla, J., Rauhe, J.C., Rosgaard Jensen, L., Luis Ruiz- Herrero, J., Villafañe, F., Rodríguez-Pérez, M.A. (2018). Evaluation of the thermal conductivity and mechanical properties of water blown polyurethane rigid foams reinforced with carbon nanofibers. *European Polymer Journal*, 108, 98–106, doi:10.1016/j.eurpolymj.2018.08.051.
6. Mucci, V.L., Hormaiztegui, M.E.V., Aranguren, M.I. (2020). Plant oil-based waterborne polyurethanes: A brief review. *Journal of Renewable Materials*, 8, 6, 579–601, doi:10.32604/jrm.2020.09455.
7. Gama, N.V., Ferreira, A. and Barros-Timmons, A. (2018). Polyurethane foams: Past, present, and future. *Materials*, 11, 10, 1841, doi:10.3390/ma11101841.
8. Szycher, M. (2013). *Szycher's Handbook of Polyurethanes*, 2nd edition. CRC Press, Boca Raton
9. Saalah, S., Abdullah, L.C., Aung, M.M., Salleh, M.Z., Biak, D.R.A., Basri, M., Jusoh, E.R., Mamat, S., Al Edrus, S.S. (2021). Chemical and thermo-mechanical properties of waterborne polyurethane dispersion derived from jatropha oil. *Polymers*, 13, 795, doi:10.3390/polym13050795.

10. Gomez-Fernandez, S., Ugarte, L., Pena-Rodriguez, C., Zubitur, M., Corcuera, M.A., Eceiza, A. (2016). Flexible polyurethane foam nanocomposites with modified layered double hydroxides. *Applied Clay Science*, 123, 109–120.
11. Bernardini, J., Cinelli, P., Anguillesi, I., Coltelli, M., Lazzeri, A. (2015). Flexible polyurethane foams green production employing lignin or oxypropylated lignin. *European Polymer Journal*, 64, 147–156.
12. Gama, N.V., Soares, B., Freire, C.S.R., Silva, R., Neto, C.P., Barros-Timmons, A., Ferreira, A. (2015). Bio-based polyurethane foams toward applications beyond thermal insulation. *Materials and Design*, 76, 77–85.
13. Kakroodi, A.R., Khazabi, M., Maynard, K., Sain, M., Kwon, O. (2015). Soy-based polyurethane spray from insulations for light weight wall panels and their performances under monotonic and static cyclic shear forces. *Industrial Crops and Products*, 74, 1–8.
14. Buffa, J.M., Mondragon, G., Concuera, M.A., Eceiza, A., Mucci, V., Aranguren, M.I. (2018). Physical and mechanical properties of a vegetable oil based nanocomposite. *European Polymer Journal*, 98, 116–124.
15. Ionescu, M., Radojcic, D., Wan, X., Shrestha, M.L., Petrovic, Z.S., Upshaw, T.A. (2016). Highly functional polyols from castor oil for rigid polyurethanes. *European Polymer Journal*, 84, 736–749.
16. Harikrishnan, G., Macosko, C.W., Choi, J.H., Bischof, J.C., Singh, S.N. (2008). A simple transient method for measurement of thermal conductivity of rigid polyurethane foams. *Journal of Cellular Plastics*, 44, 481–491.
17. Kuranska, M., Prociak, A. (2016). The influence of rapeseed oil-based polyols on the foaming process of rigid polyurethane foams. *Industrial Crops and Products*, 89, 182–187.
18. Cinelli, P., Anguillesi, I., Lazzeri, A. (2013). Green synthesis of flexible polyurethane foams from liquefied lignin. *European Polymer Journal*, 49, 1174–1184.
19. Hyuk Park J., Suh Minn K., Rae Lee H., Hyun Yang S., Bin Yu C., Yeol Pak S., Sung Oh C., Seok Song Y., June Kang Y., Ryoun Youn J. (2017). Cell openness manipulation of low density polyurethane foam for efficient sound absorption. *Journal of Sound and Vibration*, 406, 224–236.
20. Marcovich, N.E., Kuranska, M., Prociak, A., Malewska, E., Kulpa, K. (2017). Open cell semi-rigid polyurethane foams synthesized using palm oil-based bio-polyol. *Industrial Crops and Products*, 102, 88–96.
21. Olcay, H. and Kocak, E.D. (2021). Rice plant waste reinforced polyurethane composites for use as the acoustic absorption material. *Applied Acoustics*, 173, 107733, doi:10.1016/j.apacoust.2020.107733.
22. Noreen, A., Zia, K.M., Zuber, M., Tabasum, S., Zahoor, A.F. (2016). Bio-based polyurethane: An efficient and environment friendly coating systems: A review. *Progress in Organic Coatings*, 91, 25–32.
23. Carriço, C.S., Fraga, T., Pasa, V.M.D. (2016). Production and characterization of polyurethane foams from a simple mixture of castor oil, crude glycerol and untreated lignin as bio-based polyols. *European Polymer Journal*, 85, 53–61.
24. Das, A., Mahanwar, P. (2020). A brief discussion on advances in polyurethane applications. *Advanced Industrial and Engineering Polymer Research*, 3, 93–101, doi:10.1016/j.aiepr.2020.07.002.
25. Guner, F.S., Yagci, Y., Tuncer Erciyes, A. (2006). Polymers from triglyceride oils. *Progress in Polymer Science*, 31, 633–670.
26. Gurunathan, T., Arukula, R. (2018). High performance polyurethane dispersion synthesized from plant oil renewable resources: A challenge in the green materials. *Polymer Degradation and Stability*, 150, 122–132.
27. Olcay, H. and Kocak, E.D. (2020). The mechanical, thermal and sound absorption properties of flexible polyurethane foam composites reinforced with

artichoke stem waste fibers. *Journal of Industrial Textiles*, doi:10.1177/1528083720934193.

28. Zafar, F., Ghosal, A., Sharmin, E., Chaturvedi, R., Nishat, N. (2019). A review on cleaner production of polymeric and nanocomposite coatings based on waterborne polyurethane dispersions from seed oils. *Progress in Organic Coatings*, 131, 259–275, doi:10.1016/j.porgcoat.2019.02.014.
29. Liu, X., Hong, W., Chen, X. (2020). Continuous production of water-borne polyurethanes: A review. *Polymers*, 12, 2875, doi:10.3390/polym12122875.
30. Garcia Pacios, V., Colera, M., Iwata, Y., Martin-Martinez, J.M. (2013). Incidence of the polyol nature in waterborne polyurethane dispersions on their performance as coatings on stainless steel. *Progress in Organic Coatings*, 76, 1726–1729, doi:10.1016/j.porgcoat.2013.05.007.
31. Liu, T., Liu, D., Zhou, W., Ni, L., Quan, H. et.al. (2019). Effects of soft segment and microstructure of polysiloxane modified cationic waterborne polyurethane on the properties of its emulsion and membrane. *Materials Research Express*, 6, 2, 25304, doi:10.1088/2053-1591/aaea98.
32. Cakic, S.M., Stamenkovic, J.V., Djordjevic, D.M., Ristic, I.S. (2009). Synthesis and degradation profile of cast films of PPG–DMPA–IPDI aqueous polyurethane dispersions based on selective catalysts. *Polymer Degradation and Stability*, 94, 2015–2022.
33. Zhang, J., Wu, Y., Zhang, H., Yan, T., Huang, Y., Jiang, J., Tang, J. (2021). Castor oil-glycerol-based waterborne polyurethane dispersions. *Progress in Organic Coatings*, 157, 106333, doi:10.1016/j.porgcoat.2021.106333.
34. Wang, X., Zhang, Y., Liang, H., Zhou, X., Fang, C., Zhang, C., Luo, Y. (2019). Synthesis and properties of castor oil-based waterborne polyurethane/sodium alginate composites with tunable properties. *Carbohydrate Polymers*, 208, 391–397, doi:10.1016/j.carbpol.2018.12.090.
35. Fu, C., Zheng, Z., Yang, Z., Chen, Y., Shen, L. (2014). A fully bio-based waterborne polyurethane dispersion from vegetable oils: From synthesis of precursors by thiol-ene reaction to study of final material. *Progress in Organic Coatings*, 77, 53–60.
36. Zhang, P., Lu, Y., Fan, M., Jiang, P., Bao, Y., Gao, X., Xia, J. (2020). Role of cellulose-based composite materials in synergistic reinforcement of environmentally friendly waterborne polyurethane. *Progress in Organic Coatings*, 147, 105811, doi:10.1016/j.porgcoat.2020.105811.
37. Pawar, M.S., Kadam, A.S., Dawane, B.S., Yemul, O.S. (2015). Synthesis and characterization of rigid polyurethane foams from algae oil using biobased chain extenders. *Polymer Bulletin*, 73, 3, 727–741.
38. Zhang, C., Ding, R., Kessler, M.R. (2014). Reduction of epoxidized vegetable oils: A novel method to prepare bio-based polyols for polyurethanes. *Macromolecular Rapid Communications*, 35, 1068–1074.
39. Chang, C.W. and Lu, K.T. (2012). Natural castor oil based 2-package waterborne polyurethane wood coatings. *Progress in Organic Coatings*, 75, 435–443.
40. Amini, Z., Ilham, Z., Ong, H.C., Mazaheri, H., Chen, W.H. (2017). State of the art and prospective of lipase-catalyzed transesterification reaction for biodiesel production. *Energy Conversion and Management*, 141, 339–353.
41. Chen, R., Zhang, C., Kessler, M.R. (2014). Anionic waterborne polyurethane dispersion from a bio-based ionic segment. *RSC Advances*, 4, 35476–35483.
42. Wu, L., Fleischer, I., Jackstell, R., Profir, I., Franke, R., Beller, M. (2013). Ruthenium- catalyzed hydroformylation/reduction of olefins to alcohols: extending the scope to internal alkenes. *Journal of the American Chemical Society*, 135, 38, 14306–14312.

43. Turunc, O. and Meier, M.A.R. (2013). The thiol-ene (click) reaction for the synthesis of plant oil derived polymers. *European Journal of Lipid Science and Technology*, 115, 1, 41–54.
44. Hubner, S., Bentrup, U., Budde, U., Lovis, K., Dietrich, T., Freitag, A., Jahnisch, K. (2009). An ozonolysis-reduction sequence for the synthesis of pharmaceutical intermediates in microstructured devices. *Organic Process Research & Development*, 13, 5, 952–960.
45. Olcay H., Kocak E.D., Yıldız Z. (2020). Sustainability in Polyurethane Synthesis and Bio-based Polyurethanes. In: Muthu S.S., Gardetti M.A. (eds) *Sustainability in the Textile and Apparel Industries. Sustainable Textiles: Production, Processing, Manufacturing & Chemistry*. 139–156, Springer, Cham. doi:10.1007/978-3-030-38013-7_7.
46. Zhao, B., Qian, Y., Qian, X., Fan, J., Feng, Y. (2018). Fabrication and characterization of waterborne polyurethane/silver nanocomposite foams. *Polymer Composites*, 40, 4, 1492–1498, doi:10.1002/pc.24888.
47. Seol, S.M. and Kim, G.H. (2013). Effect of organoclay content on the mechanical properties of ethylene-vinyl acetate copolymer/multi-walled carbon nanotube/organoclay foams. *Polymer Composites*, 34, 5, 665–670, doi:10.1002/pc.22469.
48. VanHouten, D.J. and Baird, D.G. (2009). Generation and characterization of carbon nano-fiber–poly(arylene ether sulfone) nanocomposite foams. *Polymer*, 50, 1868–1876, doi:10.1016/j.polymer.2009.02.005.
49. Zeng, C., Hossieny, N., Zhang, C., Wang, B., Walsh, S.M. (2013). Morphology and tensile properties of PMMA carbon nanotubes nanocomposites and nanocomposites foams. *Composite Science Technology*, 82, 29–37.
50. Mao, N., Zhou, L., Ye, Z., Zheng, W., Peng, L., Li, Y. (2013). Preparation of waterborne polyurethane foam with active carbon and its adsorption for phenol in aqueous solution. *Journal of Environmental Engineering*, 139, 8, doi:10.1061/(ASCE)EE.1943-7870.0000708.
51. Alonso-Lerma, B., Larraza, I., Barandiaran, L., Ugarte, L., Saralegi, A., Corcuera, M.A., Perez-Jimenez, R., Eceiza, A. (2021). Enzymatically produced cellulose nanocrystals as reinforcement for waterborne polyurethane and its applications. *Carbohydrate Polymers*, 254, 117478, doi:10.1016/j.carbpol.2020.117478.
52. Zhang, S., Li, Y., Peng, L., Li, Q., Chen, S., Hou, K. (2013). Synthesis and characterization of novel waterborne polyurethane nanocomposites with magnetic and electrical properties. *Composites: Part A*, 55, 94–101, doi:10.1016/j.compositesa.2013.05.018.

8 Flame-Retardant Waterborne Polyurethanes

Giulio Malucelli
Department of Applied Science and Technology and local INSTM Unit, Politecnico di Torino, Viale Teresa Michel 5, 15121, Alessandria Italy

CONTENTS

8.1 Introduction 119
8.2 Thermal and Thermo-Oxidative Stability of WPUs 120
8.3 Flame Retardance of WPUs: A General Overview 121
8.4 Flame Retardance of WPUs: The Current Strategies Employed 121
8.5 Flame-Retardant WPUs Using the Additive Approach: Recent Advances 123
8.6 Flame-Retardant WPUs Using the Reactive Approach: Recent Advances 127
8.7 Conclusions and Future Perspectives 132
References 133

8.1 INTRODUCTION

Waterborne polyurethanes (WPUs) represent an interesting and up-to-date class of polymer systems that gathered much interest in academic and industrial research over the last years. Undoubtedly, WPUs constitute one of the fastest emerging and dynamic branches belonging to the chemistry of polyurethanes [1]. As clearly reported in the scientific literature, the synthesis of WPUs generally exploits the reaction between water and a polyurethane prepolymer [2]. In particular, the obtainment of WPUs is usually achieved utilizing structural modifications, which are aimed at changing the hydrophobic character of the polymer backbone with intrinsic hydrophilic functionalities [3]. Subsequently, the resulting hydrophilic WPU is emulsified and reacted with deionized water, exploiting a chain extension process [4, 5].

The "green" and low environmental impact of WPUs, and the ease of tailoring their structure based on the constituents selected for their formulation, justify their wide applications in different sectors involving coatings, adhesives, paint additives, textile dyes, primers for metals, defoamers, caulking materials, emulsion polymerization media, associate thickeners, biomaterials and pigment pastes, among a few to mention [6, 7].

DOI: 10.1201/9781003173526-8

Generally, the recipe of WPU dispersions comprises aliphatic or aromatic diisocyanates (for the build-up of hard segments), polyols (i.e. polyesters, polyethers, or polydienes, for the build-up of soft segments: their molecular weight usually ranges from 400 to 6,000 g/mol), chain extenders (both difunctional low molecular weight chain extenders and hydrophilic chain extenders), neutralizers, and water.

Despite all these interesting features, WPUs (and, in general, polyurethanes) are not intrinsically flame retarded at all; when exposed to a flame or an irradiative heat flux, they burn very easily, releasing a huge amount of smoke [8]. These issues harshly limit their use for specific applications, where flame retardance is mandatory. In this context, the chapter will summarize the recent advances concerning flame-retardant WPUs, emphasizing the current limitations and suggesting some possible progress for the next future.

8.2 THERMAL AND THERMO-OXIDATIVE STABILITY OF WPUs

Before providing an overall picture of the recent advances in flame-retarded WPUs, it is useful to examine the general thermal and thermo-oxidative stability of these polymeric systems.

Generally speaking, we have to discriminate between thermal and thermo-oxidative stability of WPUs, as the presence of air (and, in particular, of oxygen) in the testing environment may significantly affect their thermal behavior. Usually, thermogravimetric (TG) analyses can be successfully employed for investigating the effect of thermal stresses on the degradation behavior of selected WPUs. Possibly, TG analyses can be further coupled with mass spectroscopy (MS) or Fourier-transform infrared spectroscopy (FTIR) apparatuses, aiming at better elucidating the type of volatile products that originate from the thermal degradation of WPUs.

In this context, it is well documented in the scientific literature that the thermal degradation of WPUs takes place according to complicated heterogeneous processes involving many degradation reactions [9]. As a consequence, there are some difficulties in elucidating the degradation mechanisms of WPUs because of the production of several volatile products, as documented by the multiple decomposition stages usually observed in TG analyses.

Both thermal and thermo-oxidative stability of WPUs are affected by several parameters. First of all, a key role is played by the chemical structure of the polyurethane, i.e. by the types of hard and soft segments, as well as by the type of chain extender. In particular, the building blocks (i.e. diisocyanate, long- or short-chain diols) of WPUs dramatically affect the thermal behavior of the synthesized polymers. Besides, this latter may be further affected by the extent of separation into microdomains of both hard and soft segments. Further, other factors include isocyanate:hydroxyl group ratio, concentration of the catalyst, and temperature.

More specifically, during a heating-up process, the hard segments of WPUs are the first to decompose; as a consequence, their main characteristics (e.g. length, concentration, chemical structure and composition, distribution, among a few to mention) are dramatically important as far as the overall thermal stability of the polymer is concerned [10, 11]. Besides, the degradation path significantly depends on the aggregation of the hard segments, which gives rise to a pseudo-crosslinking reinforcing effect. Conversely, the amorphous domains present in the hard segments may exert a detrimental effect on the thermal stability of the whole polymer [12]. In addition, the type of structure (i.e. cycloaliphatic or aromatic) that the hard segments of WPUs are made of plays a key role. In particular, the role of these structures depends on degradation temperature; at low temperatures, cycloaliphatic structures can provide higher thermal stability than aromatic ones, as the aromatic rings weaken the interactions occurring among the hard segments, hence lowering the thermal stability at the start of the degradation process. Conversely, aromatic structures can provide WPUs with enhanced thermal behavior at high temperatures. The packing of hard segments is further affected by the type (i.e. aromatic, cycloaliphatic, or aliphatic) of the chain extender, its molecular volume, functionality, and chain length; therefore, the chain extender indirectly affects the overall thermal stability of the polymer [13–16].

Besides, the use of aromatic or aliphatic polyols in the recipes of WPUs affects the thermal stability of the resulting polymer; in particular, aromatic polyols are preferable to aliphatic counterparts to increase the thermal resistance of WPUs [17].

8.3 FLAME RETARDANCE OF WPUs: A GENERAL OVERVIEW

Generally speaking, WPUs are very combustible plastics; indeed, their Limiting Oxygen Index (LOI) ranges between 17% and 19%, particularly depending on the structure and morphology of the polymer [18, 19]. In particular, the flammability of WPUs is significantly affected by the chemical structure of the polyol; aromatic-based polyols usually provide much higher flame retardance to the resulting polymer with respect to aliphatic counterparts, because of the presence of stable aromatic rings [20, 21]. Besides, the high nitrogen content in the WPUs causes the formation of hydrogen cyanide, upon pyrolysis or the application of a flame [22]. Furthermore, the two main components of WPUs take part in the process in a different way; the solid-phase burning mainly involves the combustion of the isocyanate part of the polymer, whereas the polyol part is more active in the liquid pool combustion [23]. In the first burning step, upon the application of a flame, a yellow smoke that is an indication of the release of the isocyanate part is formed; then, this smoke further decomposes, giving rise to the formation of hydrogen cyanide and other flammable organic products. The former partially produces NOx during combustion. The second burning step mainly involves polyols, which are mostly responsible for the release of heat, CO, and CO_2.

8.4 FLAME RETARDANCE OF WPUs: THE CURRENT STRATEGIES EMPLOYED

Two classes of flame-retarded WPUs have been designed and developed so far: the first exploits the so-called reactive-type flame-retarded strategy (i.e. utilizing an "intrinsic" flame-retardant approach [24–26]), while the second uses a blending technique, i.e. an additive modification [27–29]. This latter, for which the flame retardant is incorporated into WPU by physical means, shows some drawbacks that refer to the possible migration of the blended flame retardant toward the surface of the polymer, as a consequence of either low compatibility or poor mixing. Therefore, when possible, the design, synthesis, and application of reactive-type flame retardants seem preferable to blending [30]. In general, both the described approaches show some pros and cons, which will be summarized in the following paragraphs.

The use of reactive-type flame retardants is generally very effective for providing outstanding flame-retardant properties to WPUs, but, at the same time, it promotes a decrease of the other properties (particularly referring to the thermal stability and mechanical behavior), especially when the polyurethane recipe contains high loadings of the reactive flame retardant. As an example, 2-ethyl-2-(2-oxo-5,5-dimethyl-1,3,2-dioxaphosphorinanyl-2-methylene)-1,3-propanediol was exploited as reactive flame retardant for WPUs. Though in the presence of 12 wt.% of the flame retardant in the WPU formulation the modified polymer system achieved V-0 rating in vertical flame spread tests, a significant decrease of its thermal stability, as well as of its mechanical strength, was observed [31]. In another recent work [32], phosphorus- and nitrogen-containing flame-retardant di-*N*-hydroxyethyl phosphamide was synthesized and grafted to a WPU. Unfortunately, it was not possible to exceed 26% of LOI even at high grafting levels (i.e. 9 wt.%) of the reactive flame retardant.

In another recent paper, a series of flame-retarded WPUs were synthesized, exploiting the grafting of bis(4-aminophenoxy) phenyl phosphine oxide onto the WPUs polymer backbone. Despite a remarkable enhancement of the overall flame-retardant behavior of the resulting WPUs, the chemical modification of the polymer backbone worsened the stability of the dispersion before application. Besides, the flame-retardant WPUs showed a remarkable decrease of the elongation at break, which may limit their use in specific sectors [33]. Briefly, the most significant limitation of designing reactive-type flame retardant for WPUs refers to the challenging need of harmonizing flame retardance with the overall (e.g. thermal, mechanical) behavior of the modified WPUs.

Looking at the second approach (i.e. the incorporation of the flame retardant into the WPU through blending), which is usually preferred for industrial purposes, several challenging issues have to be considered. In particular, as for every blending method, it is very important to verify the compatibility of the flame retardant additive with the WPU recipe, as compatibility plays a crucial role in the design of the flame-retarded polymer system. Limited compatibility may favor demulsification and therefore failure in the obtainment of the flame-retarded WPU. To overcome this problem, several studies have been proposed, even recently, though a certain worsening of the flame-retardant features or of the overall behavior of the modified WPUs has been highlighted [28, 29]. In this context, the incorporation of bisphosphoric-modified amino functional Fe_3O_4 and $Al(OH)_3$ nanoparticles into a WPU at 20 wt.% loading provided good flame-retardant features, but, at the same time, made the nanocomposite films opaque and worsened their thermal stability and mechanical properties [34].

Therefore, the potential of the additive approach is still being thoroughly investigated, aiming at achieving a good compromise among high flame retardance acceptable compatibility, and limited side effects that may affect the thermal stability and mechanical features of the resulting flame-retarded WPUs.

The next paragraphs will summarize the most recent achievements concerning the exploitation of the two flame-retardant approaches.

8.5 FLAME-RETARDANT WPUs USING THE ADDITIVE APPROACH: RECENT ADVANCES

Zhao and co-workers [35] synthesized a water-soluble ionic liquid, 1-methyl-3-((6-oxidodibenzo[c,e][1,2]oxaphosphinin-6-yl)methyl)-1-H-imidazol-3-ium-4-methylbenzenesulfonate, which was incorporated into a WPU formulation at different loadings (namely, 2, 4, and 6 wt.%). The obtained dispersions were found to be very stable, without showing phase separation even after more than six months. Besides, the highest flame-retardant concentration (i.e. 6 wt.%) allowed achieving V-0 rating in vertical flame spread tests, as well as a high limiting oxygen index (i.e. 27%, vs. 20% for neat WPU). As assessed by cone calorimetry tests performed at 35 kW/m^2 irradiative heat flux (Figure 8.1), this formulation exhibited a remarkable decrease in total heat release (−41.5%), peak of heat release rate (−46%), and total smoke release (−51.5%) as compared with neat WPU. Furthermore, the incorporation of the flame retardant at the selected loadings did not significantly affect the overall behavior of the obtained coatings, as far as the mechanical properties, the thermal and stability, transparency, and surface properties are considered.

Ding and co-workers [36] exploited an exfoliation method via mechanical stirring to produce mono-layered montmorillonite nanosheets in polyethyl-phosphate glycol ester; the resulting material was incorporated in a WPU formulation at different loadings (namely, 1.5, 3.0, 4.5, and 6.0 wt.%). All the formulations were found stable for at least five months, hence revealing a long shelf-life. As assessed by forced-combustion tests performed at 35 kW/m^2 irradiative heat flux (Figure 8.2), the incorporation of exfoliated montmorillonite at the highest loading (i.e. 6.0 wt.%) into WPU allowed decreasing both peak of heat release rate and smoke production rate by 41% and 7%, respectively, as compared to the WPU counterpart containing 20 wt.% of polyethyl-phosphate glycol ester only. This finding was ascribed to the thermal shielding effect exerted by the exfoliated nanoclay lamellae, which, in turn, limited the intumescence of the flame-retarded WPUs, as shown by the residues after cone calorimetry tests (Figure 8.3).

In the seeking for the design of effective flame-retardant systems for WPUs, graphene oxide has been thoroughly investigated. Bernard and co-workers [37] prepared WPU nanocoatings containing different amounts of graphene oxide (namely, 0.4, 0.8, 1.2, and 2.0 wt.%). Apart from a remarkable overall improvement of the thermal stability, mechanical and barrier properties, and weathering resistance as well, the incorporation of the nanofiller with the highest loading was also beneficial for the decrease of heat release rate (−43%) with respect to the unfilled WPU.

Pursuing this research, Du and co-workers [29] exploited an in-situ method for functionalizing graphene oxide with urethane–silica; then, the modified nanofiller

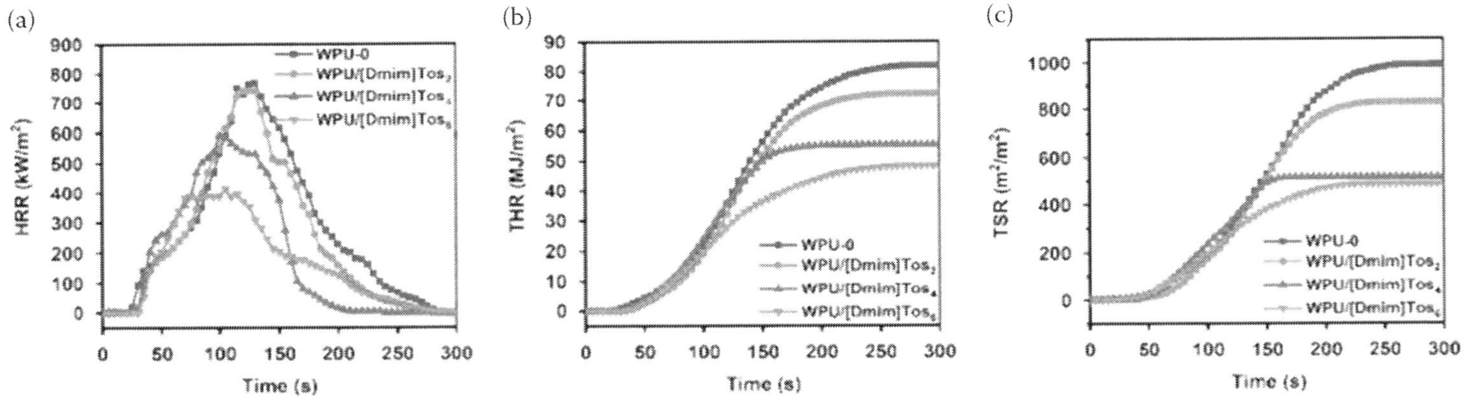

FIGURE 8.1(A) HRR, (b) THR, and (c) TSR curves of neat (WPU-0) and WPUs containing different amounts of ionic liquid flame retardant (WPU/[Dmim]Tos$_x$, where x represents the wt.% of the incorporated flame retardant). Reproduced with permission from reference [35]. Copyright (2020) Elsevier.

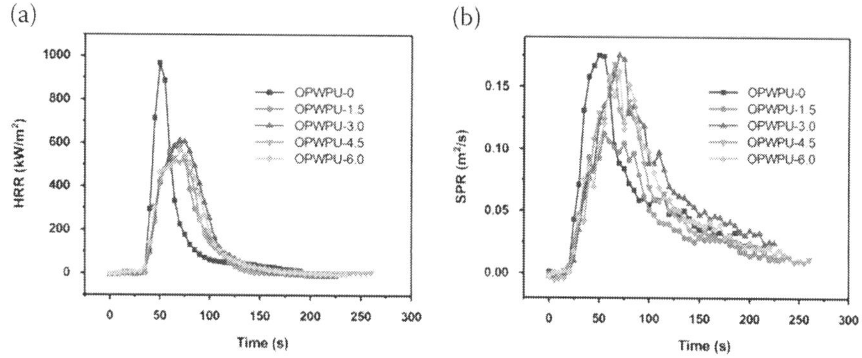

FIGURE 8.2 (A) HRR and (b) SPR curves of neat WPU (OPWPU-0, containing 20 wt.% of polyethyl-phosphate glycol ester) and WPUs containing different amounts of exfoliated montmorillonite (OPWPU-X, where X represents the wt.% of the incorporated montmorillonite). Reproduced with permission from reference [36]. Copyright (2020) Elsevier.

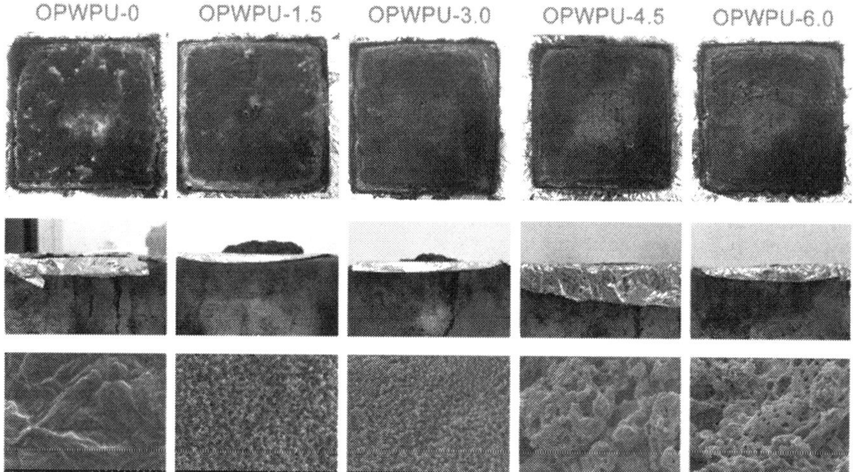

FIGURE 8.3 Photographs and SEM images of carbon residues collected after cone calorimetry tests of neat WPU (OPWPU-0, containing 20 wt.% of polyethyl-phosphate glycol ester) and WPUs containing different amounts of exfoliated montmorillonite (OPWPU-X, where X represents the wt.% of the incorporated montmorillonite). Reproduced with permission from reference [36]. Copyright (2020) Elsevier.

was incorporated in a WPU at different loadings (namely, 1, 2, and 3 wt.%). As reported in Table 8.1, 2 wt.% of functionalized graphene oxide was enough for providing acceptable flame-retardant features to WPU; besides, this formulation allowed achieving enhanced mechanical properties (either tensile strength or elongation at break increased), and water resistance.

He et al. [38] designed an environmentally friendly method for concurrently exfoliating and modifying black phosphorus (also known as black phosphorene)

TABLE 8.1
Limiting oxygen index values and vertical flame spread test rating for unfilled WPU and its nanocomposites

Sample	LOI (%)	UL-94 Rating
Unfilled WPU	17.8	Not Rated
WPU + 1 wt.% of unmodified graphene oxide	18.9	Not Rated
WPU + 1 wt.% of modified graphene oxide	20.6	Not Rated
WPU + 2 wt.% of modified graphene oxide	23.1	V-2
WPU + 3 wt.% of modified graphene oxide	23.5	V-2

through electrochemical cathodic exfoliation. More specifically, black phosphorus crystal behaved as the cathode and poly(ethylene imine) (positively charged) as either electrolyte or modifier, anchoring onto the surface of exfoliated black phosphorus nanosheets. Then, these latter were incorporated into a WPU at different loadings, namely 0.5, 1, and 2 wt.%. At the highest nanofiller content, as assessed by forced combustion tests performed at 35 kW/m^2 irradiative heat flux, peak of heat release rate, total heat release, and total smoke production, were remarkably lowered respectively by about 34%, 21%, and 22% as compared with the unfilled polymer matrix. Besides, the presence of the modified nanofiller at 2 wt.% loading turned out to enhance the tensile strength by about 71%, without impacting too much on the ductility of the polymer matrix (in fact, the elongation at break decreased by 5.6% only with respect to unfilled WPU).

Pursuing this research topic, Yin et al. [39] investigated possible synergistic effects provided by the combination of black phosphorene and hexagonal boron nitride nanosheets as flame retardants for WPUs. In particular, it was found that the concurrent presence of 0.2 wt.% of each filler promoted a significant increase of the limiting oxygen index values up to 33.8% (vs. 21.7% for unfilled polyurethane). Besides, as assessed by forced combustion tests carried out at 35 kW/m^2, peak of heat release rate and total heat release of the composite material were significantly lowered by about 51% and 24%, respectively, further demonstrating the effectiveness of the design flame-retardant system, which was found to be active both in condensed and gas phase.

Yang and Fu [34] succeeded in preparing superparamagnetic WPUs incorporating organic phosphorus-coated Fe_3O_4 and $Al(OH)_3$ nanoparticles, previously modified with sodium alendronate and therefore amino-functionalized. Three different nanoparticle loadings were employed, namely 10, 20, and 30 wt.%. The resulting nanocomposites containing 20 or 30 wt.% of filler turned out to be self-extinguishing and V-0 rated in vertical flame spread tests. Besides, LOI values for these composites were as high as 28.4% and 31.2%, respectively (vs. 20.6% for the unfilled WPU). Further, as assessed by cone calorimetry tests, the presence of the filler at the highest loading (i.e. 30 wt.%) allowed decreasing both peak of heat release rate and smoke production rate by about 58% and 64%, respectively, as compared to the unfilled polymer. These findings were ascribed to the dual physical barrier effect as a consequence of the formation of a nanoparticle network and of a char layer, which both slow down the heat and mass transfer phenomena during combustion. At variance, the obtained flame-retarded products lost their transparency and showed worsened thermal and mechanical features.

8.6 FLAME-RETARDANT WPUs USING THE REACTIVE APPROACH: RECENT ADVANCES

Wang and co-workers [40] synthesized pentaerythritol di-*N*-hydroxyethyl phosphamide, a reactive-type intumescent flame retardant, which was included in a WPU backbone. Then, the resulting flame-retarded WPU was further enhanced by the addition of tri (*N,N*-bis-(2-hydroxy-ethyl) acyloxoethyl) phosphate. A schematic illustration of the synthesis of the flame-retarded WPU is shown in Figure 8.4. The LOI of the final system achieved 25% (vs. 19.5% for neat polyurethane) and, as assessed by pyrolysis combustion flow calorimetry, the combination of the two synthesized molecules decreased peak of heat release rate and total heat release by 15.4% and 28%, respectively, as compared with the unfilled polymer. Finally, it was demonstrated that the designed flame retardant was active both in condensed (through the formation of a stable char) and gas (through the release of high amounts of nonflammable gases) phases.

Wu and co-workers [41] exploited the post-chain extension method for synthesizing flame-retardant WPUs bearing a phosphorus-containing diamine, namely, Bis (4-aminophenoxy) phenyl phosphine oxide. Figure 8.5 displays a schematic illustration of the synthesis procedure.

In particular, by increasing the loading of Bis (4-aminophenoxy) phenyl phosphine oxide, it was possible to prepare a number WPUs with different chain extension ratios, ranging from 20 to 100. The system containing the highest amount of Bis (4-aminophenoxy) phenyl phosphine oxide achieved very good fire retardant properties, as demonstrated by the V-0 rating in vertical flame spread tests, by the increase of LOI values, and by the remarkable decrease of the thermal parameters (peak of heat release rate and total heat release) observed during forced-combustion tests performed at 35 kW/m^2 irradiative heat flux (Table 8.2 and Figure 8.6). Besides, the char-forming character of the flame-retarded WPUs was witnessed by the increase of the residues at the end of the cone calorimetry tests.

Further, Zhang and co-workers [42] synthesized benzoguanamine spirocyclic pentaerythritol bisphosphonate, to be used as a chain extender in the preparation of flame-retarded WPUs. Conjugating 8 wt.% of the product with the polyurethane backbone was enough to achieve V-0 rating in vertical flame spread tests and a LOI of 27.3%.

In an extra research effort, the same group [28] prepared WPU composites using graphene oxide GO decorated with phosphazene and DOPO (i.e. 9,10-dihydro-9-oxa-10-phosphaphenanthrene-10-oxide)-based phosphonamidate; the so-obtained modified graphene oxide was covalently incorporated into WPU using an in situ polymerization approach. Three different nanofiller loadings were selected, namely

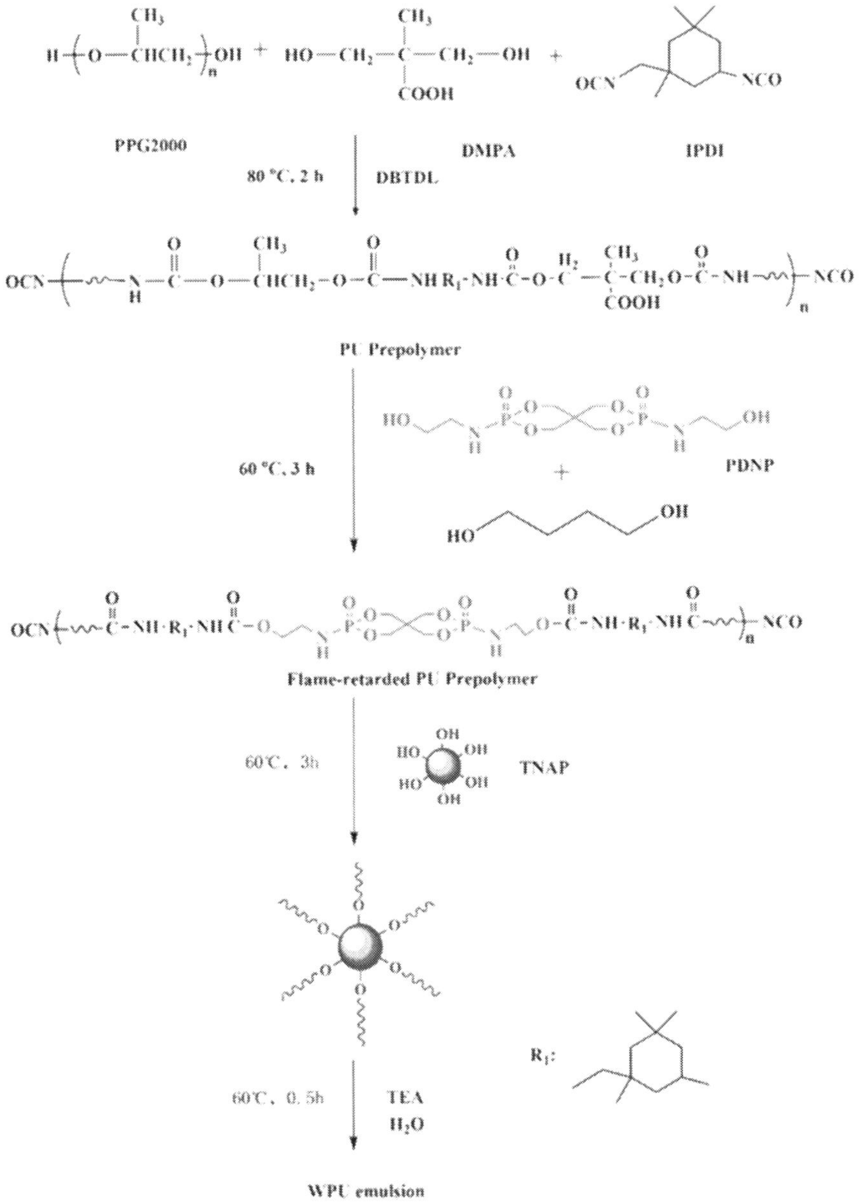

FIGURE 8.4 Scheme of the synthesis of the flame-retarded WPU. DBTDL = dibutyltin dilaurate; PDNP = pentaerythritol di-*N*-hydroxyethyl phosphamide; TNAP = tri(*N*,*N*-bis-(2-hydroxy-ethyl) acyloxoethyl) phosphate; TEA = triethylamine. Reproduced with permission from reference [40] Copyright (2019) Elsevier.

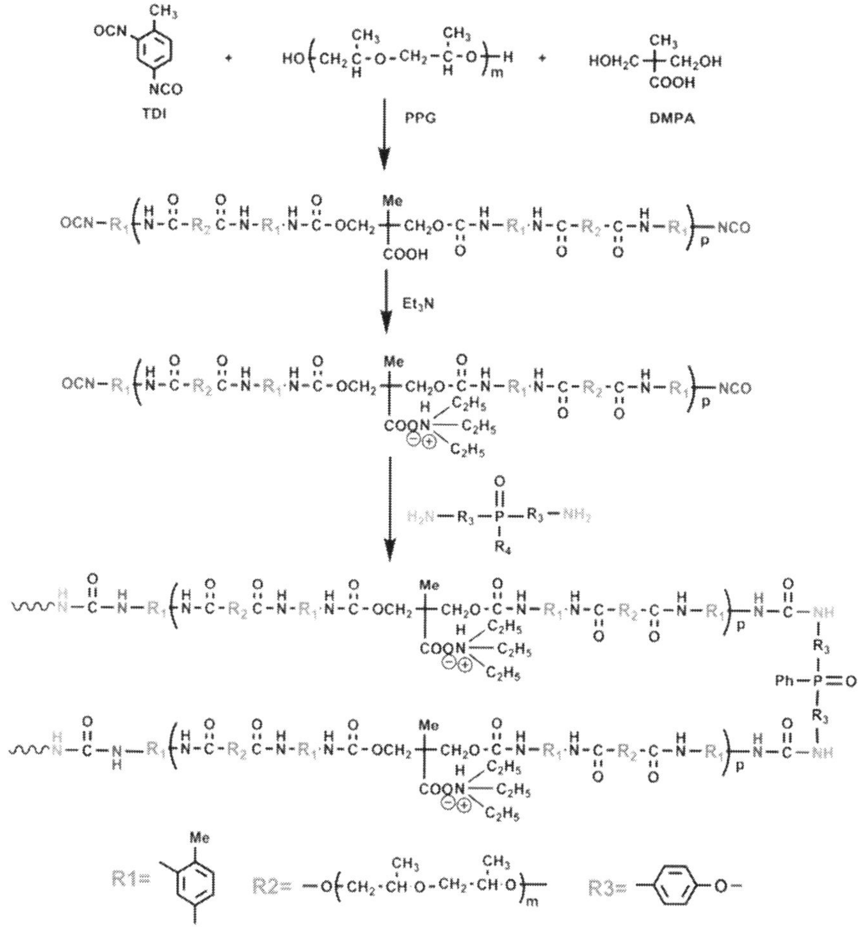

FIGURE 8.5 Scheme of the synthesis of the flame-retarded WPU with the post-chain extension of Bis (4-aminophenoxy) phenyl phosphine oxide. Et_3N = triethylamine. Reproduced with permission from reference [41] Copyright (2016) Elsevier.

0.5, 1.0, and 2.0 wt.%. Despite a limited effect on the overall flammability features of the polyurethane matrix, the modified graphene oxide was able to remarkably decrease either the total heat release or the peak of heat release by about 19% and 39%, respectively, as compared to the unfilled polymer counterpart.

Further, the nanocomposites containing modified graphene oxide showed a remarkable increase of the tensile strength (up to about 139% for the nanocomposite containing the highest loading of modified nanofiller) without significantly impacting the ductility.

An intumescent flame-retardant DOPO derivative, containing both nitrogen and phosphorus was synthesized by Wang et al. [18] and exploited as a reactive-type monomer for the preparation of flame-retardant WPUs; in particular, 3, 6, and 9 wt.% of DOPO derivative loadings were selected. As shown in Table 8.3, the presence of increasing amounts of the synthesized flame retardant promoted

TABLE 8.2
Vertical flame spread test rating and cone calorimetry data for neat WPU and for its flame-retarded modifications

Sample	LOI (%)	UL-94 Rating	Peak of Heat Release Rate (kW/m^2)	Total Heat Release (MJ/m^2)	Char Yield (%)
Neat WPU	24.1	No rating	1134	76	0
WPU modified with Bis(4-aminophenoxy) phenyl phosphine oxide (20 chain extension ratio)	26.4	No rating	953	72	3.7
WPU modified with Bis(4-aminophenoxy) phenyl phosphine oxide (40 chain extension ratio)	27.7	V-2	792	67	4.8
WPU modified with Bis(4-aminophenoxy) phenyl phosphine oxide (60 chain extension ratio)	28.2	V-2	661	61	7.0
WPU modified with Bis(4-aminophenoxy) phenyl phosphine oxide (80 chain extension ratio)	29.2	V-1	559	50	12
WPU modified with Bis(4-aminophenoxy) phenyl phosphine oxide (100 chain extension ratio)	30.1	V-0	486	44	20

an increase of the LOI values, as well as a remarkable decrease of the thermal and smoke parameters in cone calorimetry tests. Finally, it was proven that the synthesized flame-retarded WPUs were active both in condensed (through char formation) and gas (utilizing dilution effect) phases.

Pursuing this research, very recently, the same group [43] exploited an addition reaction between hydroxyl-terminated polybutadiene acrylonitrile and DOPO; the resulting product was employed as a reactive-type flame retardant in the formulation of WPUs at three concentrations, namely 0.5, 1.0, and 2.0 wt.%. The presence of increasing amounts of the DOPO derivative turned out to increase the LOI values (that achieved 26.8%—vs. 18.4% for neat WPU—when 2.0 wt.% of modified DOPO was employed), as well as to enhance the vertical flame spread rating (from not rated, for neat WPU, up to V1, for the counterpart containing 2.0 wt.% of DOPO derivative). Besides, as assessed by forced combustion tests under 35 kW/m^2 irradiative heat flux, the proposed flame-retardant strategy was very effective in decreasing either the thermal or the smoke parameters, as presented in Table 8.4.

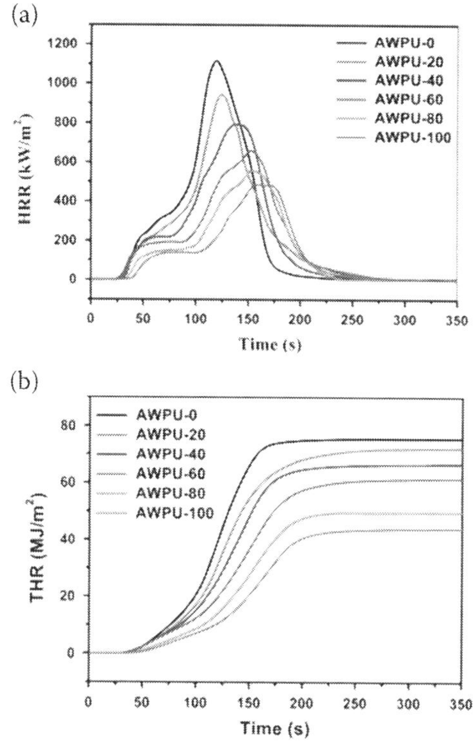

FIGURE 8.6 (A) Heat release rate (HRR) and (b) total release rate (THR) curves vs. time for flame-retarded WPUs with various chain extension ratios (irradiative heat flux: 35 kW/m^2). AWPU-0= neat polyurethane; AWPU-XX= flame-retarded polyurethane, where XX represents the chain extension ratio. Reproduced with permission from reference [41]. Copyright (2016) Elsevier.

TABLE 8.3

Limiting Oxygen Index values and cone calorimetry data for neat WPU and its flame-retarded modifications (WPU-X, where X represents the wt.% of reactive DOPO derivative)

Sample	LOI (%)	Time to Ignition (s)	Peak of Heat Release Rate (kW/m^2)	Total Heat Release (MJ/m^2)	Total Smoke Production (m^2/m^2)	Char Yield (%)
Neat WPU	18.4	35	976	64.5	7.8	1.6
WPU-3	23.2	37	949	50.1	6.5	2.3
WPU-6	27.4	39	797	47.6	6.1	5.2
WPU-9	31.0	40	577	42.4	2.6	7.4

TABLE 8.4
Cone calorimetry data for neat WPU and its flame-retarded modifications (WPU-X, where X represents the wt.% of reactive DOPO derivative)

Sample	Peak of Heat Release Rate (kW/m^2)	ΔPeak of Heat Release Rate (%)	Total Heat Release (MJ/m^2)	ΔTotal Heat Release (%)	Total Smoke Release (m^2/m^2)	Smoke Production Rate (m^2/s)	Char Yield (%)
Neat WPU	986	–	51.6	–	9.42	0.1689	0.91
WPU-0.5	791	−19.8	47.3	−8.3	8.99	0.1563	1.54
WPU-1.0	705	−28.5	41.7	−19.2	8.11	0.1479	3.12
WPU-2.0	667	−32.4	39.0	−24.4	7.44	0.1293	3.83

An interesting paper from Yin and co-workers [44] studied the possibility of working on the curing agent (being modified on purpose) in two-component WPU systems. In particular, the authors synthesized a hydrophilic curing agent combining polyethylene glycol monomethyl ether, hexamethylene diisocyanate trimer, and 2-((2-aminoethyl)amino)-ethanesulfonic acid monosodium salt. The resulting product allowed achieving a V-0 rating in vertical flame spread tests and LOI as high as 29.4%, hence showing an acceptable flame-retardant behavior.

Very recently, the up-to-date search for low environmental impact products has pushed the scientific community toward the synthesis, processing, and application of "green" flame-retardant WPUs.

In this context, some efforts toward the design of more environmentally friendly flame-retarded WPUs were made by Tabatabaee and co-workers [45]. In pursuit of this aim, a series of flame-retarded self-curable silanized WPUs was synthesized, using a commercially available oligomeric phosphonate diol (namely, Exolit® OP560), which was, together with a bio-based polyol (i.e. castor oil, aimed at increasing water resistance and hydrophobicity), exploited as a reactive-type flame retardant in the polyurethane formulation. The presence of just 8.7 wt.% of Exolit® OP560 in the WPU recipe allowed achieving V-0 rating in vertical flame spread tests, and a LOI value as high as 30.4% (vs. 26.8% for neat WPU, which, conversely, exhibited dripping phenomena). Furthermore, as assessed by pyrolysis combustion flow calorimetry, the same formulation decreased total heat release and heat release capacity by, respectively, about 6% and 32% compared to unfilled WPU. The designed flame-retarded WPUs were proven to be active both in condensed (through char formation) and gas (exploiting a dilution effect) phases.

8.7 CONCLUSIONS AND FUTURE PERSPECTIVES

At present, WPUs undoubtedly represent a very interesting class of emerging polymers within the chemistry of polyurethanes, not only because of their low environmental impact, but also for their properties that can be quite easily tailored according to the specific application they are designed for. Conversely, their broad uses can be limited severely, as they effortlessly burn when exposed to an irradiative heat flux or a flame.

This chapter has provided some clear examples of the current methodologies suitable for enhancing the fire behavior of WPUs, i.e. the reactive-type flame-retardant strategy (which exploits an "intrinsic" flame-retardant approach) and the additive approach, for which the flame retardants are incorporated into WPUs by using physical means.

Both the methodologies possess advantages and drawbacks, which at present are requiring additional research studies and developments, aiming at further improving not only the overall fire behavior of the flame-retarded products but also at minimizing the side effects on the other polymer features, particularly referring to its mechanical behavior and thermal stability.

The examples that have been elucidated in the chapter show that both the strategies, at least at an academic research level, can be successfully pursued, and in many cases, it was possible to get a good balance between the flame retardance and the other features, thus making possible the utilization of the modified WPUs for different advanced application fields.

In conclusion, despite the current limitations, some progress in the design and development of new effective flame-retardant WPUs can be expected, for the next years, even toward an improved sustainability.

REFERENCES

1. Liu SG, Zhu GQ (2007) Effects of altered reaction conditions on the synthesis of polyurethane-poly(2,2,2-trifluoroethyl methacrylate) triblock copolymer aqueous dispersion *Eur. Polym. J.* 43:3904–3911.
2. Prisacariu C (2011) *Polyurethane Elastomers: From Morphology to Mechanical Aspects*, 1st ed., Springer, New York.
3. Anderle GA, Lenhard SL, Lubnin AV, Snow GE, Tamareselvy K (2003) Plasticized waterborne polyurethane dispersions and manufacturing process. US Patent No. 6576702.
4. Nanda AK, Wicks DA, Madbouly SA, Otaigbe JU (2005) Effect of ionic content, solid content, degree of neutralization, and chain extension on aqueous polyurethane dispersions prepared by prepolymer method. *J. Appl. Polym. Sci.* 98:2514–2520.
5. Fang CQ, Zhou X., Yu Q., Liu SL, Guo DG, Yu RE, Hu JB (2014) Synthesis and characterization of low crystalline waterborne polyurethane for potential application in water-based ink binder. *Prog. Org. Coat.* 77:61–71.
6. Anıl D, Berksun E, Durmusx-Sayar A, Billur Sevinisx-Özbulut E, Ünal S (2020) *Recent advances in waterborne polyurethanes and their nanoparticle-containing dispersions*. In: *Handbook of Waterborne Coatings,* Elsevier, Cambridge, pp. 249–302.
7. Madbouly SA, Otaigbe JU (2009) Recent advances in synthesis, characterization and rheological properties of polyurethanes and POSS/polyurethane nanocomposites dispersions and films. *Prog. Polym. Sci.* 34:1283–1332.
8. Chattopadhyay DK, Webster DC (2009) Thermal stability and flame retardancy of polyurethanes. *Prog. Polym. Sci.* 34:1068–1133.
9. Scaiano JC (1989) Laser photolysis in polymer chemistry. In: *Degradation and Stabilization of Polymers*, 2nd ed., Elsevier, Amsterdam, pp. 1–43.

10. Petrovic ZS, Zavargo Z, Flynn JH, Macknight WJ (1994) Thermal degradation of segmented polyurethanes. *J. Appl. Polym. Sci.* 51:1087–1095.
11. Liu J, Ma D (2002) Study on synthesis and thermal properties of polyurethane–imide copolymers with multiple hard segments. *J. Appl. Polym. Sci.* 84:2206–2215.
12. Li Y, Gao T, Linliu K, Desper CR, Chu B (1992) Multiphase structure of a segmented polyurethane: effects of temperature and annealing. *Macromolecules* 25:7365–7732.
13. Blackwell J, Nagarajan MR, Haitink TB (1982) Structure of polyurethane elastomers: effect of chain extender length on the structure of MDI/diol hard segments. *Polymer* 23:950–956.
14. Blackwell J, Quay JR, Nagarajan MR, Born L, Hespe H (1984) Molecular parameters for the prediction of polyurethane structures. *J. Polym. Sci. Polym. Phys.* 22: 1247–1259.
15. Chuang FS, TsenWC, Shu YC (2004) The effect of different siloxane chain extenders on the thermal degradation and stability of segmented polyurethanes. *Polym. Degrad. Stabil.* 84:69–77.
16. Kendagannaswamy BK, Siddaramaiah (2002) Chain-extended polyurethanes–synthesis and characterization. *J. Appl. Polym. Sci.* 84:359–369.
17. Sarkar S, Adhikari B (2001) Thermal stability of lignin–hydroxyterminated polybutadiene copolyurethanes. *Polym. Degrad. Stabil.* 73:169–175.
18. Wang H, Wang S, Du X, Wang H, Cheng X, Du Z (2019) Synthesis of a novel flame retardant based on DOPO derivatives and its application in waterborne polyurethane. *RSC Adv.* 2019, 9, 7411–7419.
19. Cullis CF, Hirschler MM (1981) *The Combustion of Organic Polymers*, Clarendon Press, Oxford.
20. Buszard DL, Dellar RJ (1984) The performance of flame retardants in rigid polyurethane foam formulations. In: *Fire and Cellular Polymers*, Elsevier, London.
21. Buszard DL, Dellar RJ (1985) The performance of flame retardants in rigid polyurethane foam formulations. *Cell. Polym.* 4:431–443.
22. Levchik SV, Weil ED (2004) Thermal decomposition, combustion and fire-retardancy of polyurethanes—a review of the recent literature. *Polym. Int.* 53: 1585–1610.
23. Duquesne S, Le Bras M, Bourbigot S, Delobel R, Poutch F, Camino G, Eling B, Lindsay C, Roels, T (2000) Analysis of fire gases released from polyurethane and fire-retarded polyurethane coatings. *J. Fire Sci.* 18:456–482.
24. Wang S, Du X, Jiang Y, Xu J., Zhou M, Wang H, Cheng X, Du Z (2019) Synergetic enhancement of mechanical and fire-resistance performance of waterborne polyurethane by introducing two kinds of phosphorus-nitrogen flame retardant, *J. Colloid Interface Sci.* 537:197–205.
25. Sun Y, Liu C, Hong Y, Liu R, Zhou X (2019) Synthesis and application of self-crosslinking and flame retardant waterborne polyurethane as fabric coating agent. *Prog. Org. Coat.* 137:105323.
26. Wu L, Guo J, Zhao S (2017) Flame-retardant and crosslinking modification of MDI-based waterborne polyurethane. *Polym. Bull.* 74:2099–2116.
27. Neisius M, Liang S, Mispreuve H, Gaan S (2013) Phosphoramidate-containing flame-retardant flexible polyurethane foams. *Ind. Eng. Chem. Res.* 52:9752–9762.
28. Zhang P, Xu P, Fan H, Sun Z, Wen J (2019) Covalently functionalized graphene towards molecular-level dispersed waterborne polyurethane nanocomposite with balanced comprehensive performance. *Appl. Surf. Sci.* 471:595–606.
29. Du W, Jin Y, Lai S, Shi L, Shen Y, Pan J (2019) Urethane-silica functionalized graphene oxide for enhancing mechanical property and fire safety of waterborne polyurethane composites. *Appl. Surf. Sci.* 492:298–308.

30. Ranganathan T, Zilberman J, Farris RJ, Coughlin EB, Emrick T (2006) Synthesis and characterization of halogen-free antiflammable polyphosphonates containing 4,4′-bishydroxydeoxybenzoin. *Macromolecules.* 39:5974–5975.
31. Zhang P, Tian S, Fan H, Chen Y, Yan J (2015) Flame retardancy and hydrolysis resistance of waterborne polyurethane bearing organophosphate moieties lateral chain. *Prog. Org. Coat.* 89:170–180.
32. Wang S, Du Z, Cheng X, Liu Y, Wang H (2018) Synthesis of a phosphorus- and nitrogen-containing flame retardant and evaluation of its application in waterborne polyurethane. *J. Appl. Polym. Sci.* 135:46093.
33. Wu G, Li J, Chai C, Ge Z, Lin J, Luo Y (2015) Synthesis and characterization of novel post-chain extension flame retardant waterborne polyurethane. *RSC Adv.* 5:97710–97719.
34. Yang L, Fu HQ (2020) Bisphosphoric modified amino functional [xFe_3O_4-2xAl $(OH)_3$]/ waterborne polyurethane nanocomposite with superparamagnetism and flame retardancy. *Polym. Adv. Technol.* 31:338–349.
35. Zhao Z, Guo D-M, Fu T, Wang X-L, Wang Y-Z (2020) A highly-effective ionic liquid flame retardant towards fire-safety waterborne polyurethane (WPU) with excellent comprehensive performance. *Polymer* 205:122780.
36. Ding Z, Li J, Xin W, Luo Y (2020) Facile and high-concentration exfoliation of montmorillonite into monolayered nanosheets and application in multifunctional waterborne polyurethane coating. *Appl. Clay Sci.* 198:105798.
37. Bernard C, Goodwin Jr. DG, Gu X, Celina M, Nyden M, Jacobs D, Sung L, Nguyen T (2020). Graphene oxide/waterborne polyurethane nanocoatings: effects of graphene oxide content on performance properties. *J. Coat. Technol. Res.* 2020:255–269.
38. He L, Zhou X, Cai W, Xiao Y, Chu F, Mu X, Fu X, Hu Y, Song L (2020) Electrochemical exfoliation and functionalization of black phosphorene to enhance mechanical properties and flame retardancy of waterborne polyurethane. *Compos. B* 202:108446.
39. Yin S, Ren X, Lian P, Zhu Y, Mei Y (2020) Synergistic effects of black phosphorus/boron nitride nanosheets on enhancing the flame-retardant properties of waterborne polyurethane and its flame-retardant mechanism. *Polymers* 12:1487.
40. Wang S, Du X, Jiang Y, Xu J, Zhou M, Wang H, Cheng X, Du Z (2019) Synergetic enhancement of mechanical and fire-resistance performance of waterborne polyurethane by introducing two kinds of phosphorus–nitrogen flame retardant. *J. Coll. Interface Sci.* 537:197–205.
41. Wu G, Li J, Luo Y (2016) Flame retardancy and thermal degradation mechanism of a novel post-chain extension flame retardant waterborne polyurethane. *Polym. Degrad. Stabil.* 123:36–46.
42. Zhang P, He Y, Tian S, Fan H, Chen Y, Yan J (2017) Flame retardancy, mechanical, and thermal properties of waterborne polyurethane conjugated with a novel phosphorous-nitrogen intumescent flame retardant. *Polym. Compos.* 38:452–462.
43. Wang H, Wang S, Du X, Du Z, Wang H, Cheng X (2020) A novel DOPO-containing HTBN endowing waterborne polyurethane with excellent flame retardance and mechanical properties. *J. Appl. Polym. Sci.* 137:49368.
44. Yin X, Dong C, Luo Y (2107) Effects of hydrophilic groups of curing agents on the properties of flame-retardant two-component waterborne coatings. *Colloid Polym. Sci.* 295:2423–2431.
45. Tabatabaee F, Khorasani M, Ebrahimi M, González A, Irusta L, Sardon H (2019) Synthesis and comprehensive study on industrially relevant flame retardant waterborne polyurethanes based on phosphorus chemistry. *Prog. Org. Coat.* 131:397–406.

9 Synthesis, Characterization, and Applications of Smart Waterborne Polyurethanes

Ronglan Wu
Key Laboratory of Oil and Gas Fine Chemicals, Ministry of Education and Xinjiang Uyghur Autonomous Region, School of Chemical Engineering, Xinjiang University, 830046 Urumqi, Xinjiang, People's Republic of China

Wei Wang
Department of Chemistry & Center for Pharmacy, University of Bergen, 5020 Bergen, Norway

CONTENTS

9.1 Introduction 138
9.2 Synthetic Methods of SWPUs 138
 9.2.1 Acetone Process 139
 9.2.2 Pre-polymer Mixing Process 140
 9.2.3 Melt Dispersion Process 141
 9.2.4 Ketone Imide Process (Ketimine and Ketzaine Process) 142
9.3 Characterization of SWPUs 142
 9.3.1 Hydrogen Bonding Characterized by FTIR in WPUs 142
 9.3.2 Microphase Separation Characterized by DSC and DMA 144
 9.3.3 Morphology of WPU Characterized by TEM, AFM, and SEM 144
9.4 Applications of SWPUs 146
 9.4.1 Anti-Fouling WPUs 146
 9.4.2 Self-Healing WPUs 148
 9.4.3 WPU Phase Change Materials 149
 9.4.4 Shape Memory WPU 150
9.5 Conclusion 151
References 152

DOI: 10.1201/9781003173526-9

9.1 INTRODUCTION

Waterborne polyurethane (WPU), also called water-dispersing polyurethane and water-based polyurethane, is a new type of polyurethane (PU) resin with water as the dispersing medium. The synthesis of WPU requires water as a solvent, where organic solvents are either not used or used in only small quantities, which makes WPUs an environmentally friendly material. Thus, the current tendency to apply stricter environmental regulations in the chemical industry looks favorably over the synthesis of WPUs. Furthermore, WPUs exhibit properties that could be as good as those offered by other solvent-based PUs, e.g. good mechanical strength, biocompatibility, and the ability to be easily modified. Therefore, WPUs have attracted extensive attention in both scientific research and industrial applications [1].

In response to the current demand for new functional polymer materials, researchers have developed many WPUs with special physicochemical properties. Thus, functional WPUs have attracted great interest because of their unique biological, electrical, optical, and other physicochemical properties. The WPU-related technologies aim to develop new synthetic methods for WPUs, realize the functional improvement of PU products, and expand the application of smart waterborne polyurethanes (SWPUs). Here, we present the progress in synthesis, characterization, and application of SWPUs.

9.2 SYNTHETIC METHODS OF SWPUS

WPUs are block polymers that are comprised of alternate flexible and rigid chain segments that are dispersed in an aqueous phase. The main chain of WPU is formed by the polymerization of oligomer polyols (e.g. polyether polyols and polyester polyols), diisocyanates (e.g. toluene diisocyanate, and hexamethylene diisocyanate), and small molecule chain extenders (e.g. diamine or dialcohol). The high mechanical strength and rigidity of WPUs are brought about by rigid chain segments that are made up of isocyanate and extenders with a short chain. The good deformability of WPUs is due to flexible chain segments that are composed of flexible long-chain oligomer polyols.

The synthetic strategies of WPUs are divided into two methods, namely (1) external emulsification method and (2) self-emulsification method. External emulsification method refers to a method in which the pre-polymer chains contain no or only a small amount of hydrophilic chain segments that are insufficient to self-emulsify. In this case, the addition of external emulsifiers is necessary for the formation of an emulsion. In the self-emulsification method, conversely, there is no need for adding emulsifiers, because hydrophilic groups are directly introduced into the chain of PU, making it possible to form stable emulsions of PUs.

The common methods used for the preparation of self-emulsifying PUs include the acetone process method, pre-polymer dispersion method, melt dispersion method, and ketone imide method. In addition to these methods, recently several new synthetic methods have also been developed, e.g. homogeneous solution polymerization (HSP) [2], microemulsion polymerization (MEPP) [3], reversible addition–fragmentation chain-transfer polymerization (RAFT) [4], and atom transfer radical polymerization (ATRP) [5].

9.2.1 Acetone Process

The acetone process was first invented by Bayer in Germany. A typical preparation process by this method is shown in Figure 9.1 [6]. In the acetone process, first high viscosity PU pre-polymers with –NCO– as the end groups are prepared. Afterward, the PU pre-polymer is diluted and dissolved within organic solvents, where it is subjected to a chain expansion reaction. After the polymerization, water is added to the reactants to emulsify PU. Finally, the organic solvents are evaporated, resulting in the formation of emulsion droplets.

The fact that acetone is often used for this type of synthesis relies on its intrinsic properties, such as, miscibility with water, low boiling point, and being inert during the reaction. Therefore, the method is named as "acetone process". There are certain advantages of the "acetone process", such as easy production control, high

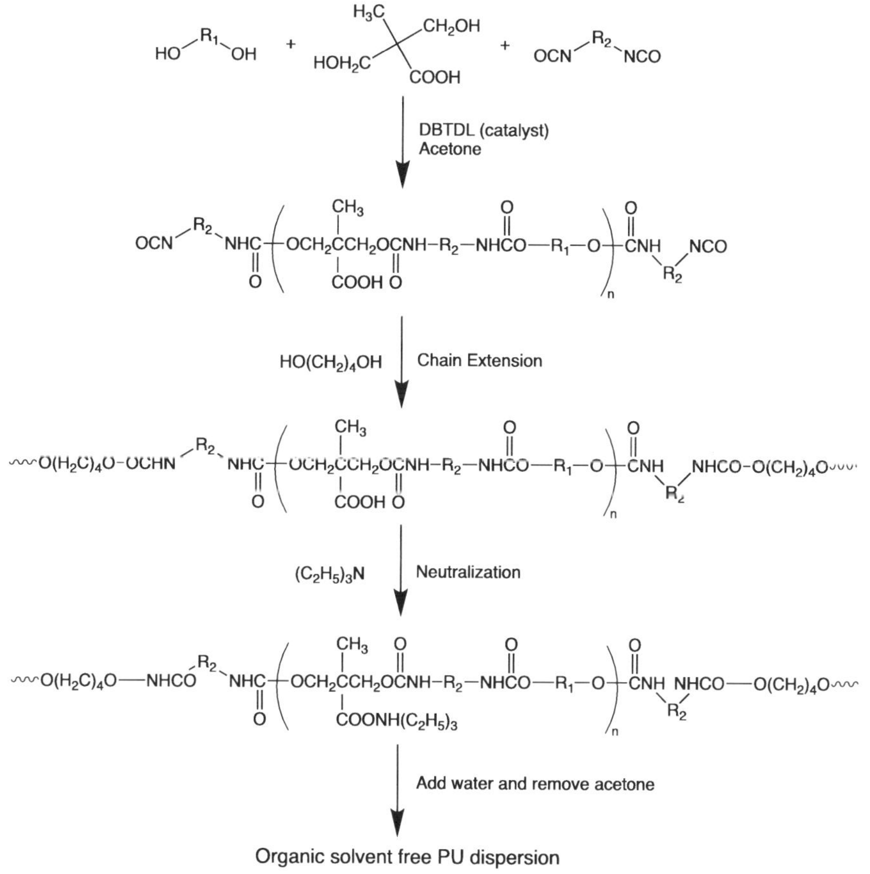

FIGURE 9.1 The reaction route for the preparation of PU dispersions by acetone process.

reproducibility, good product quality, and wide applicability. However, the process is relatively complex and consumes large amounts of organic solvents, which increases cost and brings in safety concerns.

9.2.2 Pre-polymer Mixing Process

The typical synthetic steps of the pre-polymer mixing process are presented in Figure 9.2 [7]. The pre-polymer mixing process often includes three steps. In the first step, the pre-polymer of PU is synthesized. In the second step, hydrophilic components are introduced into the pre-polymer segments. In this step, the viscosity of the pre-polymer must be adjusted under high-speed stirring. In the third step, the emulsification of PU particles and chain extension reaction occurs simultaneously through the addition of an aqueous solution that contains diamine(s).

The pre-polymer mixing method is suitable for low viscosity PU products, i.e. low molecular weight pre-polymers prepared from aliphatic and aliphatic polyisocyanates. The synthetic procedure is simple and the cost is low, making it convenient for continuous industrial production. A good example of the application of this process is provided by Saetung et al., where synthesis of natural-rubber-based WPUs has been demonstrated [8]. In this study, a series of WPUs was prepared from a renewable source, hydroxytelechelic natural rubber (HTNR), with

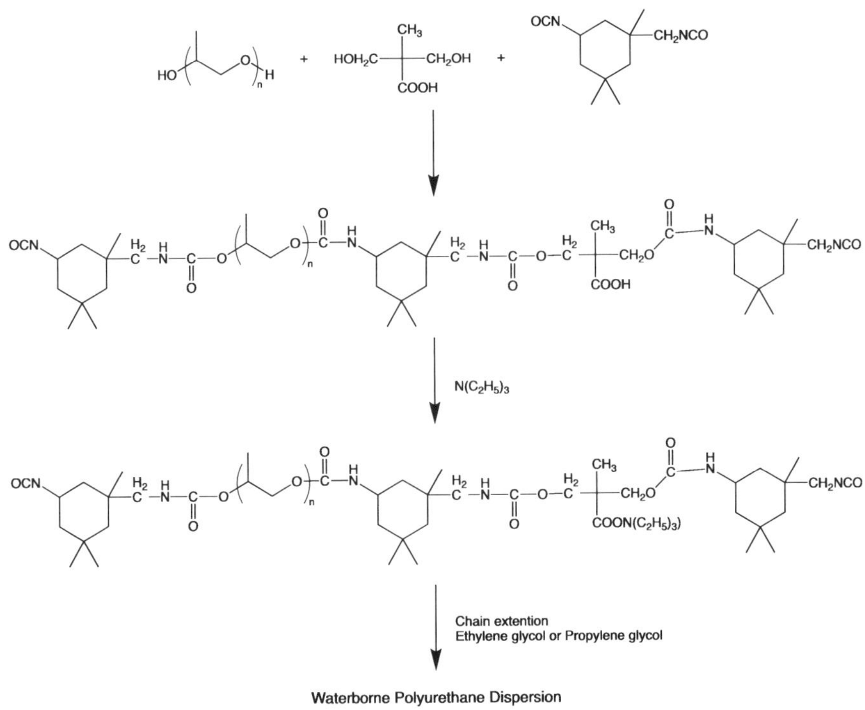

FIGURE 9.2 The reaction route for water-based polyurethane dispersion via the pre-polymer mixing process.

different amounts of dimethylol propionic acid (DMPA, 1.6–8.4 wt.%), and different molecular weights (1000–4000 g mol^{-1}), and levels of epoxide (0%–20%) of HTNR. The results indicated that the particle size of the samples decreased with an increase in the molecular weight of DMPA or HTNR. Also, increasing the content of DMPA, hard segments or epoxy segments resulted in an increase in the water uptake. The increase of DMPA content had no effect on the glass transition temperature (T_g) of WPU films.

The first steps in the acetone process and the pre-polymer mixing process are very similar. However, the two methods are noticeably different. In the acetone process, the PU pre-polymer often has a large molecular weight and a large viscosity, and the addition of solvent is necessary for diluting it. Conversely, in the pre-polymer mixing process, it is essential that the viscosity is low, because if otherwise, the dispersion step would be negatively affected. Due to the low molecular weight of pre-polymer in the latter, the method avoids the use of large amounts of solvent.

9.2.3 Melt Dispersion Process

An example of melt dispersion process is presented in Figure 9.3 [9]. In the melt dispersion process, first the polyester polyols or polyether polyols react with the tertiary amine groups and end-NCO-based groups of the pre-polymer followed

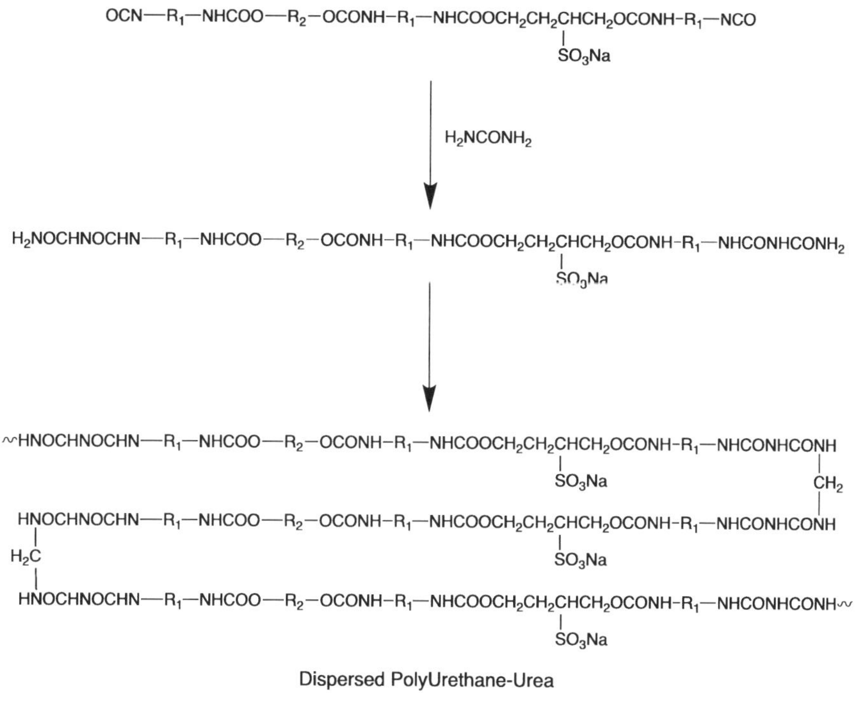

FIGURE 9.3 The synthetic route of water-based PU by the hot-melt process.

by reactions with urea to form PU biuret oligomer. Finally, the hydrophilic bis-biuret reacts with formaldehyde (hydroxymethylation). In this step, the product is in the molten state, and the molten PU is dispersed in water to form an emulsion. The melt dispersion method does not need organic solvents. The process is simple and easy to control and does not need special equipment. Therefore, the method can be easily adapted to industrial production making it promising for such a purpose. The only disadvantage of the method is that the reaction temperature is high, which makes it prone to side reactions.

9.2.4 Ketone Imide Process (Ketimine and Ketzaine Process)

The synthetic route of the ketone imide process is presented in Figure 9.4 [9]. The method is also called the Ketimine and Ketzaine Process. In this method, PU pre-polymer is mixed with a diamine-protected ketone and then is dispersed in water. The ketone imide undergoes hydrolysis and releases the free diamines and dispersed polymer particles to produce WPUs. The ketamine process is similar to the pre-polymer mixing process except that in the latter process diamine blocks are used as potential chain extenders. Under normal preparation conditions, ketone imide is almost inert to isocyanate. It hydrolyzes in contact with water to release diamines, which is followed by reactions with –NCO-based pre-polymers. This means that chain extension process and dispersion process are carried out together. Despite being simple, controlling the reaction process is not easy in this method.

9.3 CHARACTERIZATION OF SWPUS

Some methods commonly used for the characterization of morphology, structure, and properties of WPUs are described in this section together with examples from their applications in characterization of SWPUs. These methods include Fourier-transform infrared spectroscopy (IR), differential scanning calorimetry (DSC), dynamic thermomechanical analysis (DMA), X-ray diffraction (XRD), transition electron microscopy (TEM), scanning electron microscopy (SEM), and atomic force microscopy (AFM).

9.3.1 Hydrogen Bonding Characterized by FTIR in WPUs

FTIR is often used to determine vibration of chemical bonds in the wavenumber region of 400–4,000 cm^{-1}. This method has been widely used in studying the hydrogen bonding and phase behavior in PU systems, due to its extreme sensitivity to the hydrogen bond structure of substances. In WPUs, hydrogen bonding is ubiquitous and important, because the physical crosslinking network formed by it can effectively enhance the mechanical properties of WPUs. The formation of hydrogen bonds between hard segments contributes to the formation of the PU microphase. The –NH in carbamate (or allophanate) acts as an electron-donating group to form hydrogen bonding with electron-withdrawing carbonyl and ether bonds. The strength of hydrogen bonding varies between carbamate–polyether, carbamate–polyester ester, and carbamate–carbamate, and is reported to be 23.6,

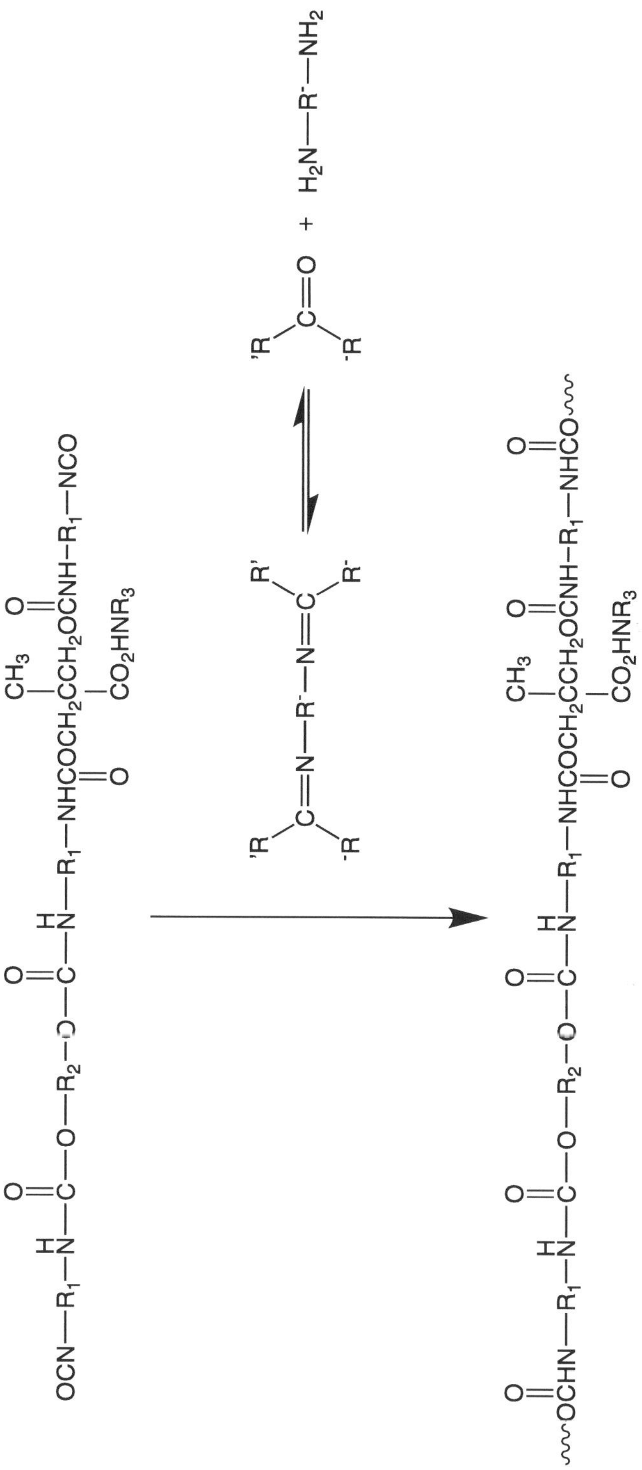

FIGURE 9.4 The synthetic route of WPU by the Ketimine-Ketazine process.

25.6, and 46.5 kJ/mol, respectively. The examples of using FTIR to analyze the change of hydrogen bonding, and consequently the phase transition in WPU systems, include the following. Guo et al. studied the change of hydrogen bonding before and after tensile fracture by FTIR [10]. Niemczyk et al. calculated the degree of phase separation of hard and soft segments of polycarbonate-based PU by monitoring the newly formed hydrogen bonding and dipole interaction through infrared spectral peak processing [11]. Lee et al. studied the change in hydrogen bonding of PU using variable-temperature infrared spectroscopy to obtain the dynamics of PU microphase separation [12].

9.3.2 Microphase Separation Characterized by DSC and DMA

DSC is a commonly used method to characterize microphase separation of polymer materials. The glass transition temperature (T_g) of the chain segment in the polymer can be determined by DSC. Generally speaking, PU exhibits two T_g values, one for soft segments ($T_{g,s}$) and one for hard segments ($T_{g,h}$). The difference between the two ($\Delta T_g = T_{g,h} - \mathrm{T}_{g,s}$) is often used as a measure of the degree of microphase separation of PU [13]. DSC can also be used for characterizing the phase separation of multi-component acrylic resins prepared by blending or copolymerization. Velankar and Cooper synthesized three series of PU that were anchored by catenarian trimethylammonium groups, which resulted from the quaternization of methane iodide [14]. DSC analysis showed that these materials exhibited different degrees of microphase separation, especially with the increase in the quaternization of the hard segments. Thus, a higher T_g of the hard segments induced a greater extent of microphase separation. Li and Shan employed DSC to study the effects of soft segment composition, DMPA content, and different –NCO/OH ratios on the microphase separation [15]. They found that the extent of microphase separation was higher in polyether-type WPUs than in polyester-type WPUs, which was consistent with earlier studies. [16] This indicates that increasing the content of DMPA and –NCO/OH ratio is beneficial to the microphase separation of WPU.

DMA is one of the effective methods to measure the T_g of polymers [17]. Similar to DSC, DMA is often used in the analysis of WPU microphase separation. In addition, DMA can be used to understand the viscoelastic behavior of WPUs, to analyze the interaction force between the hard and soft segments, and to evaluate the microphase separation within WPUs. The DMA curve measured by Comez et al. showed the dependency of the dissipation factor (tanδ) and the dynamic modulus of PU samples on the temperature [18]. When the temperature was increased, the modulus of PU decreased, where the rate of decrease reflected the degree of microphase separation in WPU. The higher the decline rate was, the less the extent of the interaction between the hard segment and the soft segment, and the greater the degree of microphase separation within WPU.

9.3.3 Morphology of WPU Characterized by TEM, AFM, and SEM

TEM is a microscopic imaging method used for studying both crystalline and amorphous WPU systems [19]. Koutsky et al. provided the first TEM images of the PU microphase separated structure, from which, it was found that the tiny segregated domains were evenly distributed in the matrix [20]. Since the extent of incompatibility between the polymer segments was large, microphase-separated domains could be readily observed using TEM. Zhang et al. used TEM to study the cross-section of WPUs prepared by polysulfide rubber. A rough surface for soft segments was indicative of the existence of phase separation between soft and hard segments. Through analyzing TEM images, the authors concluded that the introduction of polysulfide rubber reduced the hydrogen bonding between the soft and hard segments, thus enhancing the extent of microphase separation [21].

AFM is used for analyzing the size and morphology of materials in three dimensions and studying the conformation, the status of the order, and aggregation of polymer chains. Phase diagrams prepared by AFM can represent the areas of soft and hard segments. In the AFM tap mode, the hard segment regions appear in AFM images as light regions, whereas the dark regions represent the soft segment phase. Comparing the height diagrams with phase diagrams obtained by AFM, Schon et al. found that the phase diagram can better reveal the details of the phase structure of PU and clarify the microphase separated structure [22]. Yong et al. prepared a novel solvent-free WPU resin using sulfonic acid and carboxylic acid as chain extenders [23]. The results showed that the surface of the matte polymer film was extremely rough, and consisted of several spherical particles with diameters ranging from 0.8 to 3.0 μm scattered on the surface of the film. Li et al. successfully prepared a kind of low gloss WPU [24]. AFM observation of the three-dimensional morphology of WPU films showed that a large number of particles were widely distributed on the surface of WPU films (Figure 9.5), which was very important for the formation of low gloss surface morphology. In addition, the particle density was about $2.5 \times 10^5/mm^2$, which reflected the surface roughness of the film.

SEM is another technique utilized for observing the microstructure of WPU, which can be directly used to image the morphology and the structure of the sample surface. Using SEM, the surface roughness of the film can be determined. For WPU films, the surface morphology can be observed after spraying or sputtering a layer of metal film on the dry surface [25]. Yong et al. synthesized a series of low gloss waterborne polyurethane acrylate (WPUA) hybrid emulsions using soft and hard acrylate monomers with different weight ratios. SEM images showed that the surfaces of WPU films were uniform and smooth, whereas the surface of other WPUA films exhibited unevenness, indicative of the heterogeneity of the structure. In addition, by increasing the weight ratio of hard/soft acrylic monomers, the roughness of WPUA film increased. SEM equipped with energy dispersive X-ray energy dispersive spectroscopy (EDS) detectors can be used to study changes in surface chemical composition and its distribution to determine which elements contribute to surface roughness and how. Valdesueiro et al. studied the roughness of the WPU surface using an SEM microscope equipped with EDS detectors. They found that the increased roughness and subsequent loss of gloss of the coating were caused by the alumina coating particles [26].

Besides the above-mentioned methods, the particle size distribution in the aqueous PU dispersion system can be determined by dynamic light scattering (DLS) [27]. Also, the molecular weight and molecular weight distribution of WPU can be determined by gel penetration chromatography (GPC) [28].

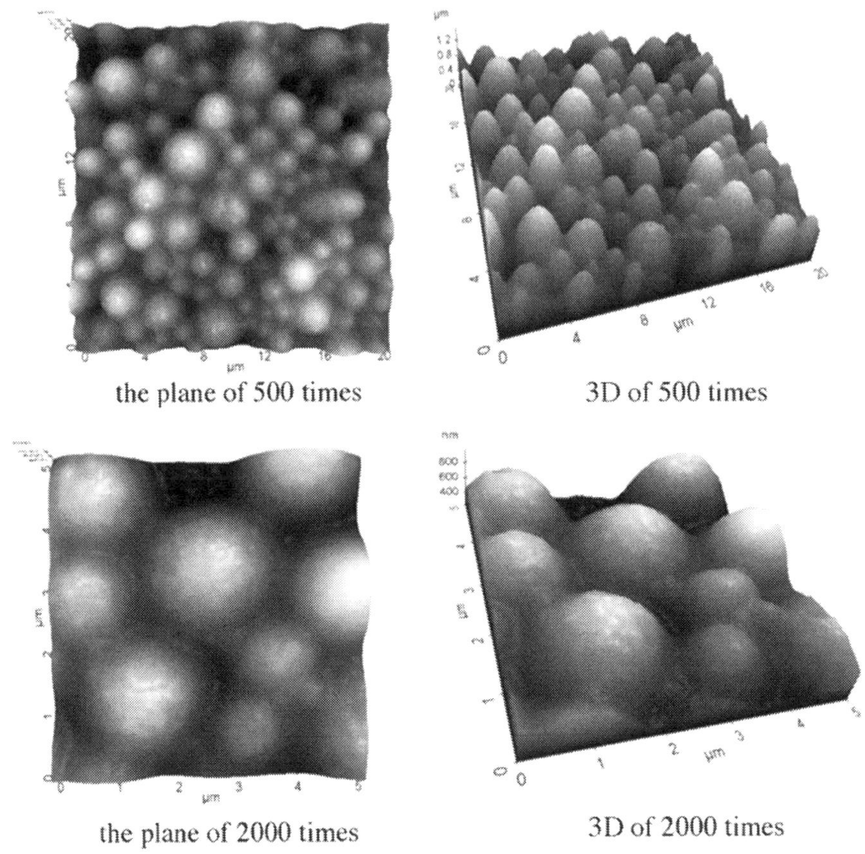

FIGURE 9.5 AFM images of WPU films [24].

9.4 APPLICATIONS OF SWPUS

Applications of SWPUs are highly dependent on the structural design during the synthesis. Matured strategies used for WPU structural design include adjusting the ratio of hard segments and soft segments, functionalization of soft segments and chemical modification of raw materials. Herein, we introduce four main applications of SWPUs, namely antifouling, self-healing, phase-change energy storage, and shape memory applications.

9.4.1 Anti-Fouling WPUs

Inspired by the superhydrophobic surface of the lotus leaf, antifouling PU surfaces can be designed by reducing the surface-free energy and increasing the surface roughness. The surface-free energy of a material is determined by its chemical composition. Typically, WPUs are modified with organosilicons or organofluorines. Silicone has both organic and inorganic properties, due to its chemical properties. The nonpolar Si–O bond in the main chain can effectively reduce the surface tension of materials and induce a remarkable degree of hydrophobicity. The incorporation of polysiloxane into the PU chain is usually achieved through the formation of a covalent bond between them.

Wen et al. used 3,3,4,4,5,5,6,6,7,7,8,8,8-tridecafluoro-1-octanol (TFO) and hexamethylene diisocyanate trimer (HDIT) to synthesize a series of fluoride WPUs (Figure 9.6) [29]. PU has low surface-free energy and strong wetting ability. A series of fluorinated WPUs with different lengths of fluorinated side chains were synthesized using different fluorine-terminated isocyanate trimers, where it was

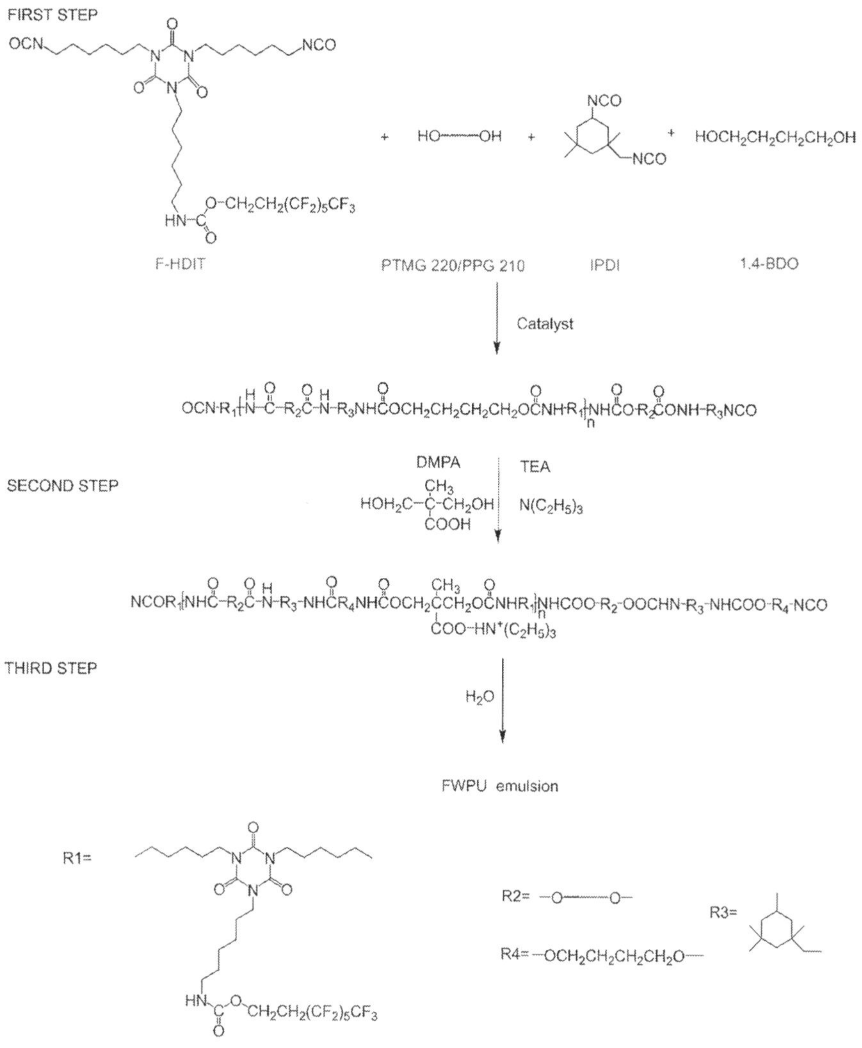

FIGURE 9.6 The synthetic route of fluorine-containing WPUS by using fluoro alcohol-terminated isocyanate trimer (F-HDIT) prepared from 3,3,4,4,5,5,6,6,7,7,8,8,8-tridecafluoro-1-octanol (TFO) and hexamethylene diisocyanate trimer (HDIT) [29].

shown that long fluorocarbon chains play a key role in the molecular design of low surface energy PU coating materials [30]. Xu et al. synthesized a CO_2-triggered hydrophobic/hydrophilic switchable WPUA containing a methyl methacrylate (MMA) unit [31]. The CO_2-induced WPUAs made stable latex particle dispersions in water and had good water resistance and mechanical properties after film formation.

9.4.2 Self-Healing WPUs

The unique microphase-separated structure of PU endows it with physical properties that can be controlled within a wide window [32], [33]. In the design of self-healing materials, PU has unique advantages, making it a research hotspot for the development of intrinsic self-healing materials in recent years [34], [35]. Due to its unique microphase separated structure, PU has two T_g values. This property could be used to adjust a temperature-sensitive self-healing property to function at lower temperatures suitable for many diverse applications. Also, the reaction of isocyanates with polyols or amines to form polyester or polyurea structures provides a large number of hydrogen bonds, which can be used for efficient self-healing without the need of other chemical bonds. The design of intrinsic self-healing PU is similar to that of other polymers. Such materials are achieved through reversible dynamic covalent bonds and reversible noncovalent bonds. For reversible dynamic covalent bonds in PU, the most studied cases are achieved through Diels–Alder reaction, disulfide bond, nitroxide radical, and acylhydrazone bond, and for non-covalent bonds, the investigated cases are hydrogen bond, metal-ligand, and π–π stacking [36–42].

Fang et al. investigated the covalent coupling reaction of novel diols containing DA bonds in the main chain of PU [43]. The WPU film was cut into two sections once WPU was exposed to heat treatment at 130°C for 30 min. Due to the Retro–DA reaction, polymer chains in the WPU film were cleaved into small molecular weight fragments. When the WPU film was exposed to a temperature of 65°C for 24 h, most of the dissociated maleimides and furans were reconnected through a DA reaction. The mechanical properties were restored due to the formation of hydrogen bonds at the breaking points. The self-healing rate of WPU was 92.5%, and the solid content of DA diol was 15.6% (Figure 9.7).

Kim et al. prepared room temperature self-healing PU elastomers by introducing 4,4′-dihydroxy diphenyl disulfide into the hard segment of PU in the form of small molecular chain extenders [44]. They discussed the effects of different isocyanates on self-healing properties. The results showed that isophorone diisocyanate (IPDI) PU had the best self-healing properties. The maximum tensile strength and toughness of IPDI PU were found to be 6.8 MPa and 26.9 MJ/m^3, respectively. The self-healing efficiency of IPDI PU can be restored up to 75% in 2 h at room temperature. In comparison to dihydroxy disulfide PU, the excellent self-healing performance of IPDI PU came from the double exchange of hydrogen bond and disulfide bond. In comparison to self-healing WPUs in which the strategy of reversible DA reaction has been employed, the PU self-healing material with disulfide bond had a lower self-healing temperature, and the prepared PU exhibited a better mechanical strength.

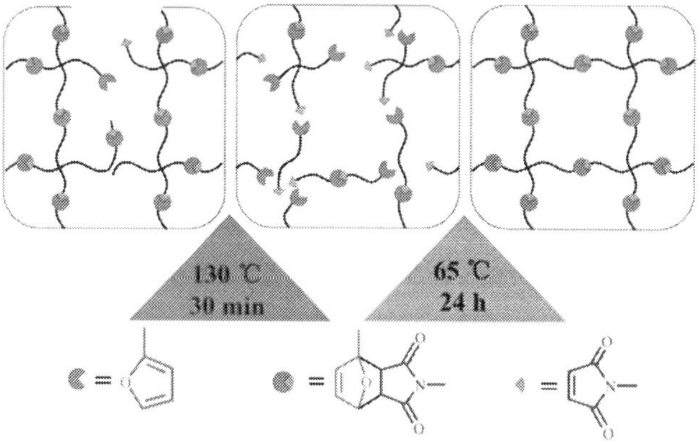

FIGURE 9.7 Schematic illustration of the self-healing mechanism of WPU through Diels–Alder reaction [43].

9.4.3 WPU Phase Change Materials

Phase change thermal energy storage (TES) is the most promising energy-saving technology, due to its high energy storage density and the isothermal characteristics of the phase change process. Waterborne polyurethane phase change materials (WPUPCMs) have been demonstrated to be effective solid–solid phase change materials for thermal energy storage. They are energy storage materials with high enthalpy and stable thermal performance. WPUPCMs are often composites of block copolymers, where alternating rigid and flexible chains are connected to each other by covalent bonds. The structure of WPUPCMs is similar to that of WPU elastomers. The hard segment is composed of chain extenders and isocyanate and constitutes the framework of WPUPCM. The soft segments are a carrier of functional groups that bring in new properties for WPUPCM.

The synthetic method for WPUPCM comprises of the introduction of hydrophilic groups into PUPCM to make it water-soluble. Du et al. prepared a thermal energy storage solid phase change material based on hyperbranched PU [45]. A series of hyperbranched PU phase change materials were prepared using hyperbranched polyester resin synthesized by dimethylolpropionic acid (DMPA) and polyethylene glycol 1,000 (PEG 1,000) and reacting with PU segments that were synthesized by IPDI and PEG 6000. A properly adjusted degree of branching ensured that the hyperbranched PU showed a good capacity for phase transition. The thermal cycle test and thermogravimetric analysis showed that the hyperbranched WPUPCM had good thermal controllability and stability, and therefore, exhibited potential application as a thermal energy storage material. Du et al. synthesized diethanolamine modified methoxy polyethylene glycol (DMPEG) by the reaction of methoxy polyethylene glycol (MPEG) with acryloyl chloride and diethanolamine [46]. Using DMPEG, 4,4-diphenylmethane diisocyanate (MDI), and water as raw materials, thermal insulation PU foams with enhanced energy storage capacity were prepared through condensation reaction (Figure 9.8). The TES capacity of WPU foams with oxidized polyethylene segments as side chains was much higher than

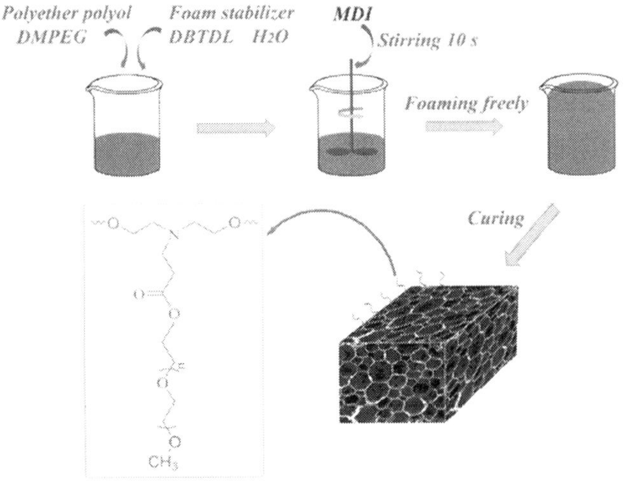

FIGURE 9.8 Schematic formation process of thermal insulation PU foams [46].

that of PU foams with oxidized polyethylene segments as the main chains. The PU foams synthesized in the experiment exhibited good thermal stability and could be used as an insulator with better energy storage capacity. Such foams featured a potential use in applications concerning temperature control and heat storage.

9.4.4 Shape Memory WPU

Shape memory polymers can remember their original shape after deformation and can restore their shape under appropriate stimuli, such as temperature, light, electric and magnetic fields, pH variations, specific ions, and enzymes (Figure 9.9) [47–49]. Stable polymer networks and reversible polymer switching transitions are the two prerequisites for the shape memory effect [46], [50]. The incompatibility of soft and hard segments in WPUs leads to microphase separation, which depends on block length, hydrogen bond, crystallization bond, and crystallization degree. A hard segment is associated with a fixed point or a freezing state that remains rigid in a temporary shape. The crystallization melting temperature of the soft section is the shape recovery temperature. The soft segment is considered to be the "molecular switch," and its melting temperature is the shape recovery temperature. The reversible phase transition is controlled by heating above and cooling below the shape recovery temperature. Lu et al. prepared a low-cost and environmentally friendly biomass polyol emulsion using castor oil as raw material, and successfully prepared castor oil-based WPU films with good shape memory (45°C heat treatment), self-healing properties (60°C, 5 min healing), and EMI shielding properties using 2-aminophenyl disulfide as chain extender [51]. Mahapatra et al. prepared strong electric response carbon nanotubes reinforced hyperbranched PU composite matrix shape-memory materials [52]. The branched structure of PU was helpful for the uniform dispersion of multi-walled carbon nanotubes (MWCNTs), and the addition of MWCNTs was beneficial for improving the mechanical properties of the

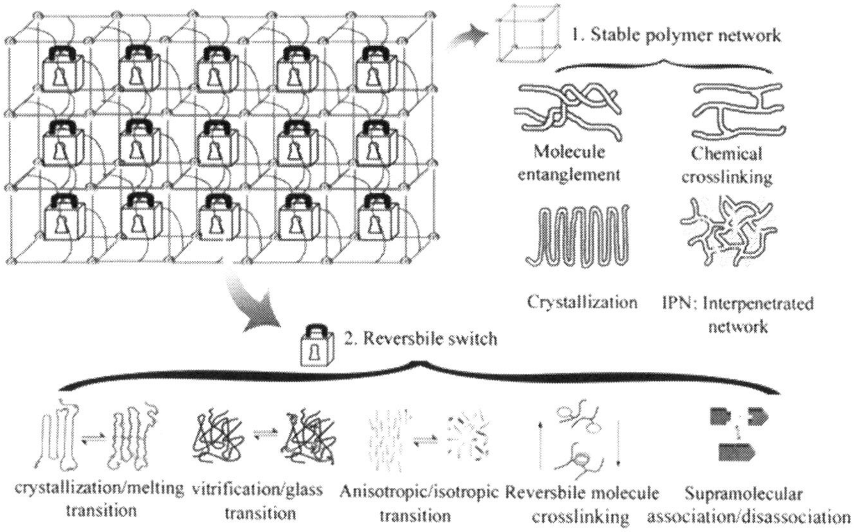

FIGURE 9.9 A stable network and a reversible switching transition are presented to illustrate the shape memory WPU [49].

polymer. The MWNTs/hyperbranched PU composites exhibited a fast recovery time of 9s in more than 98% of thermal triggering and electrically induced shape memory behavior. The carboxylic acid groups in the side chains of WPU were sensitive to pH. This could be utilized to form dimers under acidic conditions and convert carboxylic acids to carboxylic acid esters to destroy dimers under alkaline conditions. In addition, carboxylic acid dimers were affected by temperature, where they could dissociate or associate with an increase or a decrease in the temperature, respectively. Based on these properties, Song et al. developed a thermo-sensitive PU with three shape memory elements and a pH-sensitive dual shape memory element [53]. The glass transition of soft segments and the association and dissociation of carboxylic acid dimers were two switches that could control the triform memory performance. pH stimulation was achieved by association of carboxylic acid dimers in an acidic environment and their dissociation in a basic one.

9.5 CONCLUSION

SWPUs are one of the most common, widely used, and widely studied materials in the world. WPUs can be produced from a variety of diisocyanates, polyols, chain extenders, and crosslinkers, making it possible to obtain a wide range of custom materials useful in different applications. With the increasing demand on protecting the environment and human health and meeting the quality requirements of water-based materials, it is imperative to develop new synthesis and modification methods to prepare high-grade SWPUs. Functional WPUs exhibiting properties, such as antifouling, self-healing, phase change energy storage ability, and shape memory properties, are bound to become research hotspots in academic and commercial fields. Even though in this research field a great progress has been made over the past few decades, a variety of new and interesting challenges remain that should be addressed.

REFERENCES

1. H. Honarkar, "Waterborne polyurethanes: A review," *J. Dispers. Sci. Technol.*, vol. 39, no. 4, pp. 507–516, 2018, doi: 10.1080/01932691.2017.1327818.
2. W. Wang, Y. Guo, and J. U. Otaigbe, "Synthesis and characterization of novel biodegradable and biocompatible poly(ester-urethane) thin films prepared by homogeneous solution polymerization," *Polymer (Guildf).*, vol. 49, no. 20, pp. 4393–4398, 2008, doi: 10.1016/j.polymer.2008.07.057.
3. H. Wang, Y. Niu, G. Fei, Y. Shen, and J. Lan, "In-situ polymerization, rheology, morphology and properties of stable alkoxysilane-functionalized poly (urethane acrylate) microemulsion," *Prog. Org. Coatings*, vol. 99, pp. 400–411, 2016, doi: 10.1016/j.porgcoat.2016.07.005.
4. P. Wang, B. Zhang, L. Tang, L. Huang, and W. Wang, "Reversible addition fragmentation chain-transfer polymerization," *Polym. Mater. Sci. Eng.*, vol. 22, no. 5, p. 109, 2006.
5. J. Li, H. Wang, Z. Sun, R. Fan, Q. Ren, and Q. Yu, "Synthesis and characterization of linear waterborne poly (ethyl acrylate-urethane) prepared from poly (ethyl acrylate) diol via atom transfer radical polymerization," *J. Appl. Polym. Sci.*, vol. 126, no. 1, pp. 66–71, 2012.
6. A. K. Nanda and D. A. Wicks, "The influence of the ionic concentration, concentration of the polymer, degree of neutralization and chain extension on aqueous polyurethane dispersions prepared by the acetone process," *Polymer (Guildf).*, vol. 47, no. 6, pp. 1805–1811, 2006, doi: 10.1016/j.polymer.2006.01.074.
7. S. M. Cakic, J. V. Stamenkovic, D. M. Djordjevic, and I. S. Ristic, "Synthesis and degradation profile of cast films of PPG-DMPA-IPDI aqueous polyurethane dispersions based on selective catalysts," *Polym. Degrad. Stab.*, vol. 94, no. 11, pp. 2015–2022, 2009, doi: 10.1016/j.polymdegradstab.2009.07.015.
8. A. Saetung *et al.*, "Synthesis, characteristic, and properties of waterborne polyurethane based on natural rubber," *J. Appl. Polym. Sci.*, vol. 124, no. 4, pp. 2742–2752, 2012, doi: 10.1002/app.35318.
9. D. Dieterich, "Aqueous emulsions, dispersions and solutions of polyurethanes; synthesis and properties," *Prog. Org. Coatings*, vol. 9, no. 3, pp. 281–340, 1981, doi: 10.1016/0033-0655(81)80002-7.
10. Y. Guo *et al.*, "Asynchronous fracture of hierarchical microstructures in hard domain of thermoplastic polyurethane elastomer: Effect of chain extender," *Polymer (Guildf).*, vol. 138, pp. 242–254, 2018, doi: 10.1016/j.polymer.2018.01.035.
11. A. Niemczyk, A. Piegat, Á. Sonseca Olalla, and M. El Fray, "New approach to evaluate microphase separation in segmented polyurethanes containing carbonate macrodiol," *Eur. Polym. J.*, vol. 93, no. May, pp. 182–191, 2017, doi: 10.1016/j.eurpolymj.2017.05.046.
12. H. S. Lee and S. L. Hsu, "An analysis of phase separation kinetics of model polyurethanes," *Macromolecules*, vol. 22, no. 3, pp. 1100–1105, 1989, doi: 10.1021/ma00193a017.
13. P. Król, "Synthesis methods, chemical structures and phase structures of linear polyurethanes. Properties and applications of linear polyurethanes in polyurethane elastomers, copolymers and ionomers," *Prog. Mater. Sci.*, vol. 52, no. 6, pp. 915–1015, 2007, doi: 10.1016/j.pmatsci.2006.11.001.

14. S. Velankar and S. L. Cooper, "Microphase separation and rheological properties of polyurethane melts. 2. Effect of block incompatibility on the microstructure," *Macromolecules*, vol. 33, no. 2, pp. 382–394, 2000, doi: 10.1021/ma990817g.
15. R. Li and Z. Shan, "Research on structural features and thermal conductivity of waterborne polyurethane," *Prog. Org. Coatings*, vol. 104, pp. 271–279, 2017, doi: 10.1016/j.porgcoat.2016.11.027.
16. H. Benoit and G. Hadziioannou, "Scattering theory and properties of block copolymers with various architectures in the homogeneous bulk state," *Macromolecules*, vol. 21, no. 5, pp. 1449–1464, 1988, doi: 10.1021/ma00183a040.
17. Z. Wang *et al.*, "Highly stretchable and compressible shape memory hydrogels based on polyurethane network and supramolecular interaction," *Mater. Today Commun.*, vol. 17, pp. 246–251, 2018, doi: 10.1016/j.mtcomm.2018.09.006.
18. C. M. Gomez, D. Gutierrez, M. Asensio, V. Costa, and A. Nohales, "Transparent thermoplastic polyurethanes based on aliphatic diisocyanates and polycarbonate diol," *J. Elastomers Plast.*, vol. 49, no. 1, pp. 77–95, 2017, doi: 10.1177/0095244316639633.
19. W. Li, X. Jiang, R. Wu, and W. Wang, "Fast shape recovery by changing the grafting ratio in polyurethane/montmorillonite-poly(methyl methacrylate) composites," *Polym. J.*, vol. 49, no. 2, pp. 263–271, 2017, doi: 10.1038/pj.2016.98.
20. J. A. Koutsky, N. V. Hien, and S. L. Cooper, "Some results on electron microscope investigations of polyether-urethane and polyester-urethane block copolymers," *J. Polym. Sci. Part B Polym. Lett.*, vol. 8, no. 5, pp. 353–359, 1970, doi: 10.1002/pol.1970.110080508.
21. Y. Zhang, L. Shao, D. Dong, and Y. Wang, "Enhancement of water and organic solvent resistances of a waterborne polyurethane film by incorporating liquid polysulfide," *RSC Adv.*, vol. 6, no. 21, pp. 17163–17171, 2016, doi: 10.1039/C5RA24574E.
22. P. Schön, K. Bagdi, K. Molnár, P. Markus, B. Pukánszky, and G. Julius Vancso, "Quantitative mapping of elastic moduli at the nanoscale in phase separated polyurethanes by AFM," *Eur. Polym. J.*, vol. 47, no. 4, pp. 692–698, 2011, doi: 10.1016/j.eurpolymj.2010.09.029.
23. Q. Yong, F. Nian, B. Liao, L. Huang, L. Wang, and H. Pang, "Synthesis and characterization of solvent-free waterborne polyurethane dispersion with both sulfonic and carboxylic hydrophilic chain-extending agents for matt coating applications," *RSC Adv.*, vol. 5, no. 130, pp. 107413–107420, 2015, doi: 10.1039/C5RA21471H.
24. J. Li, W. Zheng, W. Zeng, D. Zhang, and X. Peng, "Structure, properties and application of a novel low-glossed waterborne polyurethane," *Appl. Surf. Sci.*, vol. 307, pp. 255–262, 2014, doi: 10.1016/j.apsusc.2014.04.022.
25. Q. Yong *et al.*, "Preparation and characterization of low gloss aqueous coating via forming self-roughed surface based on waterborne polyurethane acrylate hybrid emulsion," *Prog. Org. Coatings*, vol. 115, 2017, pp. 18–26, 2018, doi: 10.1016/j.porgcoat.2017.10.024.
26. D. Valdesueiro *et al.*, "Tuning roughness and gloss of powder coating paint by encapsulating the coating particles with thin Al 2 O 3 films," *Powder Technol.*, vol. 318, pp. 401–410, 2017, doi: 10.1016/j.powtec.2017.05.019.
27. A. Saadat-Monfared, M. Mohseni, and M. H. Tabatabaei, "Polyurethane nanocomposite films containing nano-cerium oxide as UV absorber. Part 1. Static and dynamic light scattering, small angle neutron scattering and optical studies," *Colloids Surfaces A Physicochem. Eng. Asp.*, vol. 408, pp. 64–70, 2012, doi: 10.1016/j.colsurfa.2012.05.027.
28. C. Xu, L. OuYang, Z. Cai, Y. Ren, S. Lu, and W. Shi, "Effects of polyaminosiloxane on the structure and properties of modified waterborne polyurethane," *J. Appl. Polym. Sci.*, vol. 136, no. 12, p. 47226, 2019, doi: 10.1002/app.47226.

29. J. Wen, Z. Sun, H. Fan, Y. Chen, and J. Yan, "Synthesis and characterization of a novel fluorinated waterborne polyurethane," *Prog. Org. Coatings*, vol. 131, no. January, pp. 291–300, 2019, doi: 10.1016/j.porgcoat.2019.02.029.
30. J. Wen, Z. Sun, J. Xiang, H. Fan, Y. Chen, and J. Yan, "Preparation and characteristics of waterborne polyurethane with various lengths of fluorinated side chains," *Appl. Surf. Sci.*, vol. 494, pp. 610–618, 2019, doi: 10.1016/j.apsusc.2019.07.170.
31. L. Xu *et al.*, "CO2-triggered hydrophobic/hydrophilic switchable waterborne polyurethane–acrylate with simultaneously improved water resistance and mechanical properties," *J. Coatings Technol. Res.*, 2021, doi: 10.1007/s11998-021-00476-y.
32. A. V. Menon, G. Madras, and S. Bose, "The journey of self-healing and shape memory polyurethanes from bench to translational research," *Polym. Chem.*, vol. 10, no. 32, pp. 4370–4388, 2019, doi: 10.1039/C9PY00854C.
33. U. S. Chung, J. H. Min, P.-C. Lee, and W.-G. Koh, "Polyurethane matrix incorporating PDMS-based self-healing microcapsules with enhanced mechanical and thermal stability," *Colloids Surfaces A Physicochem. Eng. Asp.*, vol. 518, pp. 173–180, 2017, doi: 10.1016/j.colsurfa.2017.01.044.
34. L. Zhang *et al.*, "Achievement of both mechanical properties and intrinsic self-healing under body temperature in polyurethane elastomers: a synthesis strategy from waterborne polymers," *Polymers (Basel).*, vol. 12, no. 4, p. 989, 2020, doi: 10.3390/polym12040989.
35. P. D. Tatiya, R. K. Hedaoo, P. P. Mahulikar, and V. V. Gite, "Novel polyurea microcapsules using dendritic functional monomer: synthesis, characterization, and its use in self-healing and anticorrosive polyurethane coatings," *Ind. Eng. Chem. Res.*, vol. 52, no. 4, pp. 1562–1570, 2013, doi: 10.1021/ie301813a.
36. L. Huang *et al.*, "Multichannel and repeatable self-healing of mechanical enhanced graphene-thermoplastic polyurethane composites," *Adv. Mater.*, vol. 25, no. 15, pp. 2224–2228, 2013, doi: 10.1002/adma.201204768.
37. P. Du *et al.*, "Diels-Alder-based crosslinked self-healing polyurethane/urea from polymeric methylene diphenyl diisocyanate," *J. Appl. Polym. Sci.*, vol. 131, no. 9, p. 40234, 2014, doi: 10.1002/app.40234.
38. Y. Xu and D. Chen, "A novel self-healing polyurethane based on disulfide bonds," *Macromol. Chem. Phys.*, vol. 217, no. 10, pp. 1191–1196, 2016, doi: 10.1002/macp.201600011.
39. C. Yuan, M. Z. Rong, and M. Q. Zhang, "Self-healing polyurethane elastomer with thermally reversible alkoxyamines as crosslinkages," *Polymer (Guildf).*, vol. 55, no. 7, pp. 1782–1791, 2014, doi: 10.1016/j.polymer.2014.02.033.
40. J. Ling, M. Z. Rong, and M. Q. Zhang, "Photo-stimulated self-healing polyurethane containing dihydroxyl coumarin derivatives," *Polymer (Guildf).*, vol. 53, no. 13, pp. 2691–2698, 2012, doi: 10.1016/j.polymer.2012.04.016.
41. Y. J. Kim, P. H. Huh, and B. K. Kim, "Synthesis of self-healing polyurethane urea-based supramolecular materials," *J. Polym. Sci. Part B Polym. Phys.*, vol. 53, no. 7, pp. 468–474, 2015, doi: 10.1002/polb.23653.
42. K. Zhu, Q. Song, H. Chen, and P. Hu, "Thermally assisted self-healing polyurethane containing carboxyl groups," *J. Appl. Polym. Sci.*, vol. 135, no. 9, p. 45929, 2018, doi: 10.1002/app.45929.
43. Y. Fang *et al.*, "Thermal-driven self-healing and recyclable waterborne polyurethane films based on reversible covalent interaction," *ACS Sustain. Chem. Eng.*, vol. 6, no. 11, pp. 14490–14500, 2018, doi: 10.1021/acssuschemeng.8b03151.
44. S.-M. Kim *et al.*, "Superior toughness and fast self-healing at room temperature engineered by transparent elastomers," *Adv. Mater.*, vol. 30, no. 1, p. 1705145, 2018, doi: 10.1002/adma.201705145.

45. X. Du, H. Wang, Y. Wu, Z. Du, and X. Cheng, "Solid-solid phase-change materials based on hyperbranched polyurethane for thermal energy storage," *J. Appl. Polym. Sci.*, vol. 134, no. 26, pp. 1–8, 2017, doi: 10.1002/app.45014.
46. X. Du, J. Li, S. Wang, H. Wang, X. Cheng, and Z. Du, "Fabrication and characterization of polyurethane foams containing phase change materials for thermal energy storage," *Thermochim. Acta*, vol. 670, no. October, pp. 55–60, 2018, doi: 10.1016/j.tca.2018.10.014.
47. C. Pan, L. Liu, Q. Chen, Q. Zhang, and G. Guo, "Tough, stretchable, compressive novel polymer/graphene oxide nanocomposite hydrogels with excellent self-healing performance," *ACS Appl. Mater. Interfaces*, vol. 9, no. 43, pp. 38052–38061, 2017, doi: 10.1021/acsami.7b12932.
48. Y. Zhang, X. Jiang, R. Wu, and W. Wang, "Multi-stimuli responsive shape memory polymers synthesized by using reaction-induced phase separation," *J. Appl. Polym. Sci.*, vol. 133, no. 24, p. 43534, 2016, doi: 10.1002/app.43534.
49. H. Meng and G. Li, "A review of stimuli-responsive shape memory polymer composites," *Polymer (Guildf).*, vol. 54, no. 9, pp. 2199–2221, 2013, doi: 10.1016/j.polymer.2013.02.023.
50. Q. Meng and J. Hu, "A review of shape memory polymer composites and blends," *Compos. Part A Appl. Sci. Manuf.*, vol. 40, no. 11, pp. 1661–1672, 2009, doi: 10.1016/j.compositesa.2009.08.011.
51. J. Lu *et al.*, "Self-healable castor oil-based waterborne polyurethane/MXene film with outstanding electromagnetic interference shielding effectiveness and excellent shape memory performance," *J. Colloid Interface Sci.*, vol. 588, pp. 164–174, 2021, doi: 10.1016/j.jcis.2020.12.076.
52. S. S. Mahapatra, S. K. Yadav, H. J. Yoo, M. S. Ramasamy, and J. W. Cho, "Tailored and strong electro-responsive shape memory actuation in carbon nanotube-reinforced hyperbranched polyurethane composites," *Sensors Actuators B Chem.*, vol. 193, pp. 384–390, 2014, doi: 10.1016/j.snb.2013.12.006.
53. Q. Song, H. Chen, S. Zhou, K. Zhao, B. Wang, and P. Hu, "Thermo- and pH-sensitive shape memory polyurethane containing carboxyl groups," *Polym. Chem.*, vol. 7, no. 9, pp. 1739–1746, 2016, doi: 10.1039/C5PY02010G.

10 Shape Memory Waterborne Polyurethanes

Arunima Reghunadhan, and Sabu Thomas
School of Energy Materials, Mahatma Gandhi University, Kottayam, Kerala, India

Jiji Abraham
Department of Chemistry, Vimala College, Ramavarmapuram, Thrissur, Kerala, India

CONTENTS

10.1 Concept of Shape Memory 157
10.2 The Thermomechanical Cycle 160
10.3 Properties of Shape Memory Materials 160
    10.3.1 The Shape Recovery Rate (R_r) 161
    10.3.2 The Shape Fixity (R_f) 161
10.4 Waterborne Polyurethanes—An Overview 161
10.5 Applications of Waterborne Polyurethane 162
10.6 Shape Memory and Polyurethanes 165
10.7 Shape Memory of Waterborne Polyurethane and its Composites 167
10.8 Applications of Shape Memory Waterborne Polyurethanes 170
10.9 Conclusions 171
References 171

10.1 CONCEPT OF SHAPE MEMORY

The shape memory effect (SME) is an interesting and attractive property of materials that makes them suitable for some specific applications, such as in robotics. Shape memory material-related inventions were started in the 1980s. When a material is heated over a particular characteristic transition temperature, the SME occurs, in which it returns to its previous size and shape. Thus, it is the restoration or regaining the shape of a body by the application of heat. Shape memory is related to plastic deformation. Primary crosslinking net sites (hard segments) memorize and determine the permanent shape, while secondary switching segments or soft segments with a transition lower the strain–stress and maintain the temporary shape in

DOI: 10.1201/9781003173526-10

the reversible transformation of shape memory polyurethanes (SMPUs) [1–3]. During the change of shape, the crystalline structure of the material undergoes a change between two states, known as austenite and martensite. In the austenite shape or structure, the phase will be highly ordered, while in the martensite, the ordered arrangement is lowered. The change from austenite to martensite happens in two stages; the grid distortion organize produces the modern structure by means of little layered disengagements, whereas the cross-section invariant shear or accommodation organize includes accommodating the unused structure by changing the encompassing austenite either by twinning (incapable to accommodate volume changes) or slip (permanent and common). The thermoelastic martensitic transition is a crystalline phase transition that causes this phenomenon [4]. Their microstructure is characterized by self-accommodating twins in this state, and the martensite is soft and readily deformed by de-twinning. The entire transformation cycle is defined as follows (Figure 10.1).

Most of the materials are renowned for single shape memory, but there are many materials that show double, triple, and multiple SMEs or shape transformation. SMPUs have a wide range of qualities, ranging from stable to volatile, to biodegradable and transitory, pliable to stiff, or soft to hard, depending on the structural pieces that comprise the SMPU [5,6]. The multiple shape memory materials are responsive toward temperature, magnetic field, light, electric field, etc.

a. One-way shape memory materials

If the strain given to the material is less than 10%, then that strain can be recovered. Thus, materials that give a shape memory only by the application of heat (temperature) are called the one-way SME. So, if a shape-memory polymer is cold (as below), it may be bent or stretched and will retain those forms until heated above the phase transition. When heated, the form returns to its original state. When the metal cools again, it retains its shape until it is distorted again (Figure 10.2).

b. Two-way shape memory materials

In this type of effect, the shape memory is prominent both at lower and higher temperatures. This type of effect is less exploited in the industry due to the difficulty in effect generation. This may also be accomplished without the use of external

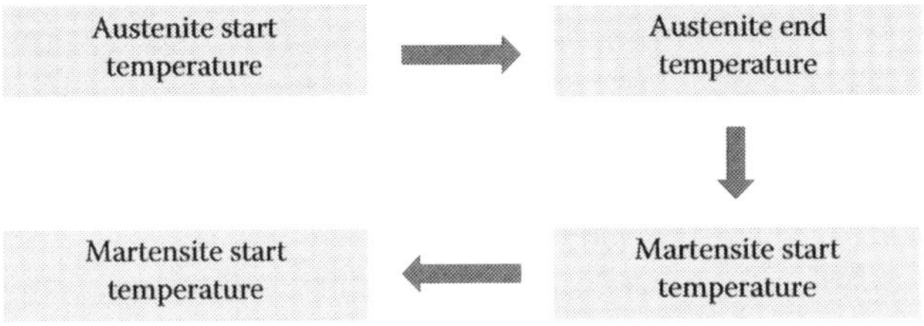

FIGURE 10.1 Schematic representation of the phase transformations during shape memory effect.

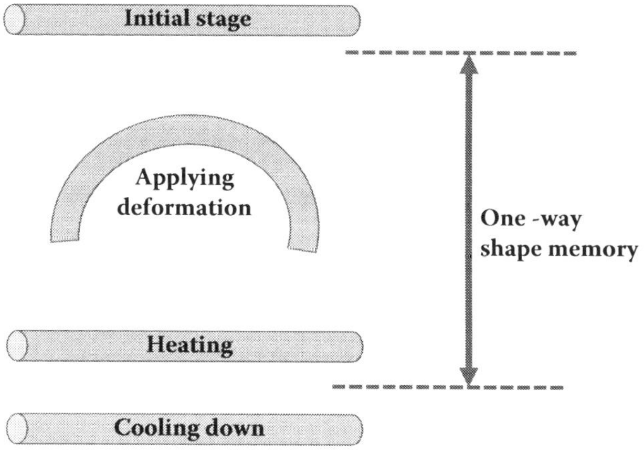

FIGURE 10.2 Schematic representation of one-way shape memory effect.

force (intrinsic two-way effect). The explanation the material reacts so differently in various instances resides in training. Training infers that a form of memory may learn to act in a given way. A shape memory alloy keeps its low-temperature shape under normal conditions but forgets the low-temperature form when heated to regain the high-temperature shape. In the high-temperature phases, it can be "taught" to "remember" to leave certain reminders of the deformed low-temperature condition (Figure 10.3).

c. Super elastic materials

Shape memory materials show a special type of behavior, often termed pseudoelasticity, which is alternatively referred to as superelasticity. The superelasticity may be a consequence of a stress-induced alteration from austenite to martensite and back once a sample is tested between zero and a finite; however, tiny strain at a continuing close temperature on top of a set temperature is typically observed because of the austenite end temperature. The sample experiences very little or no permanent deformation in such a strain cycle, giving the impression that the fabric has undergone solely elastic deformation; consequently, the term super elastic is

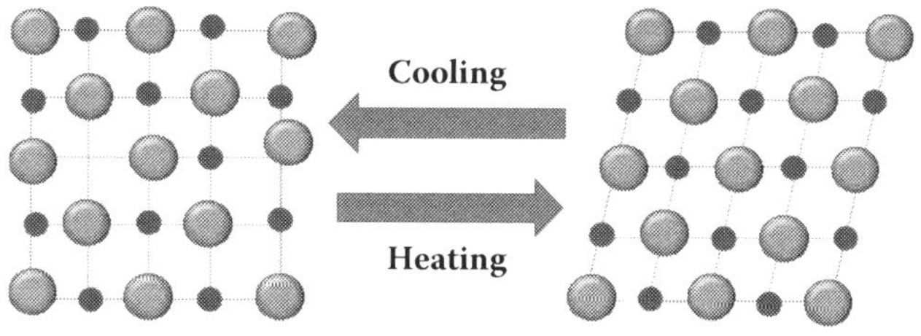

FIGURE 10.3 Schematic representation of the change in crystalline phase during two-way shape memory effect (from austenite to de twinned martensite and vice versa).

being used. The stresses suffered by superelastic materials in ordinary applications are minimal, but the rotations can be rather significant [7].

10.2 THE THERMOMECHANICAL CYCLE

A thermomechanical cycle that shows the modifications throughout the SME may be used to explain the shape memory manner in SMPUs. In the thermomechanical cycle, SMPUs are first heated at an accelerated temperature, a temperature above glass transition (T_g) with zero strain and stress. After that, SMPUs are deformed to a preferred form and correspond to strain and stress at temperatures above the glass transition temperature T_g. Strain and stress are maintained at the same time as the deformed form is fixed. Then, SMPUs are cooled to a temperature lower than T_g, where the chain segments of the substances are within the transient position. Afterward, constrain is eliminated from the polymer, and the material is stated to be in its transient form while the stress is eliminated. This manner is likewise referred to as the programming process. The thermomechanical cycle is finished when the substances are reheated to a temperature above T_g, where the pressure is changed into relief and the substances recover their authentic form. The recuperation steps may be repeated for the following cycle [8] (Figure 10.4).

10.3 PROPERTIES OF SHAPE MEMORY MATERIALS

The shape memory materials can be of several types. The main categories are shape memory metals, alloys, and polymers. The key focus of this chapter is on shape memory polymers. Depending on the materials, the parameters deciding the

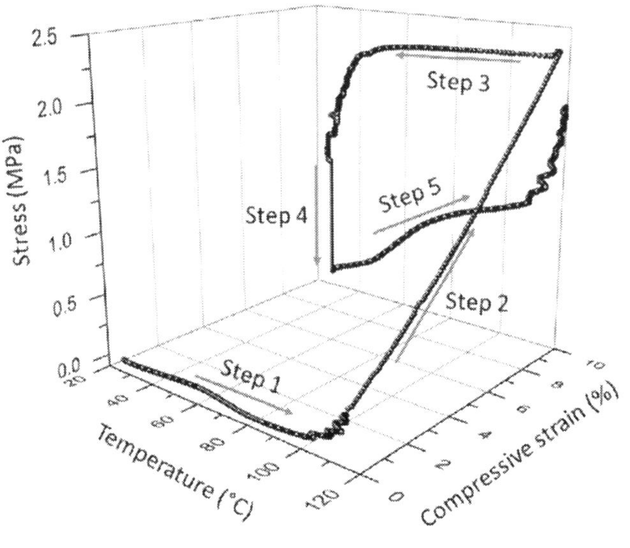

FIGURE 10.4 The thermomechanical cycle for a shape memory polymer (Reproduced with permission from reference [9]).

SME will also be different. The main parameters are shaped recovery rate and shape fixity. These two parameters represent the ability of the SME [10].

10.3.1 The Shape Recovery Rate (R_R)

Shape recovery rate or strain recovery rate refers to a material's capacity to remember its permanent shape. Shape recovery rate can be calculated from the below-given equation in which the $\in m$ represents the maximum stress given to the system and $\in_P$, the strain after recovery. N is the number of cycles.

$$R_r = \frac{\in_m(N) - \in_P(N)}{\in_m(N) - \in_P(N-1)} \times 100\%$$

10.3.2 The Shape Fixity (R_F)

The shape fixity is another important property that determines the material's efficiency to show the SME. The amount to which a temporary form is set for an SMP is known as shape fixity. It's also known as shape retention or strain fixity. As a likely result, the shape fixity is calculated as a percentage of the fixed deformation to total deformation ratio. It can be calculated from the following equation, where $\in$u is the strain at the temporary shape.

$$R_f = \frac{\in_u(N)}{\in_m(N)} \times 100\%$$

10.4 WATERBORNE POLYURETHANES—AN OVERVIEW

Polyurethanes (PUs) are polymers containing urethane (–NH(C=O) O) linkage. Monomers used are diisocyanate or polymeric isocyanate with a hydroxyl-containing group. PU is a high-performance polymer material due to its unique properties, including excellent abrasion resistance, flexibility, hardness, chemical resistance, solvent resistance, light stability, and weatherability. This polymer is widely used in various sectors like furniture and bedding, construction, automotive, footwear, coatings, and bio-medical fields. It is considered to be the fifth big plastic among the macromolecular category [11]. Accordingly, the global market for PUs is estimated to keep growing from about USD 65.5 billion in 2018 to USD 91 billion by 2026 [12–14]. Solvent-based PU is restricted because of threats from high pollution, the chemical industry, finite resources of crude oil, gradual reduction of fossil fuels, and there boom of the price of solvents. So, environmentally friendly and sustainable water-based PU has got considerable attention over solvent-based PU.

The key difference between solvent-based PUs and water-based PUs (hereafter WPU) is that water is the solvent. The water-based emulsification process results in zero toxic byproducts, such as gases, recyclability, and environmental friendliness. Moreover, some versatile properties of WPU, such as resistance to solvent, chemicals, and water, flexibility, and toughness, make it an ideal candidate for a wide range of applications. WPUs have a common structure consisting of two parts: the emulsifier and the main skeleton or backbone. Emulsifiers can be any hydrophilic materials, such as polymers. They are acting as stabilizers in water. The rigid and soft segments of WPU are comprised of ionic groups and hydrophobic units, respectively. Commonly, the diol with carboxylic, quaternary ammonium, sulfonic, etc., acts as an emulsifier. Depending on the ionic functionality added in the synthetic media, WPUs can be classified as cationomers and anionomers. The anionomers are from the reaction between isocyanates and diols containing ionic functionalities like phosphate, carboxylate, etc., while the anionomers are obtained from the reaction between the isocyanate with nitro or sulfur-containing diols.

WPUs are synthesized by many methods. Some of them are the hot-melt process, acetone method, emulsification method, Ketimine–Ketazine Process, in-situ methods, etc. The detailed discussion is not under the scope of this chapter. The WPU synthesis became greener by the so-called non-isocyanate routes. Most of these techniques use plant-derived oils such as vegetable oil. Most of the vegetable oil lacks the hydroxyl moiety to be used as polyols, and these will be modified with the addition or conversion to –OH groups. Transesterification, epoxidation followed by ring-opening polymerization, hydrogenation, coupling reaction, etc., are the available routes to introduce the hydroxyl groups. The multistep reaction of WPU dispersion synthesis can be summarized as: The NCO terminated PU pre-polymers will be produced first by the reaction of the diols from the fatty acids or glycerides [15–17]. The viscosity of the medium is adjusted by suitable solvents. The addition and co-reaction with internal emulsifying agents, which introduce ionic centers into the polymer chemical structure, is the second stage. Following this process, the PU is dispersed in water at room temperature using a high shear device. The whole process is represented schematically in Figure 10.5.

10.5 APPLICATIONS OF WATERBORNE POLYURETHANE

WPU has found applications in various sectors like plastics, the construction industry, automotive industry, medicinal science, synthetic leather, furniture, and coatings [18]. The multifunctional coating based on WPU has gotten considerable attention in both academics and industry. This is because of the fact that these coatings are free from toxic volatile organic compounds. Several studies have been reported on the application of PUs as efficient coatings. This is because of their noble flexibility at low temperature, high tensile strength, outstanding surface gloss, and good dielectric characteristics. In one study, a renewable, green, and scalable dihydroxy acid was synthesized from sunflower oil and explored as a chain extender in the synthesis of WPUs for coatings on poly (vinyl chloride) surfaces and synthetic leather [19]. Cationic waterborne polyurethane (cWPU) latex-based on epoxidized hydroxyl telechelic natural rubber (eHTNR) polyol was used for green coating applications. The fabricated films displayed good chemical resistance, thermal stability, wettability, superior hydrophilicity, and good adhesion properties [20]. UV-curable WPU acrylate

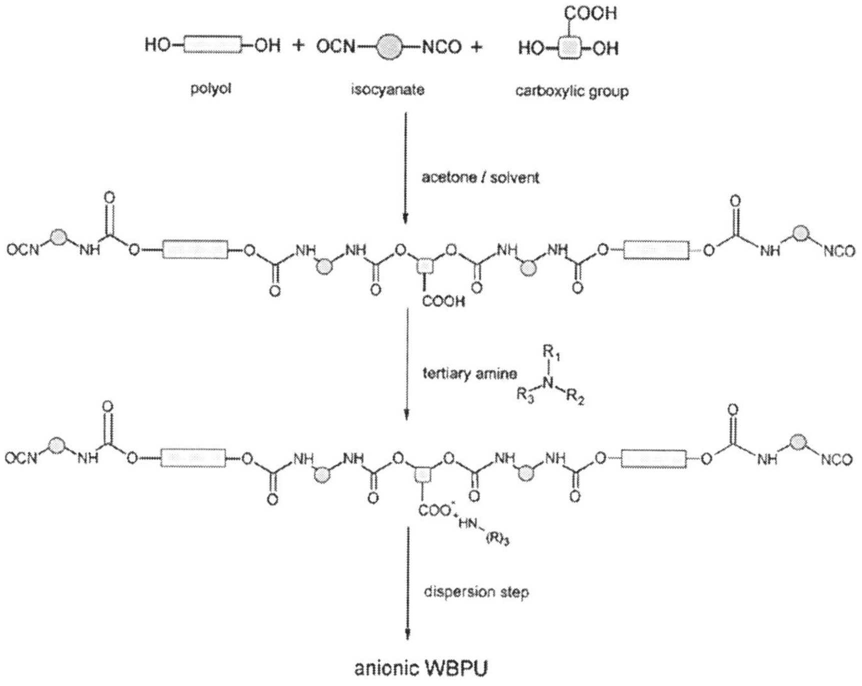

FIGURE 10.5 Schematic representation of the synthesis of waterborne polyurethane (Adapted from reference [15]).

modified with octa vinyl POSS for weatherable coating applications was also explored. Here, the demerits of WPU were overcome by the introduction of octa vinyl polyhedral oligomeric silsesquioxane (OVPOSS) [21]. WPU prepared from citric acid-based polyol, sustainable functional polyol (SFP) is used for antibacterial and breathable waterproof coating of cotton fabric. The WPU-coated fabric showed antibacterial properties ranging from 84% to 99% against *Escherichia coli* bacteria. It also showed water barrier property of 100+ cm of water pressure, and water vapor transmission rate up to 1,506 g per day, demonstrating nonporous breathable coating [22]. In another study, thermoelectric composite made from WPU and carbon nanotube was explored for use as a coating agent to give thermoelectric function to yarns [23] (Figure 10.6).

Outstanding physical properties of PU, such as good elasticity, excellent mechanical strength, and high adhesiveness, make it an ideal candidate for adhesives and finishing agents. High solid content WPU prepared from pentaerythritol (PE) was used as a leather finishing agent. These finishing agents apply to materials like clothes, shoes, and bags. Nontoxic and high value-added waterborne leather finishing agents with excellent overall performance are supposed to give intense significance to advancements in the leather industry [24].

PU-acrylic hybrid waterborne nanoparticles prepared via mini emulsion polymerization can be used as pressure-sensitive adhesives. Compared to solvent-based adhesives, WPU presents some disadvantages (e.g. water-resistance and tack/shear

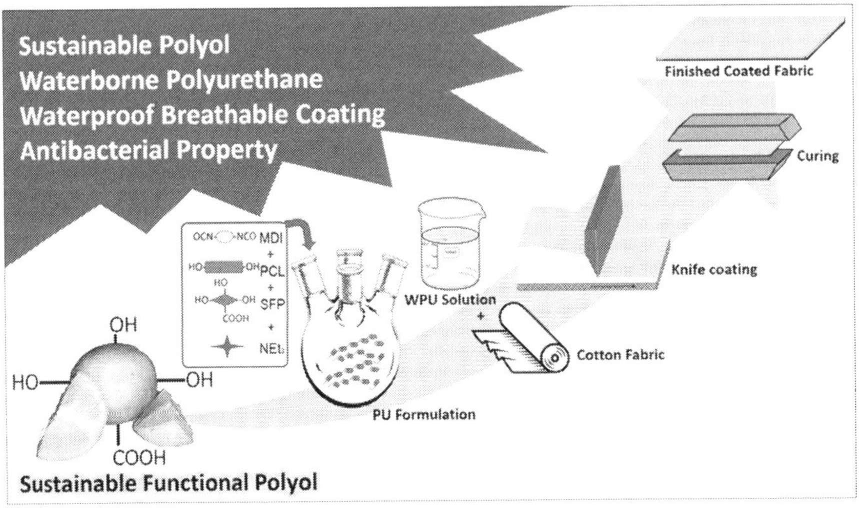

FIGURE 10.6 Waterborne polyurethane for antibacterial and breathable waterproof coating of cotton fabric (Reproduced with permission from reference [23]).

strength). Properties can be improved by the incorporation of functionalized acrylic monomer (hydroxyethyl methacrylate, HEMA). The shear resistance was considerably elevated when PU was integrated into the acrylic network, indicating an increase in film cohesion. A balance between elastic and viscous properties is required to get a higher shear holding power combined with an adequate level of tack adhesion [25].

WPU composites can be used in the medical field also. It is discovered that PU/cellulose nanofiber composites can be used as tissue engineering scaffolds or medical devices. Scaffolds are prepared via the 3D printing technique. The interface between cellulose and PU will improve rheology, biocompatibility, and degradation so the nanocomposite ink can be directly applied to 3D printing for an extensive range of applications, mainly in the construction of medical devices and tissue engineering scaffolds [26]. Cationic WPU coating modified by chitosan biguanide hydrochloride was used as a coating for medical catheters. The inhibition rate of *S. aureus* and *E. coli* reached 91% and 85%, respectively. Cytotoxicity test presented that the material had outstanding biocompatibility [27].

Nowadays, water-based ink has attracted considerable attention in the printing industry because this reduces the use of inks based on volatile organic solvents. Outstanding properties of WPU, such as high tensile and tear strength, excellent abrasion resistance, high elasticity, a range of hardness, good resistance to chemicals, and good low-temperature stability, result in applications mainly in the ink industry [28]. As compared with inks based on volatile organic solvents, WPU inks have some disadvantages like low printing speed and low water resistance. This can be improved by the use of a wide variety of PU dispersions with multifunctional post-chain extenders, including ethylene diamine (EDA), diethylenetriamine (DETA), and triethylenetetramine (TETA) [29].

10.6 SHAPE MEMORY AND POLYURETHANES

PUs are commercially the second most used polymeric material. They can have a large number of synthetic routes; hence, they are having different structures. They can be thermoplastic, elastomer, or foams. The most widely used ones are foams. These materials are also well-known for their specific property of resilience. Connecting to resilience, they are expected to show shape memory. Structurally, the PUs have consisted of hard and shape segments. This segmented structure is responsible for most of its properties, including the SME. PU offers a broad range of structure recovery, transition temperatures, retraction temperatures, intrinsic soft-hard segments, a high recoverable strain, good softening control, advantageous and adjustable physical qualities, and so on [30]. The mechanism of shape memory is dependent on the architecture and structure of the polymers. In the case of the segregated or segmented type of polymers like PU, the internal mechanism can be explained as follows. The molecular chains contain two segments called net point and molecular switches. Net points and molecular switches that are responsive to an external input are common components of appropriate polymer network topologies. In addition, the polymer must have adequate elastic deformability. The permanent shape of an SMPU is determined by the net points, which are linked by chain segments. To achieve the requisite deformability, the chain segments must enable a certain orientation, which increases with the length and flexibility of the chains. The recoiling of the chain segments allows the permanent form to be recovered. The temporary fixing of the conformation of the chain segments in the distorted shape is a need for stabilizing the temporary shape. Chemical or physical net points can determine the permanent form. Chemical net points stand for covalent bonding, and physical stands for the intermolecular interactions. Physical crosslinks are formed in a polymer whose morphology consists of at least two separated domains, such as a crystalline and an amorphous phase, using appropriate crosslinking chemistry. Domains are seen in multiphase polymers like PUs [31–33] (Figure 10.7).

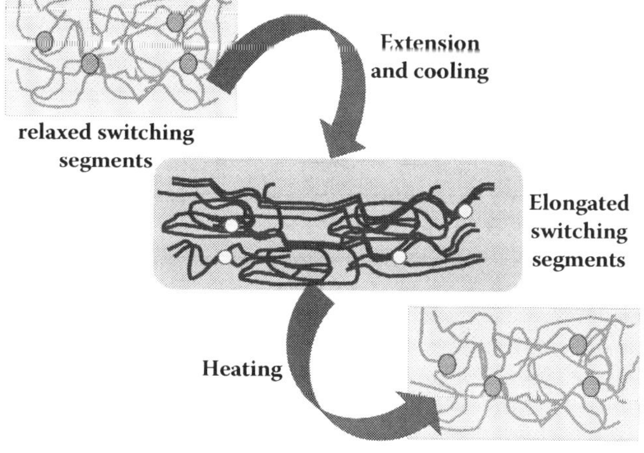

FIGURE 10.7 Mechanism of change in the structure during the shape memory effect.

The shape memory features of different forms of PUs and their composites have been explored quite often. The SMPUs were first reported by Hagoshi and Nagaya of Mitsubishi, Japan, and are the most ever applied material in the industry over these years [34]. A great number of reported researches suggest that the PUs are much efficient in terms of shape recovery and fixity [35]. Through the thermomechanical cycle tests on a PU thin film with loadings at various temperatures suggested that for loading over T_g, the strain has recovered at temperatures near the glass transition temperature, whereas for loading below T_g, it has recovered at temperatures near the glass transition midpoint temperature. Secondly, for loading over T_g, the rate of strain fixity was 98%, whereas for loading below T_g, it falls with increasing cycles. Again, except for the early cycles, the rate of strain recovery for loading exceeding T_g is 98%. Despite the variance in T_g, the thermomechanical characteristics of materials with various T_g were very similar [36]. PU foams are already well known for their resilience and were tested for shape recovery by Toubushi and coworkers. They concluded their work on the thermoplastic foam PUs that the stress is reduced and the distorted shape is stabilized by cooling the foam following compressive deformation at a high temperature. During the cooling process, stress-reduced dramatically in the temperature range below the glass transition temperature. Moreover, the original form has regained by heating the shape-fixed foam under no stress. At that point, the temperature area around T_g got a significant recovery of strain. The shape fixity ratio was 100%, whereas the shape recovery ratio was 98%. Neither of these ratios is affected by the number of cycles. Furthermore, the heating under the restriction of the fixed shape increases recovery stress. The applied maximum stress was around 80% of the recovery stress. At higher temperatures, relaxed stress does not return, and the shape distorted at high temperature was maintained under no load for 6 months without relying on the maximum strain. The original shape was then restored by heating. The original shape cannot be regained if the distorted shape is held at a high temperature. The holding conditions of strain, temperature, and time are all elements that influence shape recovery [37,38]. Segmented PU with lesser molecular weights can exhibit shape memory. PUs are known to show chemically induced and thermally induced SME. When the experiment was done by immersing PU wires in two different solvents, i.e. water and ethanol, the wires recorded a nominal T_g with a difference of 20°. The lower T_g material (immersed in water) took more time to straighten than the high T_g one (immersed in alcohol). In a solvent with hydrogen bonding ability, the glass transition temperature shows a reduction. The PU polymer gets swollen up in the water, which can be assured by spectroscopic analysis of the functionalities, like C=O and –N–H. There will be significant variation in these group vibrations. Also, when the solvent is changed from ethanol to methanol, instead of shape recovery, dissolution takes place. This experiment simply explains the phenomena of chemo-responsive SME. The changes during the experiment are represented in Figure 10.8 [39].

The PU itself and the blends and composites are also able to display shape recovery. Conductive PUs have many applications in microelectronics, aerospace, and robotics. Electroactive material fabrication is a very crucial area in smart material production. Conducting fillers and other conducting polymers can be

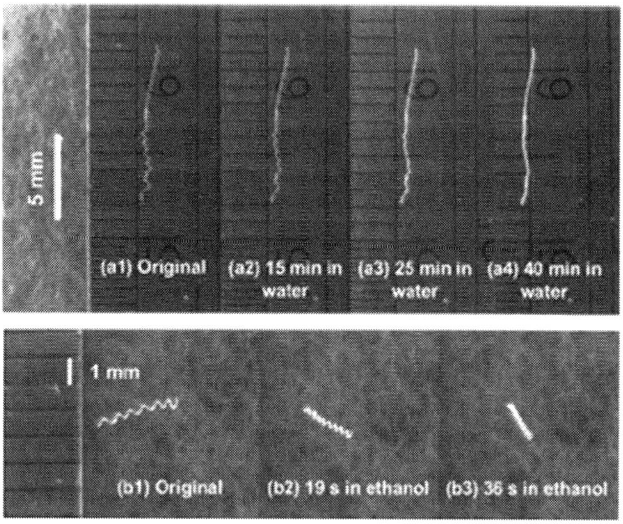

FIGURE 10.8 The shape memory effect is shown by polyurethane by the addition of different solvents (Reproduced with permission from reference [39]).

incorporated into the PU to make them more efficient. Multiwalled carbon nanotube-filled conducting PU composites showed prominent SME. The transition or the shape change from linear to bent happened in every 10-s interval [40].

Thermoset materials, even though they tend to show SME, are unsuccessful most of the time; they fail to regenerate the shape. Thermoresponsive shape memory thermoset PUs are known. It is reported that the permanent shape can be redefined or the plasticity can be introduced in the thermosets by rearranging the topology or surface by inducing transcarbamoylation. They believe that the discovery would have major benefits beyond shapeshifting because the key carbamate moiety in PUs is one of the most prevalent polymer-building elements [41].

An innovative attempt on the SMPUs could be the fabrication of fibers. Tremendous studies are available on this topic, and the SMPU fibers find application in smart textiles. The tuning of properties can be brought about by changing the fiber fabrication techniques (electrospinning, melt spinning), by surface modification chemically, and by the incorporation of filler materials. The PU molecules would be somewhat orientated in the direction of the fiber axis as a result of the spinning process [42–45]. A countless report is available in the SMPU field, but here the chapter focuses more on the WPUs, their composites, and SME.

10.7 SHAPE MEMORY OF WATERBORNE POLYURETHANE AND ITS COMPOSITES

WPU was widely accepted because of its key properties, such as nontoxicity, environmental friendliness, nonpolluting, nonflammability, etc. Since water is the solvent used in their synthesis and dispersions, they are more preferred to exhibit water-induced SMEs, but the reports on the shape memory aspects and mechanism of WPUs are negligibly less. The shape memory composites and surface modified forms are available for many applications.

WPUs can be synthesized as biodegradable by the inclusion of such polymers as a second phase. Poly-(ε-caprolactone) (PCL), polyhydroxy butyl valerate, poly (hydroxybutyrate), etc., can be blended with WPU. During the reaction, the PHB becomes the part of the soft segment in the WPU and PCL becomes the diol soft segment. Such a combination enhances the thermo-mechanical properties of the composites, and they are ready to give some SME. The shape recovery was up to 80% with a biodegradability of more than 50 in 30 days. These shape-memory materials were suggested for cardiovascular-related implants [46]. WPUs, different from solvent-based PUs, have poor dispersion and interfacial adhesion with other materials. Fillers are a very commonly employed method to enhance interfacial adhesion and relieve strain. Both organic and inorganic fillers were incorporated into WPU to make them more compatible. Silica, ZnO, titania, CNTs, clays, etc. are some of them. Most of the composites revealed enhanced properties, and some of them showed SME. The silica-based WPU composites recovered the shape during the thermomechanical cycle. For the repeated cycles at 1% functionalized silica loading, over 99% shape fixity and shape restoration were attained with minimal cyclic hysteresis [47]. Nano zinc oxide is an excellent filler for WPUs. The self-emulsified method can be employed to fabricate WPU/ZnO nanocomposite membranes with superior shape memory behavior. The shape fixity increased as a function of filler concentration and the soft segment crystallization during the cooling period corrected the shape changes caused by the high-temperature straining. The soft segment's crystallization was more flawless, which resulted in greater shape fixing. In the composites, the shape memory shown could be attributed to the strain release by the addition of ZnO. The internal tension accumulated during the shape change would be released from the composite membrane, resulting in shape recovery. The storage modulus in a high elasticity state would be improved by nanoparticles with proper dispersion in the composite membrane, and the recovery of internal stress would be increased. As a result, when the adjusted nano ZnO concentration was 0.45 wt.%, the shape recovery associated with the highest storage modulus of the membranes was at its peak [48].

Dual SMEs or multiple SMEs can be introduced in WPU films derived from caster-oil with the proper addition of dithiodiphenylamine (DTDA). Due to the increased crosslink density of PUs, the inclusion of DTDA can significantly improve the shape fixity of PU films. The Rf of the material at high temperatures increased from 29.4% to 91.9% when the DTDA ratios rose from 0% to 40% [49]. The thermomechanical changes are represented in Figure 10.9

However, increasing the number of disulfide bonds in PUs can enhance the chain mobility of the films through a dynamic reversible process, lowering the Rr of the material at high temperatures. The samples' quadruple form memories were proved by their capacity to fix three transient forms and subsequently recover them, based on the dual shape-memory phenomenon. After warming to the fixity temperature, the film precisely recovered each form and retained it for more than 30 min, demonstrating the sample's outstanding quadruple-shape memory. The changes in the films are depicted in Figure 10.10.

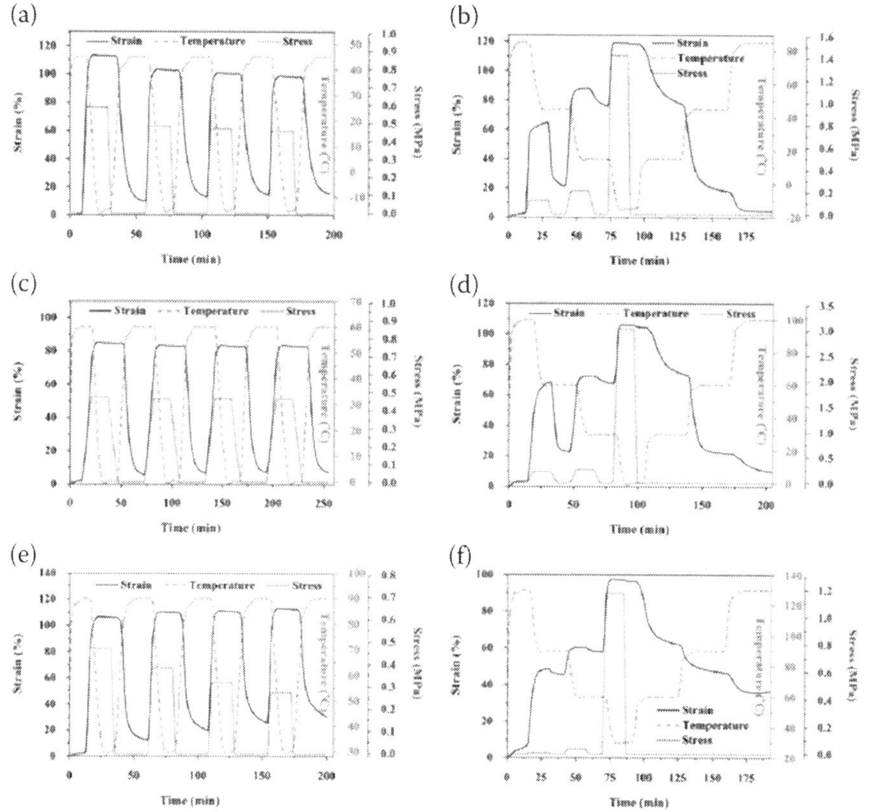

FIGURE 10.9 The shape recovery cycle of DTDA-incorporated WPU composites (Reproduced with permission from reference [49]).

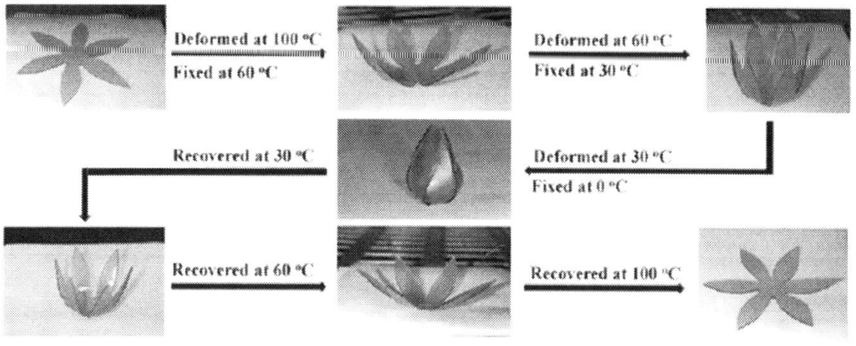

FIGURE 10.10 Demonstration of the blooming and closing of multiple shape memory composites of the DTDA incorporated WPU (Reproduced with permission from reference [49]).

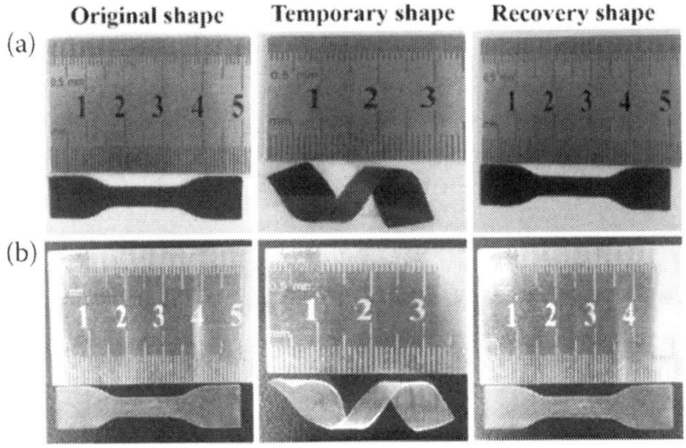

FIGURE 10.11 Shape memory effect of two different WPU composites (Adapted from reference [50]).

WPUs with both SME and recyclability are interesting; however, this combination is rare. An attempt can be made to graft or include polycaprolactone (PCL) on the end of rigid perylene bisimide (PBI), which is an organic dye, and to use it as a soft segment to make WPU. The fabricated films had microphase separation and thus revealed the ability to show SME. The shape memory experiment was done as follows. Under external stress, the dumbbell-like film was first shaped into a certain form at 25°C for 20 min [50]. The film was then placed in a refrigerator for 20 min to achieve a temporary form, which was regained after being immersed in 70°C water. The temporary spiral form restored its original shape in 9 s after being kept at 0°C for 20 min and then heated to 70°C (Figure 10.11).

10.8 APPLICATIONS OF SHAPE MEMORY WATERBORNE POLYURETHANES

WPUs and their composites are widely exploited in the coating industry and the biomedical field. Owing to their nontoxicity, biocompatibility, and recyclability, they are suitable in biomedical applications. Apart from that, the shape memory materials containing WPUs are employed in sensors, actuators, scaffolds, 3D printing inks, self-healing materials, and as matrices in different composites.

WPU scaffolds were fabricated using 3D printing technology. The basic component of the 3D printing ink for constructing bone scaffolds is water-based biodegradable SMPU. The 3D printing ink included 500 parts per million of superparamagnetic iron oxide nanoparticles to increase osteogenic induction and shape fixity, as well as polyethylene oxide (PEO) or gelatin to increase printability. Shape fixity and recovery were better in PU–PEO scaffolds than in PU–gelatin scaffolds, and shape memory characteristics in water were better than in air. They might be used as customizable replacements for bone tissue engineering in surgical procedures [51]. Medical implants can be produced from WPU surface modified with perfluoro polyether [52]. Biocompatible, water-based shape memory materials were fabricated from cellulose, and cellulose enhanced the tensile strength and modulus. The addition of bacterial cellulose enhanced the shape fixity of the films. These materials were suggested for application in the biomedical field [53]. Implantable tissue engineering can be fabricated from biodegradable WPUs, and they were fabricated through 3D printing technology. The printed products might help chondrocytes and fibroblasts stick together and proliferate. Because of the varying quantities of hydrophilic chain extenders, the printed blocks had a controlled degradability and did not gather acidic products [54]. Three-dimensional scaffolds for biomedical applications with biodegradability can be produced by freeze-drying technology. The scaffolds had an aligned and oriented porous structure in the horizontal direction. These scaffolds found their application in tissue regeneration [55].

Carbon dot grafted WPUs are efficient in the detection of Fe^{3+} ions specifically. The material was in the form of thin fluorescent films with self-healing properties. These materials can be suggested as a candidate in the category of intelligent sensing materials [56]. Self-healing materials for a variety of applications can be synthesized from nanocomposites of PUs. The composites had ZnO as the filler in the PU matrix [57]. Cellulose-based coatings materials of WPUs were produced from enzymatic cellulose derived from ancestral endoglucanase. The cellulose improved the thermomechanical properties and stability. They were applied in the paper coatings [58]. Recently, the castor oil-based WPU was converted to composites with the very advanced two-dimensional materials, the MXenes. They showed excellent compatibility and exhibited nearly 100% shape recovery. The materials were found to be apt for EMI shielding, and the flexibility suggested application in microelectronics [59]. Thus, WPUs that show SME are more relevant in advanced material-related research and industry.

10.9 CONCLUSIONS

WPUs are environmentally friendly materials without the usage of water as the solvent. The favorable properties, such as biocompatibility, biodegradability, recyclability, mechanical strength, flexibility, etc., make them widely useful in many applications. The special properties, such as SME and self-healing, make them more favorable to biomedical and smart materials. This chapter discussed briefly the general features of WPUs, and special emphasis was given to SME. The main parameters used to measure SME are shape recovery and shape fixity. The WPUs and their composites had excellent SME. Most of the reports suggested their applications in 3D printing, biomedical scaffolds, sensors, and microelectronic materials.

REFERENCES

1. E. Zharinova, M. Heuchel, T. Weigel, D. Gerber, K. Kratz, A. Lendlein, Water-blown polyurethane foams showing a reversible shape-memory effect, *Polymers (Basel)*. 8 (2016). doi:10.3390/polym8120412.
2. X. Wu, W.M. Huang, Y. Zhao, Z. Ding, C. Tang, J. Zhang, Mechanisms of the shape memory effect in polymeric materials, *Polymers (Basel)*. 5 (2013) 1169–1202. doi:10.3390/polym5041169.

3. A. Tcharkhtchi, S. Abdallah-Elhirtsi, K. Ebrahimi, J. Fitoussi, M. Shirinbayan, S. Farzaneh, Some new concepts of shape memory effect of polymers, *Polymers (Basel).* 6 (2014) 1144–1163. doi:10.3390/polym6041144.
4. J. Chen, Y. Zhang, J. Ge, H. Peng, S. Huang, Q. Yang, Y. Wen, Effect of thermo-mechanical cycling on martensitic transformation and shape memory effect in 304 austenitic steel, *Metals (Basel).* 10 (2020) 1–9. doi:10.3390/met10070901.
5. K. Strzelec, N. Sienkiewicz, T. Szmechtyk, Classification of shape-memory polymers, polymer blends, and composites, in: *Adv. Struct. Mater.*, (2020): pp. 21–52. doi:10.1007/978-981-13-8574-2_2.
6. S. Jose, J.J. George, S. Siengchin, J. Parameswaranpillai, Introduction to shape-memory polymers, polymer blends and composites: State of the art, opportunities, new challenges and future outlook, in: *Adv. Struct. Mater.*, (2020): pp. 1–19. doi:10.1007/978-981-13-8574-2_1.
7. C. Della Corte, Novel super-elastic materials for advanced bearing applications, *Adv. Sci. Technol.* 89 (2014) 1–9. doi:10.4028/www.scientific.net/ast.89.1.
8. S.A. Abdullah, A. Jumahat, N.R. Abdullah, L. Frormann, Determination of shape fixity and shape recovery rate of carbon nanotube-filled shape memory polymer nanocomposites, in: *Procedia Eng.*, (2012): pp. 1641–1646. doi:10.1016/j.proeng.2012.07.362.
9. L. Lu, J. Fan, G. Li, Intrinsic healable and recyclable thermoset epoxy based on shape memory effect and transesterification reaction, *Polymer (Guildf).* 105 (2016) 10–18. doi:10.1016/j.polymer.2016.10.013.
10. T. Tanpitaksit, M. Okhawilai, S. Rimdusit, Shape fixity and shape recovery of shape memory polymer, *J. Met. Mater. Miner.* 24 (2014) 43–47.
11. Z. Yang, W. Zhang, X. Guan, C. Zhang, Preparation, characterization, and properties of silicate/polyurethaneurea composites based on dipropylene glycol dibenzoate, *Polym. Compos.* 37 (2016) 37–43. doi:10.1002/pc.23152.
12. M.F. Sonnenschein, Introduction to polyurethane chemistry, in: *Polyurethanes*, (2021): pp. 111–132. doi:10.1002/9781119669401.ch3.
13. M.F. Sonnenschein, *Polyurethanes: Science, Technology, Markets, and Trends*, (2014). doi:10.1002/9781118901274.
14. Lepitreb, Bio-based Polyurethane (PU) Market analysis and forecast to 2020, *BioPlastics News.* (2015) 1.
15. V.L. Mucci, M.E.V. Hormaiztegui, M.I. Aranguren, Plant oil-based waterborne polyurethanes: A brief review, *J. Renew. Mater.* 8 (2020) 579–601. doi:10.32604/jrm.2020.09455.
16. M.S. Gaikwad, V. V. Kusumkar, O.S. Yemul, D.G. Hundiwale, P.P. Mahulikar, Eco-friendly waterborne coating from bio-based polyester amide resin, *Polym. Bull.* 76 (2019) 2743–2763. doi:10.1007/s00289-018-2511-y.
17. S. kumar Gaddam, A. Palanisamy, Anionic waterborne polyurethane-imide dispersions from cottonseed oil based ionic polyol, *Ind. Crops Prod.* 96 (2017) 132–139. doi:10.1016/j.indcrop.2016.11.054.
18. J.O. Akindoyo, M.D.H. Beg, S. Ghazali, M.R. Islam, N. Jeyaratnam, A.R. Yuvaraj, Polyurethane types, synthesis and applications—A review, *RSC Adv.* 6 (2016) 114453–114482. doi:10.1039/c6ra14525f.
19. H.K. Shendi, I. Omrani, A. Ahmadi, A. Farhadian, N. babnejad, M.R. Nabid, Synthesis and characterization of a novel internal emulsifier derived from sunflower oil for the preparation of waterborne polyurethane and their application in coatings, *Prog. Org. Coatings.* 105 (2017) 303–309. doi:10.1016/j.porgcoat.2016.11.033.
20. N. Sukhawipat, N. Saetung, P. Pasetto, J.F. Pilard, S. Bistac, A. Saetung, A novel high adhesion cationic waterborne polyurethane for green coating applications, *Prog. Org. Coatings.* 148 (2020). doi:10.1016/j.porgcoat.2020.105854.

21. X. Wang, Y. Hu, L. Song, W. Xing, H. Lu, P. Lv, G. Jie, UV-curable waterborne polyurethane acrylate modified with octavinyl POSS for weatherable coating applications, *J. Polym. Res.* 18 (2011) 721–729. doi:10.1007/s10965-010-9468-3.
22. I. Bramhecha, J. Sheikh, Development of sustainable citric acid-based polyol to synthesize waterborne polyurethane for antibacterial and breathable waterproof coating of cotton fabric, *Ind. Eng. Chem. Res.* 58 (2019) 21252–21261. doi:10.1021/acs.iecr.9b05195.
23. Q. Wu, J. Hu, Waterborne polyurethane based thermoelectric composites and their application potential in wearable thermoelectric textiles, *Compos. Part B Eng.* 107 (2016) 59–66. doi:10.1016/j.compositesb.2016.09.068.
24. Y. Han, J. Hu, Z. Xin, Facile preparation of high solid content waterborne polyurethane and its application in leather surface finishing, *Prog. Org. Coatings.* 130 (2019) 8–16. doi:10.1016/j.porgcoat.2019.01.031.
25. A. Lopez, E. Degrandi-Contraires, E. Canetta, C. Creton, J.L. Keddie, J.M. Asua, Waterborne polyurethane-acrylic hybrid nanoparticles by miniemulsion polymerization: Applications in pressure-sensitive adhesives, *Langmuir.* 27 (2011) 3878–3888. doi:10.1021/la104830u.
26. R. De Chen, C.F. Huang, S. hui Hsu, Composites of waterborne polyurethane and cellulose nanofibers for 3D printing and bioapplications, *Carbohydr. Polym.* 212 (2019) 75–88. doi:10.1016/j.carbpol.2019.02.025.
27. Y. Liu, Y. Zou, J. Wang, S. Wang, X. Liu, A novel cationic waterborne polyurethane coating modified by chitosan biguanide hydrochloride with application potential in medical catheters, *J. Appl. Polym. Sci.* 138 (2021). doi:10.1002/app.50290.
28. X. Zhou, Y. Li, C. Fang, S. Li, Y. Cheng, W. Lei, X. Meng, Recent advances in synthesis of waterborne polyurethane and their application in water-based ink: A review, *J. Mater. Sci. Technol.* 31 (2015) 708–722. doi:10.1016/j.jmst.2015.03.002.
29. L. Lei, L. Zhong, X. Lin, Y. Li, Z. Xia, Synthesis and characterization of waterborne polyurethane dispersions with different chain extenders for potential application in waterborne ink, *Chem. Eng. J.* 253 (2014) 518–525. doi:10.1016/j.cej.2014.05.044.
30. N. V. Gama, A. Ferreira, A. Barros-Timmons, Polyurethane foams: Past, present, and future, *Materials (Basel).* 11 (2018). doi:10.3390/ma11101841.
31. M. Behl, J. Zotzmann, A. Lendlein, Shape-memory polymers and shape-changing polymers, *Adv. Polym. Sci.* 226 (2010) 1–40. doi:10.1007/978-3-642-12359-7-26.
32. F. Pilate, A. Toncheva, P. Dubois, J.M. Raquez, Shape-memory polymers for multiple applications in the materials world, *Eur. Polym. J.* 80 (2016) 268–294. doi:10.1016/j.eurpolymj.2016.05.004.
33. C.M. Yakacki, Shape-memory and shape-changing polymers, *Polym. Rev.* 53 (2013) 1–5. doi:10.1080/15583724.2012.752745.
34. S. Hayashi, Technical Report on Shape Memory Polymers, Nagoya Research and Development Center, Mitsubishi Heavy Industries, Inc., *Nagoya, Japan* (1990).
35. W.M. Huang, Z. Ding, C.C. Wang, J. Wei, Y. Zhao, H. Purnawali, Shape memory materials, *Mater. Today.* 13 (2010) 54–61. doi:10.1016/S1369-7021(10)70128-0.
36. H. Tobushi, T. Hashimoto, N. Ito, S. Hayashi, E. Yamada, Shape fixity and shape recovery in a film of shape memory polymer of polyurethane series, *J. Intell. Mater. Syst. Struct.* 9 (1998) 127–136. doi:10.1177/1045389X9800900206.
37. H. Tobushi, D. Shimada, S. Hayashi, M. Endo, Shape fixity and shape recovery of polyurethane shape-memory polymer foams, *Proc. Inst. Mech. Eng. Part L J. Mater. Des. Appl.* 217 (2003) 135–143. doi:10.1243/146442003321673635.
38. H. Tobushi, S. Hayashi, M. Endo, D. Shimada, Shape fixity and shape recovery of polyurethane-shape memory polymer foam, Nippon Kikai Gakkai Ronbunshu, A Hen/Transactions Japan Soc. *Mech. Eng. Part A.* 68 (2002) 1594–1599. doi:10.1299/kikaia.68.1594.

39. W.M. Huang, Y. Zhao, C.C. Wang, Z. Ding, H. Purnawali, C. Tang, J.L. Zhang, Thermo/chemo-responsive shape memory effect in polymers: A sketch of working mechanisms, fundamentals and optimization, *J. Polym. Res.* 19 (2012). doi:10.1007/s10965-012-9952-z.
40. J.W. Cho, J.W. Kim, Y.C. Jung, N.S. Goo, Electroactive shape-memory polyurethane composites incorporating carbon nanotubes, *Macromol. Rapid Commun.* 26 (2005) 412–416. doi:10.1002/marc.200400492.
41. N. Zheng, Z. Fang, W. Zou, Q. Zhao, T. Xie, Thermoset shape-memory polyurethane with intrinsic plasticity enabled by transcarbamoylation, *Angew. Chemie - Int. Ed.* 55 (2016) 11421–11425. doi:10.1002/anie.201602847.
42. Q. Yang, G. Li, Spider-silk-like shape memory polymer fiber for vibration damping, *Smart Mater. Struct.* 23 (2014). doi:10.1088/0964-1726/23/10/105032.
43. F.L. Ji, Y. Zhu, J.L. Hu, Y. Liu, L.Y. Yeung, G.D. Ye, Smart polymer fibers with shape memory effect, *Smart Mater. Struct.* 15 (2006) 1547–1554. doi:10.1088/0964-1726/15/6/006.
44. Q. Meng, J. Hu, Y. Zhu, J. Lu, Y. Liu, Morphology, phase separation, thermal and mechanical property differences of shape memory fibres prepared by different spinning methods, *Smart Mater. Struct.* 16 (2007) 1192–1197. doi:10.1088/0964-1726/16/4/030.
45. Y. Zhu, J. Hu, L.Y. Yeung, J. Lu, Q. Meng, S. Chen, K.W. Yeung, Effect of steaming on shape memory polyurethane fibers with various hard segment contents, *Smart Mater. Struct.* 16 (2007) 969–981. doi:10.1088/0964-1726/16/4/004.
46. S.H. Hsu, C.T. Hsieh, Y.M. Sun, Synthesis and characterization of waterborne polyurethane containing poly(3-hydroxybutyrate) as new biodegradable elastomers, *J. Mater. Chem. B.* 3 (2015) 9089–9097. doi:10.1039/c5tb01773d.
47. S.K. Lee, S.H. Yoon, I. Chung, A. Hartwig, B.K. Kim, Waterborne polyurethane nanocomposites having shape memory effects, *J. Polym. Sci. Part A Polym. Chem.* 49 (2011) 634–641. doi:10.1002/pola.24473.
48. X. Li, S.C. Li, J.Y. Wang, Preparation and properties of shape memory water borne polyurethane/modified Nano-ZnO composites membranes, *Polym. - Plast. Technol. Eng.* 52 (2013) 1482–1487. doi:10.1080/03602559.2013.820750.
49. C. Zhang, H. Liang, D. Liang, Z. Lin, Q. Chen, P. Feng, Q. Wang, Renewable castor-oil-based waterborne polyurethane networks: Simultaneously showing high strength, self-healing, processability and tunable multishape memory, *Angew. Chemie Int. Ed.* 60 (2021) 4289–4299. doi:10.1002/anie.202014299.
50. K. Wei, H. Zhang, J. Qu, J. Wang, Y. Bai, F. Sai, Recyclable shape-memory waterborne polyurethane films based on perylene bisimide modified polycaprolactone diol, *Polymer.* 13 (2021). doi:10.3390/polym13111755.
51. Y.-J. Wang, U.-S. Jeng, S. Hsu, Biodegradable water-based polyurethane shape memory elastomers for bone tissue engineering, *ACS Biomater. Sci. Eng.* 4 (2018) 1397–1406.
52. Z. Wang, Z. Hou, Y. Wang, Fluorinated waterborne shape memory polyurethane urea for potential medical implant application, *J. Appl. Polym. Sci.* 127 (2013) 710–716. doi:10.1002/app.37862.
53. L. Urbina, A. Alonso-Varona, A. Saralegi, T. Palomares, A. Eceiza, M.Á. Corcuera, A. Retegi, Hybrid and biocompatible cellulose/polyurethane nanocomposites with water-activated shape memory properties, *Carbohydr. Polym.* 216 (2019) 86–96. doi:10.1016/j.carbpol.2019.04.010.
54. Z. Feng, D. Wang, Y. Zheng, L. Zhao, T. Xu, Z. Guo, M. Irfan Hussain, J. Zeng, L. Lou, Y. Sun, H. Jiang, A novel waterborne polyurethane with biodegradability and high flexibility for 3D printing, *Biofabrication.* 12 (2020). doi:10.1088/1758-5090/ab7de0.

55. W. Lin, W. Lan, Y. Wu, D. Zhao, Y. Wang, X. He, J. Li, Z. Li, F. Luo, H. Tan, Q. Fu, Aligned 3D porous polyurethane scaffolds for biological anisotropic tissue regeneration, *Regen. Biomater.* 7 (2020) 19–27. doi:10.1093/rb/rbz031.
56. L. Xiao, J. Shi, B. Nan, W. Chen, Q. Zhang, E. Zhang, M. Lu, Highly sensitive detection of Fe^{3+} ions using waterborne polyurethane-carbon dots self-healable fluorescence film, *Macromol. Mater. Eng.* 305 (2020). doi:10.1002/mame.201900810.
57. M.M. Rahman, Polyurethane/zinc oxide (PU/ZnO) composite-synthesis, protective property and application, *Polymers (Basel).* 12 (2020). doi:10.3390/polym12071535.
58. B. Alonso-Lerma, I. Larraza, L. Barandiaran, L. Ugarte, A. Saralegi, M.A. Corcuera, R. Perez-Jimenez, A. Eceiza, Enzymatically produced cellulose nanocrystals as reinforcement for waterborne polyurethane and its applications, *Carbohydr. Polym.* 254 (2021). doi:10.1016/j.carbpol.2020.117478.
59. J. Lu, Y. Zhang, Y. Tao, B. Wang, W. Cheng, G. Jie, L. Song, Y. Hu, Self-healable castor oil-based waterborne polyurethane/MXene film with outstanding electromagnetic interference shielding effectiveness and excellent shape memory performance, *J. Colloid Interface Sci.* 588 (2021) 164–174. doi:10.1016/j.jcis.2020.12.076.

11 Waterborne Polyurethanes for Self-Healing Applications

Masoumeh Kianfar, Shiva Mohajer, and
Mir Saeed Seyed Dorraji
Department of Chemistry, Faculty of Science,
University of Zanjan, Zanjan, Iran

CONTENTS

11.1 Introduction ... 177
11.2 Classification of SHWPU Polymers ... 178
11.2.1 Intrinsic and Extrinsic SHWPU ... 178
11.2.2 Autonomous and Non-Autonomous SHWPU ... 179
11.3 Synthesis ... 179
11.3.1 Synthesis of Intrinsic SHWPU Based on Disulfide Bonds ... 179
11.3.2 Synthesis of Intrinsic SHWPU Based on Visible-Light-Induced Dynamic Covalent Bonds ... 181
11.3.3 Synthesis of Intrinsic SHWPU Based on Diels–Alder Chemistry ... 181
11.3.4 Synthesis of SHWPU Containing H-Bond Moieties ... 182
11.3.5 Synthesis of Extrinsic SHWPU Based on Continuous Microvascular Fillers ... 182
11.3.6 Synthesis of Extrinsic SHWPU Based on Capsules ... 182
11.3.7 Synthesis of SHWPU Nanocomposites ... 183
11.4 Self-Healing of WPU Polymer Nanocomposites (Self-Healing by Nanoparticles) ... 183
11.5 Characterization ... 184
11.6 Application ... 187
11.7 Conclusion and Future Perspective ... 189
References ... 190

11.1 INTRODUCTION

Polyurethane (PU) was invented in the 1930s by Professor Otto Byer and his cow-orkers. Waterborne polyurethane (WPU) is one of the active branches of PU chemistry. They are rapidly developing. Their use has increased due to their special properties.

DOI: 10.1201/9781003173526-11

The chemistry of WPUs synthesis involves the reaction between PU pre-polymer and water [1]. It's an organic reaction in a water medium. In the formulation of WPUs, the common step to induce hydrophilicity into the PU network is structural modification. The structural modification consists of the introduction of hydrophilic groups into hydrophobic PU using ionic internal emulsifiers. Followed by the neutralizer is added to neutralize ionic groups in the polymer chains and deionized water is used to emulsify and disperse the resultant hydrophilic WPU pre-polymers along with the chain extension process. Water has a multi-role during WPU synthesis: solvent, emulsifier, and chain extender [2]. WPUs mainly consist of PU ionomers and ionic moieties. PU ionomers are prepared by reaction between polyols and isocyanate monomers.

On the other hand, the important thing to note is the damage of WPU films, which is one of the most important affecting factors on WPU lifespan. Abrasion, fatigue, impact, etc are some of the external factors which could make fractures or cracks in WPU films. If it occurs, WPU-based product performance will be affected and their service life will be extremely decreased [3–5]. To solve the problems of the WPU damages and extend their service life, reliability, and durability, it is of great significance to impart self-healing ability to WPUs [4]. With this smart feature, the damaged area is repaired without the need for detection or any type of manual intervention [6].

Today, SH WPUs are very attractive and popular among self-healing polymers since different from the contemporary self-healing polymers, which need a large amount of organic solvents to disperse. SHWPUs are dispersed in water, and they are eco-friendly and affordable [7,8]. Due to the problems mentioned above and the need to produce smart materials with higher performance, especially in waterborne polyurethanes used in various applications, it is necessary to provide smart generation of WPU with self-healing properties.

11.2 CLASSIFICATION OF SHWPU POLYMERS

11.2.1 Intrinsic and Extrinsic SHWPU

According to the self-healing mechanism used in WPUs, they are divided into intrinsic and extrinsic SHWPUs. The intrinsic SHWPU itself contains a specific molecular structure which makes it needless from the implantation of an external repair agent. Intrinsic SH WUs generally are based on either non-covalent chemistries or dynamic covalent chemistries [9–22]. The non-covalent chemistry approach uses π–π stacking, ligand–metal bonding, hydrogen bonding, or host–guest interaction stacking among other techniques [18,23], while covalent approaches use Diels–Alder reaction, radical exchange, dynamic urea bond, and trans-esterification for instance [15,24]. In this type of healing mechanism, WPU films are exposed to the external stimulus then their original function and structure are restored through reversible reactions.

The main difference between intrinsic and extrinsic SHWPUs lies in their chemistries. Extrinsic SHWPUs are based on an external healing agent in the form of capsules or vascular networks [25–27]. In both of them, the self-healing process is initiated by the external/ internal damage/rupture in the vascular networks and also in the capsules. Silica and PAN are suitable materials for making capsule shells. Recently, the development tendencies of SHWPUs have moved to intrinsic healing systems since the intrinsic SHWPUs themselves contain a specific molecular structure that makes them needless of external repair agents [10].

11.2.2 Autonomous and Non-Autonomous SHWPU

Also, there is another category for healable WPUs. They can be divided into autonomic and non-autonomic systems based on their lack of need and need for an external stimulus respectively [13,26,28,29]. In SHWPUs, when damage or rupture occurs, the filler or healing agent is automatically released and the damaged area is healed. Generally, autonomic systems are organized into encapsulation and highly dynamic supramolecular systems [25,26,30]. In contrast, in non-autonomic SHWPUs, healing the damaged area occurs after applying an external stimulus. Light and heat are the most common stimuli for this type of SHWPUs because they can be easily applied in service environments [4,12,14,16–21,23,27].

11.3 SYNTHESIS

11.3.1 Synthesis of Intrinsic SHWPU Based on Disulfide Bonds

SHWPUs based on reversible chemistry mechanisms, unlike microcapsules, can be healed repeatedly. This feature has attracted the attention of researchers to synthesis the intrinsic SHWPU. One way to produce this type of polymer is to introduce a disulfide bond into the main chain of WPU [31]. The process of preparing an SHWPU dispersion based on a disulfide bond typically involves follows steps:

1. Preparation of pre-polymer
2. Chain expansion step, which is carried out by the disulfide material
3. Neutralization step
4. Water addition and preparation of the WPU dispersion (removing solvents like acetone or methylethylketone is done just in the acetone process).
5. Preparation of film specimens

Generally, WPUs are synthesized in a 250 mL three-necked flask equipped with a mechanical stirrer, a nitrogen inlet, and a condenser. First, the polyol is melted. Then, it is mixed with a hydrophilic internal emulsifier (commonly 2-bis (hydroxymethyl) propionic acid (DMPA)) and heated at 80°C for some hours. Then, isophorone diisocyanate (IPDA) is added to the flask and stirred for 1.5 h. Subsequently, the mixture is cooled to 50°C, and dibutyltin dilaurate (DBTDL) as a catalyst is added to the mixture.

Then, the chain extender (like disulfide materials) is added gradually, and the mixture is further being stirred for another 1 h, after which the mixture is cooled to ambient temperature. The required amount of neutralizer (like TEA) is then added dropwise to neutralize the ionic groups of emulsifiers. After 30 min, for the reaction of residual NCO groups diamine (like ethylene diamine (EDA), hydrazine hydrate,

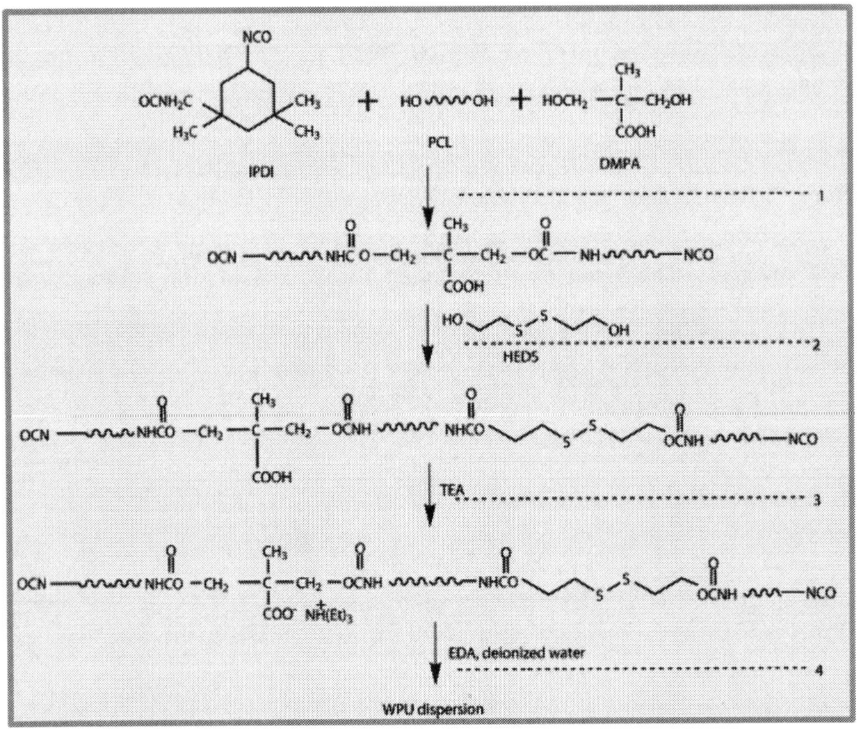

FIGURE 11.1 Elementary steps for the preparation of WPU dispersions. Reproduced from reference [31]. Copyrighting (2017) Springer.

and so on) is used. Finally, deionized (DI) water is slowly added and stirred for 40 min with an agitation speed of 1,500 rpm. The WPU emulsions are poured into a Teflon mold with appropriate dimensions and then dry under require conditions.

Wan and Chen [31] have prepared a non-autonomic self-healing-WPU by using 2-hydroxyethyl disulfide (HEDS) as a chain extender. Figure 11.1 shows the preparation steps of SHWPU.

Zhang et al. [4] prepared non-autonomic SHWPU by pre polymer mixing process. They used different contents of HEDS as a chain extender and studied its effect on the healing performance of the final films. Lee et al. [19] also synthesized non-autonomic SHWPUs by using the pre-polymer process. They used cystamine as a chain extender to prepare disulfide group-bearing WPUs. Aguirresarobe et al. [13] have prepared an autonomic self-healing-WPU by using aromatic disulfide dynamic structures (4-aminophenyl disulfide (APDS)) as a chain extender. They used the acetone process to synthesis WPU. A significant difference of this methodology from the method mentioned above is in using acetone as a solvent, which is evaporated to obtain the final WPU dispersion at the end. They also used *N*-butyl-3-aminopropyl trimethoxysilane for material cross-linking. Aguirresarobe et al. [13] synthesized self-healing waterborne poly (urethane-urea)s by using the acetone process. They used different aromatic disulfide compounds as chain extenders and made mechanically robust products. The aromatic disulfide compounds used are bis(4-hydroxyphenyl)disulfide, bis(4-aminodiphenyl)disulfide, bis[4-(3′hydroxypropoxy)phenyl]disulfide].

Huang et al. [32] synthesized a self-antiglare waterborne poly(urea-urethane) with room-temperature self-healing capability. The self-healing agents in their work were quadruple H-bond of the zigzag array and linear array and disulfide bond.

Liang et al. [20] synthesized a type of SHWPU with a disulfide bond in the main chain by using HEDS as a chain extender. Zhang et al. [21] used the three-step chain extending process to synthesis self-healing WPU. They used different chain extenders (4,4′-disulfanediyldianiline, 2,2′-disulfanediyldianiline, 4,4′-oxydianiline, triethylenetertramine) to induce intrinsic self-healing properties to the WPU.

Zhang et al. [14] made castor oil-based WPU containing reversible covalent disulfide bonds of DTDA, multiple H-bonds, and urethane bonds.

11.3.2 Synthesis of Intrinsic SHWPU Based on Visible-Light-Induced Dynamic Covalent Bonds

Today, visible-light-induced dynamic covalent bonds like diselenide, ditelluride, and aromatic Schiff base (ASB) bonds are more used to fabricate SHWPUs. Yong Jin and colleagues synthesized WPUs containing dynamic diselenide/or ditelluride/or ASB bonds along with hydrogen bonds [7,33,34]. The synthetic route is as follows:

PTMG, IPDI, and DMPA are fed in a reactor under stirring with the protection of nitrogen. Subsequently, DBTL is fed and the reaction is kept at 80°C. After that, components containing diselenide/or ditelluride/or ASB bonds are added to the mixture. When the temperature was reduced to ambient temperature, TEA dissolved in acetone is fed and reacted for 15 min. Afterward, DI water is added dropwise and vigorously stirred.

In addition to the above, chain extenders, such as coumarin [35,36], chitosan [37], and the sulfonate groups [17], were also applied successfully in the making of SHWPUs.

11.3.3 Synthesis of Intrinsic SHWPU Based on Diels Alder Chemistry

One of the most employed methods to create intrinsic self-healing features in materials is introducing Dils-Alder (DA) moieties in the polymer chain. There are two common strategies for synthesizing WPU/DA emulsion, one-pot strategy, and two-pot strategy. In one pot strategy, in the beginning, a Diels-Alder adduct (DA-diol) is synthesized by the reaction between diene components (like furfuryl alcohol) and dienophile components (like *N*-(2-hydroxyethyl maleimide)). The prepared DA diol is introduced in the WPU chains by the polymerization in the acetone process.

The two pot strategy generally involves three steps of synthesizes [11,15]:

a. Synthesis of diene component-modified WPU
b. Synthesis of dienophile component-modified WPU
c. Mixing of a and b dispersions to form WPU-containing DA moieties

Fang et al. [11] used the acetone process to construct furan modified WPU. They applied furfuryl alcohol and dihydroxy furan (F–2OH) along with other PU main materials to introduced diene components in the polymer chains. In the next step, they used *N*-(2-hydroxyethyl)-maleimide (M-OH) and dihydroxyl maleimide (M-2OH) along with other PU primary materials to construct PU chains containing dienophile components. Afterward, these two modified WPU dispersions mixed and DA bonds formed between chains as crosslinks.

11.3.4 Synthesis of SHWPU Containing H-Bond Moieties

In WPU structure, there are urethane groups that have hydrogen bond interaction together. Also, one of the ways to create a hydrogen bond between the chains is the entry of materials with this property into the WPU backbone. Materials that could cause dual or quadruple H-bond between chains.

Fan et al. [33] used ureido-4[1H]-pyrimidinone (UPy) to reinforced ditelluride-based WPU features. They synthesized 5-(2-hydroxyethyl)-6-methyl-2-aminouracil (UPy OH)) as follows: A mixture of 2-acetylbutyrolactone and guanidine carbonate was refluxed with absolute ethanol in the presence of triethylamine for 12 h. Then, the precipitate of UPy OH was obtained. The presence of UPy moieties creates physical crosslinks between chains and thus improves the mechanical and healing WPU properties.

11.3.5 Synthesis of Extrinsic SHWPU Based on Continuous Microvascular Fillers

One way to SHWPUs is to add continuous microvascular fillers in the polymeric matrix. Li et al. [26] prepared self-healing coatings by embedding electrospun fibers in WPUs. They used coaxial electrospinning to make the core(linseed oil)-shell (graphene oxide reinforced PAN) fibers. Then, WPU dispersion was sprayed on the porous fiber mats deposited on hot-dip galvanized steel substrates and dried at 40°C for 24 h.

11.3.6 Synthesis of Extrinsic SHWPU Based on Capsules

One of the common methods to create self-healing properties is to use capsules in the polymer matrix. Healing agent encapsulation could be based on the oil in water emulsion method. In this method, the first emulsion of microcapsules is prepared, then it is combined with WPU emulsion by continuously stirring.

Li et al. [25] provided SHWPU coating by using the encapsulation method. In the first step, they prepared graphene oxide microcapsules (GOMCs), containing linseed oil as the healing agent. Graphene oxide (GO) was prepared using the modified Hummers method then the aqueous solution of GO was mixed with linseed oil by high shear mixing to prepare Pickering emulsions. After that, GO microcapsules emulsions were mixed with a WPU by gentle magnetic stirring to prepare a liquid composite coating.

Mirmohseni et al. [27] made silica capsules by dynamic self-assembly of tetraethoxy orthosilicate molecules at the oil–water interface. In this method, there were two phases, aqueous and oil phases. The aqueous phase contained CTAB as the cationic surfactant, ammonia solution as the catalyst, and water. The organic phase included 2-mercaptobenzothiazole as the core of capsules and diethyl ether. The organic phase was added to the aqueous phase while stirring. Then, tetraethoxy orthosilicate (TEOS) as the precursor of shell material was added drop-wise under intense agitation. A silica network was formed between organic and aqueous phases, and capsules are precipitated and collected by centrifuging.

11.3.7 Synthesis of SHWPU Nanocomposites

Generally, WPU composites could be obtained by the physical mixing of particles such as titanium dioxide and beta-cyclodextrin nanoparticles, and collagen in WPU dispersion under constant stirring. These particles reinforce the self-healing and mechanical properties of WPU films.

Eceiza and Tercjak [29] prepared WPU nanocomposite by incorporating TiO2 nanoparticles into synthesized solvent-free waterborne poly(urethane-urea) (WPUU) based on hydrophilic poly(ethylene oxide). At first, they synthesized WPUU dispersion. In this step, diphenylmethane diisocyanate (MDI)and poly (ethylene oxide) (PEO) were added into the flask and reacted at 80°C for 2 h. After cooling to room temperature, the neutralized chain extender 2,4-diamino benzenesulfonic acid (DBSA) was added to the mixture. Finally, DI water was used to form proper dispersion. In the next step, WPU dispersion and TiO2 nanoparticles in the commercial aqueous were mixed by stirring for 24 h at 500 rpm.

Wan and Chen [38] prepared SHWPU nanocomposites by using β-cyclodextrin. First, β-CD aqueous solution was modified by blending with polyethylene glycol (PEG). They were stirred at 65°C for 1 h to obtain a homogenous solution. Then, the CD/PEG inclusion compound was prepared after the temperature decreased to 25°C. After 24 h, the formed precipitate was collected. In the next step, the modified β-CD was added to a SHWPU through a simple solution blending method.

Han et al. [39] made self-healing collagen/WPU composites. They used the collagen solution to improve WPU features. The collagen solution was created by acid extraction. In this method, collagen, which contained tissues, was dissolved in acid solution at a low temperature. Then, it was centrifugated, and its supernatant was taken. For the collagen precipitate, the supernatant was adjusted to a pH of 7–8. After that, the precipitate was washed with water several times to obtain purified collagen. In the next step, WPU and prepared collagen, which was dissolved into an acid solution, are mixed under constant stirring at room temperature for 12 h.

11.4 SELF-HEALING OF WPU POLYMER NANOCOMPOSITES (SELF-HEALING BY NANOPARTICLES)

SHWPUs face challenges, such as less mechanical stability, which limits their use in various fields [40]. One way to overcome this limitation is to use SHWPU as a nanocomposite. Nanocomposites consist of an inorganic nanometer-sized phase dispersed in a polymer. These materials are used in various fields, such as solar cells, photocatalysis, gas separation, and so on [13]. In SHWPUs, adding inorganic nano-sized particles increases their properties [41]. This result is a mixture of the properties of polymers and nanoparticles, which ultimately improves the mechanical and thermal properties [42].

The addition of different types of nanomaterials, such as Au, Ag, graphene, TiO_2, ZnO, $CaCO_3$, silica clays, Fe_3O_4, carbon nanotubes, cellulose nanocrystals, etc., to SHWPU is a useful method to enhance its mechanical properties, which does not reduce the self-healing property of this material and only increases its mechanical properties [43].

SHWPU polymer nanocomposites have enhanced healing and mechanical properties compared to pure SHWPU polymer. Díez-García et al. [43] studied the combination of TiO_2 nanoparticles reaching 40 wt.% with prepared solvent-free WPUU [waterborne poly(urethane-urea)] basis on hydrophilic PU0 [poly(ethylene oxide)] to increase the healing and mechanical efficiency of a polymer matrix. The results showed that with increasing the amount of TiO_2, mechanical and healing properties of the synthesized nanocomposite films increased and were higher than PU0. In this work, an easy and practical method for controlling the distribution of TiO_2 in WPUU for the making of nanocomposites with enhanced self-healing and mechanical features is explained.

11.5 CHARACTERIZATION

To evaluate properties of SHWPU, including physical, mechanical, etc., various analyzes have been performed. For example:

In the characterization of SHWPU composites:

Han et al. [39] prepared WPU/CO (WPU/collagen) composites in an eco-friendly, easy way. The prepared samples had properties such as vapor-stimulated healing capability, increased mechanical properties, and enhanced temperature-adaptive water vapor permeability (WVP). FTIR spectra of WPU-CO composites confirm the molecular structure of polyurethane.

Based on TEM images, WPU/CO composites contained collagen in fiber form (Figure 11.2(a1) and (b1)). In WPU/CO5.4% and WPU/CO3.8% composites, there was a 3–6 nm diameter range for single fibers (Figure 11.2(a2) and (b2)), which was related to the creation of the collagen fibril.

The size distributions indicated that the maximum amount of the collagen fibril had a diameter of 4 nm. The XRD spectra of WPU/CO and WPU composites did not show any different crystalline peaks, so all samples belong to the soft polyurethane type. The DTG and TG curves revealed that the WPU/CO composites' thermal decomposition temperature was noticeably better than that of WPU composites. Fracture cross-surface images for WPU/CO and WPU composites were shown in Figure 11.3. The results obtained from SEM images showed that WPU/CO composites do not have any layers or accumulations, which indicates a good harmony between WPU and collagen and also have a relatively smooth structure due to increased mechanical hardness of materials, but in contrast, WPU composites have a rough fracture surface.

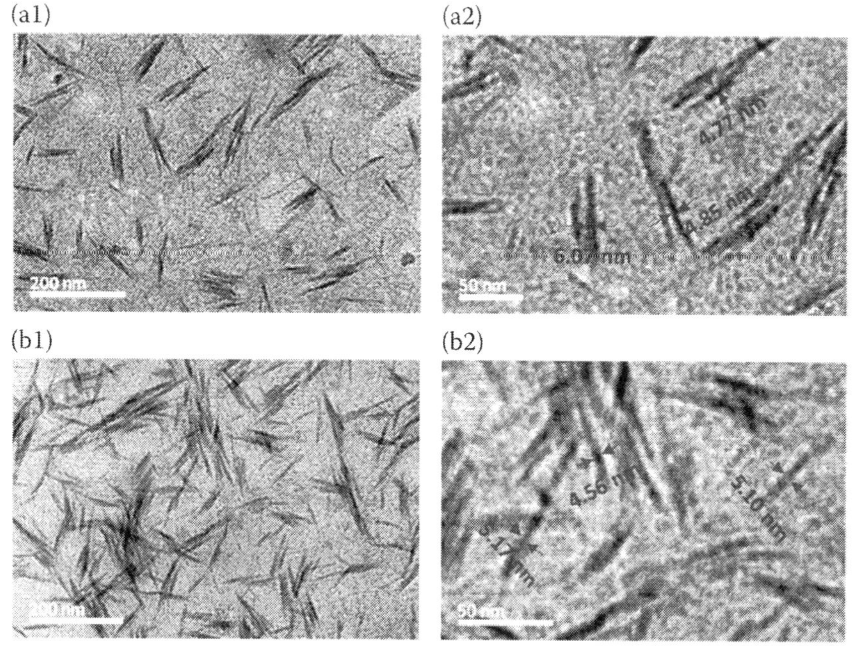

FIGURE 11.2 TEM images of WPU/CO3.8% (a1) and WPU/CO5.4% (b1); enlarged images of WPU/CO3.8% (a2) and WOU/CO5.4% (b2), with fiber diameter. Reproduced from reference [39]. Copyright (2020) Elsevier.

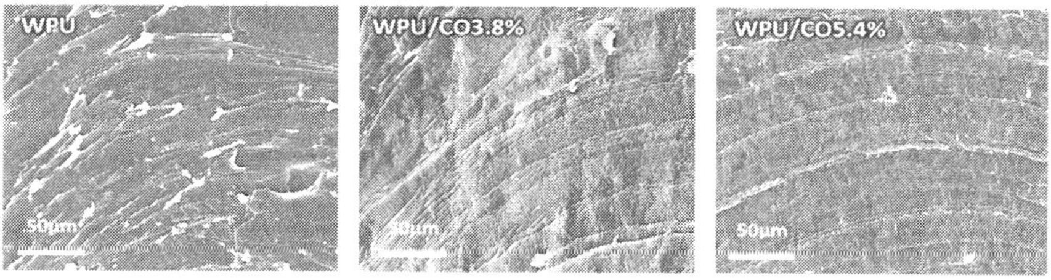

FIGURE 11.3 SEM micrograph for WPU/CO and WPU composites fracture cross-section. Reproduced from reference [39]. Copyright (2020) Elsevier.

The results obtained from comparing the tensile stress between WPU and WPU/CO composites showed that these composites have better tensile stress under unloading and loading cycles in various strains. The healing efficiency of coatings with WPU/CO composites was investigated. Increasing the amount of collagen in the range of 0% up to 5.4%, increased the self-healing efficiency of tensile stress (from 35% to 68%) and strain (from 10% to 75%) of coatings.

In the characterization of SHWPU with disulfide as a chain extender:

Wan and Chen [31] offered a novel method to prepare SHWPU from poly(e-caprolactone) glycol, isophorone diisocyanate, 2-hydroxyethyl disulfide, and 2 bis (hydroxymethyl) propionic acid. To prepare WPU with high healing property and emulsion stability, emulsions with different proportions of disulfide having chain extender and hydrophilic internal emulsifier were prepared and studied. The structure of the prepared WPU was confirmed by Raman and FTIR analysis. Raman spectroscopy was used to verify the presence of disulfide bonds whose signal was not apparent by FTIR, which showed the presence of disulfide bonds in PU5, PU3, PU7, and PU6 samples. The results showed that by increasing the molar ratio of HEDS/2-bis(hydroxymethyl) propionic acid (DMPA), the size distribution was first reduced and then increased. TGA was used to evaluate the thermal properties of WPU. The thermal stability of the WPUs was enhanced due to the exchange reaction between the disulfide bonds. GPC, DMA, DSC, polarizing microscope, and surface analysis were also carried out to determine the properties and structure of prepared WPUs. Based on the results, all synthesized WPUs have increased thermal and high dispersion stability.

In the characterization of SHWPU with dynamic aromatic Schiff base (ASB) bonds:

Based on ASB bonds, Fan et al. [34] developed a new kind of dynamic covalent bond induced by visible light, and they synthesized ASB-WPUs (WPUs with dynamic ASB bonds induced by visible light), which had appropriate storing stability. TEM showed that ASB-WPU-2 has a spherical micellar structure. The average particle size and particle size distribution of these samples were in the range of 52.5–81.2 nm, and 0.165–0.219, respectively. FTIR and ^{1}H-NMR analysis confirmed the chemical structures of the synthesized emulsions. The crystal structure of the prepared ASB-WPUs was studied by DSC experiments. The results showed that the polymers have an amorphous structure that could be maintained at room temperature. ASB-WPU polymers' amorphous structure was more confirmed using XRD. TGA was used to investigate the thermal stability of samples and revealed that with increasing ABS amount up to 37.5%, the thermal stability of the samples increased considerably. Tensile tests were used to evaluate the mechanical properties of the samples. The findings obtained from tensile tests showed that with increasing ABS amount up to 37.5%, in a short time, the tensile stress of prepared films jumped to 15.68 MPa from 8.13 MPa.

To evaluate the healing efficiency of prepared samples, ASB-WPU-2 was selected as a model sample. The cut and reconnected of this sample were irradiated with a LED lamp at various times. Figure 11.4 shows the mechanical properties recovery ratios of samples as well as the stress–strain curves of healed samples. The elongation at break and tensile stress of the healed specimens recovered 94.90% and 83.80%, respectively, after being irradiated for 24 h.

Moreover, the effect of the ASB value on self-healing efficiency was also studied. By increasing the amount of ASB from 0% to 25%, the elongation at break and tensile stress after healing under visible light for 24 h increased from 50.50% and 34.44% to 94.90% and 83.80%, while increasing up to 37.5%, these values decreased to 79.8% and 72.07%, respectively.

Figure 11.5 shows the healing procedure of prepared samples. In fact, the high-efficiency healing was caused by H-bonded interactions between the urethane groups and the imine metathesis of ASB bonds. The imine metathesis of ASB bonds is essential to achieving the final healing.

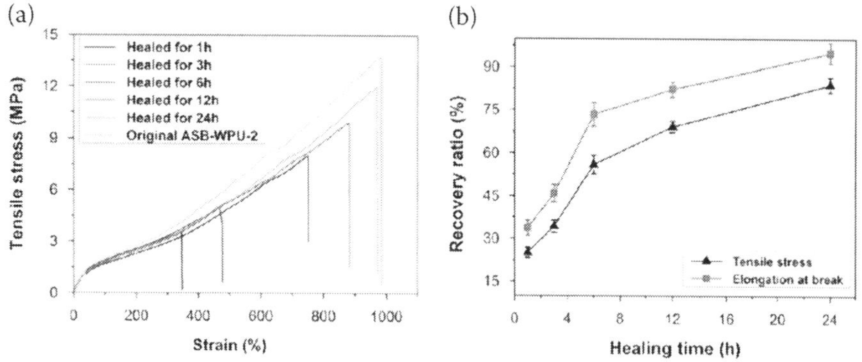

FIGURE 11.4 (a) Stress–strain curves and (b) mechanical properties recovery ratio of the cracked ASB-WPU-2 sample with various healing times. Reproduced from reference [34]. Copyright (2020) Royal Society of Chemistry.

To confirm that the prepared polymers can reprocess, ASB-WPU-2 grains were pressed into a sheet, which was then held between two plastic plates. Then, after 48 h of radiation, the opaque sheet reverted to a clear bulk with no evident crack. Besides, the tensile stress could be retained more than 75% after three reprocessing cycles.

11.6 APPLICATION

Adding self-healing properties to WPU has increased its efficiency in various fields and is also economical. SHWPU can be used in fields such as (1) leather coatings, (2) anticorrosive coatings, (3) composite conductors, and (4) flexible sensors.

1. Liang et al. [20] synthesized a kind of SHWPU with a disulfide bond by using HEDS as a chain extender to improve the scratch resistance of leather coatings. The WPU films showed a higher healing efficiency of up to 80%. These materials can completely repair the scratched coating and therefore have shown great application in textile laminating, coatings, leather/synthetic leather finishes, etc.
2. Babaei et al. [12] synthesized self-healing and anticorrosive waterborne poly (urethane-triazole) coatings in order to deposition on the metal surfaces by a cathodic electrophoretic method. The obtained coatings exhibited proper flexibility, excellent solvent resistance, very good impact resistance, favorable hardness, good corrosion resistance, and high adhesion. Hydrophobicity, big molecular cover of the metal surface through 1,2,3-triazole rings, preservation part of reactive functional groups, and the particular molecular design of coatings were the key causes for the enhanced anticorrosive and healing properties of these coatings. PUT2 (molar ratio of azide to propargyl:1 to 1.5) exhibited the ability to repair surface harm as well as proper corrosion resistance even after 2 months of soaking in NaCl.
3. Yang et al. [23] presented a molecular design method to reach high adhesion, healable, and flexible composite conductors. Specially, they prepared eco-friendly

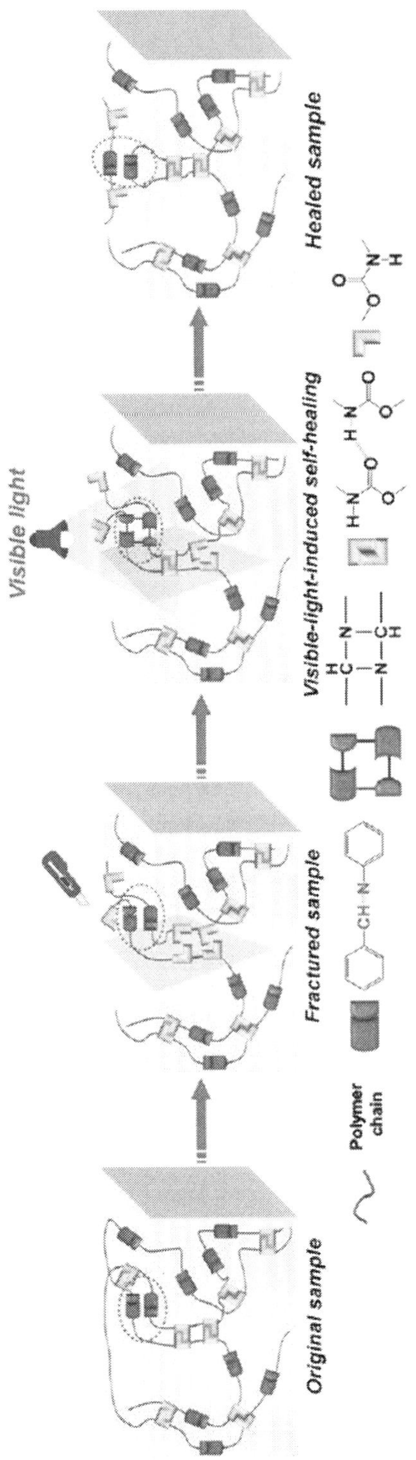

FIGURE 11.5 Schematic showing of the healing procedure of ASB-WPU polymers. Reproduced from reference [34]. Copyright (2020) Royal Society of Chemistry.

WPU-UxHy (healable UV-curing waterborne polyurethanes) using double-network design rules. The self-healing and mechanical properties of samples were investigated, and results indicated that samples have proper flexibility, great strength, and significant healing efficiency. These conductors have numerous extremely appropriate factors, such as excellent self-healing, high interfacial adhesion, and very good flexibility. Moreover, investigations offered that these conductors can be used as elastic sensors on human skin. These results clearly showed that these composite conductors are capable of producing the next generation of implantable and wearable electronics.

4. Zhang et al. [21] synthesized WPU films with a significant healing efficiency and good mechanical properties by using 22DTDA (2,2′-disulfanediyldianiline) as a chain extender. The optimal sample (P22-4: WPU film with 4 mass% of 22DTDA) after self-healing at body temperature reached a final tensile strength equal to 13.8 MPa. In this work, the significance of the blend phase in increasing self-healing efficiency and mechanical strength was also mentioned. Furthermore, ultrasound could simplify healing so that films with a final tensile strength equal to 20.3 MPa could reach 80% self-healing efficiency in strength after only 30 min of ultrasound exposure. Due to its properties, such as body temperature healing, processability, and excellent mechanical strength, WPU film has a very good capability for use in fields such as flexible sensors, electronic skin, recyclable adhesive, and smart coating.

11.7 CONCLUSION AND FUTURE PERSPECTIVE

Briefly, reduction or loss of performance of WPU film due to physical damage and the need for damage-resistant WPUs has led to the study of SHWPUs. The goal of these studies is to prepare smart WPUs with high healing efficiency, reprocessability, and mechanical properties. In this study, the synthesis methods, characterization data, and different applications of SHWPUs are discussed in detail. Importantly, the self-healing efficiency, reprocessability, and time are highlighted for various types of WPU films that have been studied recently. According to the characterization results, SHWPUs based on visible-light-induced dynamic covalent bonds could be very effective inspiration for the development of smart materials in the future. Some of their significant features are fast healing, high mechanical properties, available and cost-effective stimulus (visible light), and so on.

According to studies, SHWPU coating is one of the most suitable anticorrosion coatings for metal surfaces. The industrial application of this type of coating will have a significant impact on reducing the damage caused by metal corrosion in the future. Despite the significant advantages of SHWPU films, their low mechanical strength and reprocessing have led to extensive research in this area to improve the final properties of these films. There is also a need for more focused research to prepare them for practical application. Moreover, the challenge of improving the healing efficiency and time with the objective of effective and rapid self-healing at room temperature instead of high-temperature and heating in the furnace is still prominent.

REFERENCES

1. H.T. Jeon, M.K. Jang, B.K. Kim, K.H. Kim, Synthesis and characterizations of waterborne polyurethane–silica hybrids using sol–gel process, *Colloids Surf. A Physicochem. Eng. Asp.* 302 (2007) 559–567. doi:10.1016/j.colsurfa.2007.03.043.
2. Y. Ahmadi, S. Ahmad, Recent progress in the synthesis and property enhancement of waterborne polyurethane nanocomposites: Promising and versatile macromolecules for advanced applications, *Polym. Rev.* 60 (2020) 226–266. doi:10.1080/15583724.2019.1673403.
3. W.M. Xu, M.Z. Rong, M.Q. Zhang, Sunlight driven self-healing, reshaping and recycling of a robust, transparent and yellowing-resistant polymer, *J. Mater. Chem. A.* 4 (2016) 10683–10690.
4. M. Zhang, F. Zhao, Y. Luo, Self-healing mechanism of microcracks on waterborne polyurethane with tunable disulfide bond contents, *ACS Omega.* 4 (2019) 1703–1714.
5. B.S. Sumerlin, Next-generation self-healing materials, *Science. (80)* 362 (2018) 150–151.
6. S. Tyagi, J.Y. Lee, G.A. Buxton, A.C. Balazs, Using nanocomposite coatings to heal surface defects, *Macromolecules.* 37 (2004) 9160–9168.
7. W. Fan, Y. Jin, L. Shi, Mechanically robust and tough waterborne polyurethane films based on diselenide bonds and dual H-bonding interactions with fast visible-light-triggered room-temperature self-healability, *Polym. Chem.* 11 (2020) 5463–5474. doi:10.1039/d0py00897d.
8. W. Fan, Y. Jin, L. Shi, R. Zhou, W. Du, Developing visible-light-induced dynamic aromatic Schiff base bonds for room-temperature self-healable and reprocessable waterborne polyurethanes with high mechanical properties, *J. Mater. Chem. A.* 8 (2020) 6757–6767.
9. Y. Li, Y. Guo, W. Niu, M. Chen, Y. Xue, J. Ge, P.X. Ma, B. Lei, Biodegradable multifunctional bioactive glass-based nanocomposite elastomers with controlled biomineralization activity, real-time bioimaging tracking, and decreased inflammatory response, *ACS Appl. Mater. Interfaces.* 10 (2018) 17722–17731.
10. K. Ye, Z. Bi, G. Cui, B. Zhang, Z. Li, External self-healing coatings in anticorrosion applications: A review, *Corrosion.* 76 (2020) 279–298. doi:10.5006/3430.
11. Y. Fang, X. Du, X. Cheng, M. Zhou, Z. Du, H. Wang, Preparation of living and highly stable blended polyurethane emulsions for self-healing films with enhancive toughness and recyclability, *Polymer (Guildf).* 188 (2020). doi:10.1016/j.polymer.2019.122142.
12. N. Babaei, H. Yeganeh, R. Gharibi, Anticorrosive and self-healing waterborne poly (urethane-triazole) coatings made through a combination of click polymerization and cathodic electrophoretic deposition, *Eur. Polym. J.* 112 (2019) 636–647. doi:10.1016/j.eurpolymj.2018.10.028.
13. R.H. Aguirresarobe, L. Martin, M.J. Fernandez-Berridi, L. Irusta, Autonomic healable waterborne organic-inorganic polyurethane hybrids based on aromatic disulfide moieties, *Express Polym. Lett.* 11 (2017) 266–277. doi:10.3144/expresspolymlett.2017.27.
14. C. Zhang, H. Liang, D. Liang, Z. Lin, Q. Chen, P. Feng, Q. Wang, Renewable castor-oil-based waterborne polyurethane networks: Simultaneously showing high strength, self-healing, processability and tunable multishape memory, *Angew. Chemie - Int. Ed.* 60 (2021) 4289–4299. doi:10.1002/anie.202014299.
15. J. Aizpurua, L. Martin, E. Formoso, A. González, L. Irusta, One pot stimuli-responsive linear waterborne polyurethanes via Diels-Alder reaction, *Prog. Org. Coatings.* 130 (2019) 31–43. doi:10.1016/j.porgcoat.2019.01.008.

16. T. Li, Z.P. Zhang, M.Z. Rong, M.Q. Zhang, Self-healable and thiol-ene UV-curable waterborne polyurethane for anticorrosion coating, *J. Appl. Polym. Sci.* 136 (2019) 1–11. doi:10.1002/app.47700.
17. Y. Xiao, H. Huang, X. Peng, Synthesis of self-healing waterborne polyurethanes containing sulphonate groups, *RSC Adv.* 7 (2017) 20093–20100. doi:10.1039/C6RA28416G.
18. A.W. Grzelak, P. Boinard, J.J. Liggat, The influence of diol chain extender on morphology and properties of thermally-triggered UV-stable self-healing polyurethane coatings, *Prog. Org. Coatings.* 122 (2018) 1–9. doi:10.1016/j.porgcoat.2018.04.032.
19. D. Lee, S. Kim, D. Lee, Waterborne polyurethanes with cystamine, *Molecules.* 24 (2019) 1492.
20. F. Liang, T. Wang, H. Fan, J. Xiang, Y. Chen, A leather coating with self-healing characteristics, *J. Leather Sci. Eng.* 2 (2020) 0–4. doi:10.1186/s42825-020-0018-4.
21. L. Zhang, T. Qiu, X. Sun, L. Guo, L. He, J. Ye, X. Li, Achievement of both mechanical properties and intrinsic self-healing under body temperature in polyurethane elastomers: A synthesis strategy from waterborne polymers, *Polymers (Basel).* 12 (2020). doi:10.3390/POLYM12040989.
22. Y. Fang, X. Du, Y. Jiang, Z. Du, P. Pan, X. Cheng, H. Wang, Thermal-driven self-healing and recyclable waterborne polyurethane films based on reversible covalent interaction, *ACS Sustain. Chem. Eng.* 6 (2018) 14490–14500. doi:10.1021/acssuschemeng.8b03151.
23. Y. Yang, Z. Ye, X. Liu, J. Su, A healable waterborne polyurethane synergistically cross-linked by hydrogen bonds and covalent bonds for composite conductors, *J. Mater. Chem. C.* 8 (2020) 5280–5292. doi:10.1039/d0tc00551g.
24. W. Fan, Y. Jin, Y. Huang, J. Pan, W. Du, Z. Pu, Room-temperature self-healing and reprocessing of Diselenide-containing waterborne polyurethanes under visible light, *J. Appl. Polym. Sci.* 136 (2019) 1–12. doi:10.1002/app.47071.
25. J. Li, Q. Feng, J. Cui, Q. Yuan, H. Qiu, S. Gao, J. Yang, Self-assembled graphene oxide microcapsules in Pickering emulsions for self-healing waterborne polyurethane coatings, *Compos. Sci. Technol.* 151 (2017) 282–290.
26. J. Li, Y. Hu, H. Qiu, G. Yang, S. Zheng, J. Yang, Coaxial electrospun fibres with graphene oxide/PAN shells for self-healing waterborne polyurethane coatings, *Prog. Org. Coatings.* 131 (2019) 227–231. doi:10.1016/j.porgcoat.2019.02.033.
27. A. Mirmohseni, M. Akbari, R. Najjar, M. Hosseini, Self-healing waterborne polyurethane coating by pH-dependent triggered-release mechanism, *J. Appl. Polym. Sci.* 136 (2019) 1–12. doi:10.1002/app.47082.
28. T. Wan, D. Chen, Preparation of β-cyclodextrin reinforced waterborne polyurethane nanocomposites with excellent mechanical and self-healing property, *Compos. Sci. Technol.* 168 (2018) 55–62. doi:10.1016/j.compscitech.2018.08.049.
29. A. Eceiza, A. Tercjak, *Improvement of Mechanical Properties and Self- Healing E ffi ciency by Ex-Situ Incorporation of TiO_2* (2019). doi:10.20944/preprints201907.0017.v1
30. Y. Hua, X. Li, L. Ma, Y. Wang, H. Fang, H. Wei, Y. Ding, Self-healing mineralization and enhanced anti-corrosive performance of polyurethane CaCO3 composite film via β-CD induction, *Mater. Des.* 177 (2019) 1–10. doi:10.1016/j.matdes.2019.107856.
31. T. Wan, D. Chen, Synthesis and properties of self-healing waterborne polyurethanes containing disulfide bonds in the main chain, *J. Mater. Sci.* 52 (2017) 197–207.
32. H. Huang, W. Zhou, Z. Zhong, S. Peng, X. Peng, Self-antiglare waterborne coating with superior mechanical robustness and highly efficient room-temperature self-healing capability, *Prog. Org. Coatings.* 146 (2020). doi:10.1016/j.porgcoat.2020.105717.

33. W. Fan, Y. Jin, L. Shi, W. Du, R. Zhou, S. Lai, Y. Shen, Y. Li, Achieving fast self-healing and reprocessing of supertough water-dispersed “living” supramolecular polymers containing dynamic ditelluride bonds under visible light, *ACS Appl. Mater. Interfaces.* 12 (2020) 6383–6395. doi:10.1021/acsami.9b18985.
34. W. Fan, Y. Jin, L. Shi, R. Zhou, W. Du, Developing visible-light-induced dynamic aromatic Schiff base bonds for room-temperature self-healable and reprocessable waterborne polyurethanes with high mechanical properties, *J. Mater. Chem. A.* 8 (2020) 6757–6767. doi:10.1039/c9ta13928a.
35. C.S. Wong, N.I. Hassan, M.S. Su'ait, M.A. Pelach Serra, J.A. Mendez Gonzalez, L.A. Granda, K.H. Badri, Photo-activated self-healing bio-based polyurethanes, *Ind. Crops Prod.* 140 (2019) 1–16. doi:10.1016/j.indcrop.2019.111613.
36. R.H. Aguirresarobe, L. Martin, N. Aramburu, L. Irusta, M.J. Fernandez-Berridi, Coumarin based light responsive healable waterborne polyurethanes, *Prog. Org. Coatings.* 99 (2016) 314–321. doi:10.1016/j.porgcoat.2016.06.011.
37. D. Il Lee, S.H. Kim, D.S. Lee, Synthesis of self-healing waterborne polyurethane systems chain extended with chitosan, *Polymers (Basel).* 11 (2019). doi:10.3390/polym11030503.
38. T. Wan, D. Chen, Preparation of β-cyclodextrin reinforced waterborne polyurethane nanocomposites with excellent mechanical and self-healing property, *Compos. Sci. Technol.* 168 (2018) 55–62.
39. Y. Han, Y. Jiang, J. Hu, Collagen incorporation into waterborne polyurethane improves breathability, mechanical property, and self-healing ability, *Compos. Part A Appl. Sci. Manuf.* 133 (2020) 105854.
40. Z.P. Zhang, M.Z. Rong, M.Q. Zhang, Polymer engineering based on reversible covalent chemistry: A promising innovative pathway towards new materials and new functionalities, *Prog. Polym. Sci.* 80 (2018) 39–93.
41. Z. Zhong, S. Luo, K. Yang, X. Wu, T. Ren, High-performance anionic waterborne polyurethane/Ag nanocomposites with excellent antibacterial property via in situ synthesis of Ag nanoparticles, *RSC Adv.* 7 (2017) 42296–42304.
42. L. Zhai, Y. Wang, F. Peng, Z. Xiong, R. Liu, J. Yuan, Y. Lan, Synthesis of TiO_2–SiO2/waterborne polyurethane hybrid with amino-siloxane terminated via a sol–gel process, *Mater. Lett.* 89 (2012) 81–85.
43. I. Díez-García, A. Eceiza, A. Tercjak, Improvement of mechanical properties and self-healing efficiency by ex-situ incorporation of TiO_2 nanoparticles to a waterborne poly (urethane-urea), *Polymers (Basel).* 11 (2019) 1209.

12 Waterborne Polyurethane for Biomedical Applications

Abbas Mohammadi, Mahtab Eslamieh, Negar Salehi, and Saman Abrishamkar
Department of Chemistry, University of Isfahan, Isfahan, I.R. Iran

CONTENTS

12.1 Introduction 193
12.2 Waterborne Polyurethanes 195
12.3 Antibacterial Waterborne Polyurethanes 195
12.4 Synthesis Methods 195
12.5 Incorporation of Antibacterial Nanostructures 195
12.6 Combination With Antibacterial Polymers 197
12.7 Incorporation of Antibacterial Monomers 198
12.8 Drug Loading 198
12.9 Polymer Surface Modification 199
12.10 Wound Healing 199
12.11 WPU as a Wound Dressing 201
12.12 WPUs in Drug Delivery 202
References 207

12.1 INTRODUCTION

The infections caused by microorganisms and their prevalence have always been a matter of concern [1]. Every year, thousands of people die from bacterial infections due to the strong resistance of bacteria against old antimicrobial agents [2]. Microbes can survive and multiply under wet conditions. Since the survival of living organisms is achieved by controlling the growth of microbes, including bacteria and fungi, it is essential to kill or control microbes using antibacterial agents [3]. The fight against bacteria is one of the biggest challenges with respect to medical devices, hospital equipment, and health products [4]. One of the new approaches to prevent microbial contaminations is to use polymeric antibacterial coatings, which inhibit the adhesion of a wide range of bacteria on surfaces and hence avoid the formation of biofilms in the early stages after contact with bacteria [5]. Today, antibacterial coatings are an important part of medical science. In the

DOI: 10.1201/9781003173526-12

field of medical equipment, for instance, the biggest challenge of using artificial prosthetics, catheters, and other temporary or permanent implants is microbial contamination after placed into the body and the formation of bacterial biofilm on them.

Biofilm is an accumulation of micro-organisms that produces a viscous substance called an extracellular polymer substance (EPS). The biofilm fluid contains proteins, polysaccharides, DNA, and 97% water. Biofilm has several channels to separate colonies of microorganisms. These channels are filled with water and help to circulate and dispose of waste. The first step in biofilm formation is the attachment or fixation of bacteria by Pili, Flagella, or electrostatic interaction on surfaces. These surfaces may be living or nonliving. Then, microcolonies are formed, begin to mature, form a structure, and finally separate from the surface. Once the bacteria are removed from the surface, each bacterium creates an enzyme and forms new colonies. The bacterial cells begin to produce proteins again to form flagella so that the bacteria can move to a new location. This process leads to the spread of infections [4]. Biofilm also protects microbes as an ideal environment and enables them to continue to grow under unfavorable conditions [6]. It is important to note that there is also the possibility of gene exchange in the biofilm, resulting in the formation of drug-resistant bacterial species. Consequently, the best way to stop the spread of infections and illnesses is to inhibit the formation of biofilms using antimicrobial surfaces [7]. When active antibacterial substances are deposited at the surface, the proliferation of bacteria at the surface is prevented [8].

Skin is the body's first line of defense against pathogenic microorganisms that may enter the body through wounds or skin damages. Wound healing is a complex, dynamic, and physiological response of living tissue to an injury. In general, the wound healing process is classified into four stages: homeostasis, inflammation, proliferation, and remodeling. The immune system can heal the wound, but to improve the quality of the healing, accelerate the healing process, and prevent pathogens from entering the body, wound dressings are required [9]. Targeted drug delivery is a set of activities that lead to the accumulation of drugs in a specific area of the body. The main advantage of using targeted drug delivery is to increase the therapeutic effects of the drug without inducing side effects on healthy organs, tissues, or cells. In general, a targeted drug delivery system includes a targeted drug, carrier, and ligand. Polymers are good options for use as pharmaceutical carriers.

Polyurethane (PU) is one of the most commonly used synthetic polymers, which great diversity of raw materials makes it one of the most versatile classes of polymers. PU wound dressing can provide biocompatibility, flexibility, good permeability to oxygen and carbon dioxide, and appropriate mechanical properties. However, since they have low absorption capacity and only act as a barrier against microbial invasion, they are not suitable for wounds with high exudates [10,11]. Solvent-based PUs have long been used in a number of applications. However, one of the most important problems with their use is the release of volatile organic compounds (VOCs) into the atmosphere. Therefore, environmental agencies have published instructions to tackle this problem [12].

12.2 WATERBORNE POLYURETHANES

Waterborne polyurethane (WPU) is a binary colloidal system composed of PU particles dispersed in water. Waterborne formulations have become increasingly popular as a way to reduce costs and control VOCs. WPU is a biocompatible, nontoxic, and low-cost polymer with unique film properties, including adhesion to various surfaces, chemical and abrasion resistance, high tensile strength, optimal flexibility, and water vapor impermeability [13,14]. Therefore, it has found its application in different areas, such as medical, clothing, textile, synthetic leather, adhesives, and flooring industries [2].

12.3 ANTIBACTERIAL WATERBORNE POLYURETHANES

Antibacterial PU coatings can prevent the growth of bacteria at different levels and also restrict the spread of bacterial infections. Nowadays, such coatings are used in medical devices, hospital equipment, water purification and cooling systems [8], wound dressings [15,16], leather covers [17], hospital floor coverings, food and medicine packaging, and health products [18]. Easy preparation, low price, long-term stability, good adhesion to surfaces, high tensile strength, flexibility, spray ability on surfaces, biocompatibility, and nontoxicity of WPU film have made it a viable option for antibacterial applications [4].

12.4 SYNTHESIS METHODS

Bacteria adhere strongly to the surface of the PU film and quickly develop into colonies [19]. To prevent the growth of bacteria on the surface and provide antibacterial WPU coating, researchers have proposed various strategies, such as the incorporation of antibacterial nanostructures or monomers, combination with antibacterial polymers, drug loading, and polymer surface modification. In this section, each of these methods and also studies conducted in this field will be discussed.

12.5 INCORPORATION OF ANTIBACTERIAL NANOSTRUCTURES

Today, nanotechnology has a major contribution to the development of aerospace, electronics, environment, medicine, and especially treatment and diagnosis of diseases [20]. Metal nanostructures are known to have low toxicity, good heat resistance, and antibacterial effects against gram-negative and gram-positive bacteria [21]. Recent studies indicate that the toxicity of nanostructures depends on the type of materials and their physical and chemical properties. It also can be influenced by various factors, such as cellular adsorption of nanostructures and their interaction with cells [22]. Metal nanomaterials have a surface charge and can interact with the protein and DNA of microorganisms and cause them to be inhibited. Another important antibacterial mechanism of nanostructures is the production of reactive oxygen species (ROS), which are small, unstable, and highly reactive molecules that can oxidize proteins, lipids, and DNA. One of the most challenging aspects of using nanostructures in polymers is their dispersion quality in the polymer matrix. The high surface-to-volume ratio of nanostructures increases their surface energy and consequently their tendency to agglomerate within the polymer matrix. Usually, by modifying the surface of nanostructures, the interaction between the polymer and the nanostructure is enhanced, and as a result, the dispersion of nanoparticles in the polymer matrices is improved [23].

In recent decades, silver nanostructure with low toxicity to human cells [22] and very strong antibacterial properties has been one of the most popular research topics in the field of biomedicine. The antibacterial properties of silver are arising from its high cytotoxicity by various mechanisms. in particular, the over-production of reactive oxygen species causes stress on cells and eventually cell death [22,24]. Figure 12.1 illustrates the mechanism of interaction between the silver nanostructure and the bacterial cell membranes. As indicated, different reactions can occur between the nanostructure and the cell wall of living organisms. At first, silver ions are released from silver nanostructures. The difference between the negative charge of the cell membranes in the microorganism and the positive charge of the silver ions causes the silver ions to bind to the cell surface and eventually resulting in cell death. In addition, ions released from silver nanostructures may react with thiol (–SH) groups of proteins on bacterial wall surfaces [25]. A notable thing about the addition of silver nanostructure to WPU is the color change of the polymer solution from milky to dark brown. The reduction of silver ions is achieved by using a polyunsaturated material as one of the components used in water-based PU, or materials that can reduce silver ions, such as $NaBH_4$, DMF [26], dopamine [27], and calixarenes [28].

The problem with WPU coatings containing nanoparticles obtained from the in situ reduction of silver ions is the coagulation of PU micelles. This is due to the interaction of silver ions with the carboxylate groups of PU chains, which results in crosslinking [29]. Researchers have found that this problem can be overcome by

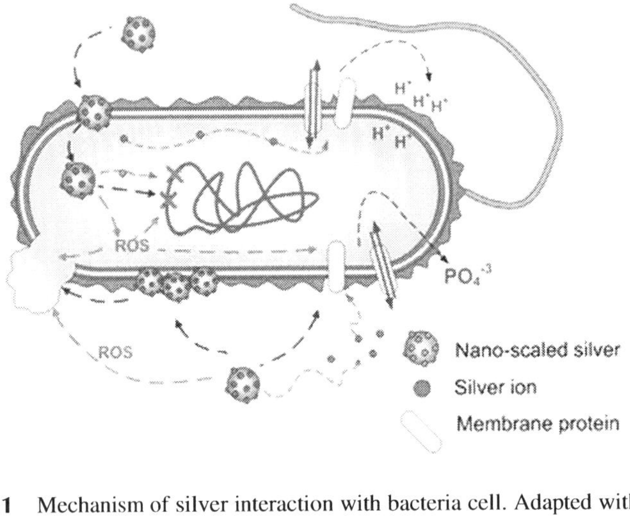

FIGURE 12.1 Mechanism of silver interaction with bacteria cell. Adapted with permission from reference [25], Copyright (2010), Springer Nature.

coordinating silver and then reducing it. Using the Schiff base ligand as a chain extender for WPU synthesis is a new method for stabilizing silver nanoparticles [2]. Cellulose nanocrystals (CCNs) have a high ability to adsorb metal cations. Due to the strong interaction between their hydroxyl group and the silver cation, a uniform distribution of the silver nanostructure in the polymer matrix is obtained. Furthermore, these interactions reduce the mobility of silver cations, prevent the formation of large particles, and stabilize silver nanostructures [30].

Zinc oxide (ZnO) nanostructure has received much attention thanks to its favorable biocompatibility, high antibacterial properties, catalytic power, anticancer properties, easy accessibility, and cost-effectiveness [18,31]. This nanostructure has dual antibacterial properties, which means it affects both gram-positive and gram-negative bacteria and halts the growth cycle of these micro-organisms [32]. ZnO nanowhiskers are synthesized using the hydrothermal method. The properties of high-performance WPU/ZnO coatings, such as mechanical strength, thermal stability as well as antibacterial effect, are all influenced by ZnO nanowhiskers [11]. Titanium oxide (TiO_2) is also another nanostructure with high photocatalytic and antibacterial properties that have been used in coatings. Titanium-containing coatings exhibit self-sterilizing, antibacterial, and nonstick properties after exposure to ultraviolet light [6].

Hybrid nanostructures enable multiple properties of the polymer to be controlled. These nanostructures, for example, SiO_2–GO hybrid, may also have a synergistic effect on the polymer properties, such as antibacterial and mechanical properties [33]. In a study by Xiong et al., WPU composites based on layered double hydroxide (LDH)/ZnO were synthesized by modification of isophorone diisocyanate (IPDI) with LDH/ZnO hybrid before polymerization. Compared to the pure WPU, the tensile strength of the resulting composites was significantly improved due to the formation of a lattice structure. They also exhibited a strong antibacterial effect against *E. coli* and *S. aureus* bacteria [34]. Layered double hydroxides are known as positively charged nanostructures with interlayer anions. The physical and chemical properties of these nanostructures are very similar to hydrotalcite [35]. LDHs have unique properties, including nontoxicity; low cost; biocompatibility; favorable thermal, mechanical, and antibacterial properties; high contact area; and good interlayer anion exchange capacity [36]. To increase the thermal and mechanical properties of WPU coatings, LDH can be introduced into WPU by in situ polymerization, after modification of isocyanate with LDH [37]. MnO_2 [38], CuO [39], rGO [40], and Fe_2O_3 [41] are other nanostructures that have been identified for antibacterial properties.

12.6 COMBINATION WITH ANTIBACTERIAL POLYMERS

The combination of the polymer matrix with antibacterial polymers leads to antibacterial properties in the final polymer. These compounds have several advantages, such as reducing environmental problems, toxic wastes, and nonvolatile materials, and also improving efficiency. For instance, chitosan is a linear biopolymer with antibacterial activity. Chitosan is the *n*-acetylated form of chitin with NH_2 and hydroxyl groups [42]. The presence of amino and hydroxyl groups in the chitosan structure makes it suitable to react with isocyanate groups [43]. However, because of its low solubility, it shows antibacterial activity only in the acidic range [44]. In the synthesis of WPU, chitosan can be used as a chain extender and improve the antibacterial activity of WPU coating [45]. In addition, a gradual increase in chitosan concentration causes a significant increase in bacterial activity due to the presence of positively charged $-NH_2$ groups that interact with negatively charged bacterial cell walls [42]. One of the main drawbacks of chitosan is that under alkaline conditions it loses its cationic nature and is no longer able to adhere to the bacterial cell wall. Therefore, it has no antibacterial activity in this pH range [46].

12.7 INCORPORATION OF ANTIBACTERIAL MONOMERS

Most antibacterial coatings are prepared by adding antibacterial agents, such as peroxides, silver nanostructures, etc. However, these antibacterial coatings have many disadvantages, such as toxicity, high cost, and unstable antibacterial activity, which limit their range of applications [47]. A novel method to decrease the viability of bacteria is the incorporation of antibacterial monomers into the polymer systems. Nitrogen, guanidine, halogen, sulfur, phosphorus, and phenol-based monomers (e.g. quaternary ammonium salt [48] and phosphonium salt [49]), and also organometallic positively charged species are inherently antibacterial. In recent decades, quaternary ammonium salts have been extensively studied due to their low toxicity and antibacterial activity for a wide range of bacteria [49]. The antibacterial mechanism of these compounds is not yet clear. However, studies suggest that this action is probably due to the binding of the ammonium group to the bacterial membrane through electrostatic interaction, which leads to membrane deformation and eventually bacterial death. One of the applications of quaternary ammonium compounds is to use them in the preparation of antibacterial WPU coatings. The antibacterial properties of quaternary ammonium compounds increase the growth and proliferation of fibroblast cells and reduce wound healing time (Figure 12.2) [50]. Recently, structures with properties similar to quaternary ammonium salts, known as quaternary phosphonium salts, have been introduced [51,52]. Several reports have demonstrated that these salts have better antibacterial activity compared to ammonium salts. In a study by Wang et al., WPU films containing quaternary phosphonium salts showed extensive antibacterial activity and greatly inhibited the growth of *S. aureus* (gram-positive bacteria) [49].

12.8 DRUG LOADING

One of the strategies to prepare antibacterial coatings is the loading of drugs into the polymer matrix. Drug-containing coatings can release the drug in a controlled manner and prevent the toxicity of the sudden drug release. Numerous studies have investigated drug loading in nanofibers prepared by electrospinning [48]. Drug loading has been widely studied by researchers in the field of membranes, wound dressings, and textiles [53]. Trapping of antibacterial drugs within these nanofibers leads to the controlled drug release and its continuation at the wound site. To

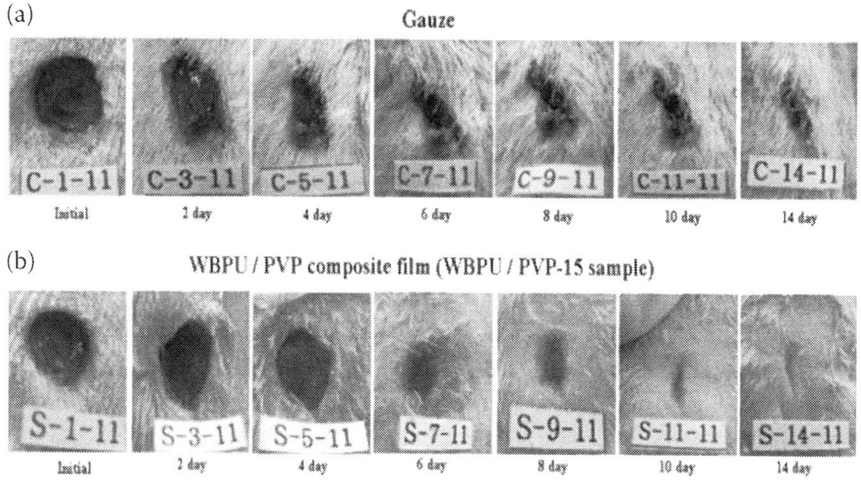

FIGURE 12.2 Comparison of wound healing effect by (a) sterile gauze and (b) wound dressed containing *n*-vinyl pyrrolidone. Adapted with permission from reference [50], Copyright (2007), John Wiley and Sons.

prepare an antibacterial coating based on WPU, Chang et al. synthesized a low molecular weight copolymer containing antibacterial ciprofloxacin drug and acrylic acid. According to the results, this water-soluble antibacterial copolymer increased the antimicrobial performance of WPU coatings. It also increased the durability of the coatings [17].

12.9 POLYMER SURFACE MODIFICATION

Surface modification methods include radiation (X-ray, electron, and gamma rays), grafting, laser, and plasma treatments. These methods are used to modify the chemical and physical properties of the surface such as surface tension, polarity, and hydrophilicity or hydrophobicity [54]. Huang et al. reported a facile method to modify chitosan under gamma rays for drug delivery. For this purpose, the NH_2 groups of chitosan were used to be grafted with other active compounds [55]. In a study by Liu et al., covalent incorporation of acryloyl chloride modified lysozyme into waterborne polyurethane-acrylic coatings (WPUA) by UV curing, represented an easy way to develop highly efficient antibacterial surfaces. These coatings were very effective in killing *S. aureus* and *E. coli* bacteria [56].

12.10 WOUND HEALING

Wound healing is a complex physiological response to physical, chemical, mechanical, and thermal injuries. Cells and matrix components collaborate to promote skin regeneration and tissue integrity during wound healing. Wound healing is a

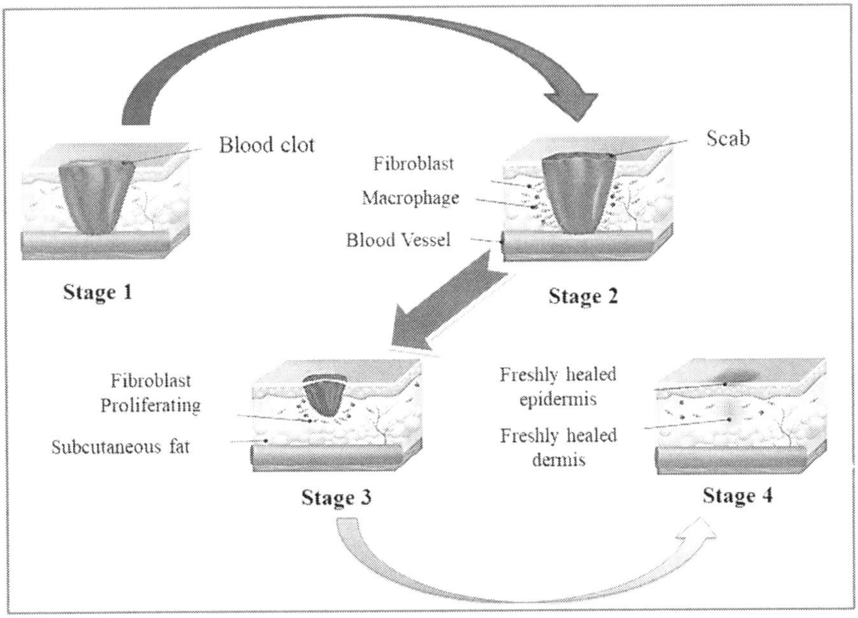

FIGURE 12.3 Wound healing stages. (1) Homeostasis stage, (2) Inflammation stage, (3) Proliferation stage, (4) Remodeling. Adapted with permission from reference [57], Copyright (2020), Elsevier.

dynamic process with four overlapping stages: homeostasis, inflammation, proliferation, and remodeling (Figure 12.3) [57].

If the healing process deviates from its normal pattern, healing may stop in the inflammatory phase. In this case, inflammatory mediators, such as reactive oxygen species, reactive nitrogen species (RNS), cytokines, as well as large numbers of neutrophils and macrophages, accumulate. Mechanical and chemical injury, burns, secondary infections, and complications from other diseases, such as diabetes, may all cause defects or delays in the healing process. The wound can be protected with a nontoxic, semi-permeable, mechanical stress-resistant, and germ-resistant dressing to facilitate the healing process [10]. Gauze, bandages, lint, plasters, and cotton wool are traditional wound dressings that are used in the early stages of wound healing to prevent bleeding and protect the wound from contact with the external environment [57,58]. One of the main drawbacks of these dressings is that they need to be replaced regularly. Traditional dressings are generally appropriate for clean and dry wounds with minor exudates. They also can be used as secondary dressings. Traditional dressings cannot retain moisture in the wound environment. Therefore, they have been replaced by modern dressings in recent decades. Modern dressings are designed to cover the wound surface, prevent dehydration, and protect the wound environment from bacterial penetration. These dressings are usually based on synthetic polymers and are available in a variety of forms, including film, foam, hydrogel, hydrocolloids, and gauze [58].

12.11 WPU AS A WOUND DRESSING

For wound healing applications, WPU may be used alone or in combination with other polymers and nanoparticles [9,10]. In general, WPU scaffolds made of polyethylene glycol (PEG) swell in an aqueous medium due to the lack of crosslinking. The main reason for phase-separated morphology is the strong hydrogen bond between the hard segment (urethane and urea) and the soft segment (polyethylene glycol). The soft segment content of hydrogels influences their water absorbability [9]. Yoo and Kim developed a series of WPU hydrogels based on polyethylene glycol to achieve optimal wound dressing. The resulting WPU hydrogels showed good potential as new wound dressing materials. This is because of their high moisture content that prevents wound dehydration and thus improves skin regeneration [59]. It was observed that with increasing healing time, the wound size is decreased for both dressings. However, the WPU hydrogel dressing performed better in terms of epithelialization than the gauze dressing. This should be attributed to the high water absorbability of WPU hydrogel [59]. In another study, Yoo and Kim investigated WPU/poly (*N*-vinylpyrrolidone) (PVP) composite films prepared by in situ polymerization in an aqueous medium. Wound healing evaluation using a full-thickness rat model experiment revealed that the wound covered with WPU/PVP film was filled with new epithelium and had no significant adverse reaction. As a result, WPU/PVP composite films can be used as new wound dressing materials to keep the wound moist and prevent dehydration [50].

As previously mentioned, preserving the moisture of the wound site has a major effect on the healing process. The excellent swelling/dehydration ability of biopolymers is one of their most valuable properties. Therefore, they can keep the wound moist. Among biopolymers, chitosan, which is present in the extracellular matrix, has been extensively studied for wound healing applications. Chitosan degradation produces harmless products, which are involved in the formation of ECM. This biopolymer also has antibacterial properties and can be used alone or in combination with other polymers as a wound dressing. In most cases, chitosan must be covalently crosslinked to maintain its structure during use. However, most of the crosslinkers used so far are toxic or have unknown fates in the body. The scarcity of healthy and biocompatible covalent crosslinkers is currently one of the most important challenges [10,60]. In the case of PU dispersions, chitosan is able to form hydrogen bonds through its amine and hydroxyl groups. Hence, the requirement for covalent crosslinkers to stabilize the structure is reduced. In a study by Bankoti et al., polyurethane dispersion (PUD) was blended with chitosan in different ratios to form macroporous hydrogel scaffolds (MHS). To investigate the wound healing process, in vivo experiment was carried out on rats by inflicting a wound on them. Tegaderm™ standard dressing was used as a control sample for this experiment. Tegaderm™ is a sterile, translucent, and waterproof dressing that acts as a barrier against external pathogens. Based on observations, the WPU/chitosan wound dressing (C7P3) adhered well to the wound bed owing to its intrinsic high protein adsorption property. Thus, it prevented the wound from being exposed to the surrounding environment [10]. On days 7, 14, and 21 the control group indicated $40 \pm 1.92\%$, $65 \pm 3.12\%$, and $82 \pm 3.91\%$ closure rates, while the C7P3 treated groups showed enhanced wound closure rates of 55 ± 2.54%, 88 ± 3.41%, and 100 ± 4.12%. Improved wound healing properties of the WPU/chitosan hydrogel scaffold may be attributed to its unique combinatorial properties, such as its high protein and water absorption capacity, interconnected porous structure, and pH-sensitive degradation [10].

Wound dressing is related to skin tissue engineering. Thickness, morphology, porosity, pore size, and interconnectivity of wound dressing are the most important parameters in tissue regeneration. It has been found that the optimal effective pore size for skin regeneration is 20–125 μm. There are many different methods to prepare wound dressings. For example, the foaming method is used for a variety of applications, such as cosmetics, food, cleaning agents, and coatings [61]. Choi et al., prepared WPU foams with macroporous structures as skin tissue engineering matrices using both foaming and freeze-drying methods. Also, they investigated the effect of a foaming agent on the foam morphology and structure. The results showed that the foaming agent causes the fabricated foams to have a macro-porous structure with interconnected pores, which allows the cells to adhere and proliferate on the foam [61]. WPU foams are the samples without a foaming agent and m-WPU foams are the morphology-modified samples, which were prepared with a foaming agent. Unlike WPU samples, m-WPU samples have a homogeneous porous structure with larger interconnected pores [61]. Hao et al. synthesized a biodegradable lysine-based waterborne polyurethane (LWPU) for soft tissue engineering applications. They reported that after 3 days of culture, macrophages adhering to the LWPU film surfaces secrete less pro-inflammatory cytokine TNF-α and more anti-inflammatory cytokine IL-10. This means that LWPUs can help macrophages to migrate to the wound healing phenotype. In addition, endothelial cells can be able to adhere to the LWPU films and proliferate. As a result, the obtained LWPUs showed great potential for wound healing and tissue repair applications [62].

Finally, it can be concluded that WPUS, alone or in combination with other polymers, nanoparticles, and other materials, are good choices for use as wound dressings for soft tissue repairing.

12.12 WPUs IN DRUG DELIVERY

Targeted drug delivery, also called smart drug delivery, is a medication administration technique in such a way that the drug concentration in certain parts of the body is higher than in other areas. A targeted drug delivery system is based on a mechanism that delivers a certain amount of drug to the diseased area of the body over a long period. This helps the body maintain the required plasma and tissue drug levels, preventing any drug-induced damage to healthy tissues. In general, drug delivery mechanisms are divided into two categories: active targeting and passive targeting. Active targeting is based on delivering a specified amount of healing agents to the diseased area. In this system, moieties such as antibodies, antibody fragments, and peptides bind to drugs so they can be detected by target tissue receptors. Drug targeting by ultrasonic energy and magnetic field are examples of active targeting. Passive targeting also called physical targeting, is focused on the preparation of a drug-carrier complex that is resistant to phagocytosis, excretion, metabolism, and other body processes, allowing it to stay in the bloodstream and be transferred to the target [63]. Recent studies in the field of smart drug delivery have led to more powerful treatments. Some drug carriers, such as soluble polymers, liposomes, cells, microcapsules, micro/nano-particles, and micelles, have been used successfully for in vivo targeted drug delivery [64]. Although there is a wide range of materials to use in the drug delivery systems, most of these materials are not bio-based. Both synthetic and natural polymers have been used for drug delivery systems, but researchers have paid more attention to bio-based synthetic polymers [65].

Polymer nanoparticles are made of either biodegradable polymers or non-biodegradable polymers and are considered potential pharmaceutical carriers. For example, supramolecular structures based on polyethylene oxide copolymers and dendrimers have been developed to deliver gene and macromolecule. Hydrogels are also capable of responding to a wide range of chemical, physical, and biological stimuli for the development of closed-loop drug delivery systems [66,67]. Another biodegradable polymer is biodegradable PU, which is synthesized by the reaction of biocompatible and biodegradable polyols with isocyanates [65]. In designing new drug carriers from polymeric materials, it is very important to consider the structure, composition, and method of synthesis. As a result, using environmentally friendly solvents in the synthesis of polymers has received much attention. WPUs are easily dispersed in water due to the presence of various ionic functional groups, such as sulfonic acid, carboxylic acid, and tertiary amine, in their structure. In drug delivery, WPUs provide more benefits than PUs, including superior biocompatibility, low toxicity, low cost, and the ability to carry hydrophobic drugs, such as anticancer medicines [65,68].

Raloxifene hydrochloride (Ral) is a drug belonging to the estrogen agonist/antagonist group and is commonly known as the selective estrogen receptor modulator (SERM). In addition to reducing the risk of breast cancer in humans, raloxifene is used to prevent and treat osteoporosis in postmenopausal women. Because of hepatic first-pass metabolism, it has low bioavailability and water solubility. To overcome these drawbacks, researchers have proposed encapsulating raloxifene in nanoparticulate drug delivery systems to increase drug bioavailability and reduce toxicity [65]. Omrani et al. developed a nanoparticulate system to control the release of raloxifene hydrochloride. They chose WPU as a suitable carrier because of its controllable biodegradability and distinct properties. One of the important factors for drug release in this system is phase separation, which is due to the formation of hard/soft segments and their weight ratio. In this study, the monomers used in the preparation of WPU resulted in the formation of a polymer with a low hard/soft segment ratio, which led to rapid degradation and drug release. WPU hydrolysis was the main driving force in the sustainable release of the drug. In other words, in the first stage, water penetrated the amorphous regions and the ester bonds underwent hydrolytic scission. In the second stage, the amorphous regions were destroyed, which caused chain scission of WPU [65]. According to in vitro drug release study, non-encapsulated raloxifene solution (Ral-S) had 80.38 ± 0.62% release in 24 h, while encapsulated raloxifene in nanoparticles (Ral-NPs) showed

94 ± 4% release in two weeks [65]. Disulfide bond with high redox sensitivity has received considerable attention for use in controlled drug delivery systems. Tumor tissue contains glutathione (GSH), a disulfide bond destroyer, 1,000 times more than blood or normal tissue. It means that the disulfide bond is stable in the bloodstream but degrades in tumor tissue. As a result, disulfide is a good candidate for use in targeted drug delivery systems [69]. Due to the presence of a redox-capacity gradient between the extracellular and intracellular milieus, micro and nanocarriers containing disulfide bonds have attracted a lot of interest. At millimolar concentrations, reducing agents like L-cysteine and glutathione can cleave disulfide bonds via thiol-disulfide exchange reactions. As a result, the nanocarrier made of disulfide bonds can facilitate intracellular transport and trigger the release of encapsulated drugs [9,15].

As reported, WPU can carry anticancer drugs, with significantly lower critical micelle concentration (CMC) value than low molecular weight surfactants. In another research, Omrani et al. synthesized a glutathione degradable WPU nanocarrier containing disulfide bonds in both hard and soft segments for redox-triggered intracellular delivery of doxorubicin (DOX). Doxorubicin is a lipophilic anticancer drug that inhibits or stops cancer cell proliferation. To divide and proliferate, cancer cells require the topo isomerase 2 enzyme. Doxorubicin acts by inhibiting this enzyme. This study was performed on two types of cells: human breast cancer cells (MCF-7) and healthy human dermal fibroblast (HDF). The results showed that DOX-loaded WPU had a high in vitro antitumor activity in both HDF and MCF-7. Furthermore, even at a high dosage of 2 mg/mL, the blank WPU nanocarriers are harmless to HDF and MCF-7 cells [15].

In recent years, pH and reduction-sensitive polymeric nanoparticles have emerged as unique kinds of nanocarriers for intracellular drug delivery. The pH of cancer cells is as low as 6.0–7.4 in the early endosome, 5.5–6.0 in the later endosome, and eventually 4.5–5.0 in lysosomes, while the pH of normal blood is around 7.4. GSH is the most prevalent reducing molecule in the body, which its concentrations in the extracellular (plasma) and intracellular (cytoplasm) environments are about 10 μM and 10 mM, respectively. Disulfide bonds are durable in the extracellular milieu, but they can be reversibly cleaved in a reductive intracellular environment. The GHS-triggered can cause polymers containing disulfide bonds to degrade quickly, whereas conventional hydrolytic degradation of some polymers in the body can take days to months. Micelles made of bio-degradable PUs with pH and redox responsive properties have garnered a great deal of attention in drug delivery. The micelle shell disintegrates and the polymer main chain breaks due to the presence of cleavable disulfide and acid linkages in the backbone of these PUs, resulting in nanocarrier disintegration [68]. In a study, Saeedi et al. developed and evaluated WPU nanoparticles for pH and redox-triggered intracellular delivery of DOX. The toxicity of blank WPU nanoparticles and the anti-cancer effect of DOX-loaded WPU nanoparticles on cells were evaluated using the MTT test. In this test, two types of cells were used: human malignant melanoma (A375) and human dermal fibroblast (HDF) cells. Cell viability after 48 h of treatment with nanoparticles at doses ranging from 0.1 to 2 mg mL^{-1}. Cell viability remained above 88% even when the WPU nanoparticles concentration was increased to 2 mg mL^{-1}.

Because of their minimal cytotoxicity to HDF and A375 cells, WPU nanoparticles are suitable as drug carriers. The resulting WPUs represented high drug loading capacity, redox and acid sensitivity, and excellent micellization properties. DOX-loaded WPU nanoparticles were found to be effective in targeting A375 human malignant melanoma cells. Furthermore, when the nanoparticles were absorbed into tumor cells, the disulfide and imine linkages in the PU backbone were broken by GSH and acid, releasing DOX into the cytoplasm quickly [68].

Breast cancer is one of the most common and serious diseases in women. Because of its rapid invasion and relapse features, an effective, immediate, and accurate treatment is highly needed. Chemotherapy is used to minimize tumor metastasis and recurrence. However, most chemotherapy drugs have limited application due to their toxicity to normal tissue and insolubility in water. Paclitaxel (PTX) is a natural diterpene alkaloid anti-cancer drug, which prevents the growth of cancer cells. One of the most important subjects in chemotherapy is how to deliver the drug to tumor tissue with minimal side effects. Local release of medicines in the lesion area is a more efficient and less dangerous treatment for solid tumors than standard chemotherapy via injection. However, choosing the right drug carrier and ensuring the controlled release of chemotherapeutic medicines are still great challenges [70]. Yin et al. combined paclitaxel-loaded phospholipid liposomes with WPU to create an encapsulated 3D scaffold for local drug release. The PTX was released in a regulated manner. First, liposomes encapsulated in the WPU scaffold was released by slowly absorbing water and destroying the WPU. After that, the PTX encapsulated in liposomes was released in water. Compared to PU scaffolds that directly encapsulated PTX, this form, known as a dual-encapsulated scaffold, provided slower initial release and higher PTX concentrations. The introduction of PTX into liposomes protected the drug from degradation in the cell-matrix and enhanced its useful life. It also increased the inhibition of tumor cells and minimized normal cell damage by releasing the drug in a controlled manner [70].

Superparamagnetic iron oxide nanoparticles (SPIO NPs) possess unique biocompatibility, magnetic, chemical, and physical properties. Therefore, they are widely used in biomedical applications, such as magnetic resonance imaging (MRI) and hyperthermia treatment. Endothelial cells of cancer tissue have irregular holes, which NPs can penetrate through them to deliver drugs or perform labeling. Since the SPIO NPs stay in the liver and spleen after entering the body, they can be used to treat liver cancer [71]. In a study by Chen et al., a green procedure was developed to encapsulate SPIO NPs and hydrophobic drugs like vitamin K3 (VK3) in water-based biodegradable polyurethane (WBPU) nanoparticles. According to the results, VK3 was effectively encapsulated by PU nanoparticles and exhibited a slow release pattern. Also, SPIO–PU NPs successfully entered the cells. Therefore, WBPU NPs can be used as carriers for SPIO and anticancer drugs [71]. PU responds differently to surrounding stimuli, such as temperature, release medium, chemical compounds, and pH. So far, many studies have been conducted on improving solvent-based pH-sensitive drug release systems. Bahadur et al. developed a waterborne polyurethane-urea (WPUU) as a long-term drug release system in response to different pHs and release media [72].

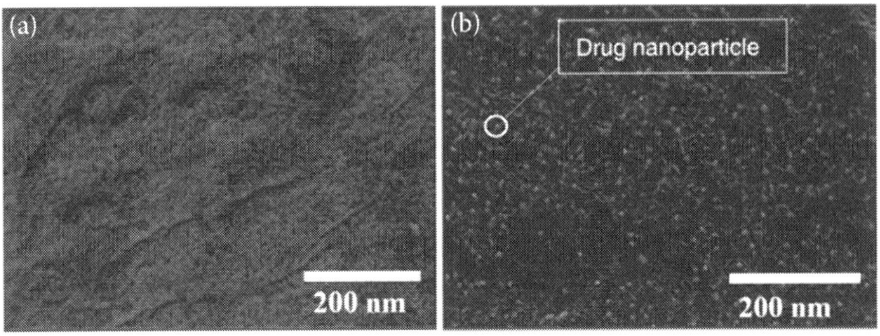

FIGURE 12.4 SEM of (a) WPUU (b) Drug-loaded WPUU. Adapted with permission from reference [72], Copyright (2017), Elsevier.

The SEM images of WPUU and drug-loaded WPUU are shown in Figure 12.4. As can be seen, WPUU has a smooth and uniform surface before loading of the cisplatin drug (Figure 12.4a), but after drug loading, the nanoscale drug is evenly distributed on the WPUU surface (Figure 12.4b) [72].

In many controlled drug delivery systems, a significant amount of drug is released immediately after the drug carrier is inserted in the release medium. This phenomenon is considered a burst release. This effect leads to uncontrolled drug release, greater cytotoxicity, higher cost, and less effectiveness. Therefore, it should be considered in designing any drug delivery system. Biodegradable PUs initially have higher swelling and drug release and finally, cause a burst release effect. Bahadur et al. used biocompatible and biodegradable methyl methacrylate (MMA) in combination with WPU to prevent burst release. For this purpose, different amounts of MMA were combined with WPU (named P1 to P5 which contain 0%, 10%, 20%, 30%, and 40% MMA, respectively). Mitomycin C was used to inject into these carriers [73]. Drug release results showed that among the MMA containing WPUs, the P2 had the highest drug release of 57.3%, while the P5 had the lowest release of 32.4% [73].

In another study, Feng et al. used chitosan combined with WPU to overcome the burst release problem. They used electrostatic interactions between chitosan and amino acid-terminated WPU to increase the bioactivity of WPU drug carriers. Here, DOX is referred to as a drug model and ultrasound method as a physical stimulant of drug release. According to an in vitro experiment, this carrier membrane had a stable and controllable drug release rate. At the same time, the membrane significantly minimized DOX toxicity in normal cells [74]. Gene therapy is a medical procedure for genetic disorders in which a new gene sequence is transferred into a patient's cells. Wu et al. used waterborne biodegradable cationic polyurethanes (WCPU) as carriers for gene transfer. WCPU NPs were not toxic at concentrations below 1000 μg/ml and could be used as a transfection chemical agent [75]. As a result, the WPU nanocarriers mentioned above can be suitable options for controlled drug/gene delivery systems.

REFERENCES

1. Wang F., Wang B., Li X., Wu Z., He Y., Song P., Wang R. Antimicrobial cationic acrylate-based hybrid coatings against microorganism contamination. *Progress in Organic Coatings*. 2020;142:105576.
2. Mohammadi A., Doctorsafaei A.H., Burujeny S.B., Rudbari H.A., Kordestani N., Najafabadi S.A.A. Silver (I) complex with a Schiff base ligand extended waterborne polyurethane: A developed strategy to obtain a highly stable antibacterial dispersion impregnated with in situ formed silver nanoparticles. *Chemical Engineering Journal*. 2020;381:122776.
3. Huang K.-S., Yang C.-H., Huang S.-L., Chen C.-Y., Lu Y.-Y., Lin Y.-S. Recent advances in antimicrobial polymers: A mini-review. *International Journal of Molecular Sciences*. 2016;17(9):1578.
4. Kenawy E-R., Worley S., Broughton R. The chemistry and applications of antimicrobial polymers: A state-of-the-art review. *Biomacromolecules*. 2007;8(5):1359–1384.
5. Mitra D., Kang E.-T., Neoh K.G. *Applications and Challenges of Smart Antibacterial Coatings. Advances in Smart Coatings and Thin Films for Future Industrial and Biomedical Engineering Applications*. Elsevier; 2020. pp. 537–556.
6. Eltorai A.E., Haglin J., Perera S., Brea B.A., Ruttiman R., Garcia D.R., Born C.T., Daniels A.H. Antimicrobial technology in orthopedic and spinal implants. *World Journal of Orthopedics*. 2016;7(6):361.
7. Marković Z., Kováčová M., Mičušík M., Danko M., Švajdlenková H., Kleinová A., Humpolíček P., Lehocký M., Marković B.T., Špitalský Z. Structural, mechanical, and antibacterial features of curcumin/polyurethane nanocomposites. *Journal of Applied Polymer Science*. 2019;136(13):47283.
8. Jiang G., Li X., Che Y., Lv Y., Liu F., Wang Y., Zhao C., Wang X. Antibacterial and anticorrosive properties of CuZnO@ RGO waterborne polyurethane coating in circulating cooling water. *Environmental Science and Pollution Research*. 2019;26(9): 9027–9040.
9. Shin E.J., Choi S.M. Advances in waterborne polyurethane-based biomaterials for biomedical applications. Novel Biomaterials for *Regenerative Medicine*. 2018:251–283.
10. Bankoti K., Rameshbabu A.P., Datta S., Maity P.P., Goswami P., Datta P., Ghosh S.K., Mitra A., Dhara S. Accelerated healing of full thickness dermal wounds by macroporous waterborne polyurethane-chitosan hydrogel scaffolds. *Materials Science and Engineering: C*. 2017;81:133–143.
11. Ma X.-Y., Zhang W.-D. Effects of flower-like ZnO nanowhiskers on the mechanical, thermal and antibacterial properties of waterborne polyurethane. *Polymer Degradation and Stability*. 2009;94(7):1103–1109.
12. Kim B., Kim T., Jeong H. Aqueous dispersion of polyurethane anionomers from H12MDI/IPDI, PCL, BD, and DMPA. *Journal of Applied Polymer Science*. 1994; 53(3):371–378.
13. Hsu S.-h., Tseng H.-J., Lin Y.-C. The biocompatibility and antibacterial properties of waterborne polyurethane-silver nanocomposites. *Biomaterials*. 2010;31(26):6796–6808.
14. Fang C., Pan S., Wang Z., Zhou X., Lei W., Cheng Y. Synthesis of waterborne polyurethane using snow as dispersant: structures and properties controlled by polyols utilization. *Journal of Materials Science & Technology*. 2019;35(7):1491–1498.
15. Omrani I., Babanejad N., Shendi H.K., Nabid M.R. Fully glutathione degradable waterborne polyurethane nanocarriers: Preparation, redox-sensitivity, and triggered intracellular drug release. *Materials Science and Engineering: C*. 2017;70:607–616.
16. Unnithan A.R., Gnanasekaran G., Sathishkumar Y., Lee Y.S., Kim C.S. Electrospun antibacterial polyurethane–cellulose acetate–zein composite mats for wound dressing. *Carbohydrate Polymers*. 2014;102:884–892.

17. Chang J., Yang G., Zheng Q., Wang Z., Xu Z., Chen Y., Fan H. Poly (*N*-acryloyl ciprofloxacin-co-acrylic acid)-incorporated waterborne polyurethane leather coating with long-lasting antimicrobial properties. *Journal of the American Leather Chemists Association*. 2017;112(01):15–22.
18. Barikani M. *Polyurethane*. Institute of Polymer and Petrochemical of Iran; 2006.
19. Francolini I., D'Ilario L., Guaglianone E., Donelli G., Martinelli A., Piozzi A. Polyurethane anionomers containing metal ions with antimicrobial properties: thermal, mechanical and biological characterization. *Acta Biomaterialia*. 2010;6(9):3482–3490.
20. Ahamed M., AlSalhi M.S., Siddiqui M. Silver nanoparticle applications and human health. *Clinica Chimica Acta*. 2010;411(23-24):1841–1848.
21. Emami-Karvani Z., Chehrazi P. Antibacterial activity of ZnO nanoparticle on gram-positive and gram-negative bacteria. *African Journal of Microbiology Research*. 2011;5(12):1368–1373.
22. Gurunathan S., Qasim M., Park C., Yoo H., Choi D.Y., Song H., Park C., Kim J.-H., Hong K. Cytotoxicity and transcriptomic analysis of silver nanoparticles in mouse embryonic fibroblast cells. *International Journal of Molecular Sciences*. 2018;19(11): 3618.
23. Mohammadi A., Barikani M., Lakouraj M.M. Biocompatible polyurethane/thiacalix [4] arenes functionalized Fe3O4 magnetic nanocomposites: Synthesis and properties. *Materials Science and Engineering: C*. 2016;66:106–118.
24. Park J.H., Kim I.K., Choi J.Y., Karim M.R., Cheong I.W., Oh W., Yeum J.H. Electrospinning Fabrication of polyvinyl alcohol)/waterborne polyurethane/silver composite nanofibre mats in aqueous solution for anti-bacterial exploits. *Polymers and Polymer Composites*. 2011;19(9):753–762.
25. Marambio-Jones C., Hoek E.M. A review of the antibacterial effects of silver nanomaterials and potential implications for human health and the environment. *Journal of Nanoparticle Research*. 2010;12(5):1531–1551.
26. Zhong Z., Luo S., Yang K., Wu X., Ren T. High-performance anionic waterborne polyurethane/Ag nanocomposites with excellent antibacterial property via in situ synthesis of Ag nanoparticles. *RSC Advances*. 2017;7(67):42296–42304.
27. Chen J., Wang Q., Luan M., Mo J., Yan Y., Li X. Polydopamine as reinforcement in the coating of nano-silver on polyurethane surface: Performance and mechanisms. *Progress in Organic Coatings*. 2019;137:105288.
28. Zhou R., Teo S., Srinivasan M. In situ formation of silver nanoparticle layer by supramolecule-directed assembly. *Thin Solid Films*. 2014;550:210–219.
29. Hasnain S., Nishat N. Synthesis, characterization and biocidal activities of Schiff base polychelates containing polyurethane links in the main chain. *Spectrochimica Acta Part A: Molecular and Biomolecular Spectroscopy*. 2012;95:452–457.
30. Liu H., Song J., Shang S., Song Z., Wang D. Cellulose nanocrystal/silver nanoparticle composites as bifunctional nanofillers within waterborne polyurethane. *ACS Applied Materials & Interfaces*. 2012;4(5):2413–2419.
31. Yang R., Du H., Lin Z., Yang L., Zhu H., Zhang H., Tang Z., Gui X. ZnO nanoparticles filled tetrapod-shaped carbon shell for lithium-sulfur batteries. *Carbon*. 2019;141:258–265.
32. Mallakpour S., Behranvand V. Nanocomposites based on biosafe nano ZnO and different polymeric matrixes for antibacterial, optical, thermal and mechanical applications. *European Polymer Journal*. 2016;84:377–403.
33. Liu L., Guo X., Shi L., Chen L., Zhang F., Li A. SiO_2-GO nanofillers enhance the corrosion resistance of waterborne polyurethane acrylic coatings. *Advanced Composites Letters*. 2020;29:2633366X20941524.
34. Xiong L., Zhang W.D., Shi Q.S., Mai A.P. Waterborne polyurethane/NiAl-LDH/ZnO composites with high antibacterial activity. *Polymers for Advanced Technologies*. 2015;26(5):495–501.

35. Tao Q., Zhang Y., Zhang X., Yuan P., He H. Synthesis and characterization of layered double hydroxides with a high aspect ratio. *Journal of Solid State Chemistry*. 2006;179(3):708–715.
36. Wang Q., O'Hare D. Recent advances in the synthesis and application of layered double hydroxide (LDH) nanosheets. *Chemical Reviews*. 2012;112(7):4124–4155.
37. Hu H., Yuan Y., Shi W. Preparation of waterborne hyperbranched polyurethane acrylate/LDH nanocomposite. *Progress in Organic Coatings*. 2012;75(4):474–479.
38. Kunkalekar R. Role of oxides (Fe_3O_4, MnO_2) in the antibacterial action of Ag–metal oxide hybrid nanoparticles. *Noble Metal-Metal Oxide Hybrid Nanoparticles*. Elsevier; 2019. pp. 303–312.
39. Azam A., Ahmed A.S., Oves M., Khan M., Memic A. Size-dependent antimicrobial properties of CuO nanoparticles against Gram-positive and-negative bacterial strains. *International Journal of Nanomedicine*. 2012;7:3527.
40. Ruan D., Zhang L., Zhang Z., Xia X. Structure and properties of regenerated cellulose/tourmaline nanocrystal composite films. *Journal of Polymer Science Part B: Polymer Physics*. 2004;42(3):367–373.
41. Chen S., Zhang S., Jin T., Zhao G. Synthesis and characterization of novel covalently linked waterborne polyurethane/Fe3O4 nanocomposite films with superior magnetic, conductive properties and high latex storage stability. *Chemical Engineering Journal*. 2016;286:249–258.
42. Fu H., Wang Y., Chen W., Xiao J. Reinforcement of waterborne polyurethane with chitosan-modified halloysite nanotubes. *Applied Surface Science*. 2015;346: 372–378.
43. Zia K.M., Zuber M., Saif M.J., Jawaid M., Mahmood K., Shahid M., Anjum M.N., Ahmad M.N. Chitin based polyurethanes using hydroxyl terminated polybutadiene, part III: Surface characteristics. *International Journal of Biological Macromolecules*. 2013;62:670–676.
44. Fei Liu X., Lin Guan Y., Zhi Yang D., Li Z., De Yao K. Antibacterial action of chitosan and carboxymethylated chitosan. *Journal of Applied Polymer Science*. 2001;79(7):1324–1335.
45. El-Sayed A.A., El Gabry L., Allam O. Application of prepared waterborne polyurethane extended with chitosan to impart antibacterial properties to acrylic fabrics. *Journal of Materials Science: Materials in Medicine*. 2010;21(2): 507–514.
46. Fu X., Shen Y., Jiang X., Huang D., Yan Y. Chitosan derivatives with dual-antibacterial functional groups for antimicrobial finishing of cotton fabrics. *Carbohydrate Polymers*. 2011;85(1):221–227.
47. Du S., Wang Y., Zhang C., Deng X., Luo X., Fu Y., Liu Y. Self-antibacterial UV-curable waterborne polyurethane with pendant amine and modified by guanidinoacetic acid. *Journal of Materials Science*. 2018;53(1):215–229.
48. Yao C., Li X., Neoh K., Shi Z., Kang E. Surface modification and antibacterial activity of electrospun polyurethane fibrous membranes with quaternary ammonium moieties. *Journal of Membrane Science*. 2008;320(1-2):259–267.
49. Wang C., Wu J., Li L., Mu C., Lin W. A facile preparation of a novel non-leaching antimicrobial waterborne polyurethane leather coating functionalized by quaternary phosphonium salt. *Journal of Leather Science and Engineering*. 2020;2(1):1–12.
50. Yoo H.J., Kim H.D. Characteristics of waterborne polyurethane/poly (N-vinylpyrrolidone) composite films for wound-healing dressings. *Journal of Applied Polymer Science*. 2008;107(1):331–338.
51. Chen Y., Tan W., Li Q., Dong F., Gu G., Guo Z. Synthesis of inulin derivatives with quaternary phosphonium salts and their antifungal activity. *International Journal of Biological Macromolecules*. 2018;113:1273–1278.

52. Anthierens T., Billiet L., Devlieghere F., Du Prez F. Poly(butylene adipate) functionalized with quaternary phosphonium groups as potential antimicrobial packaging material. *Innovative Food Science & Emerging Technologies*. 2012;15:81–85.
53. Tijing L.D., Ruelo M.T.G., Amarjargal A., Pant H.R., Park C.-H., Kim D.W., Kim C.S. Antibacterial and superhydrophilic electrospun polyurethane nanocomposite fibers containing tourmaline nanoparticles. *Chemical Engineering Journal*. 2012;197: 41–48.
54. Agnol L.D., Dias F.T.G., Ornaghi Jr H.L., Sangermano M., Bianchi O. UV-curable waterborne polyurethane coatings: A state-of-the-art and recent advances review. *Progress in Organic Coatings*. 2021;154:106156.
55. Huang W., Wang Y., Zhang S., Huang L., Hua D., Zhu X. A facile approach for controlled modification of chitosan under γ-ray irradiation for drug delivery. *Macromolecules*. 2013;46(3):814–818.
56. Liu K., Su Z., Miao S., Ma G., Zhang S. UV-curable enzymatic antibacterial waterborne polyurethane coating. *Biochemical Engineering Journal*. 2016;113:107–113.
57. Moeini A., Pedram P., Makvandi P., Malinconico M., d'Ayala G.G. Wound healing and antimicrobial effect of active secondary metabolites in chitosan-based wound dressings: a review. *Carbohydrate polymers*. 2020;233:115839.
58. Dhivya S., Padma V.V., Santhini E. Wound dressings–a review. *BioMedicine*. 2015;5(4).
59. Yoo H.J., Kim H.D. Synthesis and properties of waterborne polyurethane hydrogels for wound healing dressings. *Journal of Biomedical Materials Research Part B: Applied Biomaterials: An Official Journal of The Society for Biomaterials, The Japanese Society for Biomaterials, and The Australian Society for Biomaterials and the Korean Society for Biomaterials*. 2008;85(2):326–333.
60. Smith A.M., Moxon S., Morris G. Biopolymers as wound healing materials. *Wound Healing Biomaterials*. Elsevier; 2016. pp. 261–287.
61. Choi S., Zo S., Park G., Shin E., Han S. Preparation of waterborne polyurethane-based macroporous sponges as wound dressings. *Journal of Nanoscience and Nanotechnology*. 2020;20(8):4634–4637.
62. Hao H., Shao J., Deng Y., He S., Luo F., Wu Y., Li J., Tan H., Li J., Fu Q. Synthesis and characterization of biodegradable lysine-based waterborne polyurethane for soft tissue engineering applications. *Biomaterials Science*. 2016;4(11):1682–1690.
63. Kumar Khanna V. Targeted delivery of nanomedicines. *International Scholarly Research Notices*. 2012;2012.
64. Torchilin V.P. Drug targeting. *European Journal of Pharmaceutical Sciences*. 2000; 11:S81–S91.
65. Omrani I., Babanejad N., Shendi H.K., Nabid M.R. Preparation and evaluation of a novel sunflower oil-based waterborne polyurethane nanoparticles for sustained delivery of hydrophobic drug. *European Journal of Lipid Science and Technology*. 2017;119(8):1600283.
66. Faraji A.H., Wipf P. Nanoparticles in cellular drug delivery. *Bioorganic & Medicinal Chemistry*. 2009;17(8):2950–2962.
67. Pillai O., Panchagnula R. Polymers in drug delivery. *Current Opinion in Chemical Biology*. 2001;5(4):447–451.
68. Saeedi S., Omrani I., Bafkary R., Sadeh E., Shendi H.K., Nabid M.R. Facile preparation of biodegradable dual stimuli-responsive micelles from waterborne polyurethane for efficient intracellular drug delivery. *New Journal of Chemistry*. 2019;43(47):18534–18545.
69. Yang D., Chen W., Hu J. Design of controlled drug delivery system based on disulfide cleavage trigger. *The Journal of Physical Chemistry B*. 2014;118(43): 12311–12317.

70. Yin H., Du B., Chen Y., Song N., Li Z., Li J., Luo F., Tan H. Dual-encapsulated biodegradable 3D scaffold from liposome and waterborne polyurethane for local drug control release in breast cancer therapy. *Journal of Biomaterials Science, Polymer Edition*. 2020;31(17):2220–2237.
71. Chen Y.-P., Hsu S-h. Preparation and characterization of novel water-based biodegradable polyurethane nanoparticles encapsulating superparamagnetic iron oxide and hydrophobic drugs. *Journal of Materials Chemistry B*. 2014;2(21):3391–3401.
72. Bahadur A., Saeed A., Iqbal S., Shoaib M., ur Rahman M.S., Bashir M.I., Asghar M., Ali M.A., Mahmood T. Biocompatible waterborne polyurethane-urea elastomer as intelligent anticancer drug release matrix: a sustained drug release study. *Reactive and Functional Polymers*. 2017;119:57–63.
73. Bahadur A., Saeed A., Shoaib M., Iqbal S., Anwer S. Modulating the burst drug release effect of waterborne polyurethane matrix by modifying with polymethylmethacrylate. *Journal of Applied Polymer Science*. 2019;136(13):47253.
74. Feng Z., Zheng Y., Zhao L., Zhang Z., Sun Y., Qiao K., Xie Y., Wang Y., He W. An ultrasound-controllable release system based on waterborne polyurethane/chitosan membrane for implantable enhanced anticancer therapy. *Materials Science and Engineering: C*. 2019;104:109944.
75. Wu G.-H., Hsu S.-h.. Synthesis of water-based cationic polyurethane for antibacterial and gene delivery applications. *Colloids and Surfaces B: Biointerfaces*. 2016;146: 825–832.

13 Waterborne Polyurethanes for Tissue Engineering

D.E. Mouzakis
Hellenic Army Academy, Leoforos Eyelpidon (Varis-Koropiou) Avenue, Vari, Athens, Greece
National Hellenic Research Foundation, Vassileos Constantinou Avenue, Athens, Greece

Styliani Papatzani
Hellenic Army Academy, Leoforos Eyelpidon (Varis-Koropiou) Avenue, Vari, Athens, Greece
University of West Attica, Agiou Spiridonos, Egaleo, Athens, Greece

CONTENTS

13.1 Introduction....213
13.2 Synthesis of Waterborne Polyurethanes....215
13.3 Applications of WPU in Tissue Engineering....216
  13.3.1 2D Film Applications—Wound Healing....216
  13.3.2 Cartilage Tissue Regeneration....218
  13.3.3 Bone Tissue Augmentation....222
13.4 Epilogue—Outlook....225
References....226

13.1 INTRODUCTION

Polyurethanes (PUs) have numerous engineering applications because of their extended range of useful properties. Changes in the final properties of a PU are dictated by a few elements, e.g. the analogies of the polyol and diisocyanate, changes in the convergence of the critical step, the natural, proportion of the chain extender plus the density of the crosslinking substance. They additionally have an one-of-a-kind multiphase construction comprising softer polyol and harder isocyanate parts. The soft part is responsible for elasticity; the harder sections are providing for dimensional stability, acting as the areas of the actual crosslinking. PU is normally utilized in clinical materials in light of high inherent biocompatibility, its wide spectrum of physical-chemical properties emerging from the

DOI: 10.1201/9781003173526-13

micro-phasic division between the polymer molecular chains [1]. Improvement of the biostability, or inversely biodegradability, of PU for short- or long-term biomedicinal applications has been intensively researched in many studies. As an example, Poly(carbonate urethane) (PCU) is usually considered to be biostable and has been proved to decay in ex-vivo implantation efforts [2]. Ongoing endeavors have zeroed in on creating PUs that are biodegradable and exhibit tunable natural, mechanical, and physicochemical properties for tissue engineering [3]. PU is frequently synthesized with the help of various organic solvents, while ecological concerns have highlighted the need for a more practical engineered approach. More ecological, water-based, methods have been invented. Since ordinary PU sections show high hydrophobicity, an outer or inward emulsifier substance is utilized in order to provide PU micro-particles scattering in water. Inward structuring or self-emulsification strategy by embedding anionic hydrophilic carboxyl incorporated (COO–) sections can produce PU characterized by greater biocompatibility as well as, blood tolerance [4].

Fundamentally enzymatic, oxidative (or free-radicals oxidation), or even simpler hydrolytic, is observed for PU with respect to its process of degradation in-vivo [5]. The molecular dissolution pace of PU is dictated by the intramolecular forces, crystallinity and of course hydrophobicity, atomic weight, arrangement, and expanding conduct, just as the surrounding biochemical environment (e.g. pH) [6]. PU degradation products containing aliphatic diisocyanates have proven no signs of cytotoxicity, both in vitro and in vivo experiments [7]. Then again, aromatic diisocyanate-based PU incited more debate because the dissolution products could potentially be proven poisonous, cancer-causing, as well as mutagenic [8].

In this sense, biodegradable PU is a decent contestant material to deliver films (two-dimensional meshes also) or tissue scaffolds (three-dimensional [3D] structures porous or not) [9]. Its inherent elastic properties are similar to these of extracellular matrix (ECM), and in this sense that shall provide an appropriate environment for cells to append, multiply, and will also promote mitosis. Many biodegradable PU scaffolds utilized for tissue designing have been polymerized by organic solvent techniques from aromatic diisocyanate [10]. Likewise, water-based biodegradable PU is essentially utilized as micelles or in the form of thin films because its handling is rather difficult. Another well-known technique is freeze-drying, for producing scaffolds for tissue regeneration, being a low-temperature process basically [11]. Altering parameters like the type of polymer matrix, dissolvable and non-dissolvable proportion, cooling ramp and temperatures, lead to platforms with various permeability to cell cultures [12].

It is well known that human cartilage can withstand compressive forces by converting them into surface loads [13]. The ligament tissue scaffolds ought to give load-bearing, underlying scaffolding, and allow longer time intervals recovery and ligament growth [14]. Already suggested by some researchers, PU scaffolds with a proper mechanical response (stiffness plus strength mostly) can advance the rapid development of functional ligaments, such as the ECM [15]. Technological advancements, such as the invention of dynamic bioreactors, may additionally lead toward faster ligament reconstruction by emulating the local biological microenvironment, providing mechanical stimuli, and settling mass exchange restrictions under in vitro cultures [16].

Polymers of the PU family, with a tissue-similar rigidity as well as elasticity showing high biocompatibility, have been proposed as suitable platforms for ligament tissue engineering [17]. PU allows for tailoring of its mechanical response and also the rhythm of biodegradation of its related scaffolds, altered usually by controlling the soft fragment properties. Utilizing new trends, PUs with biodegradable properties could be prepared by an established water-based procedure and consequently manufactured in various forms of scaffolds by additive manufacturing procedures, e.g. nowadays with one of the market-available 3D printers [18]. Since mechanical stability is very important in order to support the healing tissue, elastic recovery of the 3D printed PU, when a steady strain was removed, has reached as high as 99% [18]. Cyclic loads on such 3D printer-prepared PU scaffolds have been applied using a Dynamic Mechanical Analyzer (DMA) and information subsequently obtained, after calculating elastic and damping coefficients at different cycling levels have proven high biocompatibility [19]. Mesenchymal Stem Cells (MSCs) cultivated using the PU-based, 3D printed scaffolds were found to move inside the scaffolds, populating them, before differentiating into the desirable chondrocytes. Alternatively, biodegradable waterborne PUs (WPUs) have been observed to form microspheres, for controlled drug delivery applications [20]. The amphiphilic attributes of WPUs present a large opportunity for the encapsulation of hydrophobic medications [21]. Consolidating numerous medications and delivering drugs autonomously, in a controlled manner, is a revolutionary technology, in tissue engineering biomedical applications [22]. In this chapter, the efficacy of WPUs in promoting tissue engineering applications and drug delivery is discussed. An effort is undertaken to include the most recent and significant results in this field.

13.2 SYNTHESIS OF WATERBORNE POLYURETHANES

Due to its environmental friendliness, nonflammability, the ability for film formation under room temperature conditions, wear resistance, high elasticity, and controllable structure [23], WPU is adopted in various processes and applications, such as coatings, adhesives, inks, leather finishing, composite materials, and biological materials. The existence of the hydrophilic group along with the linear structure of the WPU chain, weaknesses are present in its thermo-mechanical response, and resistance to water sorption. According to many reports, WPU performance can be improved by grafting side chain groups onto the waterborne backbone PU chain. Example gratis, because polypropylene glycol contains pendant methyl groups, there appears a reduction in the surface tension of WPU dispersed systems [24]. Attachment of long-branched fatty acid chains can promote the contact angle, adhesion, surface tension and hydrophobicity, and thermal stability and also promote the interfacial tension of WPU films, as shown by Lei et al. [25]. Zhang et al. [26] synthesized a WPU with a side chain containing phosphorus. The PU featuring the organophosphorus group incorporated in the main chain exhibits improved resistance to hydrolysis than the PU with the organophosphorus group in the side chain. Siloxane groups have been also proposed to be introduced as side chains into the WPU main chain [27]. Alkoxysilane groups, being small molecule monomers, contain silicon atoms, and compared to PDMS, they can positively affect the compatibility of WPU and siloxane [28]. Alkoxysilane has inherent good thermal stability, low-temperature flexibility, flame retardancy, high permeability, and low interfacial tension. Thus, the overall performance of the WPU-based film has also been improved [29]. Saldo et al. [30] reported on a self-crosslinking WPU-dispersed system, synthesized with (3-aminopropyl) triethoxysilane. Another diol with a triethoxy-silane incorporated in the side chain, synthesized by Michael Addition, was employed as a chain extender to prepare a WPU, as reported by Wang et al. [31] and Zhang et al. [32]. The mechanical behavior and hydrophobicity of the films were significantly enhanced after the addition of the diol with triethoxy-silane in the side chain. Long flexible side chains have a positive effect, on the silane groups, to enable migration toward the surface of the prepared films, as reported by Li et al. [33]. Surface contact angle measurements of wettability, and the mechanical response of WPUs, were also enhanced by grafting of alkoxysilane side groups. In a very recent study, Jang et al. [34], synthesized a phosphorus-containing polyol (P-polyol), which consequently reacted with isophorone diisocyanate (IPDI), to produce WPU. The addition of dimethylol propionic acid, in mixtures of P-polyol and polycarbonate diol (PCD), was afterward reacted with IPDI, to confer hydrophilicity to the WPU. Research results have proven that the mechanical strength of PU-10 (WPU with a P-polyol of a PCD molar ratio 0.1:0.9) was 16% higher in comparison to the reference P-polyol-free WPU specimens. In addition to that, the temperature of thermal decomposition of the same PU-10 sample was 27°C higher compared to the WPU reference sample.

13.3 APPLICATIONS OF WPU IN TISSUE ENGINEERING

13.3.1 2D Film Applications—Wound Healing

In the early 2000s, two interesting precursor studies were presented. With the first, the thermal degradation of aqueous PU dispersions was studied [35]. Profiles of films were cast with waterborne anionic PUs and poly(urethane-urea)s in the presence of acetone and analyzed with the help of thermal gravimetry (TGA) and differential thermogravimetry (dTG). Degradation was shown to take place in a two-stage process, in which the degradation of hard segments took place first, followed by the soft segments. However, dTG revealed phases that would have been unobserved had only TGA been considered. It was confirmed that the higher length of the soft segments and the presence of urea enhanced thermal stability, whereas the content of the hard segments also affects thermal behavior. The thermal behavior was not affected by the presence of acetone. In 2002, the second study presented swelling experiments on water-soluble diamines dispersed in aqueous PU [36]. Chain extension models were studied, and mechanical responses including tensile strength and elongation of the thin films prepared were evaluated. It was found that the PU particles could not swell with the water-soluble diamines, which were also confirmed by FT-IR studies. As discussed earlier, waterborne polyurethane dispersions (PUDs) are much preferred compared to solvent-borne PUDs, which release volatile organic compounds, both during production and use [37].

Epidermis wound healing was enhanced with the help of aqueous dispersions of PU particles and chitosan blends forming macroporous hydrogel scaffolds [38]. Chitosan (Chi), known for its antibacterial properties and improved mechanical properties, was blended with PU diol dispersion at all ratios of integral numbers starting with 10:0 and reaching 0:10. It was found that increasing chitosan content above 5:5, boosted the scaffold formation. FT-IR was used to assess the intramolecular and intermolecular hydrogen bonding, leading to conclusions for the moist environment of scaffolds, AFM to observe the morphology and surface roughness, and SEM to study the microstructure. In vivo biocompatibility and in vitro cytocompatibility was confirmed by implanting the scaffolds in albino male Wistar rats subcutaneously and by the proliferation of their fibroblast cells, respectively. Amongst a significant number of findings, higher contact angle found for the blends of Chi: PUD; 8:2, 7:3 is expected to accelerate wound healing, while the 7:3 blend also exhibited the highest elongation ratio in wet conditions due to the hydrogen bonding of PUD, demonstrated the highest swelling (without dissolution) and interconnectivity of pores and antimicrobial potential. Furthermore, it was demonstrated that the 7:3 Chi: PUD blend accelerated wound healing and wound regeneration, even compared to commercial products. A more recent study of the effect of chitosan–polyurethane hydrogel membrane (HPUC) on the healing of wounds—foot ulcers—caused by diabetes, was studied [39]. Through evaluation with FTIR, ^{13}C NMR and TGA, and ex-vivo experiments, it was found that the new membrane seeded with mononuclear bone marrow fraction cells enhanced wound healing of diabetic rat models.

Artificially crosslinked 3D systems, were acquired by joining WPU (prepared from polytetramethylene ether glycol, isophorone diisocyanate, and 2,2-bis (hydroxymethyl) butyric acid), was proven to form a bio-active extracellular network. The inclusion of small intestinal submucosa (SIS) has been proposed to create another approach for soft tissue regeneration [40]. Subsequent to portraying the design procedure, synthesis, and properties of the complex PU/SIS composites, i.e. mechanical strength, Young's modulus, and versatility of wet PU/SIS, test results were further analyzed and compared against those of crosslinked PU. Human umbilical vein endothelial cells were employed in-vitro to assess the capacity to upgrade cell attachment and multiplication in the PU/SIS scaffolds. Accordingly, the integrated PU/SIS tests showed high versatility and were equipped for upgrading cell propagation exhibiting no indications of cytotoxicity. Ex-vivo subcutaneous implantation and the ensuing testing directed following 2, 4, weeks, and up to 2 months demonstrated that satisfactory embedding, as well as a vascularization network, was formed inside the bulk of the PU/SIS composites. On the other hand, the induced occurrence of SIS advanced cell propagation and angiogenesis provided adequate signs for tissue recovery. The prepared PU/SIS composites, which have shown high bioactivity and mechanical stability, are consequently proper candidates to be utilized for soft tissue engineering applications.

Natural based protein systems have also been under the focus of investigations. Soy protein isolate (SPI) and a polyurethane prepolymer (PUP) were used to synthesize a WPU composite scaffold based on biodegradable soy protein (SWPU), by a process including substance reaction, followed by freeze–drying [40].

Variations of the SPI content (0, 10%, 30%, 50%, 70%) and its effect on the internal structure and actual properties of the resulting composite porous scaffolds were studied in detail. Results obtained, have indicated that the reactions between –NCO of PUP and also the –NH2 radicals of SPI, formed permeable, WPU composite SPI-based spongy scaffolds. Effects of estimated water uptake percentage, solvent resistance, and compression mechanical tests have proven that hydrophilicity, water absorption, and rigidity in the dry condition of the composite scaffolds was enhanced with increasing SPI content. Particularly, the rigidity (elastic modulus-compression) has risen from 0.3 to 5.5 MPa with increasing the SPI content. Biodegradability, as well as cytocompatibility of the composite WPU–SPI-based sponges, were assessed by cell cultures and ex-vivo animal implantation tests. The testing outcomes demonstrated that a specific SPI content in the scaffolds promoted attachment, development, and cell multiplication. During the 9-month long, in vivo testing the SWPU-50 sponges (content: 50% SPI) showed the least tissue inflammatory response. On the other hand, SWPU-50 scaffolds exhibited enhanced neo-angiogenesis and the highest histocompatibility. SWPU-50 composite sponge scaffolds are potentially good candidates for tissue engineering.

13.3.2 Cartilage Tissue Regeneration

PU has been employed by many researchers as a potential scaffold for tissue augmentation. Its unique elastic properties and ability to build 3D structures and scaffolds made it an excellent candidate for cartilage tissue repair and/or augmentation [41]. An articular ligament is a viscoelastic tissue with an exceptionally micro-structured 3D geometry [42] with mechano-rheological coupled behavior. Damage to this highly organized tissue is inflicted by injury, tumors, or osteoarthritis. It is severe and truly hard to repair, of the restricted inherent potential of cartilage for self-recuperating. Conventional clinical treatments, including micro-fractures, ligament allografts, arthroplasty, and infusion of bone marrow undeveloped cells (MSCs), have produced restricted success. Interestingly, PU-based tissue designing is a promising technique because of its straightforwardness and stable medication conveying capacity, along with promoting cartilage recovery. Nowadays, 3D printing technology is being widely adopted, as a promising and flexible new approach for the design (CAD) and manufacturing of almost any scaffold geometry for tissue engineering. Seeding with cells is also of paramount importance since cell-free scaffolds, supposedly do not present sufficient therapeutic efficacy, while the risk of contamination is always quite high. In a recent study [43], water-based ink for 3D printing was developed. The ink contained biodegradable PU, chemokine SDF-1, and Y27632 drug-delivering PU microspheres. The scaffolds from this specific ink were printed at low temperature (–40°C), while the scaffolds were manufactured with sequential drug release properties. PU scaffolds containing 200 ng/mL SDF-1 and 22 wt.% Y27632 of encapsulated microspheres exhibited the optimal behavior. The intrinsic 3D design approach of the scaffolds caused each of SDF-1 and Y27632 to be sequential, in vitro released, and reach the cell-culture required concentration. The Human mesenchymal stem cells (hMSCs) pre-seeded in the scaffolds exhibited remarkable

GAG deposition within only 7 days. Controlled release of SDF-1 from the PU scaffolds induced the migration of hMSCs. Also, cell-free PU/SDF-1 scaffolds were consequently implanted in rabbits with articular cartilage defects. It was shown that the scaffolds supported the potential to facilitate quick cartilage regeneration. Such 3D printed scaffolds with sequentially controlled release of drugs and substances, such as the SDF-1 and Y27632, are promising a bright future in cartilage tissue regeneration applications. In another work of the same researchers [19], water-based 3D printing, PU-materials with tailored bioactivity response, for customized cartilage tissue regeneration is presented. The "ink" for printing contained synthetic biodegradable PU in water dispersion and included also elastic nanoparticles, hyaluronan, and other bioactive ingredients such as TGFb3 or a small molecule drug (Y27632) to replace TGFb3. Elastic WPU-scaffolds were 3D-printed from the composite ink at subzero temperatures. Again, those 3D printed scaffolds promoted the self-aggregation of mesenchymal stem cells (MSCs) and showed a controlled time-dependent release profile of the bioactive substances. Differentiation of MSCs to chondrocytes was induced, and a suitable matrix for cartilage repair was produced. Moreover, it was claimed, that the design of controlled release, growth factor-free scaffolds, may prevent hypertrophy of the cartilage. Another application included nanoparticles of PU in a sponge-mimicking 3D structure obtained by freeze–drying [44]. The sponge scaffold was characterized by a porous structure. Enhanced wetting, mechanical response, degradation behavior, and by-products were evaluated. The ability to act as a scaffold for cartilage tissue regeneration, by culturing chondrocytes as well as mesenchymal stem cells (MSCs), was also evaluated. The interesting gyroid micro-structure of the scaffold may explain the excellent strain recovery (87%) and elongation compared with traditional poly(d,l-lactide) (PLA) stents. The PU scaffold was effectively seeded with chondrocytes without pre-wetting. Enhanced cell growth and secretion higher amounts of glycosaminoglycans, compared with PLA scaffolds was proven beyond doubt. In addition to that, human MSCs showed enhancement of cartilage formation and gene expression in the PU scaffold after induction was higher than that in the PLA scaffold. Based on their good hydrophilicity, elasticity and cartilage regeneration ability, the new biodegradable PU scaffold appeared more promising than other usual polymeric biodegradable scaffolds reported in cartilage tissue regeneration applications. 3D printing procedures are becoming very popular for the manufacturing of scaffolds. The methodology of polymers crosslinking via photopolymerization (Figure 13.1) can provide useful structures. In a new liquid resin preparation, the process was reported. 3D printed tailored cartilage scaffolds have been manufactured with photosensitive materials, WPU, and hyaluronic acid. The scaffolds prepared with such a procedure exhibited high cytocompatibility. The mechanical behavior of WPU resembles that of natural articular cartilages. Human Wharton's jelly mesenchymal stem cells (hWJMSCs) were cultured as shown in Figure 13.2. Excellent chondrogenic differentiation capacity was shown by hWJMSCs in that case [45].

In another study, WPU and WPU-acellular cartilage extracellular matrix (ECM) scaffolds were prepared under a low-temperature deposition manufacturing (LDM) system through water-based 3D printing. The system combined rapid deposition

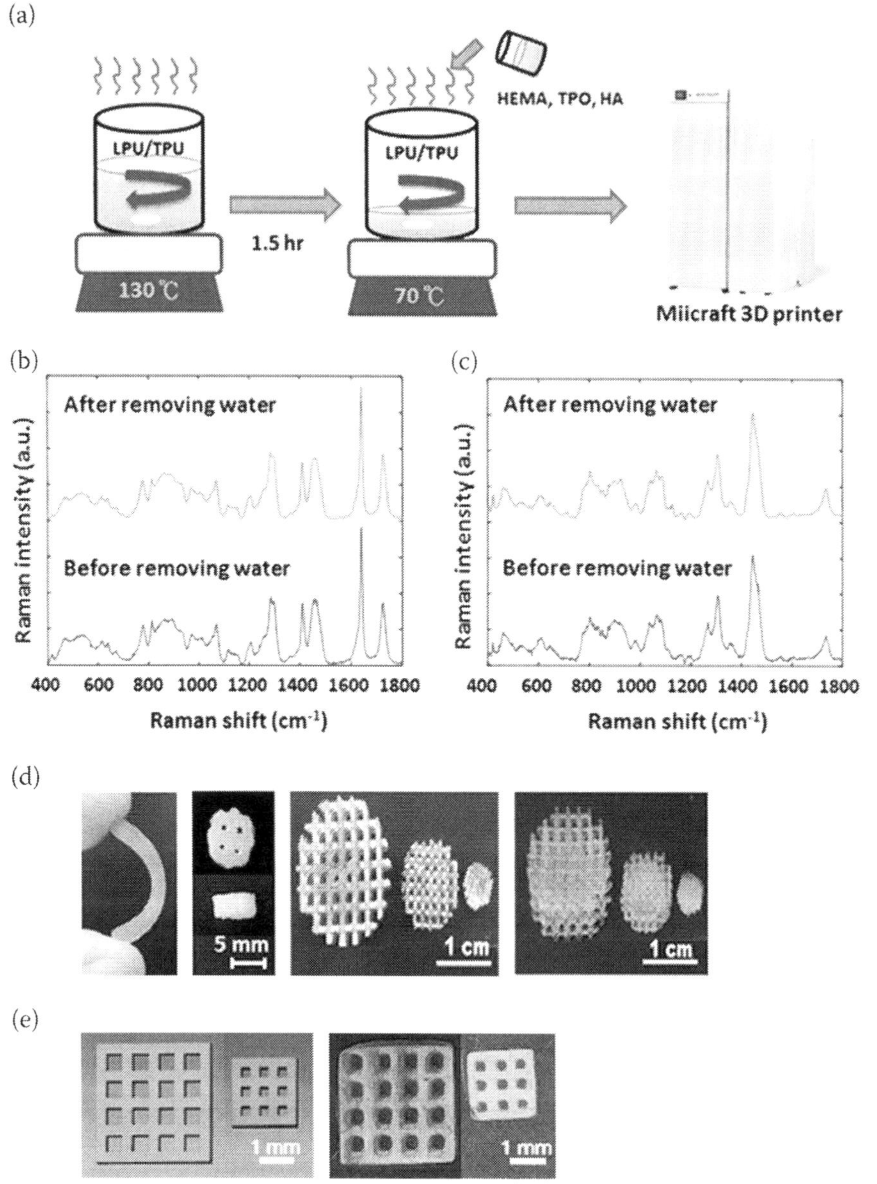

FIGURE 13.1 (A) The schematics of the manufacturing process of the water-based polyurethane photosensitive materials; The Raman spectra of the (B) water-based light-cured polyurethanes and (C) water-based thermoplastic polyurethanes with or without water removal processes; (D) The images of the printed scaffolds; (E) The images of the designed (left) and printed (right) porous lattice structures [45]. Figure 2 adapted with permission from reference [45]. Copyright Chen et al., some rights reserved; exclusive licensee [MDPI]. Distributed under a Creative Commons Attribution License 4.0 (CC BY) https://creativecommons.org/licenses/by/4.0/

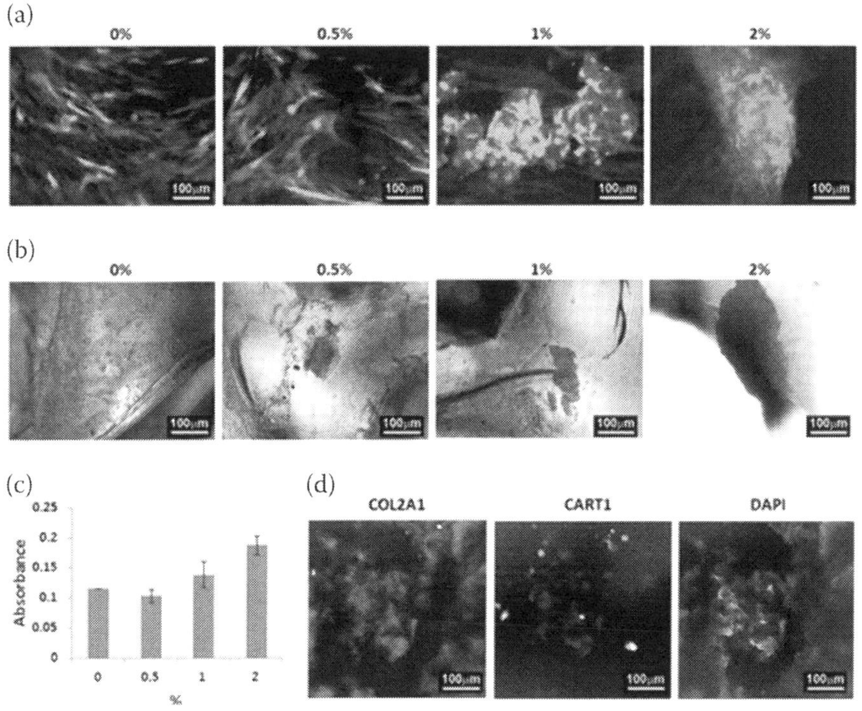

FIGURE 13.2 The (A) fluorescent and (B) Alcian blue staining images and the (C) GAG contents of the micromass cultures of WJMSCs cultured on PU/HA hybrid scaffolds for 1 day; (D) The immunofluorescence staining images of nuclei (blue), COL2A1 (red) and CART1 (green) for the micromass cultures of WJMSCs cultured on PU/HA hybrid scaffolds with 2% HA for 1 day [45]. Figure 3 adapted with permission from reference [45]. Copyright Chen et al., some rights reserved; exclusive licensee [MDPI]. Distributed under a Creative Commons Attribution License 4.0 (CC BY) https://creativecommons.org/licenses/by/4.0/

manufacturing with phase separation technologies. A layered macroporous structure was successfully created. The pore structure, hydrophilicity, and bioactive components were enhanced by adding ECM to the WPU scaffold. Cell distribution, adhesion, and proliferation, in the optimized WPU-ECM scaffold, were more enhanced than in the pure WPU scaffold. Even more significantly, the WPU-ECM stent can promote the synthesis and regulation of glycosaminoglycans (GAG) as well as collagen and cartilage-specific genes.

Shape memory cryogel systems based on WPU have also been reported as scaffolds for chondrogenic regeneration [46]. A shape-memory cryogel, based on water-based biodegradable bifunctional PU, used as a nanosized crosslinker to react with chitosan, was synthesized. Crystallinity was verified to be changed in both the crosslinker and the cryogel via WAXS employed in-situ, as martyrs, to the shape memory process, during the shape fixation and recovery procedures. The following were also discovered via in situ SAXS; (i) shape memory mechanism, (ii) the orientation of crystallinity of the cryogel and the crosslinker.

Cryogel was formed in a stripe shape. Steps of shape memory effect verification included initial deformation of the strips at 50°C to a U-shape and fixation at –20°C. At 25°C, the strip could be squeezed, whereas at 50°C in air, and it was eventually reformed to the original strip shape. Apart from air ambient, the effect of shape memory was also tested at two different temperatures in water. At 4°C, the injected cryogel recovered the U-shape in water, exhibiting elastic recovery, whereas at 37°C, it transformed to a long strip in water. Long-term cell proliferation and chondrogenic potential of the scaffolds were proven by ex-vitro human mesenchymal stem culture, grown in the shape-memory cryogels. Due to their unique injectability and cytocompatibility properties, it was inferred that injectable and expandable 3D scaffolds for tissue regeneration can offer potential applications of shape memory cryogels soon, while minimally invasive surgery may rely on these WPU systems [46].

Highly appreciated in plastic surgery applications for cosmetic or post-trauma restoration applications is cartilage-mimicking material for nasal reconstruction. Also, nasal stenting biomaterials are very useful for the treatment of nasal infections. A nasal stent equipped for forestalling attachments and aggravation is of extraordinary worth. Such a nasal stent was synthesized using a plasticized bacterial cellulose (BCG) and WPU. It additionally confronted the issue of irritation reaction and antibacterial performance. The gelation conduct of BCG safeguards the paranasal sinus mucosa; in the meantime, the WPU material with enhanced mechanical properties played a key role in supporting the thin nasal cavity. Thickness, size, and supporting power of the nasal stent could be changed to adapt to the particular geometry of the nasal cavity. Thermogravimetric analyses, contact angle, and water retention tests were applied to explore the thermal decomposition, hydrophilic, and water retention properties of the composite stents. Composites stacked with poly(hexamethylene biguanide) hydrochloride withheld their antibacterial action for more than 12 days [47]. Animal testing additionally showed that the mucosal epithelium harm due to the BCG WPU composite was minor. This new type of medication stacked nasal stent can viably address the postoperative attachments and diseases. Simultaneously, it can guarantee the strength of nasal mucosal and subsequently can offer significant clinical application prospects in treating nasal illnesses.

13.3.3 Bone Tissue Augmentation

In 2020, Yang and Wu presented results on the use of different block lengths of amphiphilic biodegradable WPU films, which were (organic) solvent-free. In addition, these films did not contain crosslinkers [48]. Hydrophilic poly(ethylene glycol) (PEG) of the soft segment was substituted by the hydrophobic poly(e-caprolactone) diol and assessed the WPU with respect to mechanical properties (tensile strength, elastic modulus, dynamic mechanical properties), thermal stability (TGA/DSC) water absorption, biodegradability (standard hydrolysis tests), and cytotoxicity (culture of bone mesenchymal stem cells). Again, both the length and chemical composition of the soft segments can be tuned for 3D printable tissue scaffolds production.

Inserting autogenous bone grafts for the healing of significant bone defects is often mandatory. However, high morbidity of donors is observed during bone harvesting, as this procedure can cause serious complications. For this, instead of bone graft, the option to use ceramics or bioresorbable/biodegradable polymers for the formation of substitute for the bone, as long as the material produced is biocompatible has been explored. Scaffolds produced by nanocomposites of PUDs grafted with collagen polypeptide/hydroxyapatite (WPU-g-CP/HAp) have been scrutinized. Initially, an emulsion of non-solvated WPU-g-CP copolymer was blended through water stage cross-connecting copolymerization between WPU pre-polymer and KH550 adjusted collagen polypeptide (CP). At that point, the WPU-g-CP/HAp was synthesized utilizing ionic precipitation for HAp nanocrystals, followed by a freeze–drying process. TEM imaging of WPU-g-CP/HAp particles in the emulsion showed aggregations of polymer nano and micro-spheres with a HAp mineralized region. FT-IR and XRD analyses affirmed that the enormous amounts of HAP microcrystalline formations were framed within the WPUg-CP/HAp nanocomposite. DMA testing showed that the WPU-g-CP/HAp had an extreme stiffness modulus (7,000 MPa) representative of the high flexibility and robustness of the composite, which was deemed as vital for the intended purpose. It was deduced after thermal gravimetric and differential thermogravimetric analyses, as well as scanning electron microscopy/X-Ray energy dispersive spectroscopy that the crystallites of HAp were well dispersed into nanocomposite, whereas stability and underlying solidity were also inferred. The matrix pattern of this novel nanocomposite underwent biodegradation testing, which verified that it can indeed be employed for bone restoring.

A similar study presented the manufacturing of 3D permeable PU frameworks, with 0, 15, 25, and 32 wt.% content of hydroxyapatite (HAp) [49]. HAp was fused into PU constituents before beginning in-situ polymerization of PU. Thickness and porosity testing of the frameworks uncovered that higher measure of HA in the platforms prompted increasing the porosity and diminishing thickness. Moreover, the higher the HAp percentage, the lower the pore size was observed to be, although the net amount of pores increased, witnessed with field emission scanning electron microscopy (FESEM) imaging. These observations were credited (i) to the interactions among HAp and PU, and (ii) to the part of HAp in pore development. Increasing the HAp content resulted into an increase in the porosity of the scaffolds, which, as expected, reduced the Elastic Modulus of the specimens. Lastly, it should be noted that simulated body liquid (SBF) was employed for bioactivity testing, which showed platforms' capacity in precipitating apatite crystals.

In the dental medicine field, periodontal regeneration has been functionally improved with guided bone regeneration (GBR) membranes. Nonetheless, only a limited number of studies on biomimetic membranes that mimic the vascularization of the periodontal ligament have been investigated. In a recent study [50], fibroblast laden, along with growth factor-2 (FGF-2), water-based PU fiber membranes by emulsion electrospinning were prepared. These could supposedly promote periodontal tissue regeneration through the vascularization of the biomimetic GBR membrane. Following the principles of green chemical synthesis technology, a biodegradable WPU was produced with the use of lysine and dimethyl-propionic acid as chain extenders [50]. Figure 13.3 depicts the manufacturing of WPU fibers

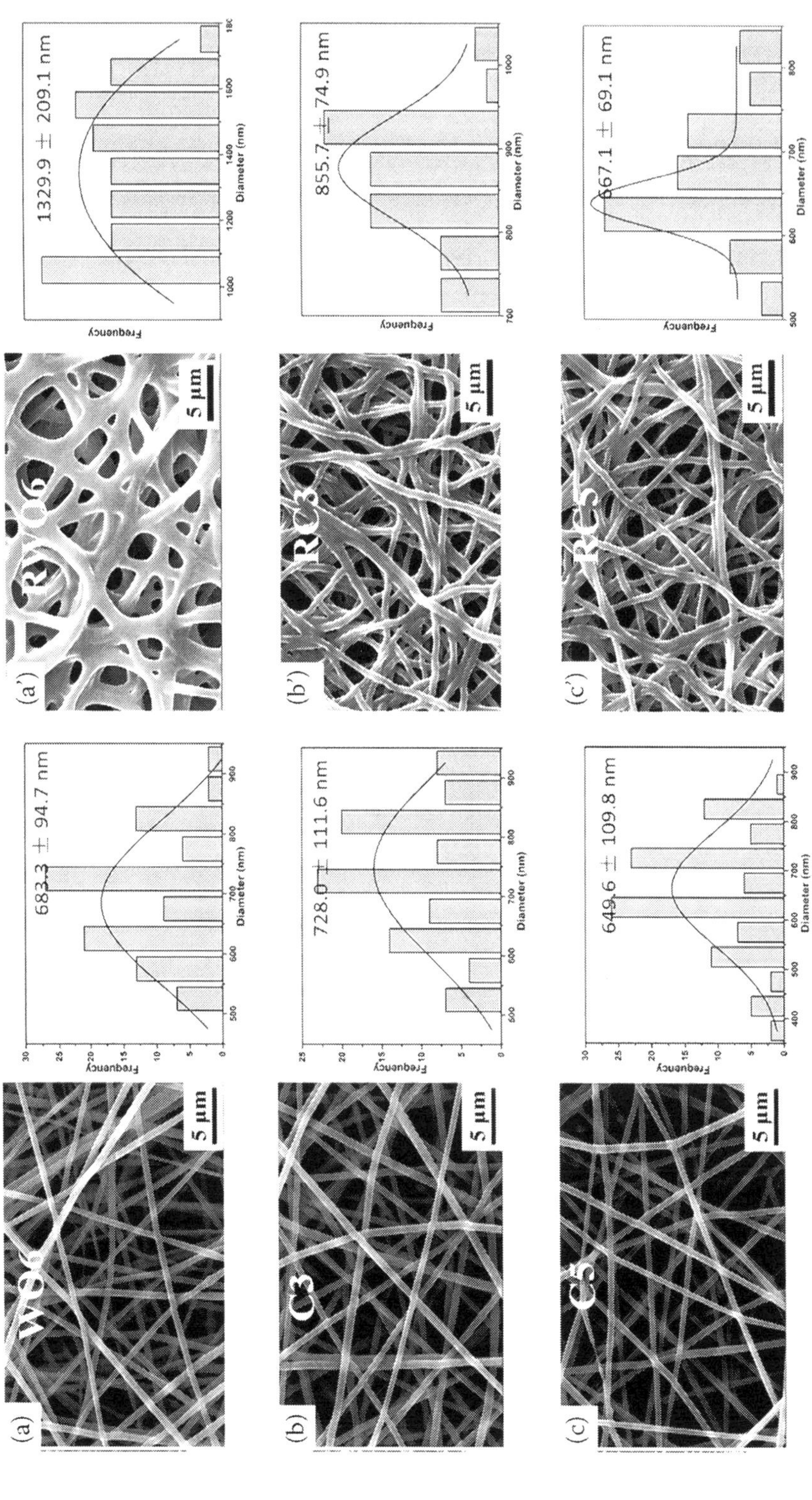

FIGURE 13.3 SEM images, diameter distribution, and average diameters (n = 100) of the WPU fibers before and after washing treatment [50]. Figure 4 adapted with permission from reference [50]. Copyright Zhang et al., some rights reserved; exclusive licensee [Oxford Academic]. Distributed under a Creative Commons Attribution License 4.0 (CC BY) https://creativecommons.org/licenses/by/4.0/

with FGF-2 by emulsion electrospinning. The results confirmed that the fiber membranes synthesized and showed controlled degradation, appropriate mechanical properties, and sustained release of growth factors. The positive immune-histochemical expression of angiogenesis-related factors was believed to promote the vascularization of cells for FGF-2 reinforced fibers. Moreover, FGF-2 reinforced fibrous scaffolds may be employed as functional GBR membranes to stimulate the formation of extraosseous blood vessels during the periodontal repair. Finally, in a water-based gas expansion polymerization [51], biomineralized PU foamy scaffolds were examined as a framework for bone tissue augmentation. Biomineralization of the foamed PU was completed by initiation on the PU surface by a two-stage mineralization system performed at different time intervals (1–4 weeks). The morphology, physicochemical, and mechanical properties of PU scaffolds were tested along with in vitro cultures with mice Bone Marrow Mesenchymal Stem Cells (BMSCs). Both untreated and biomineralized PU scaffolds exhibited homogenous morphologies and customary pore size (normal Ø = 407 μm). Crystallinity and structures of framed calcium phosphates (CaPs) layer onto the PU foam were determined by FTIR and X-beam diffraction, demonstrating the development of bonelike nano-hydroxyapatite. Biomineralization elevated the mechanical response of the treated foams as opposed to the untreated foams. Biomineralization additionally influenced the PU scaffolds' cytocompatibility, offering suitable surfaces for cell adherence and mitosis.

13.4 EPILOGUE—OUTLOOK

PU materials have always been considered for bioengineering applications; however, the toxicity due to the presence of organic solvents has recently been mitigated by the development of WPU by the substitution of the organic compounds with water as a dispersion medium. In fact, as presented in the text above, the number of scientific works on WPUs in tissue engineering is relatively limited to other biocompatible polymers or other types of solvent-based PUs. The synthesized WPU scaffolds possess significant advantages; they are considered to be environmentally friendly (solvent-free or containing low levels of organic solvents), and they are nontoxic, as they do not contain isocyanate residues. Also, they can be easily diversified for many applications as their structure and properties can be tailored and environmentally friendly fabrication methods can be adopted. The results, from research shown in the present chapter conducted by the authors, in the areas of soft and hard tissue engineering in applications prove WPU's potential in tissue engineering scaffolds. The WPU scaffolds are proposed in dental, orthopedic (bone and cartilage), as well as soft and dermal tissue regeneration or augmentation applications. They also appear to be good platforms for bio-mineralization. They have been found to promote the synthesis of important biomolecules and proteins, such as proteoglycans and glycosaminoglycans, and promote angiogenesis.

However, reading through the papers presented, one notices that the scaffolds based on WPU in spongy, fibrous, or porous form, suffer from relatively low mechanical properties, e.g. stiffness and stress retention, and require general reinforcements, which can be beneficial in the dual role: hydroxyapatite, chitosan, or other biocompatible particles. WPU appears to have an interesting future in bioengineering and tissue engineering applications. As an effect, for biomedical applications, biomaterials synthesized by water-based PUs are attracting major attention. Eventually, WPU-progressed applications can reach uncommon paths, like shape memory, self-regeneration, composites with biodegradability and biocompatibility, and so forth. Particularly, the utilization of WPU nanocomposites is limitless due to the exceptional performance of the addition of nanoparticles. In any case, tailoring the properties of such nanocomposites according to the compositions of nanomaterials is not yet as consistent as it must be. Particularly, the results behind these tissue augmentation wonders cannot be adopted beyond doubt and converted to market-level biomedical products.

REFERENCES

1. H. Goering, H. Krüger, M. Bauer, Multimodal polymer networks: design and characterisation of nanoheterogeneous PU elastomers, *Macromol. Mater. Eng.* 278 (2000) 23–35. doi:10.1002/(SICI)1439-2054(20000501)278:1<23::AID-MAME23>3.0.CO;2-P.
2. H.A. Wiggins MJ, MacEwan M, Anderson JM, Effect of soft-segment chemistry on polyurethane biostability during in vitro fatigue loading, *J Biomed Mater Res A*. 68 (2004) 668–683. doi:10.1002/jbm.a.20081.
3. S. Sartori, V. Chiono, C. Tonda-Turo, C. Mattu, C. Gianluca, Biomimetic polyurethanes in nano and regenerative medicine, *J. Mater. Chem. B*. 2 (2014) 5128–5144. doi:10.1039/C4TB00525B.
4. J. Zhu, Q. Wang, Y. Zhou, M. Mao, C. Huang, X. Bao, J. Shen, Preparation of anionic polyurethane nanoparticles and blood compatible behaviors, *J. Nanosci. Nanotechnol.* 12 (2012) 4051–4056. doi:10.1166/jnn.2012.5858.
5. X. Jiang, J. Li, M. Ding, H. Tan, Q. Ling, Y. Zhong et al., Synthesis and degradation of nontoxic biodegradable waterborne polyurethanes elastomer with poly(ε-caprolactone) and poly(ethylene glycol) as soft segment, *Eur. Polym. J.* 43 (2007) 1838–1846. doi:10.1016/j.eurpolymj.2007.02.029.
6. K. Gorna, S. Gogolewski, In vitro degradation of novel medical biodegradable aliphatic polyurethanes based on ϵ-caprolactone and Pluronics® with various hydrophilicities, *Polym. Degrad. Stab.* 75 (2002) 113–122. doi:10.1016/S0141-3910(01)00210-5.
7. P.A. Tatai, L. Moore, T. G. Adhikari, R. Malherbe, F. Jayasekara, R. Griffiths, I. Gunatillake, Thermoplastic biodegradable polyurethanes: the effect of chain extender structure on properties and in-vitro degradation, *Biomaterials.* 28 (2007) 5407–5417. doi:10.1016/j.biomaterials.2007.08.035.
8. M. Szycher, A.A. Siciliano, An Assessment of 2,4 TDA formation from surgitek polyurethane foam under simulated physiological conditions, *J. Biomater. Appl.* 5 (1991) 323–336. doi:10.1177/088532829100500404.
9. K.-C. Hung, C.-S. Tseng, S. Hsu, Synthesis and 3D printing of biodegradable polyurethane elastomer by a water-based process for cartilage tissue engineering applications, *Adv. Healthc. Mater.* 3 (2014) 1578–1587. doi:10.1002/adhm.201400018.
10. W. Ou, H. Qiu, Z. Chen, K. Xu, Biodegradable block poly(ester-urethane)s based on poly(3-hydroxybutyrate-co-4-hydroxybutyrate) copolymers, *Biomaterials.* 32 (2011) 3178–3188. doi:10.1016/j.biomaterials.2011.01.031.
11. T. Lu, Y. Li, T. Chen, Techniques for fabrication and construction of three-dimensional scaffolds for tissue engineering, *Int. J. Nanomedicine.* 8 (2013) 337–350. doi:10.2147/IJN.S38635.

12. X. Jiang, F. Yu, Z. Wang, J. Li, H. Tan, M. Ding, et al., Fabrication and characterization of waterborne biodegradable polyurethanes 3-dimensional porous scaffolds for vascular tissue engineering, *J. Biomater. Sci. Polym. Ed.* 21 (2010) 1637–1652. doi:10.1163/092050609X12525750021270.
13. R.K. Korhonen, M.S. Laasanen, J. Töyräs, J. Rieppo, J. Hirvonen, H.J. Helminen, et al., Comparison of the equilibrium response of articular cartilage in unconfined compression, confined compression and indentation, *J. Biomech.* 35 (2002) 903–909. doi:10.1016/S0021-9290(02)00052-0.
14. W.L. Grayson, P.-H.G. Chao, D. Marolt, D.L. Kaplan, G. Vunjak-Novakovic, Engineering custom-designed osteochondral tissue grafts, *Trends Biotechnol.* 26 (2008) 181–189. doi:10.1016/j.tibtech.2007.12.009.
15. S. Grad, L. Kupcsik, K. Gorna, S. Gogolewski, M. Alini, The use of biodegradable polyurethane scaffolds for cartilage tissue engineering: potential and limitations, *Biomaterials.* 24 (2003) 5163–5171. doi:10.1016/S0142-9612(03)00462-9.
16. S. Hsu, C.-C. Kuo, S.W. Whu, C.-H. Lin, C.-L. Tsai, The effect of ultrasound stimulation versus bioreactors on neocartilage formation in tissue engineering scaffolds seeded with human chondrocytes in vitro, *Biomol. Eng.* 23 (2006) 259–264. doi:10.1016/j.bioeng.2006.05.029.
17. G. Hinrichsen, *Polyurethane Handbook*, 2nd ed., Hanser Publishers, Munich, 1994. doi:10.1002/actp.1994.010450518.
18. K.-C. Hung, S. Hsu, Polymer surface interacts with calcium in aqueous media to induce stem cell assembly, *Adv. Healthc. Mater.* 4 (2015) 2186–2194. doi:10.1002/adhm.201500374.
19. K.-C. Hung, C.-S. Tseng, L.-G. Dai, S. Hsu, Water-based polyurethane 3D printed scaffolds with controlled release function for customized cartilage tissue engineering, *Biomaterials.* 83 (2016) 156–168. doi:10.1016/j.biomaterials.2016.01.019.
20. C.-Y. Lin, S. Hsu, Fabrication of biodegradable polyurethane microspheres by a facile and green process, *J. Biomed. Mater. Res. Part B Appl. Biomater.* 103 (2015) 878–887. doi:10.1002/jbm.b.33266.
21. Y.-P. Chen, S. Hsu, Preparation and characterization of novel water-based biodegradable polyurethane nanoparticles encapsulating superparamagnetic iron oxide and hydrophobic drugs, *J. Mater. Chem. B.* 2 (2014) 3391–3401. doi:10.1039/C4TB00069B.
22. M.-L. Kang, J.-E. Kim, G.-I. Im, Thermoresponsive nanospheres with independent dual drug release profiles for the treatment of osteoarthritis, *Acta Biomater.* 39 (2016) 65–78. doi:10.1016/j.actbio.2016.05.005.
23. M. Joshi, B. Adak, B.S. Butola, Polyurethane nanocomposite based gas barrier films, membranes and coatings: A review on synthesis, characterization and potential applications, *Prog. Mater. Sci.* 97 (2018) 230–282. doi:10.1016/j.pmatsci.2018.05.001.
24. M.-S. Yen, P.-Y. Tsai, P.-D. Hong, The solution properties and membrane properties of polydimethylsiloxane waterborne polyurethane blended with the waterborne polyurethanes of various kinds of soft segments, *Colloids Surfaces A Physicochem. Eng. Asp.* 279 (2006) 1–9. doi:10.1016/j.colsurfa.2005.09.027.
25. L. Lei, Z. Xia, X. Lin, T. Yang, L. Zhong, Synthesis and adhesion properties of waterborne polyurethane dispersions with long-branched aliphatic chains, *J. Appl. Polym. Sci.* 132 (2015). doi:10.1002/app.41688.
26. P. Zhang, S. Tian, H. Fan, Y. Chen, J. Yan, Flame retardancy and hydrolysis resistance of waterborne polyurethane bearing organophosphate moieties lateral chain, *Prog. Org. Coatings.* 89 (2015) 170–180. doi:10.1016/j.porgcoat.2015.09.015.
27. L. Lei, Y. Zhang, C. Ou, Z. Xia, L. Zhong, Synthesis and characterization of waterborne polyurethanes with alkoxy silane groups in the side chains for potential application in waterborne ink, *Prog. Org. Coatings.* 92 (2016) 85–94. doi:10.1016/j.porgcoat.2015.11.019.

28. T. Gurunathan, J.S. Chung, Physicochemical properties of amino–silane-terminated vegetable oil-based waterborne polyurethane nanocomposites, *ACS Sustain. Chem. Eng.* 4 (2016) 4645–4653. doi:10.1021/acssuschemeng.6b00768.
29. D.K. Chattopadhyay, A.D. Zakula, D.C. Webster, Organic–inorganic hybrid coatings prepared from glycidyl carbamate resin, 3-aminopropyl trimethoxy silane and tetraethoxyorthosilicate, *Prog. Org. Coatings.* 64 (2009) 128–137. doi:10.1016/j.porgcoat.2008.09.008.
30. H. Sardon, L. Irusta, M.J. Fernández-Berridi, M. Lansalot, E. Bourgeat-Lami, Synthesis of room temperature self-curable waterborne hybrid polyurethanes functionalized with (3-aminopropyl)triethoxysilane (APTES), *Polymer (Guildf).* 51 (2010) 5051–5057. doi:10.1016/j.polymer.2010.08.035.
31. G. Wang, G. Ma, C. Hou, T. Guan, L. Ling, B. Wang, Preparation and properties of waterborne polyurethane/nanosilica composites: A diol as extender with triethoxysilane group, *J. Appl. Polym. Sci.* 131 (2014). doi:10.1002/app.40526.
32. S. Zhang, Z. Chen, M. Guo, H. Bai, X. Liu, Synthesis and characterization of waterborne UV-curable polyurethane modified with side-chain triethoxysilane and colloidal silica, *Colloids Surfaces A Physicochem. Eng. Asp.* 468 (2015) 1–9. doi:10.1016/j.colsurfa.2014.12.004.
33. Q. Li, L. Guo, T. Qiu, W. Xiao, D. Du, X. Li, Synthesis of waterborne polyurethane containing alkoxysilane side groups and the properties of the hybrid coating films, *Appl. Surf. Sci.* 377 (2016) 66–74. doi:10.1016/j.apsusc.2016.03.166.
34. T. Jang, H.J. Kim, J.B. Jang, T.H. Kim, W. Lee, B. Seo, et al., Synthesis of waterborne polyurethane using phosphorus-modified rigid polyol and its physical properties, *Polymers (Basel).* 13 (2021). doi:10.3390/polym13030432.
35. F.M.B. Coutinho, M.C. Delpech, Degradation profile of films cast from aqueous polyurethane dispersions, *Polym. Degrad. Stab.* 70 (2000) 49–57. doi:10.1016/S0141-3910(00)00087-2.
36. I.W. Cheong, H.C. Kong, J.S. Shin, J.H. Kim, Kinetic aspects of chain extension reaction using water-soluble diamines in aqueous polyurethane dispersion, *J. Dispers. Sci. Technol.* 23 (2002) 511–518. doi:10.1081/DIS-120014019.
37. S.A. Madbouly, Waterborne polyurethane dispersions and thin films: biodegradation and antimicrobial behaviors, *Molecules.* 26 (2021). doi:10.3390/molecules26040961.
38. K. Bankoti, A.P. Rameshbabu, S. Datta, P.P. Maity, P. Goswami, P. Datta et al., Accelerated healing of full thickness dermal wounds by macroporous waterborne polyurethane-chitosan hydrogel scaffolds, *Mater. Sci. Eng. C.* 81 (2017) 133–143. doi:10.1016/j.msec.2017.07.018.
39. C. Viezzer, R. Mazzuca, D.C. Machado, M.M. de Camargo Forte, J.L. Gómez Ribelles, A new waterborne chitosan-based polyurethane hydrogel as a vehicle to transplant bone marrow mesenchymal cells improved wound healing of ulcers in a diabetic rat model, *Carbohydr. Polym.* 231 (2020) 115734. doi:10.1016/j.carbpol.2019.115734.
40. L. Da, M. Gong, A. Chen, Y. Zhang, Y. Huang, Z. Guo et al., Composite elastomeric polyurethane scaffolds incorporating small intestinal submucosa for soft tissue engineering, *Acta Biomater.* 59 (2017) 45–57. doi:10.1016/j.actbio.2017.05.041.
41. M. Wasyłeczko, W. Sikorska, A. Chwojnowski, Review of synthetic and hybrid scaffolds in cartilage tissue engineering, *Membranes (Basel).* 10 (2020). doi:10.3390/membranes10110348.
42. F. Horkay, J.F. Douglas, S.R. Raghavan, Rheological properties of cartilage glycosaminoglycans and proteoglycans, *Macromolecules.* 54 (2021) 2316–2324. doi:10.1021/acs.macromol.0c02709.
43. Y.-T. Wen, N.-T. Dai, S. Hsu, Biodegradable water-based polyurethane scaffolds with a sequential release function for cell-free cartilage tissue engineering, *Acta Biomater.* 88 (2019) 301–313. doi:10.1016/j.actbio.2019.02.044.

44. M.-C. Tsai, K.-C. Hung, S.-C. Hung, S. Hsu, Evaluation of biodegradable elastic scaffolds made of anionic polyurethane for cartilage tissue engineering, *Colloids Surfaces B Biointerfaces.* 125 (2015) 34–44. doi:10.1016/j.colsurfb.2014.11.003.
45. M.-Y. Shie, W.-C. Chang, L.-J. Wei, Y.-H. Huang, C.-H. Chen, C.-T. Shih, et al., 3D Printing of cytocompatible water-based light-cured polyurethane with hyaluronic acid for cartilage tissue engineering applications, *Materials (Basel).* 10 (2017). doi:10.3390/ma10020136.
46. C.-Y. Fu, W.-T. Chuang, S. Hsu, A biodegradable chitosan-polyurethane cryogel with switchable shape memory, *ACS Appl. Mater. Interfaces.* 13 (2021) 9702–9713. doi:10.1021/acsami.0c21940.
47. Z. Feng, M. Li, X. Jin, Y. Zheng, J. Liu, L. Zhao, et al., Design and characterization of plasticized bacterial cellulose/waterborne polyurethane composite with antibacterial function for nasal stenting, *Regen. Biomater.* 7 (2020) 597–608. doi:10.1093/rb/rbaa029.
48. Z. Yang, G. Wu, Effects of soft segment characteristics on the properties of biodegradable amphiphilic waterborne polyurethane prepared by a green process, *J. Mater. Sci.* 55 (2020) 3139–3156. doi:10.1007/s10853-019-04237-6.
49. S.A.S. Nasrollah, N. Najmoddin, M. Mohammadi, A. Fayyaz, B. Nyström, Three dimensional polyurethane/ hydroxyapatite bioactive scaffolds: The role of hydroxyapatite on pore generation, *J. Appl. Polym. Sci.* 138 (2021) 50017. doi:10.1002/app.50017.
50. C. Zhang, J. Wang, Y. Xie, L. Wang, L. Yang, J. Yu, et al., Development of FGF-2-loaded electrospun waterborne polyurethane fibrous membranes for bone regeneration, *Regen. Biomater.* 8 (2021). doi:10.1093/rb/rbaa046.
51. M. Meskinfam, S. Bertoldi, N. Albanese, A. Cerri, M.C. Tanzi, R. Imani et al., Polyurethane foam/nano hydroxyapatite composite as a suitable scaffold for bone tissue regeneration, *Mater. Sci. Eng. C.* 82 (2018) 130–140. doi:10.1016/j.msec.2017.08.064.

14 Waterborne Polyurethanes for Biodegradable Coatings

Sukanya Pradhan, Smita Mohanty, and Sanjay Kumar Nayak
Laboratory for Advanced research In Polymeric Materials (LARPM), CIPET-Bhubaneswar, Odisha, India

CONTENTS

14.1 Introduction 232
14.1.1 Waterborne Polyurethane (WPU) Dispersion-Based Coatings 233
14.1.2 Biodegradable Waterborne Polyurethane 235
14.1.3 Substrates for Screening of PU-Degrading Microbes 236
14.1.4 Polyester-PU Particle Dispersion 239
14.1.5 PU in Bulk Form 240
14.1.6 Low-Molecular-Weight (LMW) Urethane-Based Model Molecules 240
14.1.7 Fungal Biodegradation 241
14.1.8 Bacterial Biodegradation 241
14.1.9 Biodegradation of Polyester Polyurethane 242
14.1.10 Biodegradation of Polyether Polyurethane 243
14.1.11 Mechanism of Waterborne PU Degradation 243
14.1.12 Micro-Organisms Responsible for PU Degradation 244
14.1.12.1 Polyester/Polyether Degrading Microbes 244
14.1.12.2 Role of Enzymes in PU Degradation 246
14.1.12.3 Assessment of Biodegradable Waterborne PU 246
14.1.13 Monitoring Changes at Regular Intervals 246
14.1.14 Monitoring and Examining the Biodegradation Products 247
14.1.15 Monitoring the Microbial Biomass Generated After Biodegradation 247
14.1.15.1 Characterization of Physical Changes 247
14.1.15.2 Characterization of Chemical Changes 248
14.2 Conclusion 248
References 249

DOI: 10.1201/9781003173526-14

14.1 INTRODUCTION

Polyurethane (PU) is a flexible thermoset polymer comprised of organic units that are formed as a polyaddition reaction product between poly-isocyanate and polyol. It constitutes a special group of polymeric materials that differ in a variety of ways from most of the other plastic categories. They can be included in many different forms, such as paints, liquid coatings, elastomers, foams, etc., and hence, find applications in a broad range of areas, such as medical, textile, adhesives, inks, automotive, foams and cushioning, integral skin, insulation, etc. Owing to its excellent weather and chemicals resistance, flexibility, transparency, gloss, and hardness, PU has been considered one of the important classes of polymers. Additionally, it is a major component of coatings on a broad range of substrates, such as metals, plastic, glass, wood, etc. PUs are segmented block copolymers comprised of alternating hard and soft domains. The hard phase domains provide a major contribution toward attractive mechanical strength and toughness endowed with great potential reinforcements as microdomains. However, the soft segments provide flexibility to the system.

The conventional solvent-based organic coatings are employed for coating aircraft automobiles, oil tankers, ship hulls, sophisticated machines, household refrigerators, etc. due to their excellent physical barriers against aggressive environments and ability to inhibit the permeability of water, oxygen, and corrosive ions. However, the organic solvents are toxic and disturb the environmental balance. During its processing, product development, and usage process, PU also releases lots of volatile organic compounds (VOCs), directing toward rampant misuse of the resources, thereby creating environmental pollution. This particular trend has created the necessity of Green coating in order to solve the above-mentioned issues. Awareness for the protection of the environment is alarming day by day among the population, and the stringent regulations imposed by the government on the use of organic solvents that contribute to VOC emissions have forced researchers to develop eco-friendly waterborne coatings. Waterborne polyurethanes (WPUs) are being fabricated into Green high-performance coatings as well as adhesives using water as a solvent owing to their excellent adhesion, superior solvent resistance, pH stability, outstanding weathering resistance, and desirable chemical and mechanical characteristics. The major disadvantages of WPUs are low drying rate, longer curing time due to the presence of polar/ionic groups, and poor crosslinking. However, issues concerned with the overall coating performance could be vanquished by proper molecular designs and hybridization with different compatible materials. PU coatings are shown to be a promising candidate for nontoxic biodegradable coatings. In several previous reports, the antifouling properties of a different set of PUs have been extensively studied. PU polymer has to some extent inherent degradation capabilities due to the hydrolysis of ester bonds present, and it can also be easily coated on the substrate with outstanding mechanical properties. Over the past decades, many research groups have studied in detail the structure/property effects, biocompatibility, and biodegradation of a wide range of PU and its composites.

14.1.1 Waterborne Polyurethane (WPU) Dispersion-Based Coatings

In the last few decades, environmental concerns have led to the research and development of novel systems and greener routes based on the 12 principles of Green chemistry. One of the tough and tedious tasks was to fabricate waterborne systems by the inclusion of polar groups into the system to replace the conventional VOC emission contributing organic solvents by water. WPU can be synthesized with the assistance of water as a solvent or co-solvent along with an emulsifier. This type of system imparts high solid content, optimum viscosity, nonflammability properties, and curing ability at ambient temperature. In the beginning, petroleum products were the sole choice for the fabrication of WPU systems. However, the limitations of the resources as well as environmental awareness intensification have led to the development of renewable source-based monomers, including cellulose, starch, lignin, etc. along with the biobased isocyanate or non-isocyanate route for the synthesis of ecofriendly and sustainable WPU coating systems. Further, these greener materials also possess biodegradability characteristics, which could also solve the plastics pollution issues and reduce the dependence on petroleum-based polyols.

Ramanuj et al. [1], in a study, designed a dispersible coating in a two-step process derived from aqueous PU dispersions crosslinked with several modifiers. The coating formulation process of the WPU dispersion is represented in Figure 14.1. Similarly, several researchers have been reported their work on WPU,

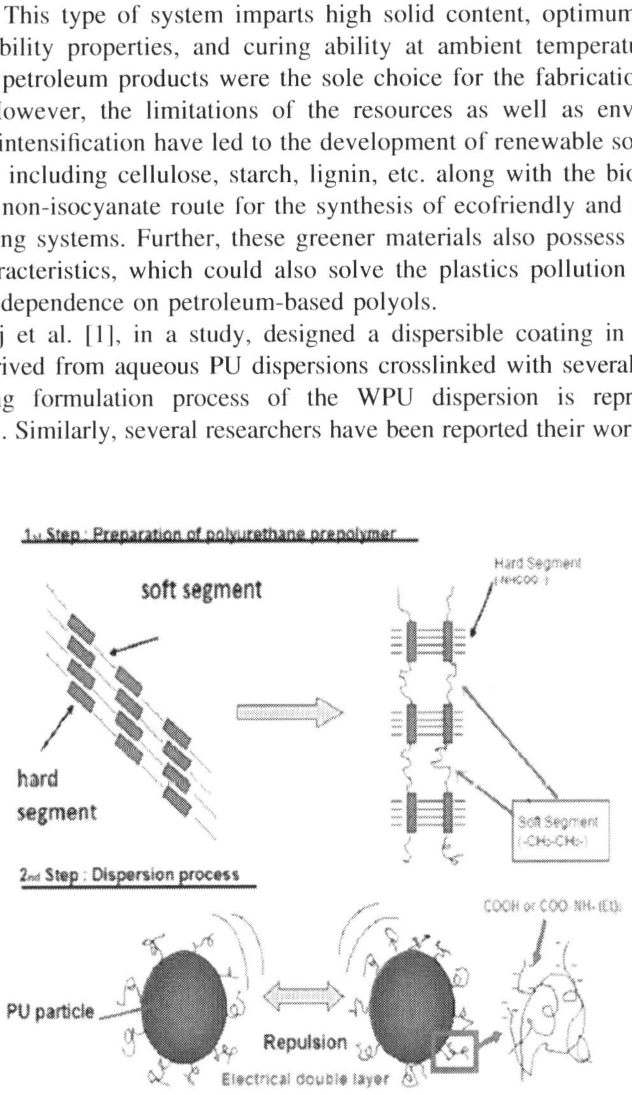

FIGURE 14.1 Schematic representation of WPU dispersion [1].

TABLE 14.1
Composition of different WPU composites for coatings and their particle average size

Sl. No	WBPU Composition	Particle Average Size (nm)	References
1	PPG/TDI/DMPA	35–225	[2]
2	PBA/PMA/PTMG/PPG/IPDI/H12MDI/TDI/ MDI/BD/TMP/HZ/DMPA	100–8000	[3]
3	PEG/PTMG/MDI/SDBS	–	[4]
4	PBA/DHA/IPDI	92	[5]
5	PPO/TDI/DMBA/APTES/SDS	–	[6]
6	PEG/HDI/LYS	–	[7]
7	MAHCSO/IPDI/HDO/DHZ	41–176	[8]
8	MAHCSO/TDI/HDO/PMDA/BPOTCDA/ HFIPDA	23–240	[9]
9	PTMG/PEG/MDI/SDBS		[10]
10	PCL/H12MDI/DMPA/BES	28–213	[11]
11	CE/PTMG/IPDI/DMPA/EDA	61.5	[12]
12	Oxymer M112/SynDD/IPDI/DMPA/HMD	81.2–139.2	[13]
13	CO/FA/IPDI	35.11 (with CO)56.11 (without CO)	[14]
14	PEG/PTMG/IPDI/GQAS/EGDE	–	[15]
15	PEG/HDI/LYS	–	[16]
16	PEG/HMDI/DMPA/MDEA/HTO	64–198	[17]
17	PCDL/CO/IPDI/DMPA/EDA/BD/THAM	50–125	[18]
18	GLY-polyols/Voranol 4701/IPDI/DMPA	–	[19]
19	PCD/PBA/IPDI/DMPA/HZ	67–84	[20]
20	PPG/IPDI/BDSA	190–320	[21]
21	PBA/DHA/IPDI/MDI/HDI	90–125	[22]
22	PTMG2000/PTMEG1000/IPDI/DMPA/ SAAS/ HZ	800–3000	[23]
23	PET/PPG/IPDI/DMPA/BD	–	[24]
24	PPG/IPDI/DMPA/TMPM/BD/EDA/ AEAPTMS/APTES	–	[25]

and a list of sources with the parameter of its composition and particle average size is represented in tabular form (Table 14.1).

In the last few decades, one of the most challenging tasks was the fabrication of waterborne products using plant-originated polyols. Most of the plant oil-based polyols do not contain inherent hydroxyl groups except castor and lesquerella oils; hence, a modified step was mandatory to incorporate polar groups to the hydrophobic backbone to reduce the VOC emissions. The modification approaches include epoxidation reactions, hydroformylation, acrylation, ozonolysis, thiol-ene coupling, transesterification, and amidation, etc. The WPU dispersion system not only provides environmental advantages but also possesses the capacity of film formation at ambient conditions owing to its high solid content, molecular weight, and greater flame retardancy properties. In general, the hydrophilicity character of the polymer is compatibilized by using an internal emulsifier, thereby cutting off the use of organic solvents. Another approach to incorporate polar groups is the use of hydrophilized polyisocyanates bearing free isocyanate groups to form a stable PU dispersion. Over the past years, high-performance renewable, resource-based PU coatings, which are being developed employing thermal/UV curing using latent agents, have attracted the researcher's fraternity working on PUs. The use of biobased isocyanates is believed to increase the biodegradability characteristics of the end products and hence, the topic is in focus for the last few years. The first biobased crosslinker for the PU system (pentamethylene diisocyanate) was developed and commercialized in 2015. Later, in 2017, this crosslinker appeared on the market in the form of its hydrophilic counterpart for its use in the development of WPU coatings. However, the overall performance of the coatings was not comparable with the conventional curing agents.

14.1.2 Biodegradable Waterborne Polyurethane

The conventional petro-based PUs developed by solvent methods are not biodegradable; release toxic VOCs due to the use of petroleum-based solvents like xylene, acetone, *N*,*N*-dimethylformamide (DMF), methyl ethyl ketone (MEK), etc. in coating formulations; and increase dependency on fossil fuels. The disposal of PU wastes is carried out through two ways: incineration or landfill. The PU products ends up in landfills and also decompose into hazardous, environment-deterioration substances. This leads to an increase in handling charges and possess a serious threat to the environment. Taking these into consideration, the scientific community has started to develop biodegradable PU coatings by incorporating specific polymeric intermediates into the system and also exploring other feasible approaches. Several studies have reported the increase in the rate of degradation by the inclusion of a variety of soft segments, such as polylactide or polyglycolic acid, polycaprolactone (PCL), polyethylene glycol (PEG), and polylactide diol, etc. The list of some commonly used diols in the synthesis of biodegradable is represented in Figure 14.2. [26]

However, an optimum balance between the final desired properties and the rate of biodegradation must be accomplished to make the modification process feasible. Further, various sources reveal the fabrication of a novel WPU system from hyperbranched polyester polyols [27,28]. In this context, the use of tannic acid, which is cheaper, widely available, and bioderived has been proved to be a potential candidate for hyperbranched WPU synthesis with excellent biodegradability and biocompatibility characteristics [29]. It is known that polyester PU is very susceptible to bacterial degradation compared to neat PU. The weight loss analysis of the resulted films showed around 40% of weight reduction on exposure to microbial degradation for six weeks, which corroborated the morphological studies obtained from SEM micrographs of waterborne hyperbranched polyurethane (WHPU),

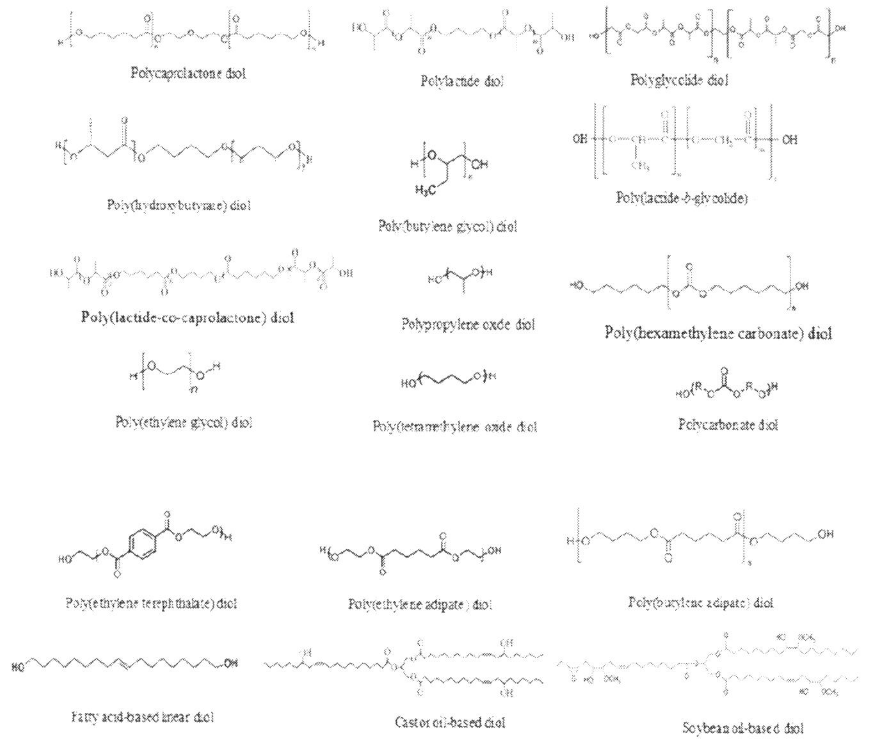

FIGURE 14.2 Some common diols used in the synthesis of polyurethane.

Adapted with permission from reference [26].

shown in Figure 14.3. In a different study [30], gelatin molecules from cold fish skin have been introduced into the WPU polymer matrix by covalent interaction, employing the sol–gel method to develop biodegradable PU coatings. The summary of other relevant degradable WPU systems highlighting the chemical composition, polymerization/formulation techniques, and degradation rate have been presented in tabulated form (Table 14.2).

The urethane linkage exists in both forms viz polyester and polyether PU. The low molecular weight urethane compounds are very susceptible to microbial degradation as they can be hydrolyzed by the action of catalyst esterase. However, the exact mechanism of the degradation of urethane bonds in PUs is unknown.

14.1.3 Substrates for Screening of PU-Degrading Microbes

The waste generated from PU gets accumulated in the environment, which poses a serious threat to ecological balance. Hence, biodegradable PUs are being considered as greener alternatives to plastics that have a wide range of applications, starting from marine to aerospace applications, as adhesives and coatings. In the early 1998s, scientists started to explore the degradability characteristics of plastics in a

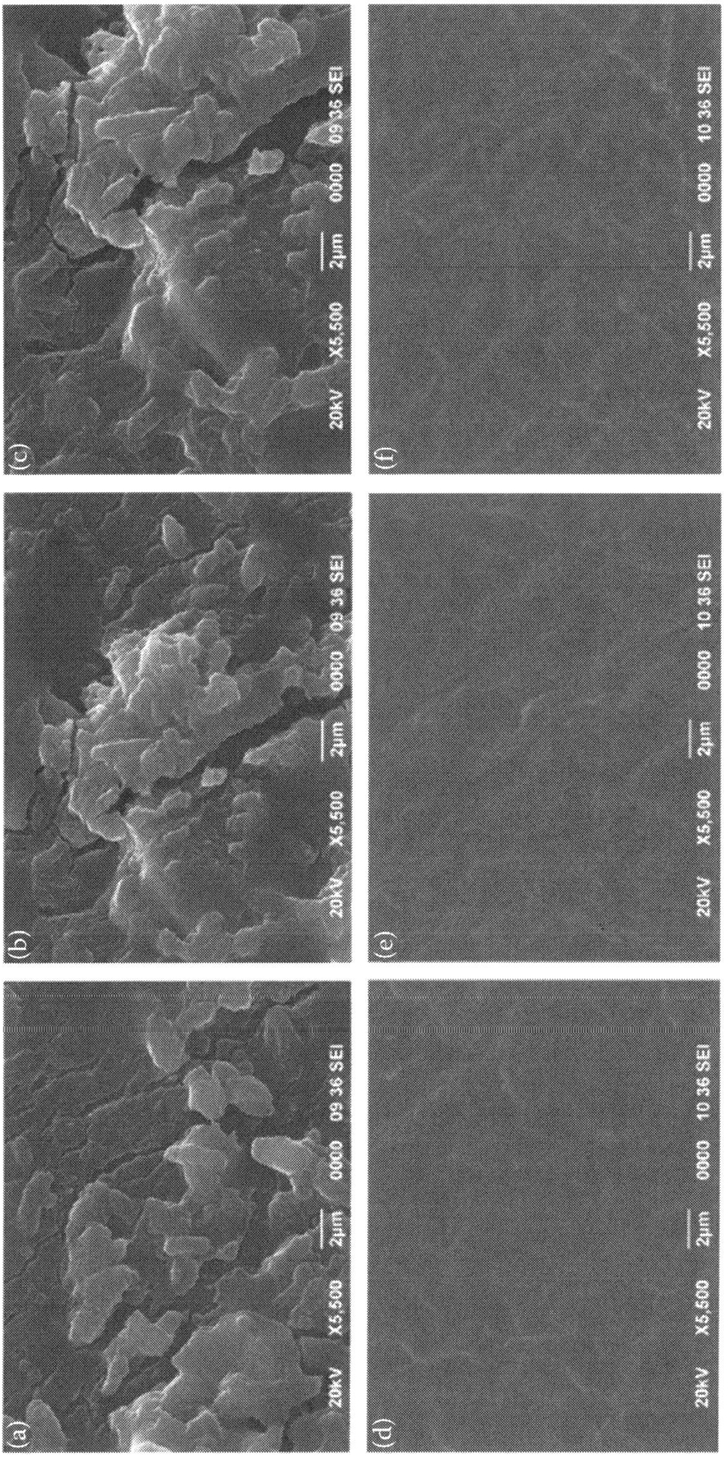

FIGURE 14.3 SEM micrographs of WHPU at different weight percentages of tannic acid.

Adapted with permission from reference (29) Copyright © 2014, American Chemical Society.

TABLE 14.2
Summary of degradable WPU systems highlighting the chemical composition, polymerization /formulation techniques, and degradation rate

Chemical Composition	Characterization Techniques	Polymerization/ Formulation Technique	Potential Applications	References
WPU/Gelatin	FTIR, DSC, DMA, Biodegradability	Sol–gel	Biomedical (Scaffold, wound dressing, tissue engineering)	[30]
Polycaprolactone diol (PCL)/, polyethylene butylene adipate diol (PEBA)	FTIR, XPS, AFM, Contact angle,	Emulsion polymerization, Spin coating	Biomaterials	[31]
Polyphenolic tannic acid, IPDI, PEG	FTIR, DSC, Radical Scavenging Assay, UV–Vis, Biodegradation	Prepolymerization, Film Casting	Biomaterials	[29]
PCL/MDI/Soybean protein	DLS, DSC, DMTA, WAXS, and hydrolytic degradation	Prepolymerization, Film Casting	Shape memory and biomedical materials	[32]
PEG/PBGA/ amino acid	Particle size, zeta potential Biocompatibility, Biodegradation performance,	Film Casting	Scaffold materials, 31 tissue engineering	[33]
Poly(3-hydroxybutyrate) (PHB) Poly(e-caprolactone) diol (PCL diol; Isophorone diisocyanate (IPDI	Shape memory behavior, Cell proliferation, Biodegradation and biocompatibility tests	Casting (Biodegradation rate-52% weight loss after 30 days)	Biodegradable elastomers and potential biomaterials for cardiovascular and other medical applications	[34]
Polypropylene oxide glycol (PPG220, Mn = 2000 Da) and dimethylolpropionic acid (DMPA), gelatinehydrosylate	FTIR, AFM, SEM, DMA, TGA contact angle and degradability testing	Emulsion polymerization (Complete degradation after 60–90 days in simulated body fluid)	Tissue engineering scaffold material	[35]
(PEG-600), citric acid, 1,4-butanediol (BD),	^{13}C NMR, chemicals resistance, UV-	Solvent-free polymerization	Transparent surface coatings	[36]

TABLE 14.2 (Continued)
Summary of degradable WPU systems highlighting the chemical composition, polymerization /formulation techniques, and degradation rate

Chemical Composition	Characterization Techniques	Polymerization/ Formulation Technique	Potential Applications	References
castor oil, glycerol, and dimer acid	aging, Biodegradation			
PCL diol, IPDI	Gelatin test, Cytotoxicity, biodegradation, anti-adhesion,	Solution polymerization Casting, Biodegradation rate (80% weight loss after four weeks)	Biomedical (Prevention of tendon postoperative adhesion)	[37]
ESO polyol (IPDI), ethylenediamine (EDA	FTIR, DMA, TGA, Hydrolytic degradation	Emulsion polymerization, molding technique, Fully degraded in water and high alkaline solution	Biomedical	[38]
Poly(e-caprolactone) diol, Isophorone diisocyanate (IPDI, 98% purity, Aladdin), dimethylol propionic acid (DMPA)	FTIr, DSC, TGA, Water absorption, Freeze–thaw stability, Cytotoxicit, Biodegradation	In-situ polymerization molding technique, Enzymatic biodegradation rate (52%, 8 weeks)	3D printing materials for tissue engineering scaffolds	[39]

microbial active environment. Since then, new strategies and standards are being developed for the safe disposal of plastics to safeguard the environment. The screening of PU-degrading microbes or enzymes can be done by using three common model substrates as polyester-PU particle dispersion, PU in bulk form (cracking, embrittlement, flaking, etc.), and low-molecular-weight urethane-based model molecules (shorter chain lengths), which are represented in Figure 14.4 [40].

14.1.4 Polyester-PU Particle Dispersion

When the Impranil (colloidal polyester-PU dispersion) acts as a substrate, the formation of clear zones could be visible in the nutrient agar medium, and the microbial colony-forming unit (CFU) can be calculated that suggests the degradability of PU. Hence, this particular screening process has wide acceptance due to its easy method and accuracy. Some other transparent substrates, such as Bayhydrol 110 and Bayhydrol 12, can also be of assistance with dyes for the clear visibility of the clear zones.

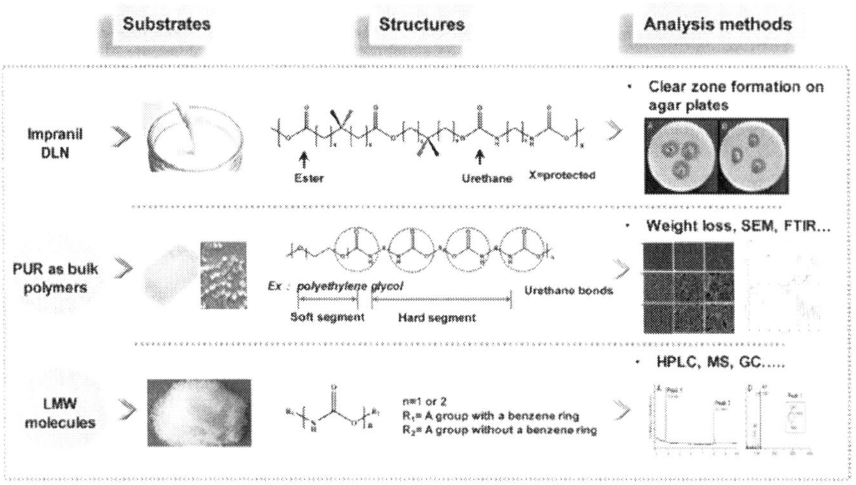

FIGURE 14.4 Commonly used substrates for identification of PU degrading microorganisms.

Adapted with permission from reference (40) Copyright © 2021 Elsevier Inc.

14.1.5 PU in Bulk Form

PU in bulk form includes the polymer in the form of foam, elastomers, thin films, etc. that have been used for screening. The degradability characteristics of the materials are examined through various analyses, such as weight loss measurement, alteration in surface morphology, changes in functionalities, molecular weight distribution, etc. Very few studies have reported the insignificant microbial degradation of PU polymers. The degradation process is very long due to the high incubation time; hence, few microbes such as *Comamonas acidovorans*, TB-35, *Aspergillus tubingensis*, etc. have been isolated and identified as responsible microorganisms for the degradation.

14.1.6 Low-Molecular-Weight (LMW) Urethane-Based Model Molecules

LMW urethane-based model molecules possess urethane bonds and are more prone to degradation; hence, they provide a clear platform for the screening of microbes and detailed degradation mechanisms. A fungal strain Exophialajeanselmei REN-11A and a bacterium Rhodococcusequi strain TB-60 have been reported to act most effectively. Due to the high rate of biodegradation of urethane bonds in the LMW model substrates, the microbial strains or enzymes isolated and identified employing this step may not necessarily be capable of breaking the same bond in PU polymers. Hence, a two-step process is highly recommended. In the first step, the screening will be carried out with both the substrates, such as LMW urethane-based model molecules and polyester-PU particle dispersion.

14.1.7 Fungal Biodegradation

PU is very prone to microbial degradation, specifically in a fungal environment, which is evidenced from several studies reported earlier. Many research groups have extensively studied the degradation of PU by the fungal community. The degradation profile mainly depends upon the type of polyols used and the isocyanates used. Different data have been reported for a variety of fungi for PU as the biodegradation analysis method employed differs for every study. The degradation process under fungal exposure is majorly due to the fragmentation of the polyol segments; hence, the intensity of the degradation is directly influenced by the structural integrity of the polyol segments, which could be either polyester or polyether type. The difference in degradation pattern could also be attributed to crosslinking density, structural orientation, and functionalities present that are prone to fungal attack, such as ester, hydroxyl, etc. Amorphous polymers can degrade easily compared to polymer with crystallinity properties; hence, the synthetic polymer with a larger degree of crystallinity has been proved to degrade to a lesser extent. Huang and Roby (1986) reported that PU with long-chain hydrocarbon backbone and polar groups are very susceptible to microbial attack as they are loosely packed and possess amorphous regions. A few reports have demonstrated and presented a comparative study on the susceptibility of polyester and polyether-type PU to fungal attack. Esterase and hydrolase are well-known enzymes for the degradation activities secreted by fungi *Chaetomium globosum* and *Aspergillus terreus*. In a different study, various fungal species such as *Curvularia senegalensis*, *Fusarium solani*, *Aureobasidium pullulans*, and *Cladosporium* sp. have been isolated and identified as responsible microrganisms for the degradation of polyester based-PU. Subsequently, *Curvularia senegalensis* has also been known to be the effective PU-degrading fungi.

14.1.8 Bacterial Biodegradation

PU degrading bacteria have been isolated and identified by various studies employing various techniques. They have been identified as effective microbes responsible for PU degradation through the use of esterase, urease, amidase, and protease enzymes. PU biodegradation is controlled to a greater extent by the influence of secreted or surface-attached enzymes and their capability to access the soft segment ester. A large-scale study reported by Kay et al. [41] identified 16 bacterial isolates to degrade polyester-PU in different mediums, such as yeast, basal, etc. The changes, such as reduction in mechanical strength and elongation, have been observed for degraded PU after several days of bacterial exposure. Further, the same research group revealed the labile ester segment of the matrix to be the main site responsible for microbial attack. Many reports published on PU microbial degradation emphasized fungal exposure; however, very few have worked on the detailed mechanism of bacterial degradation. The mechanism majorly involves the binding of microbial cells with the polymer, subsequently degrading the substrate.

14.1.9 Biodegradation of Polyester Polyurethane

Any polymers or organic compounds, when exposed to the natural environment like soil or compost, start to degrade, which means they are fragmented into organic intermediates of lower molecular weight. The fragmentation pattern of the polymers occurs either by oxidation or hydrolysis reaction, controlled by either biotic or abiotic components. These materials are available to the microbial community and become the prime carbon source. Hence, they are utilized, releasing carbon dioxide and methane in the presence of a sufficient amount of oxygen. In the assimilation stage, the organic intermediates generally use up the bacterial and fungal community as biomass carbon and ultimately go back to the soil again after the biomass is dead. The other route that is also followed for soil stabilization is humification. In this process, the organic intermediates are stabilized, known as humic substances, which become available for the soil communities as a rich source of carbon. The generalized representation of polymer in the soil environment is displayed in Figure 14.5.

The degradability of PU coating depends on both biotic and abiotic components. However, the contribution of biotic factors solely toward the degradation rate has not been extensively studied. It has also been assumed that microorganisms cannot survive exclusively on PU coatings. Comparatively, polyester-based PU is more susceptible to hydrolysis attack than polyether or any other based PU [43]. Many prokaryotic and eukaryotic organisms possess the inherent capability of degrading both polyester and polyether-based PU through hydrolysis, but fungi can hydrolyze carbamate, alkene, or other functionalities. Nakajima-Kambe et al. [44] reported a polyester PU degrading bacterium of the

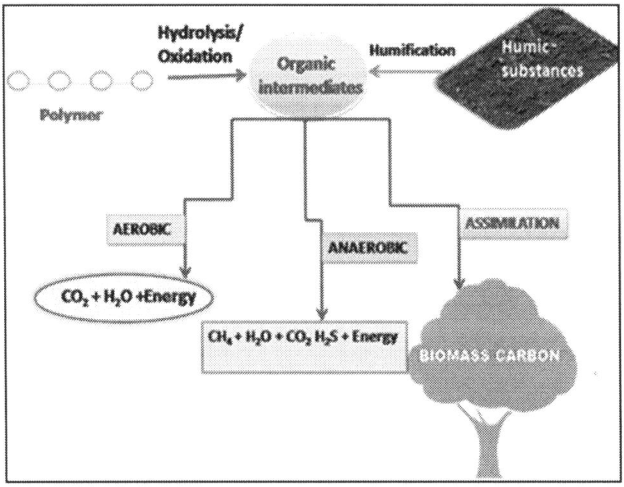

FIGURE 14.5 Schematic representation of polymer degradation in soil environment.

Adapted with permission from reference [42].

gram −ve category that has been isolated as *Comamonas acidovorans* strain TB-35, and the results revealed the metabolites to be the degraded products obtained by the cleavage of ester bonds hydrolytically.

14.1.10 Biodegradation of Polyether Polyurethane

Unlike polyester PU, polyether PU has greater crosslinking density and packed structure. Hence, it has been a tedious task to degrade completely the polyether form of PU exposed to the microbial environment. A few pieces of literature have been published that deal with the microbial degradation of polyether-based PU. Various formulations of both polyester and polyether PU have been developed and concluded the scarce susceptibility behavior of polyester PU under microbial consortium [43]. Similarly, Jansen et al.[45] reported *Staphylococcus epidermidis* strain KH11 as the degrading micro-organisms, particularly for biomer and tuftane (polyether PU). However, the degradation rate was too low, owing to the depolymerization type that directly influences the degradation pattern. The degradation pattern followed by polyester and polyether PU are endo-type and exo-type depolymerization, respectively.

14.1.11 Mechanism of Waterborne PU Degradation

The degradation mechanism of either solvent-borne PU or WPU is almost the same as it depends upon the exposure environment (aerobic, anaerobic, acidic, alkaline, enzymatic, etc.) as well the structural integrity. The actual degradation of WPU is well studied by analyzing the by-products and mineralized products. The biodegradation of PU follows either of the two mechanisms: (a) biological oxidation or (b) biological hydrolysis. The PU under polyester type is prone to hydrolytic cleavage due to the presence of ester functionalities as soft segments. However, the degradation of polyether and polycarbonate-based PU proceeds with the oxidative degradation with the abstraction of alpha hydrogen with respect to the oxygen atom of the functional groups. In the biological biodegradation system, the adherent leucocytes release reactive oxygen moieties that initiate the oxidative degradation of the soft segments. This in turn abstracts the reactive hydrogen from the methylene group of the PU, the addition of hydroxyl radical to the carbon radical forms hemiacetal, and it finally gets oxidized into the corresponding ester. The presence of an acidic environment leads to the acidic hydrolysis of the ester groups of the PU, in which the soft segments are fragmented and eventually the acid end groups with lower molecular weight are generated. The schematic representation of the mechanism is presented in Figure 14.6. In a similar way, the hard segments get degraded in which the oxygen radicals abstract the active hydrogen atom from the chain extender used in the synthesis of PU. The incorporation of hydroxyl radicals combines with the chain radical to form a highly reactive carbonyl hemiacetal that undergoes oxidative hydrolysis at the carbonyl center to form the combination of unstable adduct carbamic acid and carboxyl end groups. In the final step, the obtained carbamic acid undergoes decarboxylation to produce free amines. The detailed mechanism is shown in Figure 14.7 and Figure 14.8.

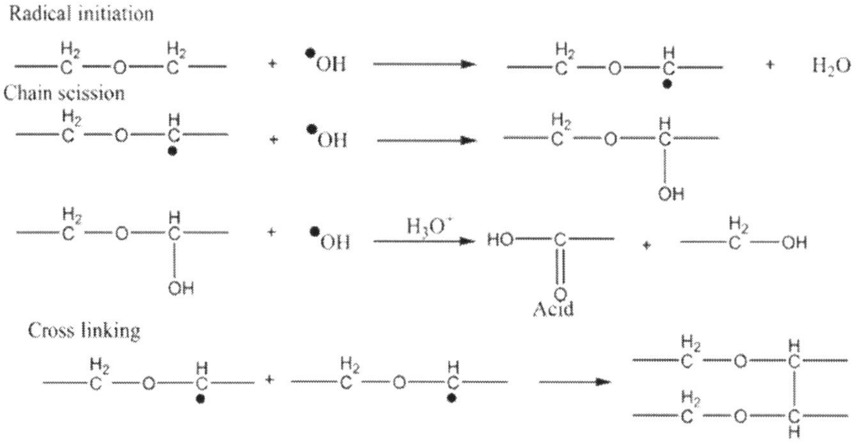

FIGURE 14.6 Biodegradation mechanism of PU soft segments of polyether type.

Adapted with permission from reference [46].

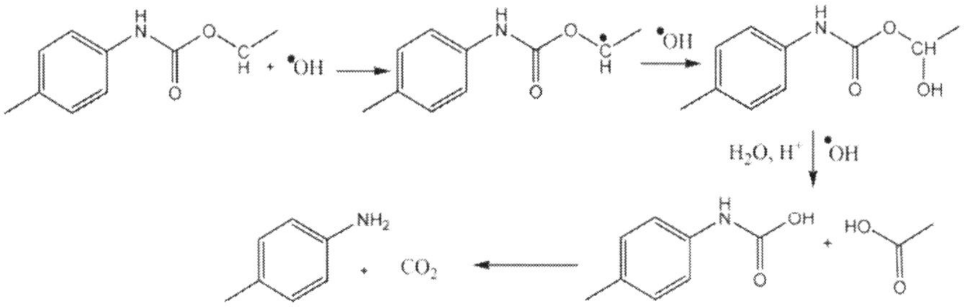

FIGURE 14.7 Biodegradation mechanism of PU hard segments of polyether type.

Adapted with permission from reference [46].

14.1.12 Micro-Organisms Responsible for PU Degradation

14.1.12.1 Polyester/Polyether Degrading Microbes

Literature sources reveal a wide range of bacteria of specific genera, such as *Comamonas*, *Pseudomonas*, *Bacillus*, *Acinetobacter*, and *Corynebacterium,* which have proved to be responsible for the biodegradation of polyester-based PU materials. The results reveal the confirmatory degradation by the appearance of amide linkage with the simultaneous disappearance of the characteristic band for ester linkages. *Comamonas* sp. strain TB-35 has been identified as a micro bacterium capable of degrading polyester PU completely into two components (diethylene glycol and adipic acid) after 7 days of exposure at ambient temperature, provided that the polymer was supplied as the sole source of carbon. Some specific bacteria, such as *Bacillus subtilis* MZA-75 and *Pseudomonas aeruginosa* MZA-85, have been reported of degrading polyester PU thin-film completely and mineralization of

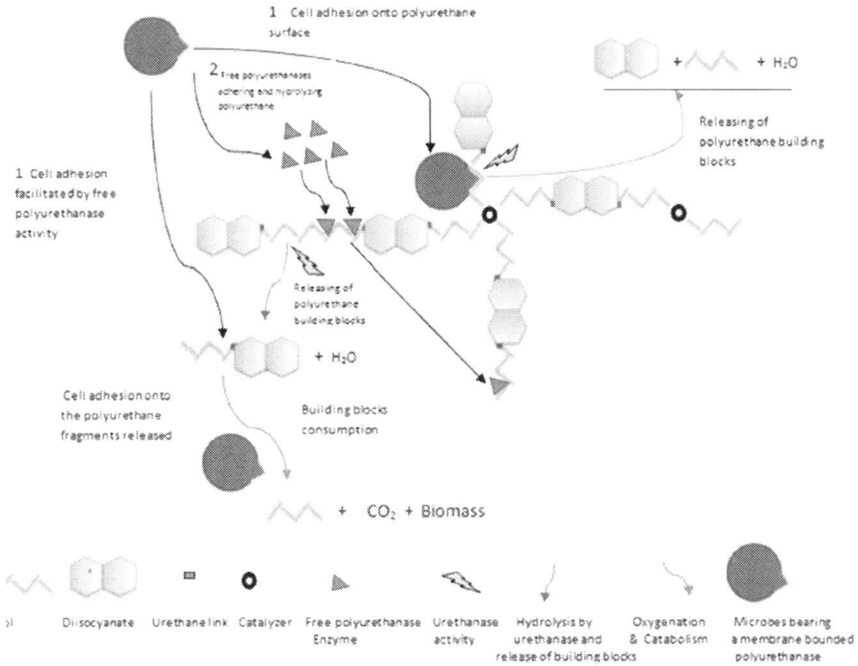

FIGURE 14.8 Mechanistic analysis of polyurethane degradation in microbial exposure.

Adapted with permission from reference [46].

the intermediates into carbon dioxide and water. The same PU film, when exposed to the simultaneous exposure of the two different strains in a mixed culture medium, experienced loss in terms of weight profile with respect to the period of 1 month of around 40% [47]. Many reports suggest the addition of nitrogen sources such as yeast extracts to the exposure medium to sustain PU biodegradation. As for most bacteria, the use of PU as a carbon and nitrogen source abolishes or decreases the level of degradation.

A very limited number of studies are focused on the degradation of PU by fungal communities. The prominent fungal genera that can be held responsible to date for the biodegradation of PU are *Aspergillus*, *Pestalotiopsis*, *Cladosporium*, *Fusarium*, and *Penicillium* [48,49]. Among these, the species of *Aspergillus genus* possesses the maximum degradability capacity for polyester-based PU in bulk form. In the Sabouraud dextrose agar culture medium, tubingensis species have been found to degrade macroscopically, as evidenced from weight loss measurement and surface degradation after three weeks of incubation. In mineral salt medium along with glucose concentration, PU could be degraded into smaller fragments within 60 days of exposure, as reported by Khan et al. [50]. In two different studies, under composting, strains of *Aspergillus* species exhibited excellent biodegradability to the tune of 60% reduction in weight for polyester PU film [51] and 20% weight reduction after 28 days of composting [52]. In another study by Cosgrove et al, the fungal colonies formed on the PU surface were identified, isolated, and compared with the neat soil using culture-based and molecular techniques. The results revealed the presence of putative PU degrading fungi in both cases. However, a few microorganisms have been isolated and identified for the degradation of polyether-based PU. Some of them are the species belonging to genera *Cladosporium*, *Aspergillus*, and *Alternaria*. Many researchers have demonstrated two novel fungal strains, namely *Alternaria solani* and *Spicaria* species, and examined the degradation potential toward polyester-based PU. A few studies have claimed several dozen endophytic fungi for the degradation of synthetic polymer polyester PU.

14.1.12.2 Role of Enzymes in PU Degradation

The microbial degradation of PU is nothing but the action of enzymes, such as esterases, ureases, proteases, and amidases, on the labile groups present, which can be easily fragmented [53]. The WPU system has been reported to be capable of degradation enzymatically in PBS (pH ¼ 7.4) with a solution mixture including PBS, $MgCl_2$, and Lipase AK with an optimum temperature of 55°C and shaking for a definite period.

Esterases are the well-known enzymes for the degradation of soft segments of PU through hydrolytic cleavage, thereby generating carboxylic and alcohol groups as end products. The other two enzymes are proteases and amidases, which are responsible for the fragmentation of peptide and amide bonds as well as the urethane linkages present in PU. The simultaneous action of enzymes amidases and esterase on the degradation behavior of PU has been well studied and showed its improved efficiency compared to enzymes when used individually. During the enzymatic action, the esterase hydrolyzes the ester linkages, and the resultant low molecular weight intermediates being susceptible to amidase; hence, the hydrolysis of urethane bonds occurs. Further, reports also reveal the hydrolytic enzymatic activity of an amidase from *Nocardia farcinica* responsible for the degradation of polyester-based PU containing polyamides (PAs) and PA-related oligomeric model substrates. Furthermore, urease is also capable of microbial degradation of polyurea-urethane polymers into amines and carbon dioxide. However, very little supporting literature is available to date as the disintegration of urea bonds is a very tough task compared to the ester bonds. The prospects in this area, therefore, deal with the exploration of screening of enzymes with "polyurethanase" activity.

14.1.12.3 Assessment of Biodegradable Waterborne PU

The biodegradation assessment methods must emphasize different aspects of the degradation process, and the choice of approach must satisfy the objectives of the test. There are numerous approaches available to assess the biodegradability of the PU coatings, which can be classified under the following categories.

14.1.13 Monitoring Changes at Regular Intervals

This is the preliminary step of analysis in which the changes in shape, size, color, and bristle formation that are visible to the naked eye are noted at a regular time. The above-mentioned changes, however, are responsible for the weight loss, loss in mechanical properties, etc.

14.1.14 Monitoring and Examining the Biodegradation Products

In an aerobic environment, energy, water, and carbon dioxide are considered to be biodegradation products for almost all polymers; hence, the release of carbon dioxide is monitored to confirm the occurrence of the biodegradation process. Further, with the known value of carbon content of the polymer under biodegradation, the amount of carbon dioxide released by the mineralization process can be determined theoretically that displays the extent of biodegradation.

The other approach applicable to polymers is to move upstream in the biodegradation process and measure the organic intermediates formed by the initial decomposition of the polymer. This process has been considered to be more specific compared to monitoring the carbon dioxide process due to the consideration of the organic intermediates in the former process, which is unique for almost every set of polymers. However, the results in this process prove the existence of biodegradation and not the extent of biodegradation as most of the organic intermediates are labile and get adsorbed by the microbial community very easily and quickly.

14.1.15 Monitoring the Microbial Biomass Generated After Biodegradation

It is known that during the biodegradation process, despite complete mineralization, some part of the biodegraded polymer is assimilated into the microbial mass that creates a new way to assess degradability characteristics. The degradation can be studied by the Petri dish test in which the microbial colonies formed on the surface of the polymer are counted by visual inspection and represented in terms of colony-forming unit (CFU). This test is carried out with the incubation of the sample in a particular or selected strain of microorganisms supplied with a different nutrient medium such as agar-agar, potato dextrose (PDA) agar, etc. for 2–3 weeks. This simple test is not used widely as the results rely on subjective judgment, and choosing specific micro-organisms is too selective to reveal the diversity found in the natural habitat.

Another method available to evaluate soil microbial mass is the chloroform-fumigation method. In this method, the cell membranes of the microbes are ruptured by employing chloroform fumigation in the soil to release biomass carbon. The released carbon dioxide is extracted directly by using potassium sulfate solution or incubated by the addition of a small amount of fresh soil. The evolution of carbon dioxide that follows the incubation can be considered an indication of the amount of the dead microbial carbon available as an energy source.

14.1.15.1 Characterization of Physical Changes

The degradation progress or the efficiency of PU coatings in the form of thin films are being evaluated by assessment of both physical and chemical changes. As there is no specific test available to date for the assessment of WPU films or coatings, a series of tests are carried out to prove and analyze the extent of the biodegradation process.

The degradation of PU films proceeds with the reduction in weight, morphological changes, deterioration in mechanical properties, alteration in hydrophilicity, molecular weight distribution, color changes, etc. Although the weight loss measurement to sample exposure time is an important and accurate technique to study the degradation behavior of polymers, it seems insignificant for most PU bulk polymers as they do not show appreciable changes in weight that have been successfully demonstrated by several researchers. The changes in weight-averaged molecular weight (Mw) and number-averaged molecular weight (Mn) also indicate the degradation behavior of the PU, which is measured by gel permeation chromatography (GPC). The reduction in mechanical properties, such as elongation and tensile strength designates the disintegration of the polymers, which is confirmed by a universal testing machine (UTM). Further, the quantitative analysis of carbon dioxide release is also considered as a potential parameter to assess the degradation rate. In addition to this, the structural changes in terms of crystalline behavior can be determined by X-ray diffraction studies, and the formation of cracks, clusters, pits, holes, and microbial colonies that indicate the degradation of polymers can be extensively studied by scanning electron microscope analysis. The surface hydrophilicity of PU polymers can be assessed by monitoring changes in water contact angle, which is an important factor for microbial adhesion on the surface of the polymer.

14.1.15.2 Characterization of Chemical Changes

The chemical changes in PU after biodegradation can be measured through spectral and GC-MS analysis. In FTIR analysis, the reduction in intensity or the disappearance of the characteristic peak, such as urethane linkage, carbonyl groups, and isocyanate bond, is the result of interest in measuring degradation. Further, GC-MS, NMR, and LC-MS techniques can be used for the identification and measurement of the metabolite products generated after the degradation of the polymer.

In addition to physical and chemical characterization, the ecotoxicity analysis of the compost/soil in which the biodegradation test is carried out can be measured by employing seed germination and plant growth technique. The percentage of germination (Germination index), root height, and shoot height are measured to evaluate the toxicity impact on the ecological balance.

14.2 CONCLUSION

PU is the most widely used polymer in the field of adhesives and coatings. However, environmental concerns and stringent regulations have limited the use of synthetic, solvent-based, nondegradable coatings. Hence, under such circumstances, the development of its potential alternative, bioderived WPU, and its commercialization has become necessary. This chapter has covered the waterborne systems, the microbial degradation of polyester, and polyether-based PU. It also provides a complete insight into the methods of biodegradation assessment.

REFERENCES

1. Narayan, R., Chattopadhyay, D. K., Sreedhar, B., Raju, K. V. S. N., Mallikarjuna, N. N., & Aminabhavi, T. M. (2006). Synthesis and characterization of crosslinked polyurethane dispersions based on hydroxylated polyesters. *Journal of Applied Polymer Science*, *99*(1), 368–380.
2. Zang, Y., Xie, H., Gong, W., Du, Z., Liu, B., & Chen, H. (2016). Migration behavior of anionic polyurethane dispersion during infiltration and redistribution in sand. *RSC Advances*, *6*(49), 43543–43550.
3. Sun, Z., Fan, H., Chen, Y., & Huang, J. (2018). Synthesis of self-matting waterborne polyurethane coatings with excellent transmittance. *Polymer International*, *67*(1), 78–84.
4. Xiao, Y., Jiang, L., Liu, Z., Yuan, Y., Yan, P., Zhou, C., & Lei, J. (2017). Effect of phase separation on the crystallization of soft segments of green waterborne polyurethanes. *Polymer Testing*, *60*, 160–165.
5. Omrani, I., Babanejad, N., Shendi, H. K., & Nabid, M. R. (2017). Preparation and evaluation of a novel sunflower oil-based waterborne polyurethane nanoparticles for sustained delivery of hydrophobic drug. *European Journal of Lipid Science and Technology*, *119*(8), 1600283.
6. Hou, Z. S., Qu, W. Q., & Kan, C. Y. (2015). Synthesis and properties of triethoxysilane-terminated anionic polyurethane and its waterborne dispersions. *Journal of Polymer Research*, *22*(6), 1–9.
7. Bahadur, A., Saeed, A., Iqbal, S., Shoaib, M., ur Rahman, M. S., Bashir, M. I., … & Mahmood, T. (2017). Biocompatible waterborne polyurethane-urea elastomer as intelligent anticancer drug release matrix: A sustained drug release study. *Reactive and Functional Polymers*, *119*, 57–63.
8. Gaddam, S. K., & Palanisamy, A. (2016). Anionic waterborne polyurethane dispersions from maleated cotton seed oil polyol carrying ionisable groups. *Colloid and Polymer Science*, *294*(2), 347–355.
9. Kumar Gaddam, S., & Palanisamy, A. (2017). Anionic waterborne polyurethane-imide dispersions from cottonseed oil based ionic polyol. *Industrial Crops and Products*, *96*, 132–139.
10. Xiao, Y., Bao, L., Fu, X., Wu, B., Kong, W., Zhou, C., & Lei, J. (2018). Effect of phase separation on water resistance of green waterborne polyurethanes: Unexpected stronger impact compared to hydrophilic segments. *Advances in Polymer Technology*, *37*(6), 1618–1624.
11. Honarkar, H., Barmar, M., & Barikani, M. (2015). Synthesis, characterization and properties of waterborne polyurethanes based on two different ionic centers. *Fibers and Polymers*, *16*(4), 718–725.
12. Patel, C. J., & Mannari, V. (2014). Air-drying bio-based polyurethane dispersion from cardanol: Synthesis and characterization of coatings. *Progress in Organic Coatings*, *77*(5), 997–1006.
13. Lokhande, G. P., Chambhare, S. U., & Jagtap, R. N. (2017). Anionic water-based polyurethane dispersions for antimicrobial coating application. *Polymer Bulletin*, *74*(11), 4781–4798.
14. Chen, R., Zhang, C., & Kessler, M. R. (2014). Anionic waterborne polyurethane dispersion from a bio-based ionic segment. *RSC advances*, *4*(67), 35476–35483.
15. Zhang, Y., He, X., Ding, M., He, W., Li, J., Li, J., & Tan, H. (2018). Antibacterial and biocompatible cross-linked waterborne polyurethanes containing gemini quaternary ammonium salts. *Biomacromolecules*, *19*(2), 279–287.
16. Bahadur, A., Saeed, A., Iqbal, S., Shoaib, M., ur Rahman, M. S., Bashir, M. I., … & Mahmood, T. (2017). Biocompatible waterborne polyurethane-urea elastomer as

intelligent anticancer drug release matrix: A sustained drug release study. *Reactive and Functional Polymers*, *119*, 57–63.
17. Ren, Z., Liu, L., Wang, H., Fu, Y., Jiang, L., & Ren, B. (2015). Novel amphoteric polyurethane dispersions with postpolymerization crosslinking function derived from hydroxylated tung oil: Synthesis and properties. *RSC Advances*, *5*(35), 27717–27721.
18. Zhang, Y., Zhou, H., Wang, L., Jiang, W., Soucek, M. D., & Yi, Y. (2017). Preparation and characterization of castor oil-based waterborne polyurethane crosslinked with 2-amino-2-(hydroxymethyl)-1, 3-propanediol. *Journal of Applied Polymer Science*, *134*(47), 45532.
19. Hu, S., Luo, X., & Li, Y. (2015). Production of polyols and waterborne polyurethane dispersions from biodiesel-derived crude glycerol. *Journal of Applied Polymer Science*, *132*(6).
20. Fuensanta, M., Jofre-Reche, J. A., Rodríguez-Llansola, F., Costa, V., Iglesias, J. I., & Martín-Martínez, J. M. (2017). Structural characterization of polyurethane ureas and waterborne polyurethane urea dispersions made with mixtures of polyester polyol and polycarbonate diol. *Progress in Organic Coatings*, *112*, 141–152.
21. Fuensanta, M., Jofre-Reche, J. A., Rodríguez-Llansola, F., Costa, V., Iglesias, J. I., & Martín-Martínez, J. M. (2017). Structural characterization of polyurethane ureas and waterborne polyurethane urea dispersions made with mixtures of polyester polyol and polycarbonate diol. *Progress in Organic Coatings*, *112*, 141–152.
22. Shendi, H. K., Omrani, I., Ahmadi, A., Farhadian, A., Babanejad, N., & Nabid, M. R. (2017). Synthesis and characterization of a novel internal emulsifier derived from sunflower oil for the preparation of waterborne polyurethane and their application in coatings. *Progress in Organic Coatings*, *105*, 303–309.
23. Yong, Q., Nian, F., Liao, B., Huang, L., Wang, L., & Pang, H. (2015). Synthesis and characterization of solvent-free waterborne polyurethane dispersion with both sulfonic and carboxylic hydrophilic chain-extending agents for matt coating applications. *RSC Advances*, *5*(130), 107413–107420.
24. Yong, Q., Nian, F., Liao, B., Huang, L., Wang, L., & Pang, H. (2015). Synthesis and characterization of solvent-free waterborne polyurethane dispersion with both sulfonic and carboxylic hydrophilic chain-extending agents for matt coating applications. *RSC Advances*, *5*(130), 107413–107420.
25. Lei, L., Zhang, Y., Ou, C., Xia, Z., & Zhong, L. (2016). Synthesis and characterization of waterborne polyurethanes with alkoxy silane groups in the side chains for potential application in waterborne ink. *Progress in Organic Coatings*, *92*, 85–94.
26. Madbouly, S. A. (2021). Waterborne polyurethane dispersions and thin films: Biodegradation and antimicrobial behaviors. *Molecules*, *26*(4), 961.
27. Asif, A., Shi, W., Shen, X., & Nie, K. (2005). Physical and thermal properties of UV curable waterborne polyurethane dispersions incorporating hyperbranched aliphatic polyester of varying generation number. *Polymer*, *46*(24), 11066–11078.
28. Asif, A., Huang, C., & Shi, W. (2004). Structure–property study of waterborne, polyurethane acrylate dispersions based on hyperbranched aliphatic polyester for UV-curable coatings. *Colloid and Polymer Science*, *283*(2), 200–208.
29. Gogoi, S., & Karak, N. (2014). Biobased biodegradable waterborne hyperbranched polyurethane as an ecofriendly sustainable material. *ACS Sustainable Chemistry & Engineering*, *2*(12), 2730–2738.
30. Lee, T. J., Kwon, S. H., & Kim, B. K. (2014). Biodegradable sol–gel coatings of waterborne polyurethane/gelatin chemical hybrids. *Progress in Organic Coatings*, *77*(6), 1111–1116.
31. Lin, Y. Y., Hung, K. C., & Hsu, S. H. (2015). Stability of biodegradable waterborne polyurethane films in buffered saline solutions. *Biointerphases*, *10*(3), 031006.

32. Madbouly, S. A., & Lendlein, A. (2012). Degradable polyurethane/soy protein shape-memory polymer blends prepared via environmentally-friendly aqueous dispersions. *Macromolecular Materials and Engineering*, *297*(12), 1213–1224.
33. Feng, Z., Wang, D., Zheng, Y., Zhao, L., Xu, T., Guo, Z., ... & Jiang, H. (2020). A novel waterborne polyurethane with biodegradability and high flexibility for 3D printing. *Biofabrication*, *12*(3), 035015.
34. Hsu, S. H., Hsieh, C. T., & Sun, Y. M. (2015). Synthesis and characterization of waterborne polyurethane containing poly (3-hydroxybutyrate) as new biodegradable elastomers. *Journal of Materials Chemistry B*, *3*(47), 9089–9097.
35. Dang, X., Li, Y., & Yang, M. (2019). Biodegradable waterborne polyurethane grafted with gelatin hydrolysate via solvent-free copolymerization for potential porous scaffold material. *Journal of the Mechanical Behavior of Biomedical Materials*, *92*, 79–89.
36. Chaudhuri, H., & Karak, N. (2020). Water dispersed bio-derived transparent polyurethane: Synthesis, properties including chemical resistance, UV-aging, and biodegradability. *Progress in Organic Coatings*, *146*, 105730.
37. Hsu, S. H., Dai, L. G., Hung, Y. M., & Dai, N. T. (2018). Evaluation and characterization of waterborne biodegradable polyurethane films for the prevention of tendon postoperative adhesion. *International Journal of Nanomedicine*, *13*, 5485.
38. Dai, Z., Jiang, P., Lou, W., Zhang, P., Bao, Y., Gao, X., & Haryono, A. (2020). Preparation of degradable vegetable oil-based waterborne polyurethane with tunable mechanical and thermal properties. *European Polymer Journal*, *139*, 109994.
39. Yang, Z., & Wu, G. (2020). Effects of soft segment characteristics on the properties of biodegradable amphiphilic waterborne polyurethane prepared by a green process. *Journal of Materials Science*, *55*(7), 3139–3156.
40. Liu, J., He, J., Xue, R., Xu, B., Qian, X., Xin, F., ... & Jiang, M. (2021). Biodegradation and up-cycling of polyurethanes: Progress, challenges, and prospects. *Biotechnology Advances*, 107730.
41. Kay, M. J., Morton, L. H. G., & Prince, E. L. (1991). Bacterial degradation of polyester polyurethane. *International Biodeterioration*, *27*(2), 205–222.
42. Pradhan, S., Mohanty, S., & Nayak, S. K. (2018). In-situ aerobic biodegradation study of epoxy-acrylate film in compost soil environment. *Journal of Polymers and the Environment*, *26*(3), 1133–1144.
43. Darby, R. T., & Kaplan, A. M. (1968). Fungal susceptibility of polyurethanes. *Applied Microbiology*, 16(6), 900–905.
44. Nakajima-Kambe, T., Onuma, F., Kimpara, N., & Nakahara, T. (1995). Isolation and characterization of a bacterium which utilizes polyester polyurethane as a sole carbon and nitrogen source. *FEMS Microbiology Letters*, *129*(1), 39–42.
45. Jansen, B., Schumacher-Perdreau, F., Peters, G., & Pulverer, G. (1991). Evidence for degradation of synthetic polyurethanes by Staphylococcus epidermidis. *Zentralblatt für Bakteriologie*, *276*(1), 36–45.
46. Mahajan, N., & Gupta, P. (2015). New insights into the microbial degradation of polyurethanes. *RSC Advances*, *5*(52), 41839–41854.
47. Shah, Z., Gulzar, M., Hasan, F., & Shah, A. A. (2016). Degradation of polyester polyurethane by an indigenously developed consortium of Pseudomonas and Bacillus species isolated from soil. *Polymer Degradation and Stability*, *134*, 349–356.
48. Magnin, A., Pollet, E., Perrin, R., Ullmann, C., Persillon, C., Phalip, V., & Avérous, L. (2019). Enzymatic recycling of thermoplastic polyurethanes: Synergistic effect of an esterase and an amidase and recovery of building blocks. *Waste Management*, 85, 141–150.

49. Magnin, A., Hoornaert, L., Pollet, E., Laurichesse, S., Phalip, V., & Avérous, L. (2019). Isolation and characterization of different promising fungi for biological waste management of polyurethanes. *Microbial Biotechnology*, 12(3), 544–555.
50. Khan, S., Nadir, S., Shah, Z. U., Shah, A. A., Karunarathna, S. C., Xu, J.,… & Hasan, F. (2017). Biodegradation of polyester polyurethane by *Aspergillus tubingensis*. *Environmental Pollution*, *225*, 469–480.
51. Mathur, G., & Prasad, R. (2012). Degradation of polyurethane by *Aspergillus flavus* (ITCC 6051) isolated from soil. *Applied Biochemistry and Biotechnology*, 167(6), 1595–1602.
52. Osman, M., Satti, S. M., Luqman, A., Hasan, F., Shah, Z., & Shah, A. A. (2018). Degradation of polyester polyurethane by *Aspergillus* sp. strain S45 isolated from soil. *Journal of Polymers and the Environment*, 26(1), 301–310.
53. Magnin, A., Pollet, E., Perrin, R., Ullmann, C., Persillon, C., Phalip, V., & Avérous, L. (2019). Enzymatic recycling of thermoplastic polyurethanes: Synergistic effect of an esterase and an amidase and recovery of building blocks. *Waste Management*, 85, 141–150.

15 Recent Developments in Waterborne Polyurethanes for Coating Applications

Verónica L. Mucci and Mirta I. Aranguren
Instituto de Investigaciones en Ciencia y Tecnología de Materiales (INTEMA), Universidad Nacional de Mar del Plata (UNMdP), Consejo Nacional de Investigaciones en Ciencia y Tecnología (CONICET), Facultad de Ingeniería, Av. Juan B Justo 4302, (7600) Mar del Plata, Argentina

Javier I. Amalvy and María E. V. Hormaiztegui
Centro de Investigación y Desarrollo en Ciencia y Tecnología de Materiales (CITEMA), Universidad Tecnológica Nacional (UTN)—Comisión de Investigaciones Científicas de la Provincia de Buenos Aires (CIC), Facultad Regional La Plata, Av. 60 y 124, (1923) Berisso, Buenos Aires, Argentina

CONTENTS

15.1 Introduction 254
15.2 Preparation Methods of Particle-Loaded WPUs 255
15.3 Viscosity of the WPU 256
15.4 Properties of WPU Coatings 257
  15.4.1 WPU Coatings That Incorporate Inorganic Particles 258
  15.4.2 WPU Coatings That Incorporate Organic Particles 260
  15.4.3 WPU Coatings That Incorporate Combinations of Fillers 260
  15.4.4 Systems of Two Components (2 K) 261
15.5 Conclusion and Future Perspectives 262
References 262

DOI: 10.1201/9781003173526-15

15.1 INTRODUCTION

The use of waterborne polymers for coating formulations has increased in the last few years due to restrictive regulations for the use of solvent-based systems [1]. As an alternative to the latter, waterborne dispersions are more friendly due to their low or zero emission of volatile organic compounds (VOCs). However, because of their usually lower performance, new approaches and strategies are focused on improving their performance [2]. From a coating point of view, the films should perform as barriers for substrate protection, and they should present good adhesion to the substrate, hardness (relevant for service life), and gloss (important for the final appearance in architectural and decorative coatings). In particular, waterborne polyurethanes (WPUs) are one of the most popular waterborne coatings, with increasing participation in the market during the last years [3]. With a share in the printing, ink, adhesive, and coating industries [4], WPU dispersions have advantages over other waterborne polymers because of their versatile structural properties and the excellent elasticity of the resulting films. The wide number of diol components (including natural oils), different isocyanates, and chain extenders available allows the development of an enormous diversity of products with final properties tailored to the type of substrate. The careful selection of raw materials, relative proportions of chemicals involved in the formulation, as well as the process of synthesis, and adjustment of the variables involved, are responsible for the characteristics of the obtained suspensions and films [5], a subject treated in other chapters of this book.

Briefly, WPUs can be considered segmented polymers, constituted by soft segments (essentially the polyol) and hard segments (formed by the isocyanate, the chain extender, and frequently the internal emulsifier). The incompatibility of these segments leads to phase separation, with a large effect on properties [6]. Usually, WPUs contains ionic centers (internal emulsifier) that allow their dispersion in water in the presence of a counterion that stabilizes the particles [7]. Anionic groups in waterborne polyurethanes contribute to adhesion to the substrate, but other specific applications require the development of cationic and bio-based WPUs [8–10]. Additionally, nonionic internal emulsifiers have also been proposed, although they usually result in dispersions less stable than those of ionic WPUs [9].

Nanotechnology is an emerging field that may have a great impact on the performance of coatings, inks, and adhesives, among other industrial products [11]. The incorporation of nanoparticles into polymeric formulations is the most widely used strategy to improve the performance of coatings. For instance, nano-silica and nano-alumina were used to obtain increased surface hardness and extended service life in coatings for industrial floors, while nano-silver was used in the formulation of hygienic paints. According to a recent market report (2021) about 90% of protective coatings are solvent-borne, 7% are water-based, and about 3% are powder systems, with a high growth expected only by 2026 [12]. Despite the relatively low share of WPUs in the market, Figure 15.1 shows the important increase in the number of publications and patents for WPUs used for coatings that occurred during the last decades [13]. The growth has been continuous, with a large increase observed during the last 15 years, evidencing the great industrial interest.

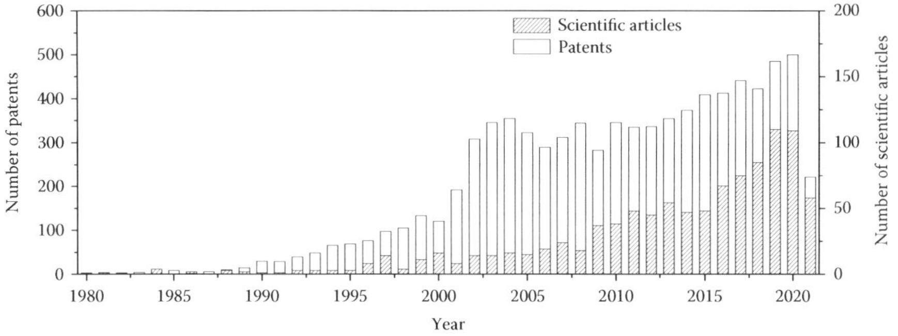

FIGURE 15.1 Number of scientific articles and patent publications on WPU for coatings between 1980 and 2021 (June). Notice the different scales. Data obtained from Scopus with the search (waterborne and polyurethane).

15.2 PREPARATION METHODS OF PARTICLE-LOADED WPUS

According to the type of particle, the kind of WPU, or the desired properties of the final material, different particle incorporation methods have been used in order to produce a homogeneous and stable dispersion of WPU composite and a coating or a film with uniform properties.

Figure 15.2 shows different ways of particle incorporation in the WPU synthetic process, for the particular case of the acetone process to obtain an anionic WPU composite.

If the particles are incorporated via an in situ method, they may react with the isocyanate and become part of the final chemical structure of the polyurethane, being covalently bonded (chemical bonding) to the polyurethane and resulting in a crosslinked structure. On the other hand, if the particles are added by sonication and/or mechanical mixing of polymer and particle dispersions, Van der Waals forces and/or hydrogen bonding interactions will develop between the particles (physical bonding) or they will be encapsulated in the polyurethane particles. For example, incorporation of cellulose nanocrystals (CNC) in a WPU, by mixing of the two water dispersions or by in situ addition of CNC aqueous dispersion, led to homogeneous distribution of the particles and reinforcing of the produced films. However, enhanced WPU–CNC interactions were obtained during the in situ incorporation of CNC [14]. The particles can also be added with the reagents used to synthesize WPUs. For example, montmorillonite (MMT) nanosheet was exfoliated

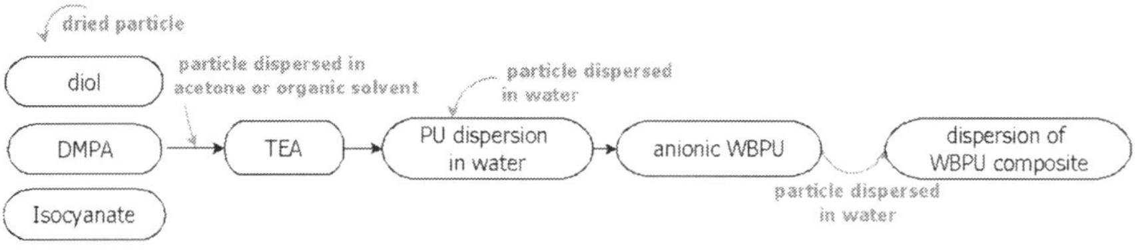

FIGURE 15.2 Different methods of particle incorporation in an anionic WPU composite.

in a polyethyl–phosphate glycol ester to synthesize a WPU with flame retardancy, via the acetone method [15].

Another strategy for incorporating particles into WPUs is to chemically modify them before incorporation. For example, a WPU nanocomposite was prepared with hydroxyl-terminated functionalized graphene (fGO) incorporated in situ. The strong filler-matrix interfacial adhesion led to extraordinary thermal, mechanical, and flame retardant properties [16]. Of course, if the particles form a stable dispersion in water, the easiest method of incorporation in WPU composite is mixing both water dispersions, in general by sonication [7,17]. For example, a CNC-reinforced coating was prepared from a castor oil-based WPU by simple blending of both water dispersions, WPU and CNC [7]. Physical mixing and chemical bonding were proposed for nanocellulose-WPU wood coating to compare the performance of both formulations [18]. The chemical bonding method resulted in the maximum overall improvement of the properties, highlighting the importance of the method of incorporation.

15.3 VISCOSITY OF THE WPU

Viscosity is a crucial property for WPUs to be applied as coatings. One important contribution made in developing WPUs, as opposed to traditional solvent polyurethane coatings, is that they can maintain a relatively low viscosity during application, even if high molecular weight polyurethanes are used in the formulation [19]. Further, tailoring of the viscosity of the dispersion can be achieved by the addition of thickeners or by water dilution [20]. Moreover, the frequently observed pseudoplastic behavior of these products is consistent with a high zero shear rate viscosity, which is needed to avoid sedimentation of the particles during storage and to ensure the stability of the dispersion.

The application of the WPU by dipping, painting, rolling, or spraying requires low viscosity to ensure easy procedure and good wetting of the surface. The following step requires a balance, since the viscosity should be high enough to prevent dripping (occurring if the viscosity is too low), but not so high as to hinder the formation of a smooth surface (leveling) [9,21]. For example, it was found that a pseudoplastic WPU with high viscosity and yield stress resulted in rugosities formed during the application, which did not level off and lead to lower gloss of the coating [22]. Several variables are intertwined to give an adequate rheological profile to these dispersions: chemical formulation, solid concentration, concentration of contraions (in the case of ionic WPU), concentration and nature of the internal emulsifier, and stirring procedure.

The total solid content affects the viscosity of the dispersion, with a "liquid–solid" transition taking place at a certain critical concentration. Comparatively, a dispersion of smaller particles shows a higher viscosity due to the relatively thicker surrounding electric double layer (compared to the size of the particles). At a fixed concentration, the number of particles is larger if the particles are smaller and the friction generated by their movement is larger than that generated by large particles. Consequently, the viscosity is higher for WPUs with small particles. The variation of the NCO/OH ratio can affect the viscosity, since high values may lead to higher contents of polar bonds formed between the polyurethane chains, leading to agglomeration and thus larger particle size [23]. Another method to obtain smaller particles is to increase the proportion of emulsifiers in the formulation, which results in enhanced transparency of the WPU films, but also a higher sensibility to humid environments [21]. Increases of emulsifier (dimethylolpropionic acid, DMPA) content (1–9 wt.%) in a WPU made from a polycarbonate diol and isophorone diisocyanate (IPDI) resulted in smaller particles. Consequently, the aspect of the dispersions changed from opaque to transparent, while viscosity increased about 26 times [24].

Some authors demonstrated that when the ionic centers were located in the soft segments, the resulting particles were smaller and the hydrophilic groups were unlocalized in the nanodroplet (instead of located on the surface, as happens when the ionic centers are part of the hard segments) [25]. Similarly, when the ionic groups are located at the chain end, the resulting aqueous dispersions also contain small polymer particles, displaying high stability and viscosity [26]. However, high concentration of ionic centers in the WPU is also responsible for the higher hydrophilicity of the films and thus for the lower protection of the substrate. Therefore, a balance must be found for the addition of the ionic emulsifier in the formulation [21]. Regarding the effect of the size distribution of the PU droplets, it has been shown that broad and bimodal distributions are effective in maintaining low viscosity at high solid content, as the small particles can fit in the interstices left between the large ones so that higher concentrations are needed to reach the liquid–solid transition [21]. Viscosity has also been used to monitor the aging of the dispersion. WPUs based on jatropha oil showed a marginal viscosity increase during the initial 9 months of storage. A stable viscosity value was reached after that time, which was still low enough to warrant usage [21].

15.4 PROPERTIES OF WPU COATINGS

The performance of waterborne coatings depends not only on the binder properties or paint formulation but also on the film quality after solvent (mainly water) evaporation. During the film formation process, latex particles approach each other and coalesce, with the size of the latex particles playing an important role in this. Thus, Patel and Kapatel [27] found that systems with large particles agglomerate earlier than systems with smaller ones, modifying the coating performance.

The selection of raw materials affects morphology. Chen et al. prepared WPU formulations using 2,2-bis(hydroxymethyl) butyric acid (DMBA) as a chain extender, and different types of diol to find that the use of the ether diol allowed for a better degree of phase separation. They found that this characteristic was the key to obtaining a good abrasion resistance without the need for adding inorganic fillers [5]. Tailoring the properties by combining the characteristics of two different macrodiols has also been considered by Li et al., who varied the ratio of polyether and polyester diols. They found that the polyester improved the compactness of the coating at the cost of increased hydrophilicity. Thus, an optimum ratio was found at which the polyether content contributed by reducing the hydrophilicity and improving corrosion resistance [28].

Renewable sources are increasingly being considered in the production of WPU. Thus, isosorbide, a green renewable polyol, has been used in the synthesis of WPU [29]. Its addition increased the degree of microphase separation and hydrogen bond interactions between hard segments in WPUs, improving mechanical properties, hardness, and chemical resistance of the coating. A frequently used renewable polyol, castor oil, was used to produce a WPU coating with damping properties, useful to solve vibration and noise problems in the design of machinery and equipment, particularly at low temperatures [30].

Another important property to consider in coatings is the hydrophilicity/hydrophobicity of the protecting films formed. The hydrophobicity of WPU has been modified by including silane derivative groups by in situ polymerization. Lyu et al. showed that the water contact angle was increased from 64° to 86° by incorporating 3-(2-aminoethylamino)propyldimethoxymethylsilane (APTS) [31]. Also, Qian et al. included a silicone derivate to increase contact angle and reduce the water penetration of the final coating [32]. An interesting UV-curable formulation was developed as a self-healable and anticorrosive paint by Li et al. [33]. In this work, trimethylolpropane diallyl ether and trimethylolpropane tris(3-mercaptopropionate) were introduced into the end of the polyurethane chains and cured by UV-radiation, leading to a crosslinked film using the reaction of the vinyl- and thiol-end-capped prepolymer. Besides changes in the chemical composition or morphology of the film, the incorporation of a second component could improve the performance of the final coating [34]. The second component could be particles of inorganic or organic nature, forming in this way composites or nanocomposites, depending on the size of the particles and the addition method. In the next section, variation of the coating properties is discussed as a consequence of particle incorporation.

15.4.1 WPU Coatings That Incorporate Inorganic Particles

Inorganic particles are most frequently used in the modification of coating properties, and silica is perhaps the most used one, mainly due to its high surface area, cost-effective production, and easy surface functionalization. Ding et al. prepared WPU/silica composites using silica without any surface modification that acted as crosslinking points forming a network structure [35]. Because of this, the thermal properties and the water-resistance of the coatings increased. Modification of the silica surface allows the preparation of coatings with specific properties. Fan et al. used modified silica, which was incorporated into a WPU using the blending method [36]. The as-prepared coating had antiglare performance, superior mechanical properties, and higher decomposition temperature compared to the neat WPU. The reduced gloss was mainly due to the light-scattering effect, and with a silica loading of 3 wt.%, the requirements of the antiglare coating were satisfied. Also, a matting coating was prepared using a SiO_2 modified with a silane derivate. The coating presented an outstanding matting performance with good storage stability and relatively low viscosity [37]. Other authors used silica modified with heptadecafluorodecyl triethoxysilane [38] for the development of a superhydrophobic coating with a water contact angle as high as of 173°.

Silica hollow capsules were used to encapsulate a corrosion inhibitor, 2-mercaptobenzothiazole, and were incorporated in a commercial WPU [39].

Although the addition of the capsules affected the coating barrier properties, it provided short-term protection against corrosion. Hollow silica microcapsules loaded with the same corrosion inhibitor and incorporated by blending were used by Mirmohseni et al. to improve the thermal and barrier properties of self-healing coatings. Their results showed that adding 1 wt.% of silica capsules enhanced the adhesion of the WPU to aluminum substrates, maintaining the improvement in barrier properties. Higher capsule concentration resulted in the coating being more prone to corrosion than the neat WPU [40]. On the other hand, Yuan et al. included a silicate obtained by sol–gel in a WPU, and the resulting coating with 4 wt.% of filler had higher mechanical properties, impermeability, and better bonding properties compared to the neat WPU [41]. Another family of silicon-containing compounds is silsesquioxane, characterized by a ratio of 1.5 between the silicon and oxygen atoms. In particular, polyhedral oligomeric silsesquioxane (POSS) was used as a reinforcing agent of a WPU used in the coating of wood [42].

Recently, green synthetic protocols have been reported to synthesize silica nanoparticles from plants and agriculture wastes, between other sources. For example, Monteiro et al. included silica obtained from rice husk ash to improve weathering resistance. In this work, the simple blending mixing process gave better results for adding the reinforcement than the in situ polymerization one. Comparison of the materials prepared with different amounts of silica (1, 3, and 5 wt.%) showed that 1 wt.% of silica produced a composite with the highest resistance to thermal degradation and accelerated weathering [43].

Probably the second most used particles for the improvement of coating properties are clays and modified clays. WPU/clay composites and nanocomposites are considered good candidates for paint formulations because of their extraordinary thermal stability, mechanical strength, and adhesive strength in addition to their superior barrier properties. For example, using MMT nanosheets, Ding et al. prepared an anti-dripping and flame-retardant coating with low gloss [44]. The coating, applied onto polyvinyl chloride leather, also improves the surface touch feeling. Incorporation of MMT modified with Ce^{3+} into a WPU was also considered as corrosion protection [45]. Salt spray test (monitoring the metal-coated samples sprayed with a 3.5 wt.% NaCl solution at 35 °C) showed no corrosion products after 768 h of exposure and an improvement of the water resistance of the coating.

Carbon nanotube (CNT) is another typical particle used to modify the coating's final properties. For instance, Wang et al. developed an anticorrosive and conductive coating by incorporation of multiwalled carbon nanotubes (MWCNTs) [46]. As the MWCNT content increased, the electrical conductivity of the electrostatic spraying coating increased; however, its corrosion resistance and bond strength showed a maximum at 0.3 wt.% MWCNT. Further increase of the filler content led to a decrease of properties because of the increased number of micro defects in the coating. Another type of particle widely used to modify coating properties is graphene oxides functionalized with chemicals such as polydopamine, polycarbodiimide, a mix of phosphoric acid, dodecylbenzenesulfonic acid, and polyaniline, to improve the anticorrosive and self-healing properties, resistance to photo-degradation, and thermo-oxidative degradation by UV irradiation, while displaying a considerable electrical conductivity of the WPU coating [13,47].

Inorganic particles with specific properties are used to prepare coatings for targeted applications. For example, Zhang et al. prepared antibacterial coatings for use in synthetic leather by ultrasonic dispersion of silver nanoparticles in WPU with a good performance against *Escherichia coli* and *Staphylococcus aureus* [48]. Zinc oxide was incorporated to reinforce the optical properties of WPU coatings [49], delaying the structural damage of the polymer, mainly in the UV region. Zhang et al. prepared a WPU doped with poly(*o*-toluidine) and a composite of poly(*o*-toluidine)-nano ZnO to inhibit metal corrosion [50]. The presence of ZnO improved the corrosion resistance, and the composite particles reduced the porosity of the coating surface.

15.4.2 WPU Coatings That Incorporate Organic Particles

The incorporation of TEMPO-oxidized cellulose nanofibers (TOCNs), a green filler, increased the pencil hardness of wood coating from B (neat WPU) up to 4 H for 5 wt.% of TOCNs, in accordance with the increase in the tensile modulus. However, a deterioration of adhesion strength between wood and the coating was observed as the TOCN content increased, due to the increase in the viscosity of the coating that resulted in the decrease of the penetration of the dispersion into the wood. The roughness of the coating composites was measured by AFM topography, and the results are in agreement with SEM images; high content of filler leads to higher roughness due to the agglomeration of TOCNs [51].

Collagen fibrils can be used to enhance the properties of WPUs. In the work of Han et al., WPU/collagen composites were used as a functional coating for applications on textile, leather, and wound dressing. Water vapor permeability was tested at two different temperatures, and the results showed that composite possessed enhanced temperature sensitivity, improving the capacity of breathability of the coated textiles [52].

15.4.3 WPU Coatings That Incorporate Combinations of Fillers

A combination of particles has also been reported. Kale et al. combined epoxy-functionalized graphene oxide with amine-functionalized silica to improve mechanical and thermal properties in coatings for leather applications [53]. A combination of copper and zinc composite oxide (CuZnO) with reduced graphene oxide (GO) was prepared by Jiang et al. [54] to produce a WPU coating for use in a circulating cooling water system with antibacterial and anticorrosive properties. WPU coatings with antibacterial and antistatic properties were developed by Mirmohseni et al. by doping reduced single-layer graphene oxide with metallic copper [55]. A similar approach combining a commercial WPU with GO modified with polytetrafluoroethylene nanopowder led to coatings with improved tribological and anticorrosive properties [56]. Significant improvement of the corrosion resistance of coatings was achieved by incorporating nano-TiO_2 modified with a silane coupling agent and GO composite particles in a commercial WPU [57] and also by using GO covered with silica nanoparticles dispersed in a commercial polyurethane/acrylic system [58]. Modifying GO with Fe_3O_4, Bai et al. [59] improved the antifriction and wear resistance of commercial WPU coatings.

15.4.4 Systems of Two Components (2 K)

Two-component paints (also referred to as 2 K paints) are systems in which a chemical reaction results in paint hardening. They were traditionally solvent-based; however, with new technology, waterborne 2 K polyurethane coatings consisting of polyisocyanate and polyol dispersion were also developed [60]. In the work of Wang et al., the WPU formulation consisted of two components: water dispersion of an OH functional acrylate dispersion and a polyisocyanate resin for which a propylene glycol diacetate was incorporated as a solvent. When this solvent was maintained at 10 wt.%, it acted as a promoter for the urethane formation between the two types of particles. The films showed a better appearance with fewer bubbles or imperfections [61]. Incorporation of octavinyl polyhedral oligomeric silsesquioxane improved pencil hardness of a 2 K-WBPU coating, according to Zeng et al. [62]. In other work, carbon steel panels were coated with 2 K-WBPU formulations (containing pigments and additives) to find the best chemical characteristics of the polyurethane to obtain good adhesion to the metal substrate [63]. The authors found that a higher OH content was preferred to promote adhesion (as compared with high NCO content). Thus, corrosion resistance was improved as shown by the results obtained using the salt spray test. According to the authors, an increased content of OH leads to better adhesion and also higher crosslinking and compactness of the formed film, which is also the reason for the improved water resistance shown by these films (Figure 15.3).

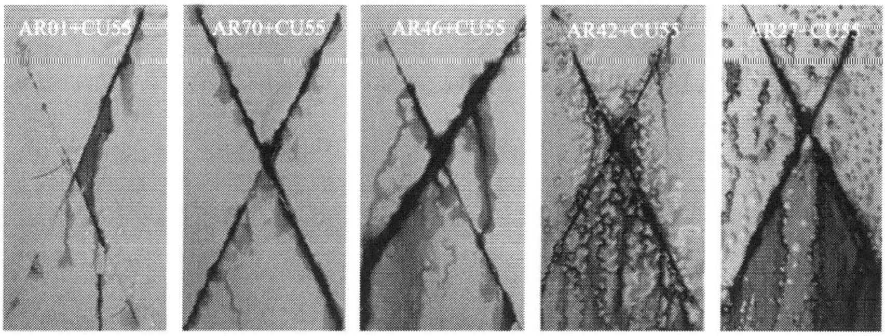

FIGURE 15.3 Images of 2 K WBPU coating based on polyacrylic polyols with different hydroxyl content. OH content increases to the left. Figure 15.3 reprinted with permission from reference [63].

15.5 CONCLUSION AND FUTURE PERSPECTIVES

This chapter aimed to present an up-to-date view of developments in the area of WPU coatings. Environmental issues are thrusting the development of these types of coatings, which despite the overwhelming number of patents and articles that have appeared in the last decades, still constitute a small percentage of the market. The WPU coatings field is estimated to grow in the next few years due to an increase of their use in several industries, such as building and automotive, mainly in Europe.

The present formulations do not provide the same range of properties as solvent-based alternatives; thus, there is still plenty of room for creative innovation in this area. It is important to keep in mind that there is not one single formulation that best fills all the requirements and that each application may require a specific combination and balance of properties. Some general, frequently shared observations highlight the effect of morphologies in neat WPUs, leading to varied transparency of the coatings, compactness of the films, and consequently mechanical as well as surface and protection-related properties (barrier, antimicrobial, etc.). Several variables are conducive to changes in these properties, and some of them were treated in detail in the text. However, it is also clear, from the dynamic of investigations in the area, that besides the chemistry of the polymers used in the paint formulations, bountiful other variables can be considered to tailor and improve the properties to adjust them to specific requirements. For example, crosslinking and the addition of nanoparticles may be a tool to improve hydrophobicity and compactness. As mentioned before, a balance may be needed to find the optimum formulation, since other properties such as adhesion or gloss may be compromised if, for example, the rigidity of the film becomes too high. The use of special monomers, the modification of the polyols, or the combination with other polymers that can co-react or produce blends with the WPU are an open field of exploration. Moreover, the pool of nanoparticles from which to select for its incorporation in the formulations is wide, so it leads to an extremely interesting way of including new and special properties in the films (antimicrobial, optical, flame retardancy, electric conductivity, etc.). The usage of renewable-based materials in the preparation of the WPU is a trend that will probably stay and grow because of the increasing pressure of international organisms and the general public involved in environmental issues.

Finally, the use of 2 K waterborne systems offers large flexibility at the moment of fitting the film properties and behavior to the particular requirement of a given application.

REFERENCES

1. T.V. Vu, T.V. Nguyen, M. Tabish, S. Ibrahim, T. Huong, T. Hoang, R.K. Gupta, T. My, L. Dang, T.A. Nguyen, G. Yasin, Water-borne ZnO/acrylic nanocoating: fabrication, characterization, and properties, *Polymer (Guildf)*. 13 (2021) 717.
2. J. Koleske, *Paint and Coating Testing Manual—15th Edition of the Gardner-Sward Handbook*, 15th ed., ASTM International, USA, 2012.

3. A. Javadi, A. Cobaj, M.D. Soucek, Commercial waterborne coatings, in: P. Zarras, A. Tiwari, M.D. Soucek (Eds.), *Handb. Waterborne Coatings*, 1st ed., Elsevier Inc., 2020: pp. 303–344.
4. F. Zafar, A. Ghosal, E. Sharmin, R. Chaturvedi, N. Nishat, A review on cleaner production of polymeric and nanocomposite coatings based on waterborne polyurethane dispersions from seed oils, *Prog. Org. Coatings.* 131 (2019) 259–275.
5. S.-Y. Chen, R.-Q. Zhuang, F.-S. Chuang, S.-P. Rwei, Synthetic scheme to increase the abrasion resistance of waterborne polyurethane–urea by controlling micro-phase separation, *J. Appl. Polym. Sci.* 138 (2021) 50561.
6. A. Santamaria-Echart, I. Fernandes, F. Barreiro, M.A. Corcuera, A. Eceiza, Advances in waterborne polyurethane and polyurethane-urea dispersions and their eco-friendly derivatives: A review, *Polymers (Basel).* 13 (2021) 409.
7. M.E.V. Hormaiztegui, B. Daga, M.I. Aranguren, V.L. Mucci, Bio-based waterborne polyurethanes reinforced with cellulose nanocrystals as coating films, *Prog. Org. Coatings.* 144 (2020) 105649.
8. N. Sukhawipat, N. Saetung, P. Pasetto, J.F. Pilard, S. Bistac, A. Saetung, A novel high adhesion cationic waterborne polyurethane for green coating applications, *Prog. Org. Coatings.* 148 (2020) 105854.
9. V.L. Mucci, M.E.V. Hormaiztegui, M.I. Aranguren, Plant oil-based waterborne polyurethanes: A brief review, *J. Renew. Mater.* 8 (2020) 579–601.
10. M.E.V. Hormaiztegui, M.I. Aranguren, V.L. Mucci, Synthesis and characterization of a waterborne polyurethane made from castor oil and tartaric acid, *Eur. Polym. J.* 102 (2018).
11. Z.W. Wicks, F.N. Jones, S.P. Pappas, D.A. Wicks, Pigments, in: F.N. Jones, M.E. Nichols, S.P. Pappas (Eds.), *Organic Coatings Science and Technology*, John Wiley and Sons Ltd, 2017: p. 512.
12. D. Gagro, A closer look at protective coatings, 1 June 2021. (2021). https://www.european-coatings.com/articles/2021/06/a-closer-look-at-protective-coatings (accessed June20, 2021).
13. D.M. Nikalin, A.S. Serdtselyubova, Y.I. Merkulova, A.A. Kozlova, Waterborne polyurethane paints and varnishes for metal surfaces: Patent review, *Russ. J. Gen. Chem.* 91 (2021) 540–545.
14. A. Santamaria-Echart, L. Ugarte, A. Arbelaiz, N. Gabilondo, M.A. Corcuera, A. Eceiza, Two different incorporation routes of cellulose nanocrystals in waterborne polyurethane nanocomposites, *Eur. Polym. J.* 76 (2016) 99–109.
15. Z. Ding, J. Li, W. Xin, Y. Luo, Facile and high-concentration exfoliation of montmorillonite into mono- layered nanosheets and application in multifunctional waterborne polyurethane coating, *Appl. Clay Sci.* 198 (2020) 105798.
16. P. Zhang, P. Xu, H. Fan, Z. Sun, J. Wen, Covalently functionalized graphene towards molecular-level dispersed waterborne polyurethane nanocomposite with balanced comprehensive performance, *Appl. Surf. Sci.* 471 (2019) 595–606.
17. B. Alonso-Lerma, I. Larraza, L. Barandiaran, L. Ugarte, A. Saralegi, M.A. Corcuera, R. Perez-Jimenez, A. Eceiza, Enzymatically produced cellulose nanocrystals as reinforcement for waterborne polyurethane and its applications, *Carbohydr. Polym.* 254 (2021) 117478.
18. L. Kong, D. Xu, Z. He, F. Wang, S. Gui, J. Fan, X. Pan, X. Dai, X. Dong, B. Liu, Y. Li, Nanocellulose-reinforced polyurethane for waterborne wood coating, *Molecules.* 24 (2019) 3151.
19. R. Li, Z. Shan, Research for waterborne polyurethane/composites with heat transfer performance: a review, *Polym. Bull.* 75 (2018) 4823–4836.
20. X. Liu, W. Hong, X. Chen, Continuous production of water-borne polyurethanes: A review, *Polymers (Basel).* 12 (2020) 1–17.

21. S. Saalah, L.C. Abdullah, M.M. Aung, M.Z. Salleh, D.R. Awang Biak, M. Basri, E.R. Jusoh, S. Mamat, Colloidal stability and rheology of jatropha oil-based waterborne polyurethane (JPU) dispersion, *Prog. Org. Coatings.* 125 (2018) 348–357.
22. Q. Yong, B. Liao, J. Huang, Y. Guo, C. Liang, H. Pang, Preparation and characterization of a novel low gloss waterborne polyurethane resin, *Surf. Coatings Technol.* 341 (2018) 78–85.
23. R.H. Patel, P.M. Kapatel, Studies on influence of the size of waterborne polyurethane nanoparticles on coating performance, *Mater. Today Proc.* 18 (2019) 1548–1555.
24. L. Hu, Z. Pu, Y. Zhong, L. Liu, J. Cheng, J. Zhong, Effect of different carboxylic acid group contents on microstructure and properties of waterborne polyurethane dispersions, *J. Polym. Res.* 27 (2020) 1–9.
25. B.K. Kim, J.S. Yang, S.M. Yoo, J.S. Lee, Waterborne polyurethanes containing ionic groups in soft segments, *Colloid Polym. Sci.* 281 (2003) 461–468.
26. S.K. Lee, B.K. Kim, High solid and high stability waterborne polyurethanes via ionic groups in soft segments and chain termini, *J. Colloid Interface Sci.* 336 (2009) 208–214.
27. R.H. Patel, P.M. Kapatel, Studies on the effect of the size of waterborne polyurethane nanoparticles on properties and performance of coatings, *Int. J. Polym. Anal. Charact.* 24 (2019) 1–9.
28. S. Li, Z. Liu, L. Hou, Y. Chen, T. Xu, Effect of polyether/polyester polyol ratio on properties of waterborne two-component polyurethane coatings, *Prog. Org. Coatings.* 141 (2020) 105545.
29. J. Hu, C. Tao, A. Yuan, J. Bao, Q. Cheng, G. Xu, Y. Huang, Effects of isosorbide on the microphase separation and properties of waterborne polyurethane coatings, *Polym.* 43 (2019) 169–180.
30. J. Shan, L. Jiang, L. Wang, H. Zhao, X. Ding, C. Zhou, Improvement of low-temperature damping performance by the control of three-dimensional network structure formed by renewable oil in modified waterborne polyurethane, *Results Mater.* 10 (2021) 100171.
31. J. Lyu, K. Xu, N. Zhang, C. Lu, Q. Zhang, L. Yu, F. Feng, In situ incorporation of diamino silane group into waterborne polyurethane for enhancing surface hydrophobicity of coating, *Molecules.* 24 (2019) 1667.
32. Y. Qian, F. Dong, L. Guo, X. Xu, H. Liu, Terpene derivative-containing silicone two-component waterborne polyurethane for coatings, *Prog. Org. Coatings.* 153 (2021) 106137.
33. T. Li, Z.P. Zhang, M.Z. Rong, M.Q. Zhang, Self-healable and thiol–ene UV-curable waterborne polyurethane for anticorrosion coating, *J. Appl. Polym. Sci.* 136 (2019) 1–11.
34. D. Anıl, E. Berksun, A. Durmuş-Sayar, E.B. Sevinis¸-Özbulut, S. Ünal, Recent advances in waterborne polyurethanes and their nanoparticle-containing dispersions, in: P. Zarras, M. Soucek, A. Tiwari (Eds.), *Handb. Waterborne Coatings*, 1st ed., Elsevier Inc., 2020: pp. 249–302.
35. X. Ding, X. Wang, H. Zhang, T. Liu, C. Hong, Q. Ren, C. Zhou, Preparation of waterborne polyurethane-silica nanocomposites by a click chemistry method, *Mater. Today Commun.* 23 (2020) 100911.
36. W. Fan, J. Wang, Z. Li, Antiglare waterborne polyurethane/modified silica nanocomposite with balanced comprehensive properties, *Polym. Test.* 99 (2021) 107072.
37. H. Ma, Y. Liu, J. Guo, T. Chai, J. Suming, Y. Zhou, L. Zhong, J. Deng, Synthesis of a novel silica modified environmentally friendly waterborne polyurethane matting coating, *Prog. Org. Coatings.* 139 (2020) 105441.
38. H. Zheng, M. Pan, J. Wen, J. Yuan, L. Zhu, H. Yu, Robust, transparent, and superhydrophobic coating fabricated with waterborne polyurethane and inorganic nanoparticle composites, *Ind. Eng. Chem. Res.* 58 (2019) 8050–8060.

39. P. Loison, V. Debout, H. Groult, J. Creus, S. Touzain, Incorporation of silica nanocontainers and its impact on a waterborne polyurethane coating, *Mater. Corros.* 70 (2019) 1884–1899.
40. A. Mirmohseni, M. Akbari, R. Najjar, M. Hosseini, Self-healing waterborne polyurethane coating by pH-dependent triggered-release mechanism, *J. Appl. Polym. Sci.* 136 (2019) 47082.
41. H. Yuan, Y. Wang, Z. Liu, S. Li, A study on the properties and working mechanism of a waterborne polyurethane-modified silicate-based coating, *RSC Adv.* 9 (2019) 26817–26824.
42. S. Wei, L. Meng, W. Liu, S. Guo, X. Zhang, Polyhedral oligomeric silsesquioxane (POSS) as reinforcing agent for waterborne polyurethane coatings on wood, *Mater. Res.* 22 (2019) e20180278.
43. W.F. Monteiro, G.M. Miranda, R.R. Soares, C.A.B. Dos Santos, M.S. Hoffmann, C.L.P. Carone, V. De Lima, M.F. De Souza, A.D. Campani, S.M.O. Einloft, J.E. De Lima, R.A. Ligabue, Weathering resistance of waterborne polyurethane coatings reinforced with silica from rice husk ash, *An. Acad. Bras. Cienc.* 91 (2019) e20181190.
44. Z. Ding, J. Li, W. Xin, G. Zhang, Y. Luo, Low gloss waterborne polyurethane coatings with anti-dripping and flame retardancy via montmorillonite nanosheets, *Prog. Org. Coatings.* 136 (2019) 105273.
45. Q. Mo, W. Li, H. Yang, F. Gu, Q. Chen, R. Yang, Water resistance and corrosion protection properties of waterborne polyurethane coating enhanced by montmorillonite modified with Ce3+, *Prog. Org. Coatings.* 136 (2019) 105213.
46. F. Wang, L. Feng, Y. Huang, G. Li, Z. Zhai, Effect of coating process on the properties of multi-walled carbon nanotubes/waterborne polyurethane anticorrosive and conductive coating, *AIP Adv.* 9 (2019) 035241.
47. F. Gao, Y. Luo, J. Xu, X. Du, H. Wang, X. Cheng, Z. Du, Preparation of graphene oxide-based polyaniline composites with synergistic anticorrosion effect for waterborne polyurethane anticorrosive coatings, *Prog. Org. Coatings.* 156 (2021) 106233.
48. X. Zhang, W. Wang, D. Yu, Synthesis of waterborne polyurethane–silver nanoparticle antibacterial coating for synthetic leather, *J. Coatings Technol. Res.* 15 (2018) 415–423.
49. P. Salazar-Bravo, D. Del Angel-López, A.M. Torres-Huerta, M.A. Domínguez-Crespo, D. Palma-Ramírez, S.B. Brachetti-Sibaja, A.C. Ferrel-Álvarez, Investigation of ZnO/waterborne polyurethane hybrid coatings for corrosion protection of AISI 1018 carbon steel substrates, *Metall. Mater. Trans. A.* 50 (2019) 4798–4813.
50. J. Zhang, Y. Li, C. Hu, W. Huang, L. Su, Anti-corrosive properties of waterborne polyurethane/poly(o-toluidine)-ZnO coatings in NaCl solution, *J. Adhes. Sci. Technol.* 33 (2019) 1047–1065.
51. D. Cheng, Y. Wen, X. An, X. Zhu, Y. Ni, TEMPO-oxidized cellulose nanofibers (TOCNs) as a green reinforcement for waterborne polyurethane coating (WPU) on wood, *Carbohydr. Polym.* 151 (2016) 326–334.
52. Y. Han, Y. Jiang, J. Hu, Collagen incorporation into waterborne polyurethane improves breathability, mechanical property, and self-healing ability, *Compos. Part A Appl. Sci. Manuf.* 133 (2020) 105854.
53. M.B. Kale, Z. Luo, X. Zhang, D. Dhamodharan, N. Divakaran, S. Mubarak, L. Wu, Y. Xu, Waterborne polyurethane/graphene oxide-silica nanocomposites with improved mechanical and thermal properties for leather coatings using screen printing, *Polymer (Guildf).* 170 (2019) 43–53.
54. G. Jiang, X. Li, Y. Che, Y. Lv, F. Liu, Y. Wang, C. Zhao, X. Wang, Antibacterial and anticorrosive properties of CuZnO@RGO waterborne polyurethane coating in circulating cooling water, *Environ. Sci. Pollut. Res.* 26 (2019) 9027–9040.

55. A. Mirmohseni, M. Azizi, M.S. Seyed Dorraji, Facile synthesis of copper/reduced single layer graphene oxide as a multifunctional nanohybrid for simultaneous enhancement of antibacterial and antistatic properties of waterborne polyurethane coating, *Prog. Org. Coatings.* 131 (2019) 322–332.
56. T. Bai, L. Lv, W. Du, W. Fang, Y. Wang, Improving the tribological and anticorrosion performance of waterborne polyurethane coating by the synergistic effect between modified graphene oxide and polytetrafluoroethylene, *Nanomaterials.* 10 (2020) 137.
57. X. Wang, X. Li, L. Long Xu, Q. Zhang, Y. Gu, Preparation and corrosion resistance of γ-aminopropyltriethoxysilane-TiO_2-GO/Waterborne Polyurethane Coating, *Int. J. Electrochem. Sci.* 15 (2020) 11340–11355.
58. L. Liu, X. Guo, L. Shi, L. Chen, F. Zhang, A. Li, SiO2-GO nanofillers enhance the corrosion resistance of waterborne polyurethane acrylic coatings, *Adv. Compos. Lett.* 29 (2020) 1–9.
59. T. Bai, Z. Liu, Z. Pei, W. Fang, Y. Ma, Tribological performance studies of waterborne polyurethane coatings with aligned modified graphene oxide@Fe_3O_4, *ACS Omega.* 6 (2021) 9243–9253.
60. S. Zhang, X. Jiang, Synthesis and characterization of non-ionic and anionic two-component aromatic waterborne polyurethane, *Pigment Resin Technol.* 47 (2018) 290–299.
61. Z. Wang, Z. Jiang, M. Zhao, Study of the effect of PGDA solvent on film formation and curing process of two-component waterborne polyurethane coatings by FTIR tracking, *Coatings.* 10 (2020) 461.
62. W. Zeng, H. Huang, L. Song, X. Jiang, X. Zhang, A novel hydroxyl polyacrylate latex modified by OvPOSS and its application in two-component waterborne polyurethane coatings, *J. Coatings Technol. Res.* 17 (2019) 181–191.
63. Q. Xu, Q. Lu, S. Zhu, R. Pang, W. Shan, Effect of resins on the salt spray resistance and wet adhesion of two component waterborne polyurethane coating, *E-Polymers.* 19 (2019) 444–452. 10.1515/epoly-2019-0046.

16 Waterborne Polyurethanes for Weather Protective Coatings

Sonalee Das
CIPET, SARP-APDDRL, Hi Tech Defence and Aerospace Park (IT Sector), Jala Hobli, Bande Kodigehalli Bengaluru, Singahalli, Karnataka, India

CONTENTS

16.1 Introduction....268
16.2 Basic Chemical Components and Synthesis of WPU Coatings....269
16.3 Classification of WPU Coatings....270
16.4 Characteristic Features of WPU....272
16.4.1 Controlled Viscosity....272
16.4.2 Controlled Thermal Stability....272
16.4.3 Auto-Oxidative Curing....273
16.4.4 Biodegradability....273
16.4.5 Low VOC or No VOC Coatings With Reduced Flammability....273
16.5 Types of WPU Dispersion....274
16.5.1 One-Component WPUDs....274
16.5.2 Two-Component WPUDs....274
16.5.3 Waterborne-Hybrid Polyurethane-Acrylic Systems....274
16.6 Application of Waterborne Polyurethane Coatings....275
16.6.1 Automotive Topcoats....275
16.6.2 Wood Coatings....275
16.6.3 Floor Coatings....275
16.6.4 Adhesives....275
16.7 Recent Trends in the Development of WPU for Weather Protective Coatings....276
16.8 Future Perspectives and Concluding Remarks....284
References....284

DOI: 10.1201/9781003173526-16

16.1 INTRODUCTION

Polyurethane (PU) is a versatile synthetic copolymer with urethane as a functional group; it has extensive applications in the fields of polymers, elastomers, fibers, roofing material, adhesives, and coatings owing to its functional properties, which include flexibility, adhesion strength, chemical resistance, durability, weather resistance, and film-forming ability [1–4]. Among the various application areas, PU coatings are the most sorted area owing to their functional properties. Traditional PU coatings are derived from toxic organic solvents, which are costly and possess health and environmental risks [5]. Hence, these organic solvents need to be replaced with nontoxic ones, such as water, which is eco-friendly and has low cost [6]. Waterborne PU (WPU) coatings have become an epicenter and a rapidly growing segment of the PU coating industry due to stringent environmental regulations, such as the Clean Air Act, and demand for low volatile organic compound (VOC) emissions, thereby imparting greener eco-friendly solutions [7]. Water used as a solvent helps to retain the original properties as well as impart additional properties i.e. easy applicability, robustness, stain resistance, flexibility, corrosion resistance, good control over viscosity, superior adhesion, faster curing, uniform texture, and nontoxicity [6–8]. Further, it reduces the dependency on toxic organic solvent-based traditional PU coatings [7].

As per the global market review, the revenue of waterborne coatings is expected to grow at a CAGR of over 5% during 2020–2026 owing to their environmental friendliness, sustainability, and easy-to-apply solutions [8]. In addition, as per the global market review, the wood-based waterborne coatings market is expected to reach over USD 4 billion by 2026 [8]. Similarly, the architectural coatings segment accounts for over 55% of the global waterborne coatings market [8]. The drastic demand for WPU coatings has accelerated due to the rapid growth in the agricultural, automotive, and industrial sectors [7,8]. Countries such as China, India, Malaysia, Taiwan, South Korea, Indonesia, and Vietnam have shown rising demand toward the development of WPU coatings for various industries, such as marine, construction, automotive, etc., to address sustainability and global warming issues [9]. The various key players in the WPU coating market include Covestro (Germany), DSM (Netherlands), BASF (Germany), R STAHL (Germany), Chemtura (the US), Dow Inc. (the US), H.B. Fuller (the US), Wanhua Chemical Corporation (China), SNP Inc. (the US), and KAMSONS Chemical Pvt. Ltd. (India) [7].

There are two production processes for WPU coatings: batch fed and continuous production [10]. The batch-fed production process has certain limitations, which include a small production scale and labor-intensiveness, thereby affecting the quality of coatings, with dispersions bearing short shelflife [10]. On the other hand, the continuous production of WPU coatings provides a highly automated method suitable for large-scale production with improved performance properties in the final product [10].

In the following sections of the chapter, we will discuss the characteristics of WPU coatings, the latest trends and development, and their applications.

16.2 BASIC CHEMICAL COMPONENTS AND SYNTHESIS OF WPU COATINGS

The basic chemical components for the formulation of WPU are diisocyanates; polyols; curatives; and specialty additives, such as amines, catalysts, and additives, as shown in the table below [11].

Polyols	Isocyanates	Specialty intermediates	Curatives
• PPG based polyols	• HDI	• Polyaspartic acid	• DETA
• Polyether polyols	• HDI trimer	• Dimethylol Propionic acid (DMPA)	• Ethylene diamine
• PTMEG based polyols	• IPDI		
	• IPDI trimer		
• Polyester polyols	• $H_{12}MDI$		
• Acrylic polyols	• MDI		

WPU molecules with block structures consist of repeating units of hard and soft segments. The hard segment comprises polar groups, i.e. carbamates or ureas arises from the reaction of isocyanate with polyol and chain extender [10]. The polar groups exhibit strong intermolecular forces, resulting in close packing of hard segments, forming a rod-like structure [10]. The soft segment consists of polyether, polyester polyols, or low-molecular-weight polyol segments crosslinked with hard segments [12]. Polyether-type WPUs with flexible ether bonds have good hydrolysis resistance, moisture permeability, and flexibility [10]. On the other hand, polyester-based WPUs with strong polar ester bonds have poor hydrolysis resistance with good mechanical performance, weather resistance, antifungal properties, and low-temperature elasticity [10]. Diols are mostly aliphatic and aromatic diols with low molecular weights, while the diamines are primarily aromatic amines [10]. Catalysts are often tertiary amines/organo-stannic compounds [10].

The poly-addition reaction of excess isocyanate with polyol results in the formation of PU pre-polymer with a low molecular weight [10]. Thereafter, the pre-polymer reacts with amine chain extenders, diols to produce a polymeric chain. Figure 16.1 depicts the flow chart of the reaction for the synthesis of WPU dispersion (WPUD).

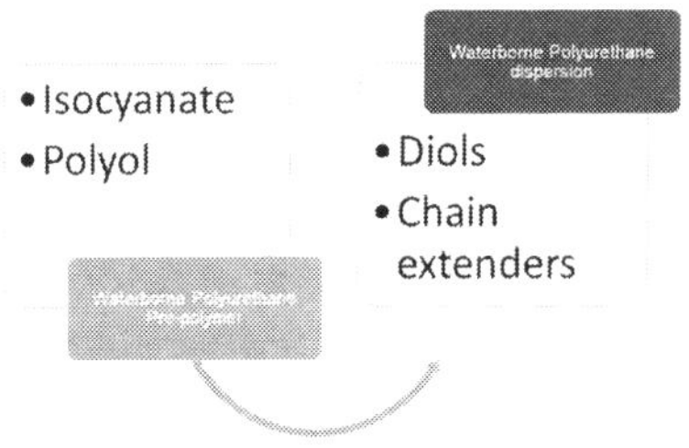

FIGURE 16.1 Flow chart of the reaction for the synthesis of WPU dispersion.

16.3 CLASSIFICATION OF WPU COATINGS

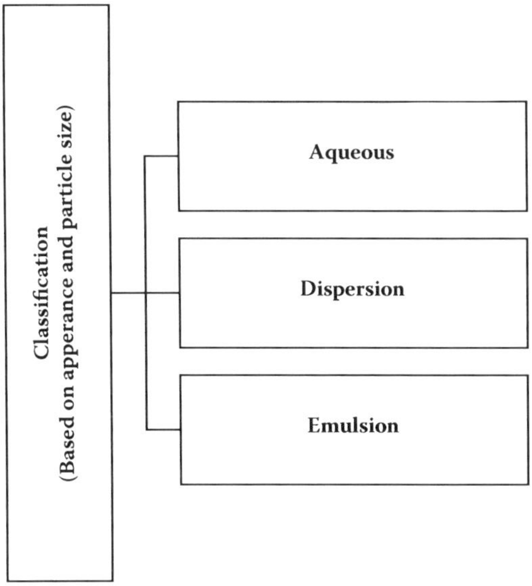

- Aqueous solution is colorless and transparent, with a particle size less than 1 nm [10].
- Dispersion solution is slightly turbid, white, and translucent, with a particle size ranging from 1 to 100 nm [10].
- Emulsion solution is white and turbid, with a particle size greater than 100 nm [10].

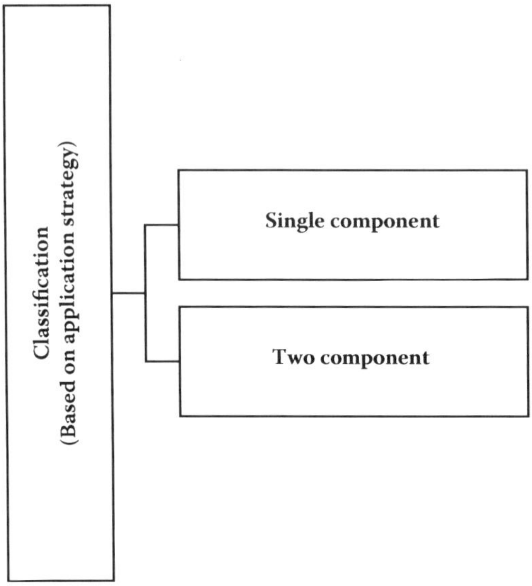

- Single components are used without crosslinkers to achieve desired properties [10].
- Two components generally require crosslinkers [10].

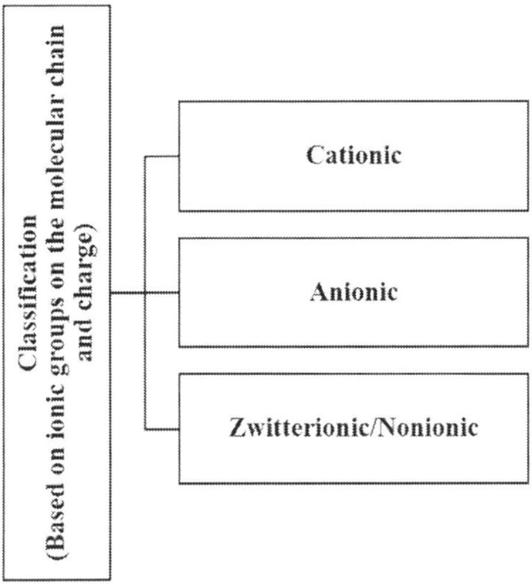

- Cationic WPUs constitute isocyanate pre-polymers with building blocks comprised of tertiary amines, quaternized with a protonic acid/alkylating agent to form water-soluble polymers [11].
- Anionic WPUs are synthesized from polyols containing carboxylic acid and sulfonic acid groups [11].
- Water-emulsifiable PU of nonionic type are formed by incorporating hydrophilic, nonionic building blocks [11].

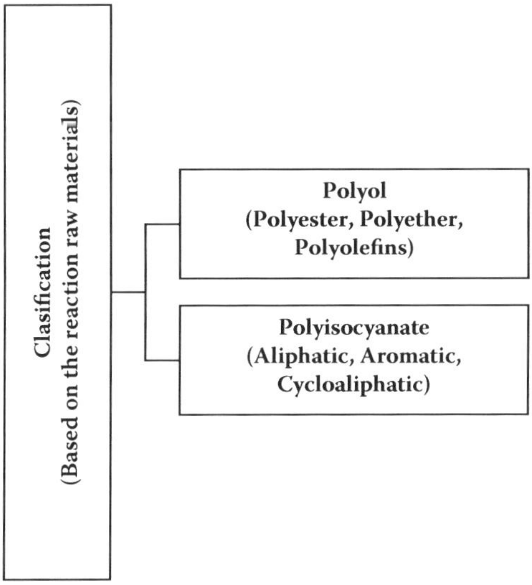

16.4 CHARACTERISTIC FEATURES OF WPU

16.4.1 Controlled Viscosity

The viscosity of WPU is quite low and can be manipulated by adding water and water-soluble thickening agents. Further, the other factors that affect the viscosity of WPU include external polymer thickening agents, core–shell structure, NCO/OH molar ratio, and emulsion particle size [13]. Apart from the above-mentioned factors, the solid content, molecular weight of PU resin, crosslinker, have a negligible effect on the viscosity of WPUs; however, they tend to increase the molecular weight and branching of PU, thereby improving its cohesive strength [14,15]. With the increase in NCO/OH molar ratio, the polar bond content increases, resulting in the aggregation of dispersed particle size and an increase in viscosity [10]. The higher the free radical density in the main chain and side chains of WPU, the greater the viscosity of the resulting emulsion [10].

16.4.2 Controlled Thermal Stability

WPUs are composed of alternating hard and soft segments with significantly different chemical properties. Both the hard and soft segments exhibit thermodynamic incompatibility due to different glass transition temperatures (T_g) [10]. The hard segments exhibit T_g above room temperature and impart rigidity, whereas the soft segments exhibit T_g below room temperature and impart flexibility to the system [10].

The thermal decomposition of WPU initiates with the hard segment, wherein its chemical composition, content, length, and structure dictate the thermal stability of WPUs [10]. The hard segments usually exhibit an ordered/semi-crystalline structure in the initial stage of thermal decomposition, thereby enhancing the thermal stability of WPUs [10].

On the other hand, the soft segments, i.e. polyether/polyester polyol, address the second stage of thermal degradation of WPUs. With an increase in soft segment content and with fewer side chains, a more regular structure is produced, thereby improving the thermal stability of the resulting WPUs [16]. Further, the length and molecular weight of soft segments also dictate the thermal stability of WPUs. The greater the length and higher the molecular weight of the soft segments, the greater the thermal stability of the resulting WPUs [10]. Polyester polyol-based WPUs exhibit greater cohesive energy and can easily form hydrogen bonds (H-bonds) with the amino group compared to polyether polyol [17]. Hence, the heat resistance of polyester polyol-based WPUs is better compared to the polyether ones.

On the other hand, an increase in NCO/OH molar ratio increases the concentration of hard segments [10]. As a consequence, H-bonding increases, increasing the crosslinking density [10]. The increase in crosslinking density of the polymer results in an increase in T_g and an overall improvement in the thermal stability of WPUs [18].

16.4.3 Auto-Oxidative Curing

WPUs that do not contain –NCO groups in the molecular chain are mostly cured by cohesive and adhesive forces, i.e. hydrogen bonds and van der Waals forces generated by polar groups, i.e. carbamates/urea within the molecule [10,19]. Further, WPU dispersions can be crosslinked via the auto oxidative process, especially in the absence of accelerated energy cure, i.e. external heat, with better hardness and good resistance properties compared to solvent-borne/high solid materials, making it an idle choice for floor coatings [20].

16.4.4 Biodegradability

WPUs are composed of alternating hard and soft segment chains, as discussed in the earlier sections. The hard segment and soft segment – owing to thermal incompatibility and different chemical groups – tend to phase separate, and as a consequence, the hard segments form a micro-domain structure within the soft segment, leading to micro-phase separation and materials with a broad range of properties [10]. Based on the mixture composition and reactant's nature, the soft and/or hard segments can be arranged in either an amorphous disordered or crystalline ordered domain, as depicted in Figure 16.2 [10,21,22]. This micro-phase separation structure tends to be quite similar to that observed in the case of bio-films, which tend to impart bio-degradable behavior to the WPUs [10]. The chemical structure of the hard and soft segments and NCO/OH molar ratio may dictate the biodegradability, thermal stability, and functional properties of the WPUs [10,23].

16.4.5 Low VOC or No VOC Coatings With Reduced Flammability

WPUs exhibit low VOCs in the final coating with good surface hardness and high-performance characteristics [20]. Acrylic emulsions can also impart low VOC

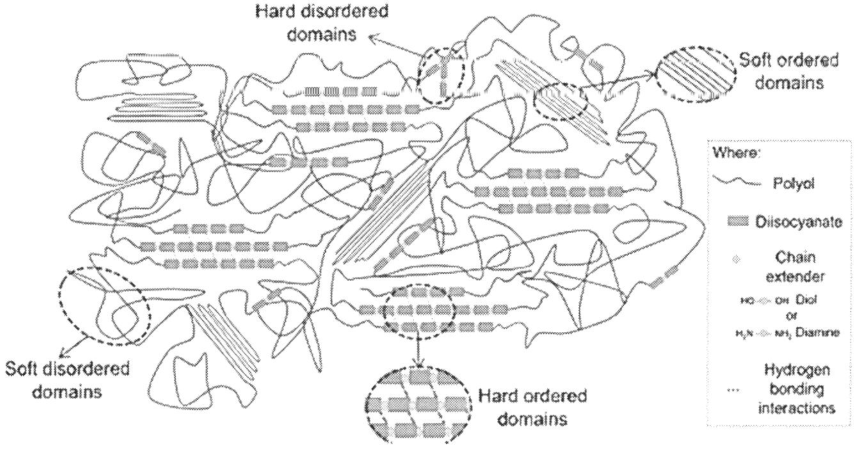

FIGURE 16.2 Schematic representation of interactions and the resulting ordered and disordered micro-domains. Adapted with permission from reference [24], [2021], Polymers.

emission coatings at low or room temperature; however, they tend to have lower performance targeted applications [20]. In the case of solvent resistance, urethanes have properties that acrylics don't have, and vice versa. Hence, blending WPUs with acrylic emulsions can provide a blend of optimum cost and performance characteristics for a particular application. High solids or 2 K WPUDs can impart low VOCs, high hardness, and high-performance characteristics at ambient temperature, but it takes longer for curing time to achieve appreciable properties [20]. WPUDs also exhibit reduced flammability properties, which makes them an ideal choice for wood flooring [20].

16.5 TYPES OF WPU DISPERSION

WPUDs involve the dispersion of PU particles in water. The various techniques involved in achieving complete dispersion include the incorporation of a carboxylic acid moiety, such as carboxylated diol (i.e. dimethylolpropionic acid, DMPA), which serves as an internal emulsifier within the PU backbone [25]. These emulsifiers are then co-reacted with polyol and isocyanate and thereafter neutralized with a base/tertiary amines/ammonium hydroxide, to promote the dispersion of the PU particles in water [25]. The various types of WPUDs are discussed below.

16.5.1 One-Component WPUDs

This involves the reaction of isocyanate pre-polymer with a protecting group. In this type of WPUD, there is a choice of polyol and isocyanate; NCO/OH molar ratio controls the coating properties [25]. The PUs can be dispersed in water in the presence of DMPA, neutralizing agents, such as TEA or ammoniam, to form water-dispersible PUs [25]. Thereafter, the dispersed WPUs can be coated and air-dried to release ammonia or TEA, forming a dry film coating [25].

16.5.2 Two-Component WPUDs

This involves the curing of isocyanate-terminated pre-polymers with polyamines, i.e. ethylene diamine and diethylenetriamine with faster curing as compared to isocyanate termination with water [25]. Hydroxy-terminated PU dispersions are used in this formulation involving chain extension and crosslinking with isocyanates, such as 1,6-hexane diisocyanate trimers [25]. The dispersion of PUs is achieved by incorporation of carboxylated diol like DMPA, followed by neutralization with ammonia hydroxide/tertiary amines/metal bases [25]. Urethane catalysts can also be used to accelerate the reaction between polyol and isocyanate. Hydroxy-functional PU dispersions are formulated using blocked isocyanate, which thermally de-blocks the isocyanate after application and baking, to form crosslinked PU dispersions [25].

16.5.3 Waterborne-Hybrid Polyurethane-Acrylic Systems

Waterborne-hybrid polyurethane-acrylic systems (WPUAs) are formulated by combining PUs with hydroxyl functional polyacrylates, known for their good weatherability, chemical resistance, and good appearance characteristics [25]. High-quality coatings can be produced by crosslinking hydroxy-functional acrylates with blocked isocyanates or aliphatic isocyanates [25]. Also, end-cap chemistry can be used to encapsulate isocyanate pre-polymer/DMPA/polyol with hydroxyethyl methacrylate (HEMS), followed by water dispersion using triethylamine (TEA) and subsequent copolymerization with methylmethacrylate/butyl acrylate/other acrylic monomers [25].

16.6 APPLICATION OF WATERBORNE POLYURETHANE COATINGS

Owing to their wide performance characteristics, WPUDs find enormous application in the area of automotive coatings, industrial coatings, plastic coating, weather protective coatings, and corrosion protective coating.

16.6.1 Automotive Topcoats

2 K WPUD is in huge demand for automotive clear coat applications owing to its ease of application, fast drying, high gloss, appreciable performance characteristics, good adhesion, chemical and mechanical resistance, low VOCs, and good weatherability properties [11]. Aliphatic polyisocyanate-based waterborne two-component PUs exhibit better weatherability features compared to aromatic ones.

Although WPUD technologies are gaining immense interest as primer and basecoat in the automotive industry, they find limited use due to water evaporation, water sensitivity of hardeners, and surface defects, i.e. pinholes/microfoams [26].

16.6.2 Wood Coatings

Solvent-borne waterborne systems find extensive application in the field of wood coatings owing to short dry times, good sandability, good build, good pore-filling capacity, excellent lifting resistance, good aesthetic results, improved mechanical strength, chemical and weathering resistance, high durability, and extremely short working times with low VOC emission [11].

16.6.3 Floor Coatings

WPU is applied as a top coat over a functional layer for abrasion resistance, chemical resistance, and weather resistance. [27]. 2 K WPU with low solid content can act as an excellent primer coat as it provides superior substrate penetration [28]. 3 K WPU imparts excellent chemical and abrasion resistance [28]. Recently, various binders and additives have been formulated to develop durable and high-quality floor coatings with finishing similar to oil-based coatings.

16.6.4 Adhesives

WPUDs have been used for the formulation of heat resistant, fast-drying, good bonding strength, eco-friendly adhesive with low VOC content [29]. WPUD

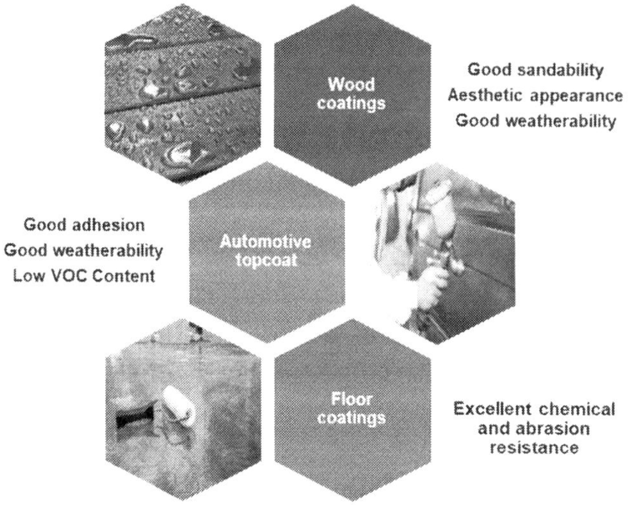

FIGURE 16.3 Application of waterborne coatings.

adhesives find wide application in the field of footwear products, paper products, construction, plastic processing, and automobile decoration [29].

16.7 RECENT TRENDS IN THE DEVELOPMENT OF WPU FOR WEATHER PROTECTIVE COATINGS

Bramhecha et al. [30] developed antibacterial and waterproof breathable WPU functionalized by graphene for near-infrared (NIR) shielding and UV protective cotton fabric. Different characterization studies, such as Fourier transform infrared spectroscopy (FTIR), mechanical, thermal, morphological, physical, and functional properties of the coated fabric, have been studied. The inclusion of graphene within the WPU has led to the increase in water vapor transmission rate (WVTR) from 861.51 g/24 h m^2 to 962.60 g/24 h m^2. This is due to the formation of the tortuous pathway for the flow of water vapor through the polymer matrix boundaries. The water contact angle of the developed waterproof fabric was found to be in the acute range due to the presence of PEG 200 and the carboxylic acid group in the WPU backbone. This imparts breathable nature to the fabric developed and selectively hydrophilic continuous film coating with possible interface geometry, depicted in Figure 16.4 below.

The doping of WPU with graphene resulted in a significant enhancement in the UV protection factor (UPF) values of the cotton-coated fabric. WPU cotton-coated fabric with 0.025 wt.% loading of graphene exhibited an UPF value of 135.08, which increased to 219.23 with 5 wt.% loading of graphene. The results obtained prove the UV protection efficiency of graphene due to the presence of free delocalized π orbital electrons or aromatic conjugation in graphene plates. The coated fabric exhibited antibacterial properties above 99% against *Escherichia coli* and *Staphylococcus aureus* bacterial colonies, which might be

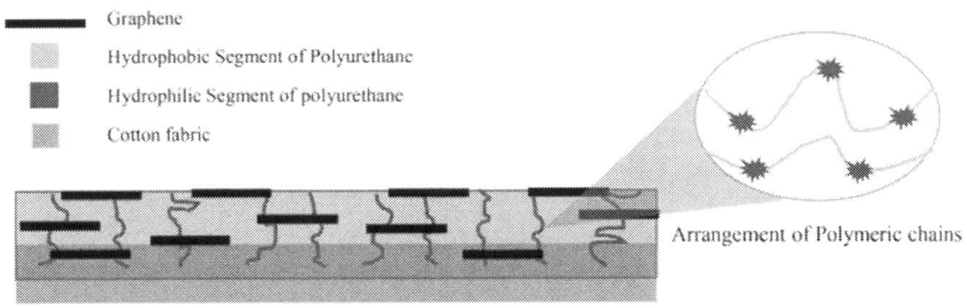

FIGURE 16.4 Interface geometry of the coated fabric time. Adapted with permission from reference [30], [2021], Carbon Trends.

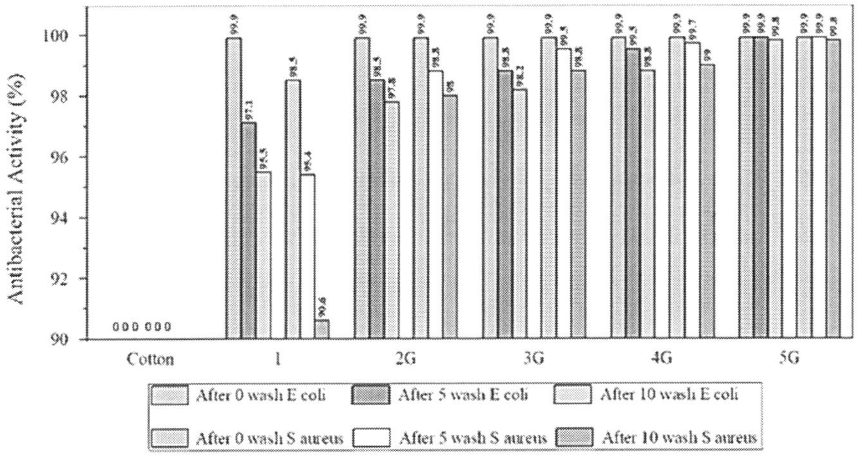

FIGURE 16.5 Coated samples' antibacterial activity. Adapted with permission from reference [30], [2021], Carbon Trends.

due to the antibacterial property and strong adherence of graphene within the PU matrix, as shown in Figure 16.5.

The coated fabric composites also exhibited NIR resistance of up to 90% of the incident, with an increase in graphene concentration, indicating the ability of the coated samples to block the NIR radiations. For the wavelength range from 2,100 to 2,400 nm, the transmittance was reduced by 10% with an increase in graphene concentration. This was because graphene can dissipate the incident energy due to the presence of a large number of sp^2-hybridized free electrons. Further, the uniform distribution of graphene throughout the polymer matrix owes to strong H-bonding in between the carboxylic and hydroxyl group of WPU and edge-hydroxyl groups of graphene. Thus, from the results obtained, the authors concluded that the graphene-functionalized PU formulation can be used as multifunctional coatings for protective textiles.

Wang et al. [31] studied the preparation and corrosion resistance of AKT–WPU coating. The authors demonstrated an inorganic–organic composite modification of nano-TiO_2 by using aluminum sulfate octahydrate and γ-aminopropyltriethoxysilane (KH550) to obtain AKT–WPU coating. The authors investigated the water resistance, heat resistance, and corrosion resistance properties of the developed AKT–WPU composite coatings. As compared to neat WPU coating, the water absorption of the AKT–WPU composite coating reduces significantly owing to the presence of AKT nanoparticles, which can fill the gap of the aqueous PU emulsion, thereby hindering the entry of corrosive ions. With the increase with the addition of AKT, the water absorption of the coating initially increases and then decreases, since a small amount of AKT nanoparticles can fill the void incompletely and cause corrosive ions to migrate through the PU matrix. Further, the agglomeration of excess AKT in the aqueous PU emulsion forms larger voids and accelerates the corrosion of corrosive ions to the substrate. Electrochemical impedance spectroscopy and polarization curves have been used to investigate the corrosion resistance properties of the AKT–WPU composite coatings. The authors observed that the addition of AKT improves the corrosion resistance significantly compared to the unreinforced ones. This was because of the inclusion of AKT nanoparticles, which can reduce the pores and gaps inside the PU coating, thereby improving the coating's compactness and hence overall corrosion resistance. This observation holds good for 0.7 wt.% addition of AKT; beyond it, the corrosion resistance decreases due to the agglomeration of particles that create a path for the penetration of electrolyte solution into the coating, causing corrosion. The authors observed that the resistive modulus of the composite coating with 0.7 wt.% AKT was $5.05 \times 10^6\ \Omega$. Also, the corrosion current density was 5.21×10^{-9}, about two-fold orders less in magnitude compared to neat WPU coatings. Further, the AKT-reinforced WPU coatings exhibited corrosion inhibition efficiency of 99.54%. Further, scanning electron microscopy studies have been carried out to study the morphological features of coating subjected to a long-term immersion test. It was observed that the AKT-reinforced WPU coating remains intact with no surface change, even after 360 h of soaking.

Liu et al. [32] investigated the preparation and performance optimization of two-component WPU hydroxy acrylic resins with hydroxy acrylate aqueous dispersion for application in the locomotive coating. The authors studied the mechanical properties and storage stability of the locomotive coatings by subjecting them to QUV exposure for 1,000 hrs. Scanning electron microscopy analysis was also carried out to investigate the morphological features of the exposed and unexposed coating surface. The authors observed that the coatings exhibited excellent hardness and good chemical resistance. Moreover, the weather resistance of the locomotive topcoat indicated high gloss and high weather resistance by adding 0.3 wt.% of ultraviolet light absorber, i.e. UV2 200 F.

Ozgenc et al. [33] evaluated the weathering resistance of waterborne acrylic and alkyd-WPU based coatings with HALS, UV absorber, and bark extracts on wood surfaces. The bark extracts were used as functional additives and acted as natural photo stabilizers to provide protection and inhibit UV degradation in waterborne acrylic- and alkyd-based PU coatings. A comparative analysis was carried out in between the coating systems containing bark extracts with control coating groups containing ultraviolet light absorbers (UVA) of the hydroxyphenyl-*s*-triazine (HPT) for acrylic and nonbasic amino ether (NOR) hindered amine light stabilizers

(HALS) for alkyd. ATR–FTIR spectroscopy was used to investigate the chemical structure of acrylic and alkyd WPU coating systems. Color changes and surface roughness parameters were evaluated to access the weathering resistance of coating systems containing extracts. Alkyd-based WPU coatings reinforced with bark extract exhibited the lowest color change (ΔE). Further, the alkyd WPU coating with the UV absorber also had a smoother surface as compared to those reinforced with bark extracts. However, color stabilization of the acrylic coatings with UV absorber was found to be much better than the acrylic coatings with extracts. The authors concluded that the waterborne wood coating systems reinforced with tree bark extracts have excellent potential to substitute the commercial UV absorbers as a natural product.

Kong et al. [34] studied the influence of nanocellulose-reinforced polyurethane for waterborne wood coating. The authors aimed to use PU in order to lower the VOC content and address the sustainability issues related to global warming with low-temperature flexibility, excellent acid and alkali resistance, excellent solvent resistance, and superior weather resistance. Nanocellulose was derived from biomass resources by the oxidation of 2,2,6,6-tetramethylpiperidine-1-oxyl (TEMPO). Thereafter, the obtained nanocellulose (different concentrations ranging from 0.1 to 0.4 wt.%) was used to reinforce the WPU via two methods: chemical addition and physical blending. Different characterization studies, such as tensile strength, elongation at break, hardness, and abrasion resistance, have been carried out to investigate the influence of chemical/physical blending of nanocellulose with WPU on the properties of the comprehensive coating. The authors observed that reinforced WPU coatings exhibited improvement in mechanical, abrasion resistance, and other coating properties as compared to the unreinforced ones due to the reinforcing effect of well-dispersed nanocellulose that could further promote the compatibility of WPU with nanocellulose. Further, the authors also observed that maximum improvement of coating properties was observed for the chemical grafting method compared to physical blending due to improvement in the compatibility of nanocellulose with PU film. This was due to the fact that in the case of physical blending, WPU and nanocellulose could form a hydrogen bond between the carbamate group and the hydroxyl/carboxyl group. On the other hand, in the case of chemical grafting, WPU and nanocellulose could form a chemical bond from the carbamate group via the chemical reaction of the isocyanate group and the hydroxyl/carboxyl group. In the case of chemical grafting, the formation of chemical bonds could improve the interfacial compatibility and bonding strength between the WPU and nanocellulose, compared to physical blending. The tensile strength (TS) and tensile elongation at break (TE) of the chemical grafted nanocellulose/WPU film improved to 58.7%, and 55%, respectively, compared to neat ones. Similarly, the glossiness of chemical grafted nanocellulose/WPU was higher compared to the physical blended ones due to better compatibility between WPU and nanocellulose via the chemical addition method. The hardness result also indicated an improvement for reinforced WPU compared to the neat ones due to the reinforcing effect arising from uniform dispersion of the nanocellulose within the WPU. Both the physical and chemical grafted nanocellulose/WPU indicated 6.98% and 14.05% hardness improvement compared to the control ones, as shown below.

It is imperative from the results obtained that chemical grafted nanocellulose/WPU exhibited better results compared to physical ones due to better compatibility issues. A similar observation was also observed for the abrasion test results, indicating an improvement of 7.4% and 3.45% mass loss for physical and chemical grafted nanocellulose/WPU (Figure 16.6).

Wu et al. [35] investigated the enhanced corrosion resistance of WPUs containing sulfonated graphene/zinc phosphate (SG-ZP) composites. The precipitation method has been used to prepare the SG-ZP composite, which is used as a functional filler for improving the anti-corrosive property of WPU coating. The inclusion of sulfonated graphene as a carbon matrix improved the dispersion of ZP, thereby preventing the agglomeration effectively. Further, the introduction of SG within ZP improved the corrosion resistance of the coatings. The synthesized spherical ZP has a larger specific surface area in the nano-meter range as observed through SEM studies, as shown below (Figure 16.7).

Electrochemical impedance spectroscopy (EIS) and potentiodynamic polarization studies (PPS) have been conducted to investigate the corrosion resistance properties of the developed WPU-reinforced coatings. The authors observed that the WPU composites had better corrosion resistance properties and exhibited the best anticorrosion property with the lowest corrosion current density (I_{corr}) of 0.4252 $\mu A/cm^2$ and maximum coating resistance (I_{CR}) value of $10.937 \times 103\ \Omega\ cm^2$ at a loading of 0.5 wt.% SG–ZP nanofiller (Figure 16.8).

Wu et al. [36] investigated the development of hydrophobic, transparent WPU-amino polydimethylsiloxane (PU-DN 100) composites prepared from an aqueous sol–gel process for corrosion protective coatings. The authors observed that FTIR and crosslink density result in an increase in DN100 content; the hydrogen bonding decreased, but chemical crosslinking increased. The higher the content of DN 100, the greater the interference for the formation of ordered hydrogen bonding. Lower content of DN 100 favored the formation of strong hydrogen bonding. The extent and strength of hydrogen bonding dictate the formation of strong chemical crosslinking and interaction between the chemical groups. The presence of PDMS creates hydrophobic and lowers the roughness due to its low surface energy, favoring a higher contact angle with a value of 93.9°. Hence, when the composite coating surface is exposed to salt spray test, it exhibits good corrosion resistance and excellent weatherability properties.

Rahman et al. [37] studied the influence of functionalized multiwalled carbon nanotubes (MWCNT) on the weather degradation and corrosion resistance of WPU coatings using three different concentrations of 0.5, 1.0, and 2.0 wt.% f-MWCNT. Potentiodynamic polarization (PDP) and X-ray photoelectron spectroscopy (XPS) analyses have been used for characterizing both exposed and unexposed surfaces of the coatings. XPS studies were carried out to investigate the degree of degradation by calculating the carbonyl content and surface morphology. It was observed from XPS analysis that the neat WPU coating exhibited higher carbonyl content, indicating slight degradation and corrosion protection. However, the inclusion of f-MWCNTs resulted in significant improvement of both the degradation and corrosion protection efficiencies of the coatings. The authors observed that 2.0 wt% f-MWCNT mixed with WPU indicated maximal

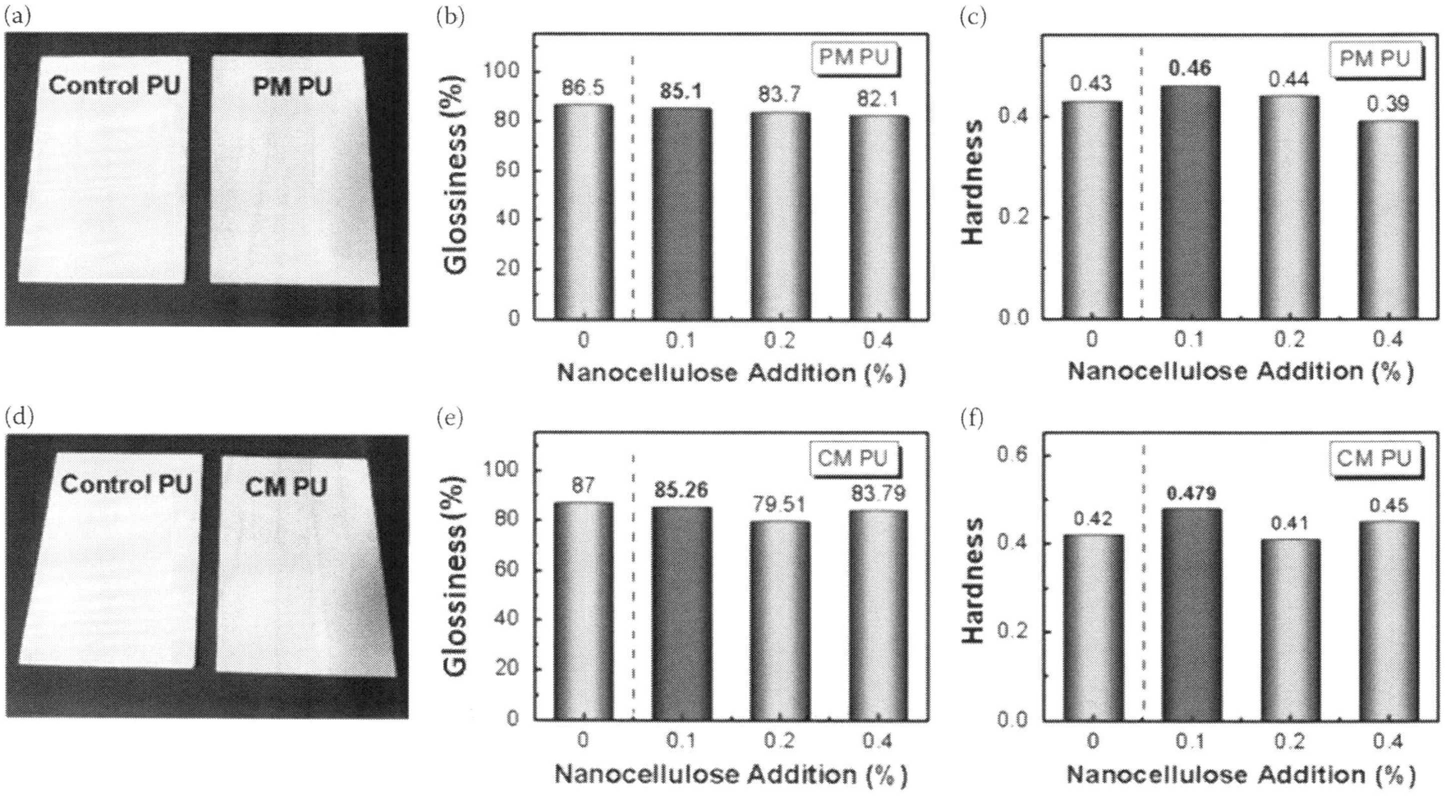

FIGURE 16.6 Comparison of glossiness and hardness of neat and reinforced nanocellulose composite samples. Adapted with permission from reference [34], [2019], Coatings.

FIGURE 16.7 SEM images of (a) ZP and (b) SG–ZP composites. Adapted with permission from reference [35], [2019], Progress in Organic Coatings.

degradation compared to all other coatings. The reason behind exception weathering and corrosion protection for coatings loaded with f-MWCNT was due to the fact that f-MWCNTs acted as a UV absorber, hindering the chain scission of urethane/urea groups, thereby decreasing the coating degradation. PDP studies conducted for the exposed and unexposed coatings also corroborated with the XPS results, indicating complete damage of the neat WPU coatings compared to the reinforced ones. The results obtained further supported the fact that WPU/f-MWCNT coatings had better weathering and corrosion protection compared with the neat WPU coating due to the presence of MWCNTs.

Bhargava et al. [38] developed a WPU elastomer-based high reflectivity coating and investigated the ultraviolet, water, and thermal aging properties for 1,000 h. The coating durability of the samples was tested every 200 h. The authors used ATR-FTIR spectroscopy to monitor the chemical changes occurring during the aging process. Further, UV–vis with integrating sphere was utilized to investigate the change in diffused reflectance. Surface morphological changes were investigated by using an optical microscope and scanning white light interferometry (SWLI). The authors observed a decrease in the intensity of functional groups related to –CONH, –CH, –CO, and C–O–C using FTIR studies, as shown below. Further, the FTIR spectra indicated the appearance –NH group attributed to chain scission of the PU

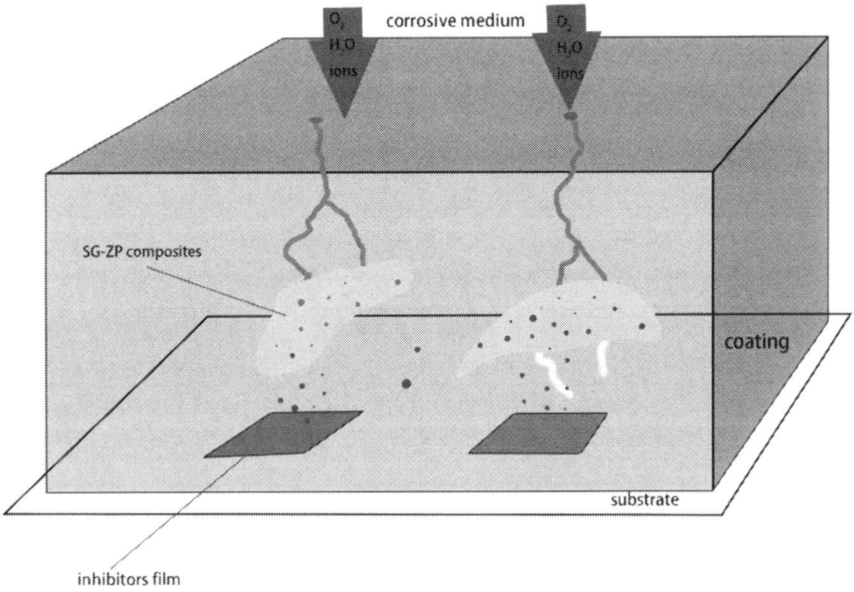

FIGURE 16.8 Corrosion inhibition mechanism of SGZP/ WPU coating. Adapted with permission from reference [35], [2019], Progress in Organic Coatings.

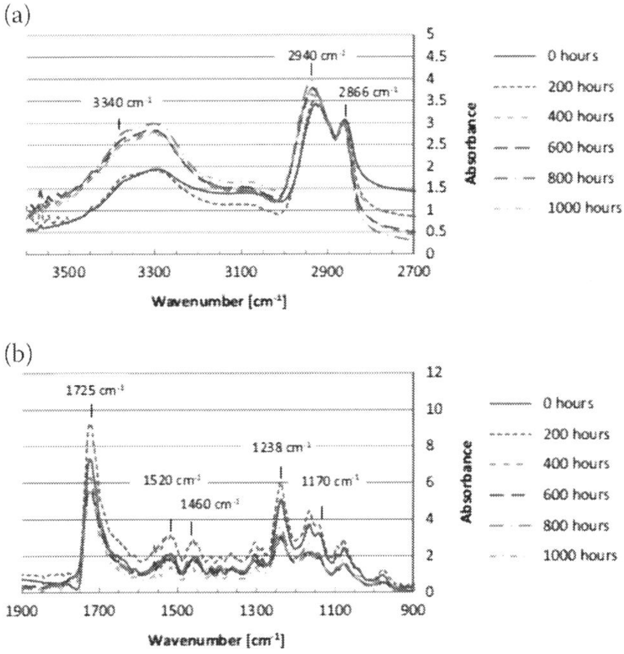

FIGURE 16.9 ATR-FTIR measurements indicating the influence of UV exposure on the relative magnitude of peak areas for the coatings. Adapted with permission from reference [38], [2019], Progress in Organic Coatings.

binder. Water and thermal aging studies indicated a degradation mechanism via penetration of water. Reflectively reduced in all aging processes due to physical defects caused by different aging mechanisms (Figure 16.9).

16.8 FUTURE PERSPECTIVES AND CONCLUDING REMARKS

Waterborne coating systems are eco-friendly and provide a green sustainable alternative to solvent-borne coatings, resulting in a reduction in harmful emissions and reduced volatile content. It is imperative to form the above discussion and findings that WPU coatings have found profound applications in the field of weatherproof, corrosion-resistant coatings, construction, wood, and automotive industries; however, much R&D work remains to be carried out in the future. WPU coatings with switchable properties should be developed, which can be incorporated into multilayer laminates. Further, the action of hybrid nano-fillers and hyperbranched structures and their interaction effects on the properties of WPU coatings needs to be explored to accelerate and widen its application. Further steps need to be developed for effective recovery of feedstock and recycle/re-use for developing new products for industrial applications. Hence, we hope that the present chapter has covered and provided in-depth information to the scientific community regarding the recent findings of WPU coatings, their types and applications, and future perspectives for developing novel sustainable and green materials for the various realm.

REFERENCES

1. B. Ghosh, and M. W. Urban, Self-repairing oxetane-substituted chitosan polyurethane networks, *Science*, 323, 2009, 1458–1460.
2. Z. L. Yang, D. A. Wicks, C. E. Hoyle, H. T. Pu, J. J. Yuan, D. C. Wan, and Y. S. Liu, Newly UV-curable polyurethane coatings prepared by multifunctional thiol- and ene-terminated polyurethane aqueous dispersions mixtures: Preparation and characterization, *Polymer*, 50, 2009, 1717–1722.
3. J. H. Shi, X. Han, and K. L. Yan, A novel bio-functional finishing agent for wool based on waterborne polyurethane mixed with chitosan, *Textile Research Journal*, 84, 2014, 1174–1182.
4. K. A. Chaffin, A. J. Buckalew, J. L. Schley, X. Chen, M. Jolly, J. A. Alkatout, J. P. Miller, D. F. Untereker, M. A. Hillmyer, and F. S. Bates, Influence of water on the structure and properties of PDMS-containing multiblock polyurethanes, *Macromolecules*, 45, 2012, 9110–9120.
5. A. Dalmais, C. A. Serra, Z. Q. Chang, M. Bouquey, and R. Muller, Microfluidic-assisted synthesis of waterborne and solvent-free urea-crosslinked polydimethylsiloxane microparticles, *Macromolecular Materials Engineering*, 299, 2014, 698–706.
6. T. Zhang, W. J. Wu, X. J. Wang, and Y. P. Mu, Effect of average functionality on properties of UV-curable waterborne polyurethane-acrylate, *Progress in Organic Coatings*, 68, 2010, 201–207.
7. https://www.wboc.com/story/43987262/waterborne-polyurethane-market-by-application-end-use-industry-region-global-forecast-to-2025.
8. https://www.globenewswire.com/news-release/2021/05/17/2230841/0/en/The-global-waterborne-coatings-market-by-revenue-is-expected-to-grow-at-a-CAGR-of-over-5-during-2020-2026.html.

9. https://www.marketsandmarkets.com/Market-Reports/waterborne-waterbased-coatings-market-205422792.html.
10. X. Liu, W. Hong, and X. Chen, Continuous production of water-borne polyurethanes: A review, *Polymers*, 12, 2020, 2875–2891.
11. K.-L. Noble, Waterborne polyurethanes, *Progress in Organic Coatings*, 32, 1997, 131–136.
12. A. Kultys, M. Rogulska, and H. Głuchowska, The effect of soft-segment structure on the properties of novel thermoplastic polyurethane elastomers based on an unconventional chain extender, *Polymer International*, 60, 2011, 652–659.
13. H. Li, Y. Liu, N. Sun, Z. Xu, Y. Li, and S. Jiang, Unique viscosity mutation of multi-generation hyperbranched waterborne polyurethane, *Journal of Polymer Engineering*, 34, 2014, 605–609.
14. Y. Lu, and R. C. Larock, Soybean-oil-based waterborne polyurethane dispersions: Effects of polyol functionality and hard segment content on properties, *Biomacromolecules*, 9, 2008, 3332–3340.
15. L. Bao, H. Fan, Y. Chen, J. Yan, T. Yang, and Y. Guo, Effect of surface free energy and wettability on the adhesion property of waterborne polyurethane adhesive, *RSC Advances*, 6, 2016, 99346–99352.
16. L. H. Bao, Y. J. Lan, and S. F. Zhang, Synthesis and properties of waterborne polyurethane dispersions with ions in the soft segments, *Journal of Polymer Research*, 13, 2006, 507–514.
17. D. K. Chattopadhyay, and D. C. Webster, Thermal stability and flame retardancy of polyurethanes, *Progress in Polymer Science*, 34, 2009, 34, 1068–1133.
18. B. J. Vincent, and B. Natarajan, Waterborne polyurethane from polycaprolactone and tetramethylxylene diisocyanate: Synthesis by varying NCO/OH ratio and its characterization as wood coatings, *Open Journal of Organic Polymer Materials*, 14, 2013, 37–42.
19. J. L. Stanford, and R. F. T. Stepto, A study of intramolecular reaction and gelation during non-linear polyurethane formation, *Polymer International*, 9, 2010, 124–132.
20. https://www.lubrizol.com/Coatings/Blog/2020/03/Advantages-of-WB-Polyurethanes
21. L. T. J. Korley, B. D. Pate, E. L. Thomas, and P. T. Hammond, Effect of the degree of soft and hard segment ordering on the morphology and mechanical behaviour of semicrystalline segmented polyurethanes, *Polymer*, 47, 2006, 3073–3082.
22. A. Saralegi, L. Rueda, B. Fernández-D'Arlas, I. Mondragon, A. Eceiza, and M. A. Corcuera, Thermoplastic polyurethanes from renewable resources: Effect of soft segment chemical structure and molecular weight on morphology and final properties, *Polymer International*, 62, 2012,106–115.
23. Z. Yang, and G. Wu, Effects of soft segment characteristics on the properties of biodegradable amphiphilic waterborne polyurethane prepared by a green process, *Journal of Material Science*, 55, 2020, 3139–3156.
24. A. Santamaria-Echart, I. Fernandes, F. Barreiro, M. A. Corcuera, and A. Eceiza, Advances in waterborne polyurethane and polyurethane-urea dispersions and their eco-friendly derivatives: A review, *Polymers*, 13, 2021, 409–439.
25. https://www.gantrade.com/blog/the-chemistry-of-waterborne-polyurethane-coatings.
26. J. V. Koleske, R. Springate, and D. Brezinski, Additives Reference Guide, 2013.
27. https://www.constrofacilitator.com/water-borne-polyurethane-coatings-for-industrial-flooring.
28. F. Zhang, Innovative Polyurethane Coating Solutions for Construction, 2016.
29. M. V. Navarro-Bañón, M. M. Pastor-Blas, and J. M. Martín-Martínez, Water-based chlorination treatment of SBS rubber soles to improve their adhesion to waterborne

polyurethane adhesives in the footwear industry, *Journal of Adhesion Science Technology*, 19, 2005, 947–974.
30. I. Bramhecha, and J. Sheikh, Antibacterial and waterproof breathable waterborne polyurethane functionalised by graphene to develop UV and NIR-protective cotton fabric, *Carbon Trends*, 4, 2021, 100067.
31. X. Wang, L. Hou, X. Ling-Long, L. Xiong, H. Jiang, and Wen-jie Zhou, Preparation and corrosion resistance of AKT-waterborne polyurethane coating, *International Journal of Electrochemical Science*, 15, 2020, 1450–1464.
32. H. Liu, B. Zhijie, W. Zhong, W. Xianming, W. Yong, G. Xiangxin, and Zhongyu Cai, Preparation and performance optimization of two-component waterborne polyurethane locomotive coating, *Coatings*, 10, 2020, 10010004.
33. Ö. Özgenç, S. Durmaz, S. Şahin, and İ. H. Boyaci, Evaluation of the weathering resistance of waterborne acrylic- and alkyd-based coatings containing HALS, UV absorber, and bark extracts on wood surfaces, *Journal of Coatings Technology and Research*, 2, 2020, 461–475.
34. L. Kong, D. Xu, Z. He, F. Wang, S. Gui, J. Fan, X. Pan, X. Dai, X. Dong, B. Liu, and Y. Li, Nanocellulose-reinforced polyurethane for waterborne wood coating, *Molecules*, 24, 2019, 3151.
35. Y. Wu, S. Wen, K. Chen, J. Wang, G. Wang, and K. Sun, Enhanced corrosion resistance of waterborne polyurethane containing sulfonated graphene/zinc phosphate composites, *Progress in Organic Coatings*, 132, 2019, 409–416.
36. Y. Wu, P. Guo, Y. Zhao, X. Liu, and Z. Du, Hydrophobic, transparent waterborne polyurethane-polydimethylsiloxane composites prepared from aqueous sol-gel process and applied in corrosion protection, *Progress in Organic Coatings*, 127, 2019, 231–238.
37. M. R. Mohammad, R. Suleiman, and H. D. Kim, Effect of functionalized multiwalled carbon nanotubes on weather degradation and corrosion of waterborne polyurethane coatings, *Korean Journal of Chemical Engineering*, 34, 2017, 2480–2487.
38. S. Bhargava, M. Kubota, R. D. Lewis, S.G. Advani, A. K. Prasad, and J. M. Deitzel, Ultraviolet, water, and thermal aging studies of a waterborne polyurethane elastomer-based high reflectivity coating, *Progress in Organic Coatings*, 79, 2015, 75–82.

17 Recent Developments in Waterborne Polyurethanes for Corrosion Protection

Felipe M. de Souza,
Muhammad Rizwan Sulaiman, and Ram K. Gupta
Department of Chemistry, Kansas Polymer Research Center,
Pittsburg State University, Pittsburg, KS, USA

CONTENTS

17.1 Introduction 287
17.2 Classification of Corrosion Inhibitors 289
17.3 Recent Developments in Waterborne Polyurethanes as a Corrosion Inhibitor 290
17.4 Conclusion 299
References 300

17.1 INTRODUCTION

Corrosion is a process of degradation of materials that can occur via environmental, chemical, or biological processes under a corrosive environment. The corrosion process begins with the oxidation of a material, which involves electron loss. Therefore, an electron transferring pathway is required to initiate the corrosion. An excellent example of grasping the corrosion understanding would be the corrosion cell. The cell comprises a pair of electrodes, such as an anode and cathode, an electrolyte, and a conductive electron pathway (Figure 17.1). Daniel cells are corrosion cells with a zinc anode and a copper cathode. Zinc metal (anode) has greater oxidative potential than copper (cathode). The difference in oxidation potential results in electron transfer via external wire from anode to cathode. As a result, $Zn^0_{(s)}$ undergoes an oxidation reaction and transforms into $Zn^{2+}_{(aq)}$ after losing two electrons, known as corrosion, whereas $Cu^{2+}_{(aq)}$ in the electrolyte undergoes a reduction reaction and transforms into $Cu^o_{(s)}$, which gets deposited on the cathode [1]. It is a spontaneous process, and the goal of employing it in batteries is to utilize the electrical energy produced through it, but mostly, the corrosion process occurs in

DOI: 10.1201/9781003173526-17

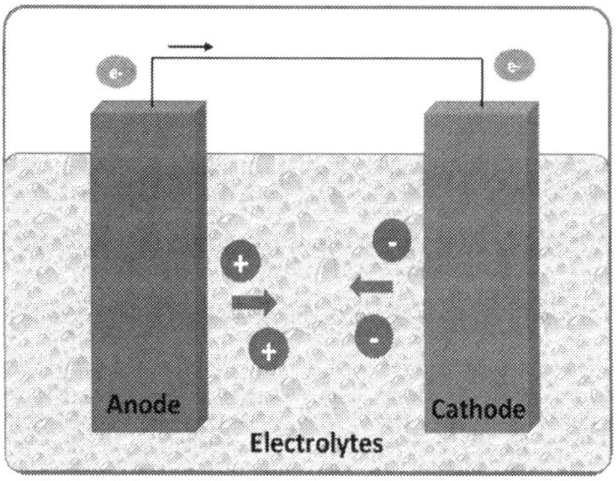

FIGURE 17.1 Schematic of a corrosion cell.

unwanted ways. For instance, metal cans are corroded when exposed to air, resulting in leakage and deterioration of the internal materials; pipelines lose transportation capacity due to corrosion; degradation of industrial equipment results in shutdown; or decay of ship bottoms might result in the boat sinking. The typical corrosive mediums are air, water, brine, gases (e.g. H_2S, SO_2), acids, and bases. In general, metals exposed to any corrosion medium are prone to corrosion without the need for other metals, as in the case of the Daniel cell. A general corrosion process is demonstrated in Figure 17.2. The moisture level in air or dissimilarity in oxygen-level concentrations works like an electrolyte and promotes metal corrosion. Likewise, the pipeline offers a similar situation, where the difference in the external and internal environment promotes corrosion.

Corrosion is a global challenge, and a tremendous amount of work is being done to overcome this problem. However, this challenge can be addressed by employing good strategies to avoid corrosion. Therefore, searching for potential ways of preventing corrosion can be beneficial for the global economy. Corrosion retardant had a whopping market of $7.4 billion in 2019 and will be improving by around 3.8% each year until 2027. The elevation in the corrosion retardant market is due to the employment of renewable methodologies and environmentally friendly products. The creation of cheap and environmentally friendly corrosion-retarding coating is required to avoid losses. The coating should have the strength to withstand a harsh environment and possess essential characteristics, such as high mechanical and thermal durability. The starting material for coatings is important and relies on its application and on the material it is being applied to.

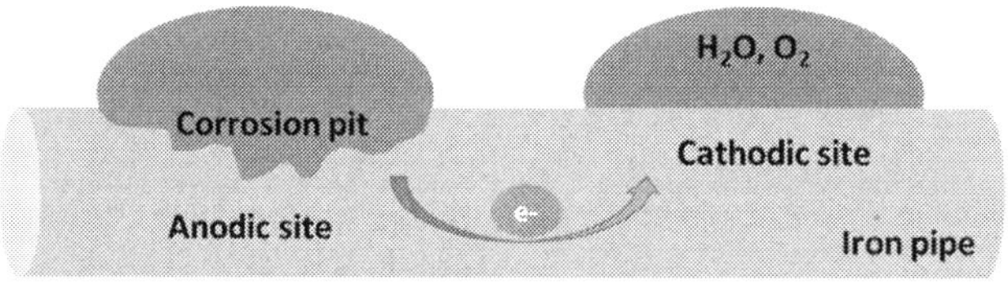

FIGURE 17.2 General schemes for corrosion.

17.2 CLASSIFICATION OF CORROSION INHIBITORS

Corrosion inhibitors can be categorized into two main groups: organic and inorganic inhibitors. The inorganic inhibitors are further classified into anodic and cathodic inhibitors. Besides these inhibitors, some inhibitors change the surroundings to restrict corrosion. These inhibitors are called environmental conditioners, which can also be categorized into scavengers and biocide inhibitors [1]. The anodic inhibitors are metal oxides covering the anode metal produced by chemical reaction or undergo natural oxidation with air. The oxide layer formed naturally is known as the native oxide layer, which prevents corrosion due to its inactive character [2]. Various metals, such as Al, Zn, Mg, Cu, Cd, Ti, Ag, Si (semimetal), and Sn, can undergo passivation with their corresponding oxides. The oxide covering is typically in the range of a few nanometers for efficient protection. The oxide covering thickness plays a vital role in productive corrosion protection. If the thickness is more than required, the coating can get cracked and may get inefficient potection [2]. However, the employment of surface passivating materials has also been done for various implementations. These include the passivation of titanium-based materials with zinc alloys utilized in semiconductors, aluminum oxides with silicon-containing covering in solar cells, and application of the dielectric coating in electronic components.

In the case of cathodic inhibitors, various polyphosphates can be applied. The adherence between polyphosphates and the cathode comes from the electrostatic pull. The electrostatic pull is produced between electrons on the cathodic surface and the positive-charged colloidal suspension, produced by the mixture of metal ions and polyphosphates, which creates a protective covering. Polyphosphates offer some drawbacks, for instance, the formation of complexes with iron, which facilitate dissolution rather than protection. Furthermore, polyphosphates must be introduced within the pH range of 6.5–7.5; otherwise, polyphosphates can be transformed into orthophosphate via the hydrolysis process, which results in losing the protective characteristics [3]. Therefore, polyphosphates cannot be employed as a covering in ships, platforms, and similar applications because of the sea's high pH.

Some corrosion inhibitors are produced by combining anodic and cathodic inhibitors to attain desired properties for effective protection. For instance, chrome and zinc-containing salts, when used separately, create an ineffective covering; however, the combination of these two salts with the polyphosphates enhances the inhibition characteristics and lessens the inhibitor amount usage. Moreover, some inorganic inhibitors, such as chrome-containing coating, have a detrimental effect on the environment and decrease their performance in alkaline media. Inhibitors derived from organic sources, i.e. amines and amides, are diverse and can be employed as anodic as well as cathodic inhibitors. These types of inhibitors work based on adsorption. As an instance, amines contain electron pairs, which can lead to interaction with the anode, and therefore are known as cationic inhibitors. In contrast, sulfonates carry a partial positive charge, which attaches to the cathodic surface and results in bond development between the metal surface and organic inhibitors, eliminating H_2O from the particular zone. This process is known as chemisorption, which restricts anodic as well as cathodic reactions and provides protection to the metal from corrosion. Other methods for inhibiting oxidation include the removal of O_2 from the region with the help of scavengers. The O_2 expedites the oxidation rate by undergoing a reaction with cathodic electrons, leading to charge depolarization, which eventually results in corrosion. Frequently employed scavengers include $(NH_4)_2SO_4$, Na_2SO_3, and N_2H_4.

17.3 RECENT DEVELOPMENTS IN WATERBORNE POLYURETHANES AS A CORROSION INHIBITOR

Technological advancements in the field of coating led to the advent of new materials to serve the purpose. Among these materials, polyurethane-containing coating acquired tremendous attention due to its low price, high efficiency, and range of synthetic procedures. Recently, waterborne polyurethane (WPU) films have garnered enormous attention due to increasing environmental concerns, which can significantly minimize volatile organic component (VOC) usage. WPU costs less due to the elimination of high-priced organic solvents and offers a clean approach by utilizing nontoxic materials, such as vegetable oils, for its preparation. In general, polyurethanes are produced by the reaction between polyisocyanates (–N=C=O) and polyols (–OH), which leads to the urethane bond (–N=C(O)–O–) [4]. The effectiveness of the coatings can be identified by various characteristics, such as resistance to chemicals and temperature, mechanical strength, stability with radiation, coating appearance, and synthetic approach. The coating characteristics can be controlled by employing different starting materials. For example, methylene diisocyanate (MDI) can be used as starting material, which possesses hard aromatic rings that enhance the mechanical properties of the coating. Simultaneously, it has low UV resistivity, which turns it into pale-colored material and loses its dyed color over time. However, isophorone diisocyanate (IPDI) possesses aliphatic chains that are responsible for high UV resistivity and offer excellently homogenous dye colors. However, due to flexible linear chains, their mechanical properties are extensively compromised.

Similarly, functionality and structural variation also play a significant role in determining polyurethane coating properties. For instance, acrylic polyols offer an excellent appearance and remain stable when exposed to UV radiation. In contrast, polyester-containing polyols hold higher mechanical properties due to their relatively rigid chain segments. Furthermore, the polyol/diisocyanate ratio also influences the coating properties. Surplus polyol usage results in improved elasticity and attachment to the surface while diminishing its strength against chemicals (water, acids, alkalis, solvents). Similarly, excessive utilization of diisocyanate extends the drying time and offers low surface bonding with enhanced mechanical properties and superior chemical resistance. Some of the efficient WPU-based materials for coating and corrosion shielding are discussed below.

WPUs with novel properties can be produced by a variety of techniques. Adding graphene in WPU is one of these methods to produce coatings with superior barrier characteristics and chemical durability that ultimately enhance the material's corrosion resistivity. The low dispersibility of graphene is one of the major drawbacks of graphene, which can be resolved by introducing hydrophilic functional groups. Hydrophilic functionalization of graphene also reduces its resistance against water, solvents, alkalis, and acids [5,6]. The most feasible approach to deal with low dispersibility is to crosslink the WPU with graphene that reduces the hydrophilic effect of polar groups and improves penetration of graphene's nanostructures in WPU. Recently, the approach of employing polycarbodiimide (PCD) as a crosslinker for graphene oxide and dispersing the crosslinked graphene into WPU has garnered immense interest [7–9]. In this reaction, the proportion of diisocyanate, crosslinker, and functionalized graphene is a significant parameter to avoid conglomeration and produce a thick coating. The corrosion shielding scheme for these coatings can be visualized in Figure 17.3. The testing was performed by using a salt-based corrosion spray on the coated metal samples. The appearance of the sample was analyzed after 72 and 120 h of spraying. The sample WPG-8.75 (PCD to GO ratio is 8.75) offered superior corrosion resistance and did not show any decay. Other samples include pure WPU and WPU with graphene, which showed substantial decay. The enhanced crosslinking and appropriate component ratio resulted in greater barrier properties and avoided ion penetration into the metal. Furthermore, the passive nature of graphene helped in resisting a violent environment [7].

Polydopamine (PDA) is another material that shows effective anticorrosion properties. Its monomer, dopamine, is commonly known as a neurotransmitter in the human body, which plays a major role in emotions related to pleasure and reward. Dopamine can undergo self-polymerization under mild conditions, and its polymeric form provides great protection against corrosion. PDA displays effective corrosion protective behavior due to its highly packed, chemically stable structure and ability to adhere to many substrates [10–12]. The combination of polydopamine with graphene oxide (GO) could lead to an expressive increase in corrosion protection due to the synergism between them. Zhao et al. fabricated a WPU/GO-PDA composite through a facile sonication process using a buffer solution at room temperature [13]. This process led to the self-polymerization of dopamine as well as its covalent grafting over the GO's surface. Since PDA has several hydroxyl groups in its structure, it improves the dispersibility of the GO in water along with its

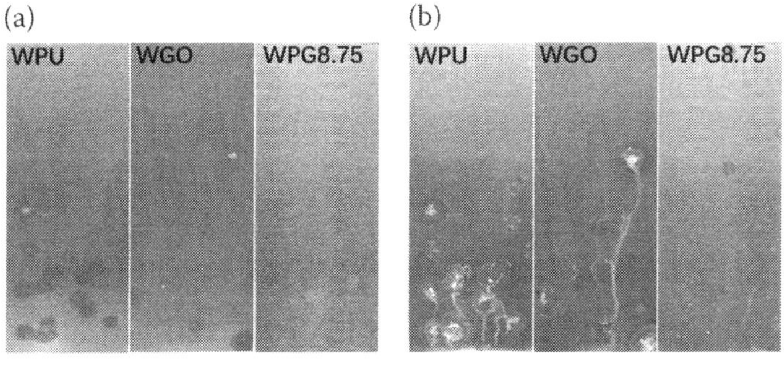

FIGURE 17.3 Salt spray test for the samples after exposure during (a) 72 h and (b) 120 h. "Adapted with permission from reference [7]. Copyright (2020) Elsevier."

adhesiveness. Also, the aromatic rings present in PDA allow the formation of π–π stacking interaction, which prevents the agglomeration of GO's nanosheets. Even though PDA improves the dispersibility in water after the composite coating was applied to a metal surface, the contact angle increased from 74.2° to 82°, suggesting a more hydrophobic behavior. The introduction of PDA into GO's nanosheets likely allowed it to arrange properly due to the decrease in agglomeration. Also, since the GO nanosheets were relatively larger than PDA, it was able to properly coat the substrate. The composite coating presented a denser and uniform distribution, which provided a more effective barrier effect against corrosive agents as well as permeation of water. The schematics for this mechanism are shown in Figure 17.4.

Metal oxides have extensively been employed for corrosion resistance produced via passivation of corresponding metals or by mixing constituents to form composites. The metal oxides are favorable as coating components because of their high resistivity toward corrosion and UV light. However, one restriction for using metal oxides is their limited applicable concentration since higher metal oxide concentration enhances the barrier properties of the material but decreases the abrasion, impact, and scratch resistivity of the material; increases flexibility; restricts easy detachment; and improves viscosity. These challenges can be resolved by employing nanosized metal oxides, such as Al_2O_3, $CaCO_3$, ZnO, TiO_2, and SiO_2 [14–16]. Some of the favorable characteristics obtained by employing nanosized materials are superior specific area, high electron density, and nanosized structures, which enable them to be dispersed uniformly throughout the coating. One important advantage of nanosized materials involves similar effectiveness in barrier effect while being used in less quantity than macro materials, retaining even rheological properties. Corrosion is an electrochemical operation, so electrochemical testing offers essential insight into the coating's corrosion resistance approach. For this reason, electrochemical impedance spectroscopy (EIS) is a crucial characterization that informs about adhesive characteristics of the coating, regions vulnerable to corrosion, delamination, defects, and resistance of the coating material.

Due to the low price and excellent durability, aluminum is extensively utilized as a coating. However, the creation of micro-sized cracks in aluminum expedites the deterioration process [17], which can be addressed by transforming it into its oxide

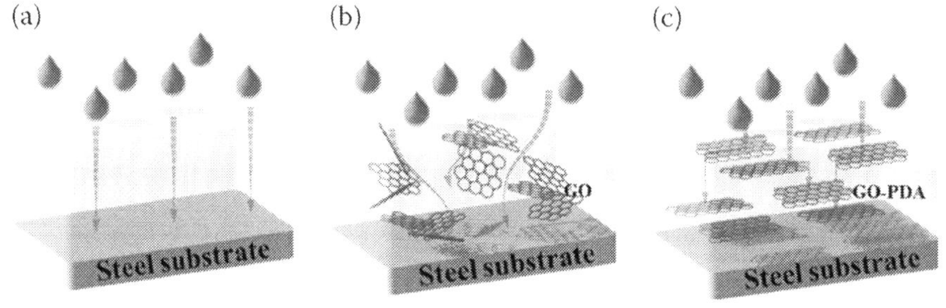

FIGURE 17.4 Schematic for the barrier effect mechanism for the (a) bare steel substrate, (b) GO-coated substrate, and (c) GO-PDA steel substrate. "Adapted with permission from reference [13]. Copyright (2019) Elsevier."

(Al_2O_3) that allows it to be mixed in a strong hydrophobic resin. The special water-repellent characteristics of Al_2O_3 allow it to be employed in various coating applications, such as a self-cleaning, anti-icing, low friction, corrosion coating. The application of nanoparticles for coating is another way for corrosion protection. However, the loosely bonded nanoparticles dissociate from the surface and lead to degradation in the long run. They can also conglomerate, leading to reduced coating efficiency. To address these drawbacks, WPUs with chemically bonded nanoparticle coating have been prepared. For example, functionalized TiO_2 nanoparticles can link to the carboxyl group hanging to the polymer chain that is already present in WPU [16]. Nanostructures are usually functionalized with hydroxyl or amine groups, which helps them link with the diisocyanates in WPU to produce urethane or urea bonds [18].

Seawater contains an extreme corrosion medium, which demands efficient coating to tolerate those conditions. WPU can be an efficient candidate for this application with the addition of sufficient additives. $CaCO_3$, due to its superior corrosion resistivity, has been investigated as a nanofiller in WPU to enhance corrosion protection in ships [19,20]. The coating composite offered excellent adhesion via mineralizing the substrate with –OH and –NH functional groups that are abundantly present in polyurethanes. The substrate was mineralized in vitro, meaning that the $CaCO_3$ was layered naturally on the top of the coating in the marine medium. This technique promotes the stability of the coating due to broadly available $CaCO_3$ in the sea and introduces self-healing ability in the coating because of a higher quantity of Ca^{2+} and CO_3^{2-} ions in the sea. The WPU was synthesized by using crown-shaped β-cyclodextrin (β-CD) [21]. The β-CD complexed with Ca^{2+} ions initially adsorbed on the surface and CO_3^{2-} ions of $CaCO_3$ coating. The influence of mineralization with the increase in β-CD quantity was observed by immersing different coatings in marine water for 10 days, as shown in Figure 17.5.

Over the years, conductive polymers (CPs) emerged as essential polymeric materials due to their modifiable electron conductance, excellent optical and chemical properties, which enable them to be employed in various applications. These materials, such as polyaniline, polypyrrole, and polythiophene, have also been applied in polyurethanes to develop corrosion-resistant coating [22]. One of the primary benefits of employing CPs is their superior miscibility with the polyurethane matrix, which occurs due to the intermolecular attraction that helps chains penetrate and prevent accumulation. The mixing technique used for this reason is the interpenetrating polymer network that involves one blending polymer to be crosslinked. This crosslinking firmly interlinks the polymers in a network and offers a homogenous mixture for coatings with enhanced characteristics, such as barrier effect, thermal and mechanical resistance, and superior dielectric constant [23]. The improved dielectric constant of the material leads to higher Helmholtz double-layer capacitance and exhibits excellent protection against corrosion [24,25]. Furthermore, the introduction of CPs offers free charges to the material, which allows Maxwell–Wagner–Sillars surface polarization that further enhances the dielectric permittivity and conductance [25,26].

Polyaniline (PANI) is a renowned CP with various applications. It presents aromatic rings, which provide high rigidity and mechanical properties, and –NH–

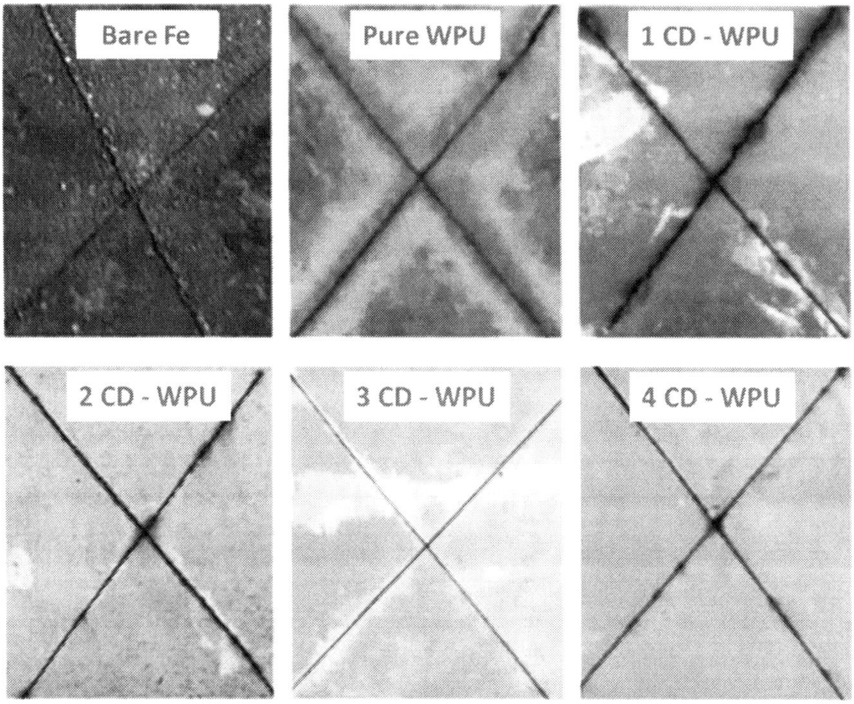

FIGURE 17.5 Immersion test of mineralized film for different coating in artificial marine water for 10 days, where the number represents the weight percentage. "Adapted with permission from reference [21]. Copyright (2019) Elsevier."

groups that bring about excellent mixability by interlinking with urethane via H-bonding [27–29]. A mechanically strong coating can be developed through this technique, but a phase separation between the mixing polymers is also noticed in a few cases, which can be prevented by grafting the polymers chemically. The grafting technique offers a homogenous framework, which can be produced by employing polyurethane prepolymer with isocyanate termination and PANI [30]. The influence of adding hydroxy functional groups in WPU by using poly(*meta*-aminophenol) was investigated. The corrosion resistance of the coating was observed to be improved significantly, with a corrosion rate of 1.28×10^{-4} mm/yr and corrosion voltage of −349.8 mV. The corrosive spray test of the synthesized coating was also carried out, demonstrating the significantly delayed deterioration of the steel coated with the composite polymer, while the pure PU started deteriorating after 72 h. When applied in mild steel, polyaniline can reduce and allows the iron to oxidize. This reduction and oxidation resulted in the formation of an inactive film of iron oxide, which adds more resistance to corrosion. Hence, introducing CPs over the surface of a metallic substrate provides this passivation process, at which a thin metal oxide layer is formed between the substrate and the CP. However, dispersing CP such as PANI into the polyurethane matrix is not a simple task as PANI has low solubility in water. To address this condition a research group synthesized an electroactive WPU coating that chemically incorporated aniline trimer by taking advantage of its amine groups that can react with isocyanate to form a urea bond [31]. Through this approach, the researchers were able to introduce similar properties to those related to neat PANI, such as the facile redox process that forms a passive layer of metal oxide over the substrate with improved dispersibility through hydrogen bonding. The synthesis of this electroactive coating is demonstrated in Figure 17.6. After the application of electroactive WPU coating, the corrosion rate decreased from 0.328 to 0.244 mm/year.

Similarly, other CPs such as polypyrrole (PPy) can also be used in coatings for corrosion protection. Previous studies demonstrated satisfactory properties of thin PPy films that were electropolymerized over copper and zinc substrate, which yielded satisfactory anticorrosion properties [32,33]. As scientific research progresses, new approaches and techniques are developed, which often combine two or more materials to deliver a final product with enhanced properties, such as composites. A composite tung oil-based WPU coating with PPy doped with CeO_2 nanoparticles was developed [34]. The morphology of the coating was studied and found to be very smooth on carbon steel with improved corrosion protection behavior. The synthetical scheme for the synthesis of the composite coating is explicated in Figure 17.7. The coating obtained through this work not only presented anticorrosion properties but also flame retardancy, high hydrophobicity, and thermal stability up to 200°C. Tafel plots are commonly used electrochemical analysis to reveal the anticorrosion properties of coatings providing corrosion potential (E_{corr}), polarization resistance (R_p), corrosion current density (I_{corr}), and corrosion rate. The highest E_{corr} and lowest I_{corr} were observed, which was attributed to the combination of doping/dedoping process of PPy as well as the compact passivation layer formed through CeO_2 that captures Fe^{3+} cations and converts in Fe_xCeO_y alloy. Other factors, such as hydrophobic nature of coating, high redox potential, small size (around 40 nm) of nanoparticles, and formation of a compact and smooth surface, provided strong adhesion to the metal surface as well as efficiency barrier effect [35,36].

Nanomaterials have also been introduced in the WPU to improve corrosion protection. These nanomaterials include nano-clay, CNTs, silver, boron nitride, and various self-healing materials. Clay is a phyllosilicate that contains Na, Al, Ca, Si, Mg, O, and H_2O in its framework. Na^+-montmorillonite is a common type of clay with a superior aspect ratio of the sheet-like structure of about 1 × 200 nm (thickness × length). The clay has a layered framework with edges made up of tetrahedral silica layers. These edge-encapsulated octahedral structures can be composed of Al or Mg hydroxide. The Na^+ and Ca_2^+ cations within the layers bring about excellent hydrophilicity of the clay [37]. The cations enhance the barrier, thermal, and mechanical properties of the coating and also provide flame retardancy to the material [38–41]. The exceptional framework of the clay is developed when the layer intercalates or exfoliates from the inorganic to the organic phase. Such events enhance layer interaction and preventions percolation, which results in improved barrier properties and restricts ions and metal surface interaction [42]. An emulsion polymerization process of water-dispersed-polyaniline with clay has been reported [42]. The synthetic procedure involves dispersing

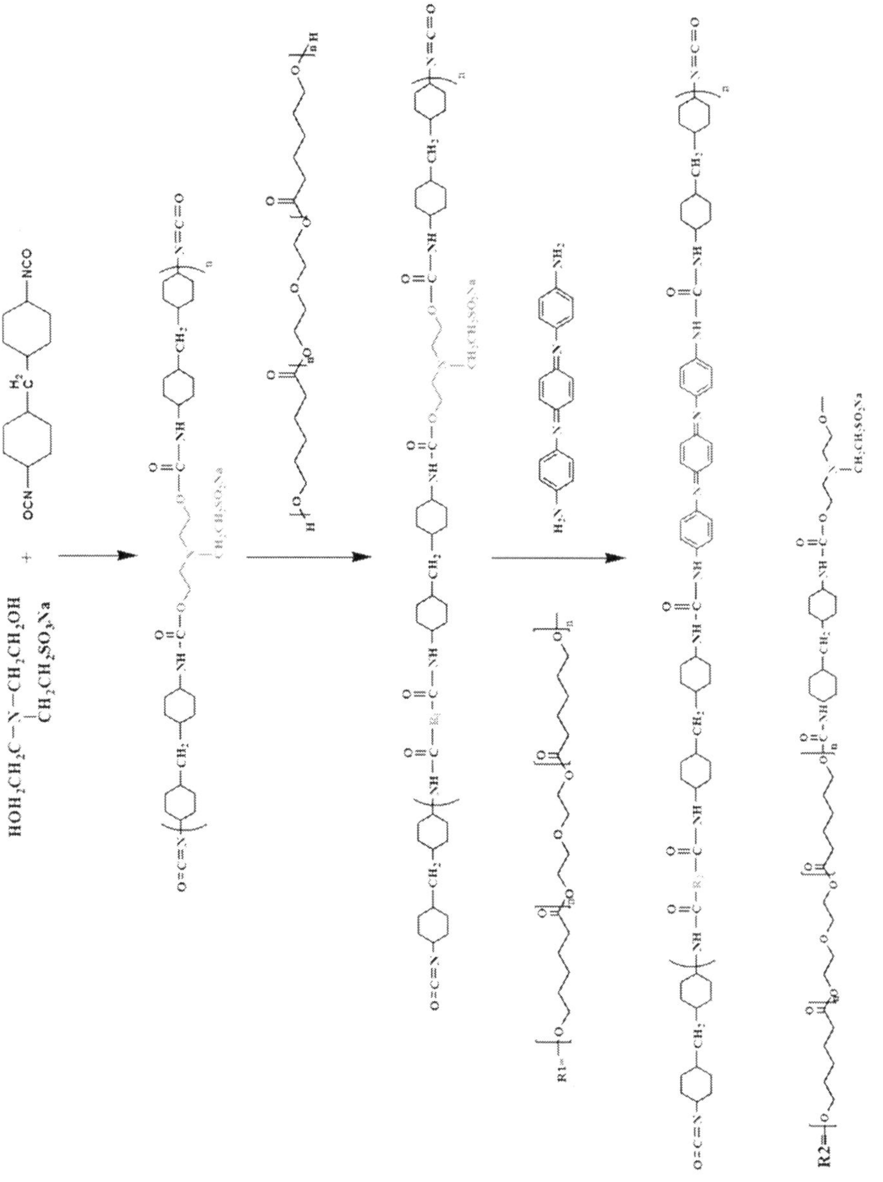

FIGURE 17.6 Synthesis of an electroactive WPU coating incorporated with a trimer aniline to add enhance the corrosion resistance properties along with water dispersibility. "Adapted with permission from reference [31]. Copyright (2013) Elsevier."

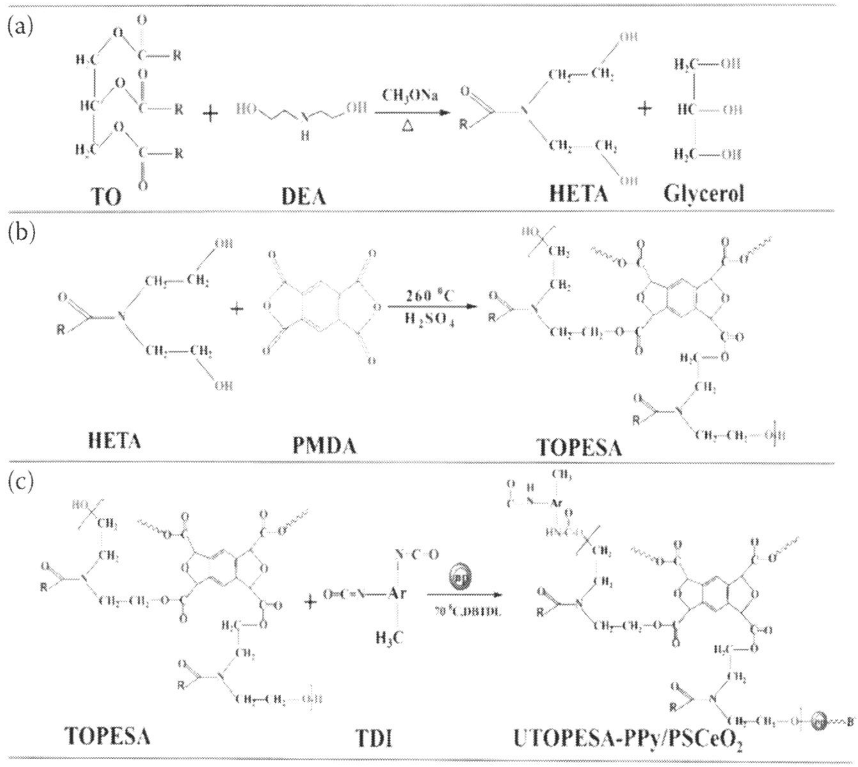

FIGURE 17.7 Synthesis scheme for (a) *N,N*-bis(2-hydroxyethyl) tung oil amide (HETA). (b) Polyesterification reaction between HETA and pyromellitic dianhydride (PMDA) to obtain TO-polyesteramide (TOPESA). (c) Polyaddition reaction between TOPESA and TDI along with PPy-CeO_2 nanoparticles to obtain the coating named as UTOPESA-PPy/PSCeO_2. "Adapted with permission from reference [34]. Copyright (2020) Elsevier."

dodecylbenzene sulfonic acid and aniline in H_2O, producing a uniform DBSA-aniline complex. Then, water dispersed Na^+-montmorillonite clay was mixed with the prepared complex, and the emulsion polymerization reaction started with the help of an initiator, for example, ammonium persulfate or ferric chloride [42]. A lower concentration, such as 1 wt.%, is even found to be offering efficient corrosion protection [42,43].

Boron nitride is another material that can be used to make a WPU composite for corrosion protection. It has a hexagonal framework like graphene with excellent mechanical and barrier properties with superior stability against the acidic and alkaline environment, making them efficient in corrosion protection application [44–46]. The difference between graphene and boron nitride is that boron nitride is an insulator and has excellent thermal durability that can withstand a temperature of around 800 °C [47,48]. Furthermore, unlike graphene, the notable intermolecular attraction is present in the planar direction instead of between layers, which results in excellent barrier properties compared to graphene [49,50]. Boron nitrate is also employed as a lubricant, which offers substantial enhancement in reducing abrasion in PU coating [51]. Due to these characteristics, boron nitrate employment to make composite is considered a good approach to obtain corrosion resistance. However, the low dispersibility of boron nitrate in water restricts its application, which can be overcome by functionalizing it with hydrophilic groups [52]. For example, sodium hydroxide-treated boron nitride produced hydroxylated boron nitride, which allowed homogeneous dispersion in coatings. It is observed that the a small amount of boron nitrate (about 2 wt.%) in WPU could result in excellent barrier properties, enhanced abrasion, and thermal stability.

Carbon-derived nanomaterials, including carbon black, graphene, graphene oxide, and carbon nanotubes (CNTs), offer strong stability against harsh acidic and alkaline environments along with other favorable properties, such as superior surface area and thermal and mechanical properties [53,54]. These characteristics enable them to be used as corrosion-resisting materials in the coatings. Some investigations indicated that the multiwalled CNTs offered a significant hydrophobic improvement to coatings because of piled-up layers of CNTs that resist water penetration through the coating [55,56]. Moreover, only a small amount of multiwalled CNTs were required to offer superior mechanical properties along with scratch and UV resistivity [55,56]. Despite these characteristics, CNTs have high surface energy that results in CNT accumulation, which reduces coating corrosion-resisting efficiency. For addressing this problem, polydopamine (PDA) can be used to improve adhesion between substrate and coating material, as discussed above. Similarly, poly(2-butylamine) can also be employed as a dispersant to avoid CNT accumulation [57]. PDA functionalized CNTs have been investigated as effective coating material produced via a simple process [58]. The PDA-CNTs were utilized as a filler in polyurethane for corrosion protection. The PDA was added to develop –OH and $-NH_2$ groups to mitigate intrinsic CNT hydrophobicity. Furthermore, it also helps in creating a crosslinked framework that can be efficient in avoiding water penetration through the coating to the substrate. Additionally, the insulated nature of PDA precludes the development of the galvanic system, where conductive CNTs can function as a cathode, metal as an anode, and both encourage galvanic corrosion.

Nanosized particles have been utilized as fillers to protect the material from corrosion. The nanosized particles obstruct the micropores of the coating material, which decreases the free space and enhances the barrier effect against ions and moisture [59]. The polymer materials adsorb the metal nanoparticles efficiently because of the high surface area of nanoparticles. Despite all the mentioned characteristics, their efficiency in WPUs is not as good as metal oxides, particularly for chlorine-containing corrosive materials. However, Ag nanoparticles mixed in high solid polyurethane (HSP) demonstrated a significant stable coating. It is assumed that the Ag^{2+} cations interact with the carbonyl group to develop stable complexes in the crosslinked framework in HSP and help reduce the chlorine interaction. However, the corrosion protection characteristics remain almost the same as pure polyurethane [60]. Furthermore, WPU can be functionalized to obtain self-healing characteristics, which can offer various benefits, such as minimum maintenance, long service life, and excellent corrosion response [61,62].

Microencapsulation is an efficient method to attain self-healing characteristics. The microcapsules made up of polyurethane, phenol, or urea-formaldehyde, and silica capsules are promising to attain efficient self-healing coating. At the same time, the low viscosity epoxies, inorganic inhibitors, aliphatic diisocyanates, and vegetable oils can be employed as a healing agents. In one investigation, toluene diisocyanate and 1,4 butanediol were used as starting materials [63]. Isophorone diisocyanate, an aliphatic isocyanate, was used as a healing agent, which reacts with environmental moisture upon breakage, creates urea linkages, and offers UV resistivity. Self-healing properties are highly desirable in several areas as materials with such features can recover from physical and chemical damage, hence providing long-term protection. It can be handy for the case of oil pipelines or drilling platforms that can be difficult sites for maintenance. Self-healing materials can be extrinsic or intrinsic. The first consists of encapsulating a flaxseed oil, reactive monomer, or corrosion inhibitor in small hollow spheres that can be blended or interpenetrated into the coating [64]. Hence, when the coating is damaged these healing agents are released to repair the area. However, such composites are not able to perform multiple healing sessions nor be applied as extra thin coatings due to the required particle size to efficiency self-heal. On the other hand, intrinsic self-healing occurs through reversible chemical bonds, such as Dies–Alder bonds, siloxanes, metal coordination, and others [65,66]. Due to the reversibility of the chemical bonds, this type of self-healing material can perform several cycles of cracking and healing before it degrades. Hence, it is a viable tool for industrial applications. Yet, manufacturing of these materials can be challenging as they often require organic solvent and complex synthesis, which may increase the overall cost for large-scale processes. To find an alternative to these drawbacks, one possible way is using compounds that contain disulfide bonds (–S–S–), which do not impose challenging synthesis. Also, self-healing properties can be triggered by UV-light and thermal curing, among other ways [67]. Other paths to introduce these type of properties were studied by Shahabadi et al. [68], who manufactured lignin that was noncovalently bonded with graphene, named lignin modified graphene (LMG) composite. The LMG was blended with a WPU dispersion in the desired quantity. Different from the reversible bond mechanism for self-healing, the process with LMG occurs through viscoelastic recovery, which is the movement of two interfaces created after a cut toward each other to reestablish the surface. This process is effective since it can take place in around 150 s of exposure to infrared radiation. Aside from self-healing properties, the composite also presented UV resistance and conductibility of 0.276 S/m which prompted applications as an antistatic coating along with corrosion protection.

17.4 CONCLUSION

The WPU coatings are very efficient and can be used in various applications, particularly for corrosion protection. For this reason, researchers are investigating WPU-based coatings to achieve properties that have not been developed yet. WPU offers excellent properties that help save metal surfaces exposed to the corrosive environment and provide stability to the material. Since the elimination of chromium-based coating due to environmental concerns, the introduction of polyurethane allows a safe, stable, and environmentally friendly way to obtain better corrosion-resistive coating that can be modified to achieve desirable characteristics. Therefore, WPU is considered a remarkable polymer group that has proven its efficiency and is of great significance for the economy worldwide.

REFERENCES

1. Ahmad Z (2006) *Principles of corrosion engineering and corrosion control.* Elsevier
2. McCafferty E (2010) Passivity. In: *Introduction to corrosion science.* Springer, pp 209–262.
3. Uhlig HH, Triadis DN, Stern M (1955) Effect of oxygen, chlorides, and calcium ion on corrosion inhibition of iron by polyphosphates. *J Electrochem Soc* 102:59.
4. Tersac G (2007) Chemistry and technology of polyols for polyurethanes. Milhail Ionescu. Rapra Technology, Shrewsbury, UK. *Polym Int* 56:820–820.
5. Zhu D, Van Ooij WJ (2004) Corrosion protection of metals by water-based silane mixtures of bis-[trimethoxysilylpropyl]amine and vinyltriacetoxysilane. *Prog Org Coatings* 49:42–53.
6. Liu M, Mao X, Zhu H, Lin A, Wang D (2013) Water and corrosion resistance of epoxy-acrylic-amine waterborne coatings: Effects of resin molecular weight, polar group and hydrophobic segment. *Corros Sci* 75:106–113.
7. Cui J, Xu J, Li J, Qiu H, Zheng S, Yang J (2020) A crosslinkable graphene oxide in waterborne polyurethane anticorrosive coatings: Experiments and simulation. *Compos Part B Eng* 188:107889.
8. Posthumus W, Derksen AJ, van den Goorbergh JAM, Hesselmans LCJ (2007) Crosslinking by polycarbodiimides. *Prog Org Coatings* 58:231–236.
9. Hesselmans LCJ, Derksen AJ, van den Goorbergh JAM (2006) Polycarbodiimide crosslinkers. *Prog Org Coatings* 55:142–148.
10. Singer F, Schlesak M, Mebert C, Höhn S, Virtanen S (2015) Corrosion properties of polydopamine coatings formed in one-step immersion process on magnesium. *ACS Appl Mater Interfaces* 7:26758–26766.
11. Yang N, Yang T, Wang W, Chen H, Li W (2019) Polydopamine modified polyaniline-graphene oxide composite for enhancement of corrosion resistance. *J Hazard Mater* 377:142–151.
12. Bernsmann F, Ball V, Addiego F, Ponche A, Michel M, Gracio JJ de A, Toniazzo V, Ruch D (2011) Dopamine–melanin film deposition depends on the used oxidant and buffer solution. *Langmuir* 27:2819–2825.
13. Zhao Z, Guo L, Feng L, Lu H, Xu Y, Wang J, Xiang B, Zou X (2019) Polydopamine functionalized graphene oxide nanocomposites reinforced the corrosion protection and adhesion properties of waterborne polyurethane coatings. *Eur Polym J* 120:109249.
14. Yang LH, Liu FC, Han EH (2005) Effects of P/B on the properties of anticorrosive coatings with different particle size. *Prog Org Coatings* 53:91–98.
15. Chen X, Yuan J, Huang J, Ren K, Liu Y, Lu S, Li H (2014) Large-scale fabrication of superhydrophobic polyurethane/nano-Al_2O_3 coatings by suspension flame spraying for anti-corrosion applications. *Appl Surf Sci* 311:864–869.
16. Charpentier PA, Burgess K, Wang L, Chowdhury RR, Lotus AF, Moula G (2012) Nano-TiO_2/polyurethane composites for antibacterial and self-cleaning coatings. *Nanotechnology* 23:425606.
17. Abedi Esfahani E, Salimijazi H, Golozar MA, Mostaghimi J, Pershin L (2012) Study of corrosion behavior of arc sprayed aluminum coating on mild steel. *J Therm Spray Technol* 21:1195–1202.

18. Behniafar H, Alimohammadi M, Malekshahinezhad K (2015) Transparent and flexible films of new segmented polyurethane nanocomposites incorporated by NH_2-functionalized TiO_2 nanoparticles. *Prog Org Coatings* 88:150–154.
19. Liu X, Zhou Y, Pei C (2018) Mimetic biomineralization matrix using bacterial cellulose hydrogel and egg white to prepare various morphologies of $CaCO_3$. *CrystEngComm* 20:4536–4540.
20. Tom S, Jin H-E, Heo K, Lee S-W (2016) Engineered phage films as scaffolds for $CaCO_3$ biomineralization. *Nanoscale* 8:15696–15701.
21. Hua Y, Li X, Ma L, Wang Y, Fang H, Wei H, Ding Y (2019) Self-healing mineralization and enhanced anti-corrosive performance of polyurethane $CaCO_3$ composite film via β-CD induction. *Mater Des* 177:107856.
22. Chandrasekhar P (1999) *Conducting polymers, fundamentals and applications.* Springer
23. Lee DS, Kang WK, An JH, Kim SC (1992) Gas transport in polyurethane-polystyrene interpenetrating polymer network membranes. II. Effect of crosslinked state and annealing. *J Memb Sci* 75:15–27.
24. Guiffard B, Seveyrat L, Sebald G, Guyomar D (2006) Enhanced electric field-induced strain in non-percolative carbon nanopowder/polyurethane composites. *J Phys D Appl Phys* 39:3053–3057.
25. Mishra RK, Abraham J, Kalarikkal N, Jayanarayanan K, Joseph K, Thomas S (2017) Chapter 8—conducting polyurethane blends: rRecent advances and perspe*ctives.* In: Thomas S, Datta J, Haponiuk JT, Reghunadhan ABT-PP (eds.), Polyurethane polymers: Blend interpenetrating polymer networks. Elsevier, pp 203–231.
26. Putson C, Jaaoh D, Muensit N (2016) Large electromechanical strain at low electric field of modified polyurethane composites for flexible actuators. *Mater Lett* 172: 27–31.
27. Rodrigues PC, Lisboa-Filho PN, Mangrich AS, Akcelrud L (2005) Polyaniline/polyurethane networks. II. A spectroscopic study. *Polymer (Guildf)* 46:2285–2296.
28. El Faydy M, Galai M, Touhami ME, Obot IB, Lakhrissi B, Zarrouk A (2017) Anticorrosion potential of some 5-amino-8-hydroxyquinolines derivatives on carbon steel in hydrochloric acid solution: Gravimetric, electrochemical, surface morphological, UV–visible, DFT and Monte Carlo simulations. *J Mol Liq* 248:1014–1027.
29. Gurunathan T, Rao CRK, Narayan R, Raju KVSN (2013) Polyurethane conductive blends and composites: Synthesis and applications perspective. *J Mater Sci* 48: 67–80.
30. Karmakar HS, Arukula R, Thota A, Narayan R, Rao CRK (2018) Polyaniline-grafted polyurethane coatings for corrosion protection of mild steel surfaces. *J Appl Polym Sci* 135:45806.
31. Huang H-Y, Huang T-C, Lin J-C, Chang J-H, Lee Y-T, Yeh J-M (2013) Advanced environmentally friendly coatings prepared from amine-capped aniline trimer-based waterborne electroactive polyurethane. *Mater Chem Phys* 137:772–780.
32. Fenelon AM, Breslin CB (2002) The electrochemical synthesis of polypyrrole at a copper electrode: corrosion protection properties. *Electrochim Acta* 47:4467–4476.
33. Pournaghi-Azar MH, Nahalparvari H (2005) Zinc hexacyanoferrate film as an effective protecting layer in two-step and one-step electropolymerization of pyrrole on zinc substrate. *Electrochim Acta* 50:2107–2115.
34. Bakshi MI, Ahmad S (2020) In-situ synthesis of synergistically active ceria doped polypyrrole oleo-polyesteramide hybrid nanocomposite coatings: Corrosion protection and flame retardancy behaviour. *Prog Org Coatings* 147:105778.
35. Qiang Y, Guo L, Zhang S, Li W, Yu S, Tan J (2016) Synergistic effect of tartaric acid with 2,6-diaminopyridine on the corrosion inhibition of mild steel in 0.5 M HCl. *Sci Rep* 6:33305.

36. Rawat NK, Ahmad S (2018) Synergistic effect of nanosize and irradiation on epoxy/conducting poly(o-phenyldiamine) nanospheres composite coatings: Synthesis, characterization and corrosion protective performance. *Mater Chem Phys* 204: 282–293.
37. Chang K-C, Lai M-C, Peng C-W, Chen Y-T, Yeh J-M, Lin C-L, Yang J-C (2006) Comparative studies on the corrosion protection effect of DBSA-doped polyaniline prepared from in situ emulsion polymerization in the presence of hydrophilic Na^+-MMT and organophilic organo-MMT clay platelets. *Electrochim Acta* 51:5645–5653.
38. Wang Z, Pinnavaia TJ (1998) Nanolayer reinforcement of elastomeric polyurethane. *Chem Mater* 10:3769–3771.
39. Tyan H-L, Liu Y-C, Wei K-H (1999) Thermally and mechanically enhanced clay/polyimide nanocomposite via reactive organoclay. *Chem Mater* 11:1942–1947.
40. Lan T, Kaviratna PD, Pinnavaia TJ (1994) On the nature of polyimide-clay hybrid composites. *Chem Mater* 6:573–575.
41. Yeh J-M, Liou S-J, Lin C-Y, Cheng C-Y, Chang Y-W, Lee K-R (2002) Anticorrosively enhanced PMMA–clay nanocomposite materials with quaternary alkylphosphonium salt as an intercalating agent. *Chem Mater* 14:154–161.
42. Yeh J-M, Yao C-T, Hsieh C-F, Lin L-H, Chen P-L, Wu J-C, Yang H-C, Wu C-P (2008) Preparation, characterization and electrochemical corrosion studies on environmentally friendly waterborne polyurethane/Na^+-MMT clay nanocomposite coatings. *Eur Polym J* 44:3046–3056.
43. Yeh J-M, Liou S-J, Lai C-Y, Wu P-C, Tsai T-Y (2001) Enhancement of corrosion protection effect in polyaniline via the formation of polyaniline–clay nanocomposite materials. *Chem Mater* 13:1131–1136.
44. Meng W, Huang Y, Fu Y, Wang Z, Zhi C (2014) Polymer composites of boron nitride nanotubes and nanosheets. *J Mater Chem C* 2:10049–10061.
45. Li LH, Chen Y (2016) Atomically thin boron nitride: Unique properties and applications. *Adv Funct Mater* 26:2594–2608.
46. Lin Y, Connell JW (2012) Advances in 2D boron nitride nanostructures: Nanosheets, nanoribbons, nanomeshes, and hybrids with graphene. *Nanoscale* 4:6908–6939.
47. Li LH, Cervenka J, Watanabe K, Taniguchi T, Chen Y (2014) Strong oxidation resistance of atomically thin boron nitride nanosheets. *ACS Nano* 8:1457–1462.
48. Bernard S, Salameh C, Miele P (2016) Boron nitride ceramics from molecular precursors: Synthesis, properties and applications. *Dalt Trans* 45:861–873.
49. Joshi MD, Goyal A, Patil SM, Goyal RK (2017) Tribological and thermal properties of hexagonal boron nitride filled high-performance polymer nanocomposites. *J Appl Polym Sci* 134.
50. Gao C, Guo G, Zhang G, Wang Q, Wang T, Wang H (2017) Formation mechanisms and functionality of boundary films derived from water lubricated polyoxymethylene/hexagonal boron nitride nanocomposites. *Mater Des* 115:276–286.
51. Xing T, Mateti S, Li LH, Ma F, Du A, Gogotsi Y, Chen Y (2016) Gas protection of two-dimensional nanomaterials from high-energy impacts. *Sci Rep* 6:35532.
52. Li J, Gan L, Liu Y, Mateti S, Lei W, Chen Y, Yang J (2018) Boron nitride nanosheets reinforced waterborne polyurethane coatings for improving corrosion resistance and antifriction properties. *Eur Polym J* 104:57–63.
53. Zhao Y, Zhao S, Guo H, You B (2018) Facile synthesis of phytic acid@attapulgite nanospheres for enhanced anti-corrosion performances of coatings. *Prog Org Coatings* 117:47–55.
54. Vosgien Lacombre C, Trinh D, Bouvet G, Feaugas X, Mallarino S, Touzain S (2017) Influence of pigment on the degradation of anticorrosion polymer coatings using

a thermodynamic analysis of electrochemical impedance spectroscopy data. *Electrochim Acta* 234:7–15.
55. Wernik JM, Meguid SA (2014) On the mechanical characterization of carbon nanotube reinforced epoxy adhesives. *Mater Des* 59:19–32.
56. Asmatulu R, Mahmud GA, Hille C, Misak HE (2011) Effects of UV degradation on surface hydrophobicity, crack, and thickness of MWCNT-based nanocomposite coatings. *Prog Org Coatings* 72:553–561.
57. Cui M, Ren S, Qiu S, Zhao H, Wang L, Xue Q (2018) Non-covalent functionalized multi-wall carbon nanotubes filled epoxy composites: Effect on corrosion protection and tribological performance. *Surf Coatings Technol* 340:74–85.
58. Cai G, Hou J, Jiang D, Dong Z (2018) Polydopamine-wrapped carbon nanotubes to improve the corrosion barrier of polyurethane coating. *RSC Adv* 8:23727–23741.
59. Shi X, Nguyen TA, Suo Z, Liu Y, Avci R (2009) Effect of nanoparticles on the anticorrosion and mechanical properties of epoxy coating. *Surf Coatings Technol* 204:237–245.
60. Akbarian M, Olya ME, Mahdavian M, Ataeefard M (2014) Effects of nanoparticulate silver on the corrosion protection performance of polyurethane coatings on mild steel in sodium chloride solution. *Prog Org Coatings* 77:1233–1240.
61. Koh E, Kim N-K, Shin J, Kim Y-W (2014) Polyurethane microcapsules for self-healing paint coatings. *RSC Adv* 4:16214–16223.
62. Yuan YC, Rong MZ, Zhang MQ, Chen J, Yang GC, Li XM (2008) Self-healing polymeric materials using epoxy/mercaptan as the healant. *Macromolecules* 41: 5197–5202.
63. Yang J, Keller MW, Moore JS, White SR, Sottos NR (2008) Microencapsulation of isocyanates for self-healing polymers. *Macromolecules* 41:9650–9655.
64. Kumar A, Stephenson LD, Murray JN (2006) Self-healing coatings for steel. *Prog Org Coatings* 55:244–253.
65. Wang Z, Yang Y, Burtovyy R, Luzinov I, Urban MW (2014) UV-induced self-repairing polydimethylsiloxane–polyurethane (PDMS–PUR) and polyethylene glycol–polyurethane (PEG–PUR) Cu-catalyzed networks. *J Mater Chem A* 2:15527–15534.
66. Kötteritzsch J, Stumpf S, Hoeppener S, Vitz J, Hager MD, Schubert US (2013) One-component intrinsic self-healing coatings based on reversible crosslinking by diels–alder cycloadditions. *Macromol Chem Phys* 214:1636–1649.
67. Xu WM, Rong MZ, Zhang MQ (2016) Sunlight driven self-healing, reshaping and recycling of a robust, transparent and yellowing-resistant polymer. *J Mater Chem A* 4:10683–10690.
68. Seyed Shahabadi SI, Kong J, Lu X (2017) Aqueous only, green route to self healable, UV-resistant, and electrically conductive polyurethane/graphene/lignin nanocomposite coatings. *ACS Sustain Chem Eng* 5:3148–3157.

18 Waterborne Polyurethane for Electrically Conductive Coating

Sheraz Iqbal and Tauqir A. Sherazi
Department of Chemistry, COMSATS University Islamabad, Abbotabad Campus, Abbottabad, Pakistan

Tahir Rasheed
Interdisciplinary Research Center for Advanced Materials, King Fahd University of Petroleum and Minerals, Dhahran, Saudi Arabia

Muhammad Bilal
School of Life Science and Food Engineering, Huaiyin Institute of Technology, Huaian, China

CONTENTS

18.1 Introduction 306
18.2 Synthesis of Polyurethane 306
18.3 Classification of Polyurethane 307
18.4 Waterborne Polyurethane 308
18.5 Advantages of Waterborne Polyurethane 308
18.6 Conductive Coating of WPU and CNT 309
18.7 Conducting Coating of Graphene Oxide/Waterborne Polyurethane 309
18.8 WPU Coating Materials Containing Polyaniline and Other Polymers .. 311
18.9 Conductivity of Coated Composite of WPU/GNs 312
18.10 Conductive Coatings of MWCNT/WPU 314
18.11 Conductivity of WPUs Coated With Metal Oxide 314
18.12 Thermal Conductivity of WPUs 315
18.13 Conclusion 317
References 317

DOI: 10.1201/9781003173526-18

18.1 INTRODUCTION

Polyurethane (PU) is a polymer composed of organic units joined by carbamate linkage. It was first synthesized by a German chemist, Dr. Otto Van Bayer, and his co-worker in 1937 while working in IG Farben, a German chemical company. PU is a plastic material, which exists in various forms. It can be tailored to be either rigid or flexible and is the material of choice for a broad range of end-user applications. There are various types of PUs, which differ in appearance and quality from each other. They are used in diversified products, such as coatings, adhesives, shoe soles, mattresses, and foam insulation. However, the basic chemistry of each type is essentially the same. Widespread use of PUs was first seen during World War II, when they were utilized as a replacement for rubber, which at the time was expensive and hard to obtain. During the war, other applications were developed, largely involving coatings of different kinds, from airplane finishes to sports clothing and varieties of other accessories.

By the 1950s, PUs were being used in adhesives, elastomers, and rigid foams, and in the latter part of the same decade, flexible cushioning foams similar to those used today were developed. Subsequent decades saw many further developments, and today we are surrounded by numerous materials made up of PU in our everyday lives. Although PU is a product that most people are not overly familiar with, as it is generally "hidden" behind covers or surfaces made of other materials, it would be hard to imagine life without PUs [1]. In the years 1952–1954, a variety of polyester-polyisocyanate was synthesized by Bayer and his co-workers. It was gradually replaced by polyether polyols because of cost-effectiveness, ease of handling, good hydrolytic stability, and many other advantages of polyester polyols over the poly (tetramethylene ether) glycol (PTMG), which was introduced by DuPont in 1956. Later, in 1957, BASF and Dow industries started producing chemicals such as polyalkylene glycols [2] 4,4′-diphenylmethane diisocyanate (MDI), and ethylenediamine, which were the starting material for PU. It took almost a decade in the transition of PU from rigid PU foams (polyisocyanurate foams-1960) to flexible PU foams (1967), with the involvement of several blowing agents. In 1969, high-performance PU materials were developed through reaction injection molding (RIM) technique. Due to this technology, advancements were made in reinforced reaction injection molding (RRIM) for the production of high-performance PU material. In 1983, the first plastic-body automobile became functional in the United States. Recently, the world of PU has come through a long passage like the preparation of PU hybrids, non-isocyanate PU, and PU-based composites, which are being applied successfully in a wide range of advanced applications [3].

18.2 SYNTHESIS OF POLYURETHANE

PUs are polymers made by combining diisocyanates like toluene diisocyanate (TDI), methylene diphenyl diisocyanate (MDI), and polyols. There are hundreds of different types of PUs, and each is made in a slightly different way. PU is one of the traditional polymers most commonly formed by treating di- or polyisocyanate with polyol; both these reagents that are used to develop PU contain one or more functional groups per molecule. PU can also form through the polymerization of the carbamate unit [4]. The major chemicals for PU synthesis are isocyanate, like methylene diphenyl diisocyanate (MDI) and toluene diisocyanate (TDI). They are mixed to form long chemical structures because they have few places to bind with each other. PUs are generally used to make flexible products like seat cushions, mattresses, and sealants. Polymeric MDI produces a more reliable structure, generally used to make rigid products like picnic coolers, foam insulation, and wood boards. Resins are mainly composed of polyols, which construct the backbone of the polymer and also determine the physical properties of the product, such as mechanical strength, hardness, etc. Besides the main ingredients or precursors of the reaction, the reaction condition or additional additives are also important, such as catalysts, which control the rate of chemical reaction; surfactants, which determine the structure and influence the flow; pigments, which determine the color; flame retardants, which make it safer; adhesion promoters, which make it sticky; and finally blowing agents, which help determine the density and foaming action. The isocyanate and polyols combine under various conditions or in the presence of different reagents to form unique products, ranging from flexible to rigid PU.

PU acquired importance due to its many applications. PU nowadays is used as an insulating material that cools in summer and warms in winter to avoid energy loss, such as in refrigerators. It is also used in extreme environments like spacesuits and diving suits in cold water. One of the most important uses of PU is in furniture. More importantly, PU can be recycled easily; it is environmentally friendly and eco-friendly. A class of PU called spandex is quite elastic and flexible; thus, it is used in socks, undergarments, and many sports-related items [3,5]. Isocyanate affects PU because of its chemical nature. PU exists in various forms, like foam, elastomers, paints, coatings, adhesives, and elastomeric plastics. Commonly used PU films are flexible, semi-rigid, rigid, and integral. Flexible PU is used in bedding, furniture, automotive interiors, underlying carpets, and packaging. These flexible PUs are light, durable, supportive, and comfortable. In contrast. rigid PU is used in insulation material, mainly in walls and roofs. Isocyanate is a compound that is reactive and has a lower molecular weight; moreover, it causes irritation to the eyes and skin and also causes breathing problems. That's why it should be handled properly [6]. The most commonly used isocyanate for the synthesis of PU is TDI in the manufacture of flexible PU films. It has a fast gelation rate but a poisonous effect. The other isocyanate used in the manufacture of rigid and some flexible PU foams is MDI [7].

18.3 CLASSIFICATION OF POLYURETHANE

PUs are classified into the following classes:

- Flexible PU foam
- Rigid PU foam
- Coating, adhesives, sealants, and elastomers
- Thermoplastic PU
- Binders
- Waterborne polyurethane (WPU)

18.4 WATERBORNE POLYURETHANE

WPU is a polymer that disperses in water and has a low odor and no free isocyanate, which makes it safe and allows it to be used in multiple applications. In other words, WPU is a class of coatings and adhesive materials that uses water as the primary solvent. In addition to offering high performance in many modern applications, the characteristics such as low odor and absence of residual isocyanate make PU a safer potential alternative material for many applications [8]. In today's lifestyle, the coatings and adhesives industries are important because of the development of high-performance and eco-friendly polymers like WPU. Thus, WPU is considered to be a potential candidate due to its excellent performance and environmental compatibility, compared to solvent-based PUs. Materials synthesized from WPUs are increasingly significant since they can deliver much better results in modern applications. Researchers and manufacturers are working to deliver high-performance products that are not only cost-effective but also environmentally friendly, too [9]. The application of materials based on WPU is dependent upon its physical properties, like viscosity, solidity, wetting properties, etc. Flexible and rigid substrates, such as flooring, leather, metal, fabric, and paper, are the systems that prefer to be cured at ambient conditions, that is, air-dried or baked coatings.

18.5 ADVANTAGES OF WATERBORNE POLYURETHANE

- The volatile organic compounds (VOCs) in solvent-free PU dispersions (PUDs) are much lower compared to WPU dispersions (WPUDs) that contain a co-solvent system.
- On sensitive substrates, solvent-free PUDs are minutely aggressive and can be damaged by solvents.
- WPUDs are slow drying if they contain co-solvent because the most common co-solvents are much slower to evaporate water than solvent-free PUDs.
- Accordingly, environmentally friendly and user-friendly free PUDs are even more important than WPUDs that contain co-solvent.

WPUs have a number of applications in many fields. Despite all these applications, the most important application of WPU is conductive behavior. WPU is used as a conductive agent in practical applications, shows reliable performance, and is stable compared to PUs [10]. WPU formulation is green synthesis with no emission of any VOCs, which has attracted attention from scientists because of environmental considerations. In the past few decades, WPU has received considerable attention because it is used as a substrate in the coatings and adhesives industries. The application of WPU includes automotive, packing, construction, transportation, electronics, textiles, etc. For the synthesis of stable PU ionomers, minimum ionic content is required, which is an important feature of WPU over conventional PU [11]. A large number of techniques is available in which WPU is used as a conductive agent, depending on which type of species is required. Formation of the ion, which can be an anion or cation, forms an anionic PUD or cationic PUD. It is also possible to prepare a nonionic PUD used as a conductive agent. These ionic

PUDs are synthesized by using materials that will produce an ethylene oxide or other chain pendant from the main polymer backbone. Anionic PUDs are the most common ones available commercially. Some of the modifiers, like dimethylol propionic acid (DMPA), contain two hydroxy groups as well as a carboxylic acid group, which is a perfect example of cationic PUD [12].

18.6 CONDUCTIVE COATING OF WPU AND CNT

The majority of raw materials and petrochemical products are flammable and corrosive. The sparks originating from turboelectric charging can easily allow petrochemicals to catch fire abruptly, which may cause severe damage. To avoid such a loss the conductive coatings can play a vital role by avoiding the formation of electrostatic sparks by means of discharging static electric current [13,14]. Therefore, anticorrosive as well as conductive coatings have been widely used in the petrochemical industry. However, in the past, the majority of conductive coatings were oil based and had safety problems because during the construction process they can produce flammable organic solvents that can catch fire easily. Additionally, the content of conductive agents that already exist in conductive coatings is very high. Generally, it should be almost 10 wt.%. When the amount of conductive agent in the conductive coating is higher, it will generate interfaces between the polymer matrix and conductive mediums after the curing process [15]. These practical issues can be solved when WPU is used as a polymer matrix due to its safe and reliable ability with multiwalled carbon nanotubes (MWCNTs) used as a conductive medium because of its properties like anticorrosive and electro conductibility. MWCNT/WPU dispersion is prepared through magnetic stirring or ultra-sonication in such a way that low content of MWCNTs as a conductive agent is dispersed in the WPU matrix, which is sprayed on a substrate that is a steel rod by means of electrostatic spraying, resulting in a thin layer on the substrate with good anticorrosive and conductive abilities [13]. For good atomization of material, electrostatic spraying is used, which also prevents MWCNTs from agglomeration. The effect of electrical conductivity, as well as corrosion resistance due to MWCNT content, was that similar adhesive strength of the WPU conductive coating synthesized by electrostatic spraying was reported. These results provide a theoretical clue for preparing water-based conductive coating with low content of conductive mediums through electrostatic spraying [16]. Figure 18.1 portrays the use of WPU for conductive coatings.

18.7 CONDUCTING COATING OF GRAPHENE OXIDE/ WATERBORNE POLYURETHANE

Graphene oxide (GO) is the most often used nano-filler for waterborne coatings due to its outstanding mechanical and physical properties, very good dispersibility in aqueous medium, and cost-effectiveness compared to graphene [17]. The performance of WPU used as a nano-coating formulated with different GO loadings can be effective due to its conductive properties. When the degree of dispersion or adhesion for GO was evaluated, GO with WPU showed good conductive properties

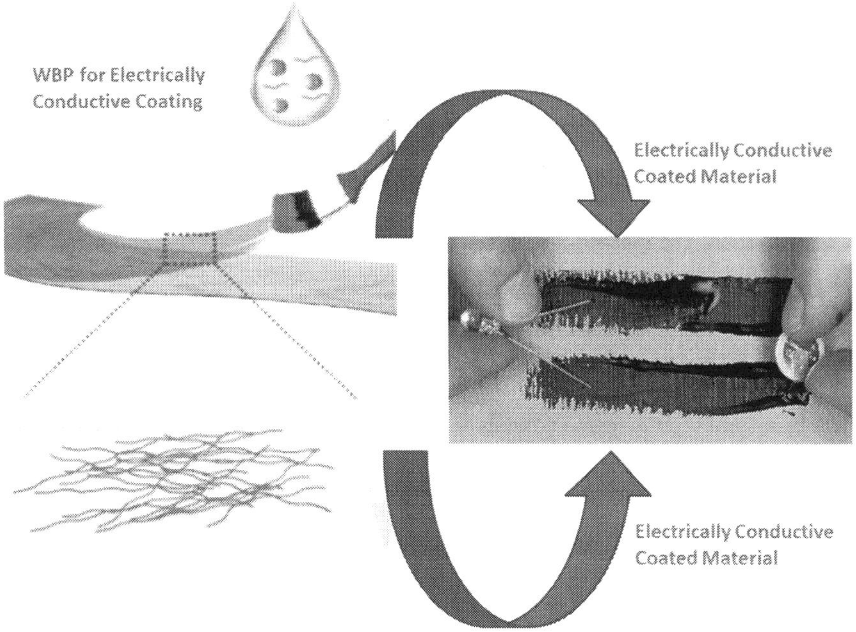

FIGURE 18.1 Representation of waterborne polyurethane used for conductive coatings.

in addition to mechanical properties [9,18]. Higher content of loadings between 1.2% and 2% in GO increased the conductivity of WPU by 200%–300%, respectively, and thermal conductivity was also increased by 38% of the composite. The composite of GO with WPU is a suitable choice to increase the conductive properties of WPU. Figure 18.2 illustrates the schematic synthesis of composite WPU/GO.

Graphene is a monolayer of carbon with a hexagonal lattice. It is a wonder material known to have exceptional properties like 2,630–2,965 m^2/g of specific surface area [20]. It shows an electrical conductivity of 10 S/cm with a modulus of 10 GPa and shows outstanding thermal conductivity of 3,000 W/mK [21]. Due to its exceptional properties, graphene is an effective material to be used in diverse applications, mostly in electronics such as sensors, supercapacitors, and batteries. In the recent few decades, mostly inorganic nanomaterials function as fillers in the fabrication of polymer/inorganic composites, which are conductive in nature and have many potential applications like electromagnetic shielding, corrosion resistance, and antistatic coatings. WPU comprises a minute amount of organic solvent, which is why it has very little environmental effect. The majority of WPU is used in adhesives and coatings [18], but WPU has some conductive properties as well. For electronic equipment, such materials are selected with unique conductivity because it is of prime importance in such devices [22]. WPU/graphene composites were prepared through solution mixing, and electrical conductivity reaches its peak value of 5.47×10^{-4} S/cm with a loaded mass of graphene of 5.0 wt.%. However, the composite prepared by solution mixing

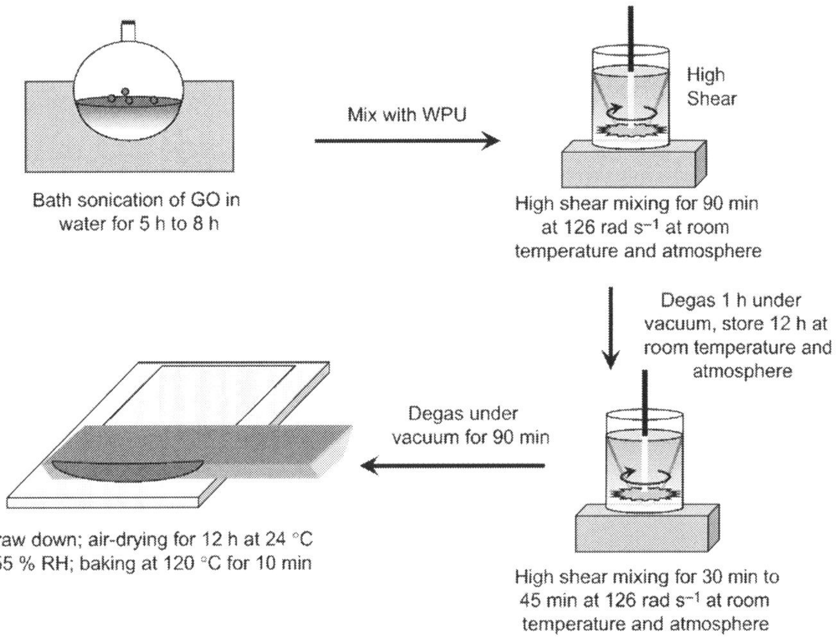

FIGURE 18.2 Schematic synthesis of composite WPU/GO. Reproduced from reference [19] with permission. Copyright © 2019, American Coatings Association.

technique usually has lower electrical conductivity compared to composites prepared by in situ intercalation method, but properties like dispersion and stability of the graphene solution are not effective [23]. Different techniques have been practiced for the development of composites of WPU with structural carbons. Hydrothermal or aqueous carbonization was used for the conversion of GO into graphene to improve dispersion of graphene. Polyvinylpyrrolidone is used to form a stable solution of graphene with higher concentration. WPU/graphene composites were prepared by mixing graphene solution with WPU [24]. The electrical conductivity in terms of S/cm of WPU was improved sufficiently by incorporation of graphene. When the loaded mass of graphene increased up to 4.0 wt.%, conductivity of 8.31×10^{-4} S/cm was reported.

18.8 WPU COATING MATERIALS CONTAINING POLYANILINE AND OTHER POLYMERS

Since successful preparation of conducting polymers like polyacetylene in 1977 by Shirakawa et al., these polymers have been used in many applications like batteries, electrochromic devices, electromagnetic interference (EMI) shielding, and sensors, and also to replace metallic conductors in a large variety of electrical devices. That is why electrical conducting polymers got huge interest due to their use in potential applications in this diverse field [25]. Among various electrical conducting polymers, polyaniline (PANI) showed an electrical conductivity that exists in the range of 10^{-10} to 10^{-2} S/cm; it is used for flexible LED and antistatic materials. PANI is under study because it is cost–effective, environmentally friendly, stable, and easily polymerize [26]. Jeevanandas et al. prepared PANI/PC and reported the electrical conductivity of 5.7×10^{-5} to 4.7 10^{-2} S/cm. Ho et al. [27] synthesized PANI/PU blends which showed excellent electrical conductivity of 0.570–23.50 S/cm and stable enough to show 2.11–5.55 Mpa of tensile strength. PANI/SBS blends show 10^{-2} to 100 S/cm of conductivity and were synthesized by Leyva et al. Roichman et al. synthesized PANI/PS blends composite and reported electrical conductivity between 10–15 and 100 S/cm [28]. Rupali groups synthesized PANI/PVA a conducting polymer composite show 10–100 S/cm of electrical conductivity. WPUs are nonflammable, nontoxic, and environmentally friendly. Moreover, WPUs are used in versatile commercial applications, such as adhesives and coatings for many substrates. However, now that researchers use WPUs as conducting polymers and they have importance in electrical conduction applications, the conducting polymer like PANI-DC has been prepared through oxidative polymerization in an aqueous system. The conductivity of WPU can be increased in several ways. Conductive polymers can play a vital role by increasing the conductivity of WPUs by preparing an aqueous dispersion of poly(aniline-dodecyl benzene sulfonic acid complex) (PANI-DC) viscosity (η) 1.30 dL/g using aniline as a monomer [29] and by using dopant dodecylbenzene sulfonic acid (DBSA) and ammonium peroxydisulfate (APS) as an oxidant. The electrical conductivity of the composite was 0.7 S/cm, which is a reliable result. Blended films of the resultant composite, that is, WBPU/PANI-DC, fabricate with different ratios of weight through solution casting and visualization.

For the development of coating materials with good conductive property, composite formation is one important and suitable technique. Composites of the conducting polymer with WPU, like (WPU)/PANI-DBSA complex (PANI-DC) (100–66/0–34 wt.%), as a result of solution blending could form an aqueous dispersion of WPU and PANI-DC by physical treatments like magnetic stirring or ultrasonication. Conductivity and other physical properties increase.

18.9 CONDUCTIVITY OF COATED COMPOSITE OF WPU/GNS

Electrical conductivity and EMI are considerable problems in structural development in sensitive electronic devices like optoelectronic devices, spacecrafts, and densely packed systems with diverse applications [30], but recently the main focus of scientists for the development of EMI shielding materials has been finding materials with the following characteristics: inexpensive, flexible, lightweight, and easy to process. Mostly metal-based composite is used due to unique electrical conductivities in EMI shielding materials. However, some of the problems can also be associated with such metal-based materials. Mostly, they are bulky as well as difficult to process. In electroplating vacuum deposition approaches for EMI shielding applications, metal-based composite materials can deposit on lightweight materials forming a homogeneous layer on the substrate, but some of the problems can also be associated with such coating, such as oxidation, resistance, corrosion, and difficulty processing. Composite-based conductive polymers are also a substituent material to metal-based

EMI shielding materials. The composites made from conductive polymer have the following advantages: lightweight, corrosion-resistant, flexible, easy to handle, and suitable for EMI shielding applications like flexible electronic devices as well as automobiles. The shielding efficiency of composite is associated with the following properties: intrinsic electrical conductivity, filler content, and aspect ratio. In graphene nanosheets (GNs), 2D hexagonal lattices of carbon atoms arrange in a honeycomb structure and are extremely lightweight nanomaterials with importance because of their mechanical, electrical, and thermal conductivities [25]. Therefore, GN/polymer is considered a proposed composite material for researchers developing high EMI shielding properties in addition to conductive channels, which enable the polymer matrix to become electrically conductive with an effective EMI. But the problem associated with the GNs is compatibility and aggregation in the polymer matrix, so it is not an easy task to achieve promising results. Mainly, electrical conductivity can be removed through the homogeneous dispersion of GNs and the polymer matrix. To improve the compatibility of GNs and polymer matrices, surface changes called modification of GN is a reliable method that can be achieved in many ways. Mainly, covalent and noncovalent interactions or modifications are important. Covalent modification is done through grafting of organic functionality on the surface of GN, which also increases the interactions between GNs and the polymer matrix. In contrast to covalent modification, a noncovalent modification involves van der Waals interaction. The use of a noncovalent modification without any structural changes of GN via adsorption of a cationic surfactant can improve compatibility, but interaction that develops through noncovalent modification is weak, and the dispersion is unstable during the preparation of polymer composites. In contrast with noncovalent modification, covalent modification of GNs is a better option because, due to such modification, strong interaction is developed between the polymer matrix and GNs. The development of such polymeric materials as WPU is not only environmentally friendly but also exhibits desirable properties for global researchers due to its mechanical and electrical performance. Substitution of hydrophilic groups, such as sulfonic or carboxyl, make homogeneous dispersion in an aqueous solution and make WPU negatively charged. In addition, the soft and hard segments of WPU cannot be changed and can be easily handled and modified to use in many applications. Flexible composites can be synthesized from the GN, and WPU shows excellent electrical performance and electromagnetic shielding. GNs are covalently modified through free radical polymerization reaction with aminoethyl methacrylate (AEMA; AEMA-GNs). Moreover, the composite of AEMA-GNs shows desirable compatibility with the WPU matrix. In addition to the grafting, some functional groups, like sulfonated, show electrostatic attraction, caused by AEMA-GNs uniformly scattered in the matrix. Such distribution of AEMA-GNs enables GNs to produce electrically conductive channels. The electrical conductivity and EMI shielding performance of AEMA-GN/WPU with different ratios of AEMA segments were incorporated into the matrix. GN loading of 5 vol.% shows promising conductivity (approximately 43.640 S/cm). Sheng et al. fabricated GNs through free radical polymerization, a covalent modification that was mainly processed to prepare a composite of aminoethyl-methacrylate-grafted GNs (AEMA-GNs), and studied the important effects of functional group morphologies and electrical conductivities.

18.10 CONDUCTIVE COATINGS OF MWCNT/WPU

A class of eco-friendly coatings, such as WPU in aqueous medium, is an interesting material for researchers, and it usually replaces solvent-based PU because it has environmentally friendly characteristics. However, WPU has poor mechanical strength and conductivity, which is why it may restrict the use of WPU in many applications. Therefore, the modification of WPU is necessary to meet the requirements of harsh conditions. These modifications can be done in many ways; researchers have previously shown that the fabrication of organic/inorganic nanocomposites by the reinforcement of nanoparticles to polymer matrix not only increases physical properties like conductivity, but also enhances mechanical properties of polymers. Carbon materials like MWCNTs are known for their performance due to sp^2 hybridized structure so the reinforcement of MWCNTs in WPU as matrix the conductivity and mechanical properties of WPU-coated materials sufficiently increase. The crosslinking reaction will proceed between the polar and nonpolar groups that are present on the molecular chain of WPU, during the curing groups like –OH adsorbed on the surface. Moreover, the fibrous structure of the MWCNT mechanical property of the composite coating will also increase. The conductive coating of the composite can be synthesized through electrostatic spraying (ES), in which a nozzle acts as a spray gun that generates high-voltage discharge because such high-voltage discharge establishes the electrostatic field between the metal substrate and gun. The liquid coating ejected from the nozzle bears negative charges. Under the influence of electrostatic attraction, it is attracted to the substrate that bears a positive charge. The coating could be more effective if atomization is achieved by using compression for the transportation of liquid coating and also avoiding the agglomeration of conductive filler [13].

Some additional importance of ES is that it will overcome the problem of overloading of coating material and will provide uniform distribution, which cannot be achieved through brushing (BrC). The nanocomposite MWCNT/WPU material can be fabricated through ES not only to improve the dispersity of filler but also to promote antistatic, mechanical properties of the WPU coating.

18.11 CONDUCTIVITY OF WPUS COATED WITH METAL OXIDE

As we discuss, there are many ways to improve the EMI shielding and electrical conductance of WPU by fabricating the composite. Just like the carbonaceous materials are reinforced into the polymer matrix to improve certain properties, some metal-based composites can play a vital role by enhancing conductivity and other properties [4]. WPU composites with FeCo alloy are decorated with reduced GO (FeCo/rGO) nanosilver (Ag) to form a layer arrangement [31]. The FeCo/rGO layer acts as an absorption layer that is generated through natural sedimentation proceeds at the bottom of the WPU film, and the ultra-thin Ag layer acts as a highly efficient shielding layer over the top surface to maintain good electrical properties. The obtained composite that is Ag/rGO@FeCo/WPU film (10 wt.% rGO@FeCo and Ag layer with a thickness of 500 nm the schematic procedure for the fabrication) is given in Figure 18.3.

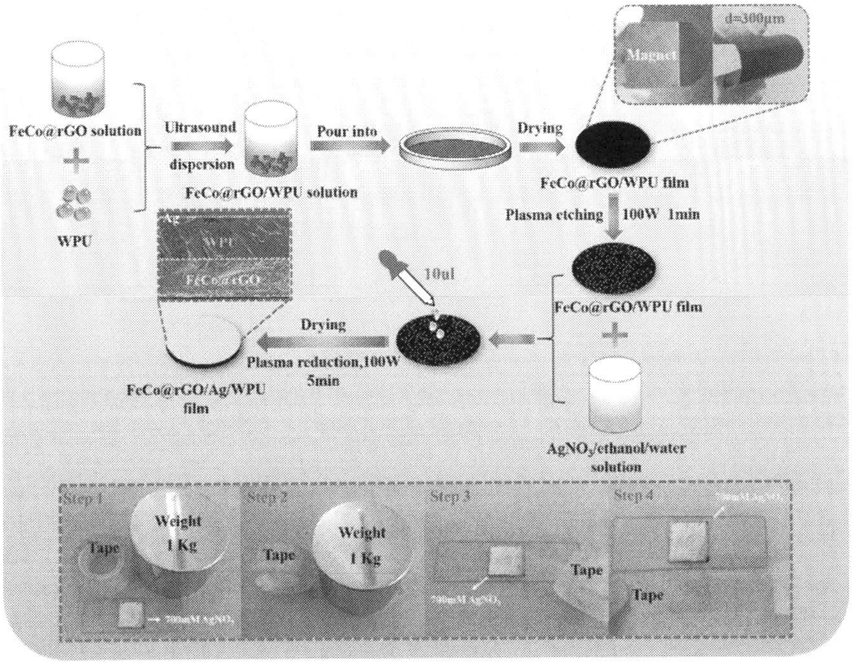

FIGURE 18.3 Synthesis of Ag/rGO@FeCo/WPU film. Reproduced from reference [32] with permission. copyright© 2018 Elsevier B.V.

Recently, with an increase in demand for portable electronic devices, various EMI shielding materials have been developed due to their excellent electrical conductivity and the compatibility of metallic oxide with conducting polymers [33]. Metal and metal-based composite with conducting polymer is also a vital choice for researchers for the fabrication of materials with outstanding electrical conductivity and effective EMI shielding performance. This specific structural layer has the ability to be selective material. Ag/rGO@FeCo/WPU composite film shows excellent EMI shielding performance as well. Figure 18.4 shows the schematic description of electromagnetic microwave dissipation in the Ag/rGO@FeCo/WPU composite film.

18.12 THERMAL CONDUCTIVITY OF WPUS

WPUs have been moving toward multifunctional WPUs that are used in developing numerous products and have many applications because of their ease of availability and eco-friendly nature. Despite their use in many applications, some of the miniaturization drawbacks regarding heat, such as heat dissipating and heat aging, need to be resolved [34]. The organic polymers, because of their excellent performance as substitutes for metal substrate, can be used in electronic devices. However, many of the organic polymers do not show thermal conduction for the improvement of thermal conductivity of composites. Many researchers rely on

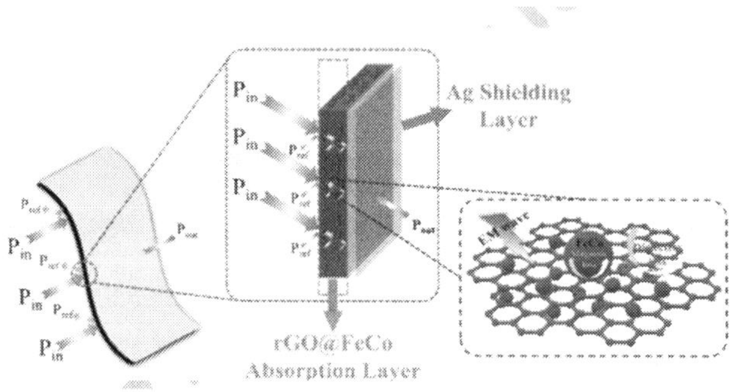

FIGURE 18.4 Schematic description of electromagnetic microwave dissipation in the Ag/rGO@FeCo/WPU composite film. Reprinted from reference [32] with permission. Copyright© 2018 Elsevier B.V.

the hybrid method. The conductivity and compatibility of hybrid methods include physical and chemical modification with noncovalent and Vander Waals interactions, as discussed earlier in the chapter. WPU has microcrystal structures consisting of short- and long-range ordering, which facilitate the thermal conductivity of the material. Because electronic products run for a long time, they will generate heat, which in turn causes adverse effects, such as heat aging, lifespan reduction, or even financial loss.

In addition, the reliability of these devices depends upon operating temperature. The lifespan of these devices may be reduced through small fluctuations in operating temperature (10–15°). The thermal conductivity of metallic materials is much higher and brings some limitations, like increased weight and corrosion [35]. Therefore, to improve the lifespan and weight issues of products, special products are needed that possess thermal conductivity, are lightweight, and dissipate heat under high frequency. Composite materials of conducting polymers are cheap, lightweight, and anticorrosive, and they have attracted more attention recently because of their outstanding merits over metallic materials. For instance, most organic polymers show poor thermal conduction, which is why they are limited in certain fields. Global demand is for materials with better performance in regard to heat transfer and that fulfill the requirements of heat dissipation. The thermal conductivity of polymer materials can be improved in two aspects according to researchers. Synthesizing polymers that possess a high degree of crystallinity increases the transportation of phonon. Materials that have a higher degree of polarization and consist of conjugate radicles increase the pathway for heat transfer, like polyacetylene (PA), polypyrrole (PPY), and polyaniline. The addition of fillers into the polymer matrix increases heat transfer performance, but the filler used must be conductive. Lower content of fillers in the polymer matrix leads to lower performance and vice versa [36]. WPU bears excellent physical properties, including abrasion high tensile strength, tear resistance, and solvent resistance. The development of conductive polymers like WPU has received attention, especially in the coating and adhesive industries. In addition, WPUs have many advantages, which have already been discussed. WPUs can effectively be used in electronic devices as coatings for electrical conductance and electromagnetic shielding. These properties of WPUs can be developed by the involvement of filler. Nanosized conductive fillers and the properties of composite materials depend upon the loading of the filler, such as graphene monolayer of graphite, and have found applications as the key nanofiller in carbon-based nanofillers to strengthen the resulting nanocomposite and affect the thermal conductivity. Carbon in the form of graphite, graphene, and CNT hybrid with WPUs enhances thermal conductivity in thermal engineering. The schematic diagram of how the thermal conduction of WPUs can be increased follows. Conducting polyaniline is doped by DBSA from chemical oxidation. Conducting polyaniline is doped with DBSA and then coupled with the WPUD results in the formation of WPU/PANDB, which is a conductive blend coating, which is then applied on poly (ethylene terephthalate) (PET) film and dried to produce WPU/PANDB conducting blend film. This film is used for multiple applications, like thermal and electrical conduction.

18.13 CONCLUSION

The conductivity of materials can be increased via coating of WPU. The coating is done through various techniques like facile latex technology, BrC method, electrospraying, etc. The conductivity of the composite depends upon the dispersion of the WPU and also the amount of the conductive filler like MWCNT, SWCNT, GO, and many other conductive polymers. Lower wt.% of filler with excellent dispersion shows excellent conductivity. Furthermore, composites exhibited desirable mechanical properties for practical applications, especially for stretchable conductive devices.

REFERENCES

1. Wen, T.-C. and Y.-J. Wang, Application of experimental design to the conductivity optimization for waterborne polyurethane electrolytes. *Industrial & Engineering Chemistry Research*, 1999. **38**(4): p. 1415–1419.
2. Chen, C.-H., et al., Fabrication and characterization of water-based polyurethane/polyaniline conducting blend films. *Surface and Coatings Technology*, 2013. **231**: p. 71–76.
3. Chattopadhyay, D.K. and K. Raju, Structural engineering of polyurethane coatings for high performance applications. *Progress in Polymer Science*, 2007. **32**(3): p. 352–418.
4. Gurunathan, T., et al., Polyurethane conductive blends and composites: synthesis and applications perspective. *Journal of Materials Science*, 2013. **48**(1): p. 67–80.
5. Akindoyo, J.O., et al., Polyurethane types, synthesis and applications–a review. *Rsc Advances*, 2016. **6**(115): p. 114453–114482.
6. Rojek, P. and A. Prociak, Effect of different rapeseed-oil-based polyols on mechanical properties of flexible polyurethane foams. *Journal of Applied Polymer Science*, 2012. **125**(4): p. 2936–2945.
7. Song, Y., et al., Effect of isocyanates on the crystallinity and thermal stability of polyurethanes. *Journal of Applied Polymer Science*, 1996. **62**(5): p. 827–834.
8. Ma, X.-Y. and W.-D. Zhang, Effects of flower-like ZnO nanowhiskers on the mechanical, thermal and antibacterial properties of waterborne polyurethane. *Polymer Degradation and Stability*, 2009. **94**(7): p. 1103–1109.

9. Hu, J. and F. Zhang, Self-assembled fabrication and flame-retardant properties of reduced graphene oxide/waterborne polyurethane nanocomposites. *Journal of Thermal Analysis and Calorimetry*, 2014. **118**(3): p. 1561–1568.
10. Wen, T.-C., et al., The effect of DMPA units on ionic conductivity of PEG–DMPA–IPDI waterborne polyurethane as single-ion electrolytes. *Polymer*, 1999. **40**(14): p. 3979–3988.
11. Guo, Y.-h., et al., Properties and paper sizing application of waterborne polyurethane emulsions synthesized with isophorone diisocyanate. *Progress in Organic Coatings*, 2014. **77**(5): p. 988–996.
12. Zhou, X.-d., Y.-l. Jian, and G.-b. Hu, Synthesis and application of waterborne blocked polyurethane. *Textile Auxiliaries*, 2007. **5**.
13. Wang, F., L. Feng, and M. Lu, Mechanical properties of multi-walled carbon nanotube/waterborne polyurethane conductive coatings prepared by electrostatic spraying. *Polymers*, 2019. **11**(4): p. 714.
14. Wu, Y., et al., Ultralight graphene foam/conductive polymer composites for exceptional electromagnetic interference shielding. *ACS Applied Materials & Interfaces*, 2017. **9**(10): p. 9059–9069.
15. Tian, Y., et al., Carbon nanotube/polyurethane films with high transparency, low sheet resistance and strong adhesion for antistatic application. *RSC Advances*, 2017. **7**(83): p. 53018–53024.
16. Wang, F., et al., Effect of coating process on the properties of multi-walled carbon nanotubes/waterborne polyurethane anticorrosive and conductive coating. *AIP Advances*, 2019. **9**(3): p. 035241.
17. Tian, K., et al., N-doped reduced graphene oxide/waterborne polyurethane composites prepared by in situ chemical reduction of graphene oxide. *Composites Part A: Applied Science and Manufacturing*, 2017. **94**: p. 41–49.
18. Lei, L., et al., Preparation and properties of amino-functional reduced graphene oxide/waterborne polyurethane hybrid emulsions. *Progress in Organic Coatings*, 2016. **97**: p. 19–27.
19. Bernard, C., et al., Graphene oxide/waterborne polyurethane nanocoatings: Effects of graphene oxide content on performance properties. *Journal of Coatings Technology and Research*, 2020. **17**(1): p. 255–269.
20. Mirmohseni, A., M. Azizi, and M.S.S. Dorraji, Facile synthesis of copper/reduced single layer graphene oxide as a multifunctional nanohybrid for simultaneous enhancement of antibacterial and antistatic properties of waterborne polyurethane coating. *Progress in Organic Coatings*, 2019. **131**: p. 322–332.
21. Zhang, W., et al., Exfoliation and defect control of graphene oxide for waterborne electromagnetic interference shielding coatings. *Composites Part A: Applied Science and Manufacturing*, 2020. **132**: p. 105838.
22. Ding, J., et al., Electrical conductivity of waterborne polyurethane/graphene composites prepared by solution mixing. *Journal of Composite Materials*, 2012. **46**(6): p. 747–752.
23. Luo, J., et al., Improved corrosion resistance based on APTES-grafted reduced sulfonated graphene/waterborne polyurethane coatings. *Journal of Coatings Technology and Research*, 2018. **15**(5): p. 1107–1115.
24. Nan, B., et al., Covalently introducing amino-functionalized nanodiamond into waterborne polyurethane via in situ polymerization: Enhanced thermal conductivity and excellent electrical insulation. *Colloids and Surfaces A: Physicochemical and Engineering Aspects*, 2020: p. 124752.
25. Kwon, J.-Y., E.-Y. Kim, and H.-D. Kim, Preparation and properties of waterborne-polyurethane coating materials containing conductive polyaniline. *Macromolecular Research*, 2004. **12**(3): p. 303–310.

26. Pei, Q. and X. Bi, Electrochemical preparation of electrically conducting polyurethane/polyaniline composite. *Journal of Applied Polymer Science*, 1989. **38**(10): p. 1819–1828.
27. Kazantseva, N.E., et al., Magnetic behaviour of composites containing polyaniline-coated manganese–zinc ferrite. *Journal of Magnetism and Magnetic Materials*, 2004. **269**(1): p. 30–37.
28. Tian, Z., et al., Recent progress in the preparation of polyaniline nanostructures and their applications in anticorrosive coatings. *RSC Advances*, 2014. **4**(54): p. 28195–28208.
29. Hsiao, S.-T., et al., Effect of covalent modification of graphene nanosheets on the electrical property and electromagnetic interference shielding performance of a water-borne polyurethane composite. *ACS Applied Materials & Interfaces*, 2015. **7**(4): p. 2817–2826.
30. Kausar, A., Emerging research trends in polyurethane/graphene nanocomposite: A review. *Polymer-Plastics Technology and Engineering*, 2017. **56**(13): p. 1468–1486.
31. Rasheed, T., et al., Catalytic potential of bio-synthesized silver nanoparticles using Convolvulus arvensis extract for the degradation of environmental pollutants. *Journal of Photochemistry and Photobiology B: Biology*, 2018. **181**: p. 44–52.
32. Zhu, H., et al., Layered structural design of flexible waterborne polyurethane conductive film for excellent electromagnetic interference shielding and low microwave reflectivity. *Applied Surface Science*, 2019. **469**: p. 1–9.
33. Bilal, M., et al., Photocatalytic degradation, toxicological assessment and degradation pathway of CI Reactive Blue 19 dye. *Chemical Engineering Research and Design*, 2018. **129**: p. 384–390.
34. Li, R., J.T. Loontjens, and Z. Shan, The varying mass ratios of soft and hard segments in waterborne polyurethane films: Performances of thermal conductivity and adhesive properties. *European Polymer Journal*, 2019. **112**: p. 423–432.
35. Liu, X., et al., Preparation and properties of waterborne polyurethanes with natural dimer fatty acids based polyester polyol as soft segment. *Progress in Organic Coatings*, 2011. **72**(4): p. 612–620.
36. Zhang, S., et al., Thermal and crystalline properties of water-borne polyurethanes based on IPDI, DMPA, and PEBA/HNA. *Journal of Applied Polymer Science*, 2007. **103**(3): p. 1936–1941.

19 Waterborne Polyurethanes for Electrical Applications

V. Dhinakaran, P.M. Bupathi Ram, S. Narain Kumar, M. Tharun Kumar, K.P. Manoj Kumar, and M. Varsha Shree
Centre for Applied Research, Chennai Institute of Technology, Chennai, 600069, India

CONTENTS

19.1 Introduction....321
19.1.1 Electrical Applications of Waterborne Polyurethanes....322
19.1.1.1 Electric Vehicle....322
19.1.2 Urethane Potting Compounds....323
19.1.3 Thermoelectric Generator....323
19.1.4 Electro-Conductive Para-Aramid Knit....324
19.1.5 Stretchable Conductors....324
19.1.6 Stretchable Pressure Sensor....324
19.1.7 Wearable Electric Heaters....325
19.1.8 Electronic Fabric....325
19.1.9 Strain Sensing....326
19.1.9.1 Stretchable Electronic Skin....326
19.1.10 Thermotherapeutic Devices....327
19.1.11 Electromagnetic Interference Shielding (EMI)....327
19.1.12 Recycling of Waterborne Polyurethanes....329
19.2 Future Possibilities....330
19.3 Concluding Remarks....330
References....331

19.1 INTRODUCTION

Polyurethanes are distinct materials that belong to the polymeric group of materials with material characteristics different from several plastic materials. These materials find their application in various fields, such as liquid layers, paints, coatings, elastomers, foams, nonconductors, connective tissues, etc. [1,2]. The method of aqueous polyurethane dispersions was developed in the late 1960s. Waterborne polyurethane (WPU) dispersion is easily defined as the binary colloidal system where the particles

DOI: 10.1201/9781003173526-19

of the polyurethane will be dispersed in the continuous aqueous medium. This basic idea is easily executed through the integration of hydrophilic monomers with ionic functionality, along with sulfonate, carboxylate, or quaternary ammonium groups. These ionic groups are called internal emulsifiers and polyurethane ionomers [3]. The application of WPUs depends on the following major reasons: at first, administration highlighted the decrease in solvent emission into the atmosphere. Second, solvent prices are too high. On the other hand, WPU's quality has made it highly suitable for a broad spectrum of applications [4]. Since these WPUs are environmentally friendly, they do not pollute the environment. They are nonflammable, are nontoxic, and don't pollute the water. As only water evaporates during this process, it does not cause any harm to the environment. These materials provide ideal properties, such as durability, toughness, and extreme elasticity properties, making them an alternative for metals, plastics, and rubber in various engineering applications [5]. Their other applications include biomedical, building and construction, automotive, textiles, etc., due to their extreme properties, such as hardness, elongation, stiffness, strength, and modulus. These materials can be synthesized from various sources and applied to several applications according to their requirements as well as placed in groups such as thick, thermoplastic, soft, binders, waterborne, sealants, coatings, adhesives, and elastomers [6]. This chapter deals with electrical applications of WPU.

19.1.1 Electrical Applications of Waterborne Polyurethanes

19.1.1.1 Electric Vehicle

Covestro Taiwan has introduced a range of aqueous laminating material solutions for single or two-component WPU coating systems, including Bayhydrol® dispersion and Bayhydur® polyisocyanate crosslinkers, to support motorcycle manufacturers in greening. Primers, base coats, topcoats, clear finishes, and various coat levels can be used. They are available in the car. They can also lower emissions of VOCs by up to 80% and increase the safety of the industrial environment. Moreover, Covestro's newest WPU coating solution is highly reactive and significantly decreases energy usage that reduces carbon dioxide emissions during production, compared to previous waterborne solutions. Being easy to spray, being appropriate for a broad variety of work circumstances, and having blister film-free thickness are features of double WPU coating technology. The texture and external design of the motorbike yield amazing results, and the overall application efficiency of the solvent-borne system is comparable. Nick Sun, head of coatings, adhesives, specialties (CAS) of Covestro Taiwan, said, "Water technology has a major role for many sectors towards sustainability and 'going green.'" In the development of electric cars, apart from the problem of air pollution, solutions still require creating greater battery capacity and longer cruising distance. Therefore, it is essential to develop lighter carriages and components. The polyurethane soft foam of Covestro is applied to motorcycle seats for improved comfort, and TPU may also be utilized for handlebars, tires, and other soft plastic applications. Polycarbonates (PCS) are appropriate for light systems, battery packs, and indoor and outdoor electric vehicles. Thermoplastic composites also have features such as being lightweight and having great toughness when utilizing PCS as a foundation material. They may also be used for interior and external electric vehicles, battery external components, and other manufacturing sectors and may deliver goods with a distinctive design quality [7].

19.1.2 Urethane Potting Compounds

Potting compounds of urethane and polyurethane for electronics combine mechanical strength and flexibility at working temperatures below 125°C. This implies that they sit between silicone and epoxy potters and offer a low-temperature option to other formulations when flexibility is required. The optimum option for electronic assembly is electrical potting compounds, which provide strong, flexible, mechanical protection for sensitive components and cable. The potting of electronic components needs high-quality materials that can resist tough circumstances. In cases where conventional protectors may be utilized, electronic assemblies can be used.

19.1.3 Thermoelectric Generator

The wearable difficulties of existing organic foil generators might be solved using a fabric-based thermoelectric generator that demonstrates promising use in flexible and self-powered electronic devices through the harvest of body heat to create energy. However, thermoelectric yarn with a thermoelectric function is essential to making a wearable fabric generator, since yarns are the fundamental fabric structure composition units that may be processed in the thermoelectric fabric generator as thermoelectric legs. Thus, a novel thermoelectric coating material consisting of WPU and multiwalled carbon nanotubes (MWCNT) composite poly(3,4-ethylenedioxythiophene): polystyrene sulfonate (PEDOT:PSS) was created to provide the thermoelectric function for yarns. The composite thermoelectric characteristics with various components have been studied. An optimum electrical conductivity of ~13826 S/m and ~10 mV/k could be achieved at room temperature, and the power factor could be ~1.41 mW m^{-1} K^{-2}. Thus, at normal temperature in the following laminating experiment with commercial cotton and polyester yarns, the produced composite was successfully coated. The long continuous polyester filament with a superior coating layer is more suited as a coating substrate compared with the staple cotton filament. Our next TEG research will be based on the manufactured yarns. The characterization of composite films reveals that increased MWCNT/PEDOT/PSS ratios, the inclusion of highly conductive DMSO doped PH1000, as well as greater MWCNT concentration, can improve the composite power factors. The optimal formula for the doped sample was found with an electrical conductivity of approximately 13,826 S/m, a Seebeck ratio of approximately 10 mV/K, and a power factor of approximately 1.41 mW m^{-1} K^{-2} at room temperature, which was based on 20 wt.% MWCNT. The sample had 1:4 ratios of MWCNT to PH1000 [8]. This water-based composite displays satisfactory thermoelectrical performance and good processability in contrast to organic solvent-based thermoelectric polymers.

19.1.4 Electro-Conductive Para-Aramid Knit

A para-aramid electro conductive knit was produced by dip-coating in a polyurethane graphene/WPU composite to be used as a protective apparel heating element for durability. The knit of para-aramid was dipped into a composite solution of 8 wt.% graphene/WPU for up to five cycles. The electric, particularly electrical, heating performance increased the number of cycles from 1 to 5 as a result of electro conducting textiles by the number of dip covering cycles. The five-coat sample with the best heating and electrical efficiency was hot-pressed at 100, 120, 140, and 160°C to improve its characteristics. The whole surface of the sample was filled after heat pressing with graphene/WPU composite and a smooth surface indicating superior electrical and electric heating performance than the five-coat sample. At 140°C and a surface resistivity and capacity of 7.5 × 104 Ω/Sq and 89.4 pF, respectively, the greatest performance was indicated. When a voltage of 50 V was applied, the surface temperature reached 54.8°C. The surface temperature was 54.8°C when the 50 V voltage was applied, and the sample temperatures were stable less than 20 min, while the temperature was sustained 60 min without any fluctuation. Through direct dip-coating and stitching, hot-pressing conditions were, therefore, optimal at 140°C, with five coatings applied to para-aramid gloves. The touchscreen and heating characteristics were identical to the sample of the cloth coated. The para-aramid electro-conductor knit is predicted to be applicable for high-durability protective and multifunctional garments.

19.1.5 Stretchable Conductors

There are many techniques for producing stretchable conductors. Existing stretchable conductors include electronic conductors, e.g. metal nanoparticles (NPs), Ag NWs, Ag flakes, fractal Ag nanostructures, Cu NWs, carbon nanotubes (CNTs), graphene, serpentine-shaped metallic wires, conductive polymers, and their composites. These conductive components are often used as fillers and arranged in the elastomer matrix while combining the structure design. For instance, we fabricated a flexible and stretchable conductor by embedding fractal-structured Ag particles into a PDMS substrate, which could stretch up to 100%. A layer-by-layer mounting procedure and a vacuum-aided flocculation approach are used to construct two stretchy PU conductors with Au NP's. Both composites show high conductivity and stretchability. As stretchy and very conductive interconnects, fluid metal interconnected connections embedded in flexible polymer substrates are employed. The manufacturing process is extremely straightforward, and the active devices in preparation have a high fill factor. Set zigzag structures are utilized to make stretchy conductive routes, whose high degree of stretching is the synergistic effect of the interpenetration of polymer gel and Ag-NP conductive networks and zigzag structures [9].

19.1.6 Stretchable Pressure Sensor

The major stretchable pressure sensors are resistivity, piezoelectricity, capacitance, and piezoresistive function; each of these sensing methods has its unique characteristics based on the active materials employed and device architecture. Extendable substrates are commonly utilized in the manufacture of the pressure sensor-type force collector that uses force collectors to detect stresses (or deflections), due to their force exerted in an area (pressure). However, hysteresis may usually be seen with PSR-based pressure sensors. Different techniques and ideas for the construction of high sensitivity stretchable pressure sensors are therefore used. A resistive sensor has been produced using a polydimethylsiloxane (PDMS) compressible substratum with PEDOT: PSS/polyurethane dispersion (PUD), polymer electrode micro-pyramid characteristic arrays and tensile arrays. The pressure leads to geometry changes, and with observed sensitivity of 10.3 kPa^{-1}, the composite electrode may spread up to 40% strain. In situations, including body movement such as bending and stretching, the stretchy array of extremely sensitive stretchable pressure sensor devices can concurrently sense both pressure and the form of an item. A resistive tactile sensor ultra-therapy active-matrix array has been created, for example, which can crumble like paper, stand up to 230% on pre-extended elastic substrates, and work at high temperatures or in aquatic conditions.

19.1.7 Wearable Electric Heaters

Stretchable or even wearable electric heaters with uniform heating behavior are regarded as next-generation electronic devices, which have been extensively studied for personal thermal management and healthcare purpose. In this work, highly stretchable electrothermal heaters were developed by using composites of intrinsically conductive poly(3,4-ethylenedioxythiophene): poly (styrene sulfonic acid) (PEDOT:PSS), elastomeric waterborne polyurethane (WPU), and reduced graphene oxide (rGO). rGO was mixed into the PEDOT: PSS/WPU blends to improve the temperature uniformity because rGO has high thermal conductivity while the polymers have very low thermal conductivity. The PEDOT:PSS/WPU/1 wt.% rGO composite film exhibits an electrical conductivity of 18.2 S cm^{-1} and elongation at a break of 530%. The electrothermal performances of the polymer heaters were investigated with respect to the applied voltage, tensile strain, and the voltage on/off cycling process. The heater had a consistent heating performance under repeat voltage on and off cycles, and a tensile load of up to 30% remains almost constant. They may be conveniently fixed to human skin, e.g. on the wrist, and even under mechanical disturbance, they have a consistent and steady heating profile. The composites WPU/PEDOT:PSS/rGO can be utilized for wearable and long-term thermotherapy applications because of their exceptional stretchability, biological compatibility, and desired electrical and thermal conductivity [10].

19.1.8 Electronic Fabric

The fabric includes just 800 μm separate electroluminescent (EL) components. It consists of three kinds of yarn: transparent, conductive, ionic-lubricated polyurethane yarn; commercial silver-plating, conductive zinc-luminescent thread; and ordinary cotton or polyester yarn. The yarn consists of three different threads. Whenever two types of conductive yarn meet, an EL unit is located. The threads become blue when linked to a low alternating voltage. The team recommended that a tiny number of copper or manganese be added to the sulfide of the phosphorus to modify the color. The most widely utilized transparent conductivity was indium tin oxide (ITO) and silver nanowire transparent electrodes. However, that ITO was produced by sputtering the magnetron, which is difficult to deposit on the surface of the fiber constantly, and the thin silver nanowire cover is exposed to frictions when weaving. "The extremely uniform and transparent weaving fibre-packed ionic-liquid-doped polyurethane gel could flex to match the luminous warp surface and create a steady surface contact during tissue," Peng writes. "An elastic surface contact resulted in a curved surface electric field as consistent as that obtained by planar devices, which made it easy to create a homogeneous and steady textile display."

19.1.9 Strain Sensing

Strain sensing is one of the key components of the intelligent fabric that may be reflected via externalized electrical signals. Using acidified carbon nanotubes as the conductive layer (a-CNTs), and washable polyurethane (WPU) as an adhesive based on foam finishing technique, the strain sensing functional fabric was created. The results indicated that acidification introduces numerous hydroxyl and carboxyl groups into CNTs surfaces, which enhances their aqueous resolution dispersibility. The a-CNTs and anionic WPU blended film showed good conductive characteristics. With a ratio of 0.13%–0.15% and a corresponding tensile sensing sensitivity up until 55.2, the fabric had the lowest resistivity, at 9:1. Leading materials should have a high degree of stretchability and adherence to the material while retaining strong electric conductivity. Wearable textile strain sensors are difficult. A glue that assures the integrity of the prepared strain sensor enables the CNTs to be attached firmly to the cloth. Because of the large amine and carbonyl groups, polyamide fibre is hydrophilic. Only a large number of hydroxyl groups and carboxyl groups are available after treatment of carbon nanotubes with mixed acid on a carbon tube surface, which can continue to experience grafting or surface modifying reactions using other reagents to enhance the dispersibility and solubility of carbon nanotubes. The acidified CNT may readily enter the material and absorb it via foaming. Polyamide fiber, on the other hand, offers an excellent wear strength and resistance that also assures the sensor's stability. A-CNT as the conductive layer and anionic WPU as a dispersant and foaming adherence were used to produce this functional structure with high strain sensing sensitivity, broad strain range, and stable performance.

19.1.9.1 Stretchable Electronic Skin

An elastomeric matrix is required for a stretching electronic skin (e-skin) to be used under varied situations. Excellent and balanced characteristics, including elasticity, water-proofing, toughness, and self-healing, are therefore required. However, optimizing them simultaneously is exceedingly difficult and frequently conflicting. A single-pot synthesis was developed here to include a polyurethane (BS-PU-3) that has a hard, hydrophobic soft segment and a dynamic disulfide bond. Unlike the normal two-pot reaction, BS-PU-3 obtained through the one-pot method owned a higher density of self-healing points along the main chain and a faster self-healing speed, which reached 1.11 μm/ min in a cut-through sample and recovered more than 93% of virgin mechanical properties in 6 h at room temperature. Moreover, its considerable toughness as an e-skin matrix of 27.5 MJ/m^3 ensures its endurance, even though the tensile pressure could endure up to 324% without crack propagation with a 1 mm diameter (half the entire width) of a typical dumbbell specimen. The form, microstructure, and conductivity of the BS-PU-3 and BS-PU-3 stretchable electronics were extremely stable after 3 days of water absorption using polybutadiene as the soft segment, which demonstrates their super-waterproof feature. An e-skin demo was built, and self-healing was validated in terms of pressure and mechanical and electrical characteristics. The One-Pot Synthesis technique makes the hard segment heterogeneous length distribution and a larger density of self-healing sites efficient, which then enhances the self-healing speed and strength of the room temperature. The hydrophobic polybutadiene is successful in generating a good water-resistant e-skin as a soft segment.

19.1.10 Thermotherapeutic Devices

A thermochromic and thermotherapeutic, ultra-stable, stretchable device is used as both a substratum and a binder, in an intra-stretchable PDMS elastomer utilizing strain-responsive Ag NW networks and thermochromic colors. Spontaneous patterning of AgNWs onto PDMS-surface-patterned wrinkles enables the electric performance of stretchy instruments such as electrodes or active components to be controlled. This method may be expanded by utilizing noble metal nanowires to produce oxidation-resistant appliances. In addition, a thermochromic film can experience color change under various external stresses on a stretchy stress sensor with the same current biases. Its great sensitivity and extensibility allow it to connect with the live tissue in an adaptable manner. In several applications such as a user-interactive motion detector and a thermotherapy device, the device placed on the finger and the wrist joint is expected. At different joint flexing levels, the gadget changes its temperature and colour, efficiently checking the heat transmission to the muscles, ligaments, and tendons, and prevents skin burns. Integrating the stretched pressure sensor and a play warmer enables further application in rehabilitation patients to increase the extensibility of damaged tissue. Electronic interactive skin (e-skin) with a distinctive output offers a huge potential for interactions between people and machinery and medical applications. Despite progress with user interactive e-skins, progress is uncommon with visual user interactive therapeutic e-skins. A user-interactive thermotherapeutic device is claimed to be manufactured by mixing thermochromic composites with strain-replicating silver nanowire networks on surface energy-patterned micro-wrinkles. Electrical resistance generated by applied mechanical strain may readily regulate the color and heat of the gadget [11].

19.1.11 Electromagnetic Interference Shielding (EMI)

Electromagnetic interference (EMI) shielding is a kind of electromagnetic wave emission that is applied in most electronic appliances. It usually appears in the

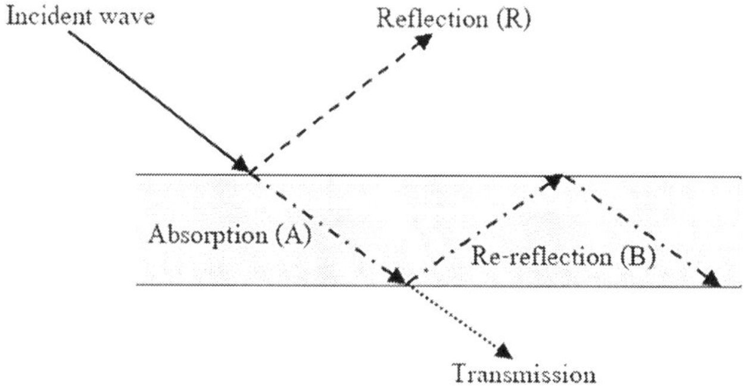

FIGURE 19.1 Schematic of the four main mechanisms of EMI shielding materials (Reprinted from reference [13], under the terms of the Creative Commons CC-BY 4.0, https://creativecommons.org/licenses/by/4.0/).

form of radio waves. Electromagnetic interference is obtained as a by-product of electronic devices. Electromagnetic interferences are also a kind of electromagnetic wave that affects human health. EMI shielding is used to control the level of electromagnetic radiation that is attained by blocking the electromagnetic waves. EMI shielding has a vital role in the protection of the environment [12]. Since these electromagnetic waves collide, it leads to the formation of four different mechanisms, as shown in Figure 19.1. Among the four mechanisms, the finest mechanism is reflection or absorption. This mechanism depends on the materials of shielding and the nature of the electromagnetic waves. Reflection is the most important mechanism among the four of them.

The reflection of electromagnetic waves from the surface of the materials is the most fundamental interaction that takes place when the electromagnetic waves collide. The usage of the conductive materials proves a beneficial way for the reflection mechanism. The next type of mechanism is the absorption mechanism, which occurs when the electromagnetic waves launch into the material. This electromagnetic wave absorption occurs when the electric or magnetic dipoles convert the energy into heat, which is present in the waves. The following mechanism of electromagnetic waves is the multiple reflections that occur when the waves get reflected from the surfaces. The efficiency of this mechanism is improved by using porous materials. This is because of its increased surface area. The above mechanism can be increased by passing the electromagnetic waves into the shielding material, leaving out attenuation.

Based on the research in the electromagnetic radiation field, EMI shielding has a vital role in shielding from electromagnetic radiation and development of materials due to the harmful effects on electronic appliances as well as on human beings. The most crucial material for EMI shielding is WPU composites. WPU composites have achieved great importance due to their high electrical conductivity [13]. The characteristics of the EMI shielding materials for electronic devices are that the cost of the materials is cheap, the weight of the materials is low, and they are also resistant to corrosion. The suggested shielding materials are WPUs and polymers. The WPUs are widely used as EMI shielding materials as they are pure and have excellent resistance to corrosion along with cost-effectiveness, are both chemically and thermally stable, and are suitable for processing. The WPUs are advantageous over metal-based EMI shielding materials and can be applied to several other applications.

19.1.12 Recycling of Waterborne Polyurethanes

WPU products are attracted toward various applications, where the economic and environmentally friendly characteristics of the materials are very much concerned. As a result of this concern, the recovery and recycling of WPU materials have been given great significance to overcome other commercially used environmentally friendly materials. Also, from the economical point of view, the process is found to be efficient and beneficial in reducing the waste material produced, which may be recycled and used as scrap. Commonly, there are four ways by which the recycling process are done:

- Advanced chemical and thermochemical recycling
- Mechanical recycling
- Energy recovery
- Product recycling

These techniques add to the advantages of fabrication and utilization of WPU materials. The recycling process of the materials requires physical treatment, and the other methods, thermochemical and chemical recycling methods, require chemical treatment to generate feedstock chemicals for industry. Also, during energy recovery, waste materials undergo partial or complete oxidation, which consequently produces electricity and gaseous fuels [14]. The fact that the by-products produced during the recycling process are also nonhazardous and can be disposed of with ease without causing damage to the environment. The major recycling and energy regeneration techniques of PUs are chemical, mechanical, and thermochemical recycling. The recycling of WPU foams can be done by regrinding them into powder that can be used for multiple purposes. The processes used for this purpose include the following:

- Compression moulding
- Adhesive pressing
- Flexible foam bonding

Another main application of WPU powder is the fabrication of doormats and tires, where the WPU granules are coated with a binder, followed by a curing process under pressure and heat. Pump and mother housing are examples of such compression moulding techniques, which are done under high heat and pressure. During the energy recovery, the maximum intensity of electricity can be produced by burning the WPU completely [15]. The thermochemical and chemical recycling processes comprise many chemical reactions, on which the process is dependent.

- Hydrogenation
- Pyrolysis
- Hydrolysis
- Glycolysis

Due to its hard and semi-rigid structure, the process of recycling polyurethane materials is economical and practical. As a result, recycled WPU may be utilized to effectively make quarter panels, wheel covers, steering wheels, bumper covers, and cores in motor vehicles, as well as other household and industrial parts. Because of recycling and recovery, WPU material is more ecologically safe and economical compared with other traditional materials.

19.2 FUTURE POSSIBILITIES

WPU coatings and adhesives have become constantly fixed in the market in several fields. They are expected to get high on top of moderate growth within the future. Their exceptional ability to be tailored to satisfy numerous requirements is consistently yielding new and effective solutions. Market profit of these products has some key factors, such as high quality to a standard system, appearing to have a gradual increase in the service life of coatings and thus longer conservation cycles, dominant variability in alignment, enabling the construction of environmental familiarity, huge solids, costless solvents, and waterborne systems. It is feasible with WPU coatings to add new operations like self-curing, accessibility to cleaning, easy touch effects, substantial performance in processing, operation, and dominant to saving of components and efficiency; these provide economic profit. Future demand for reusable basic materials for the alignment of adhesives and coatings may occur. On the other side, due to WPUs quality level, classical oil systems placed on plant oil like alkyd resins are being increasingly recovered by modern oil, such as crude oil systems. Currently, a return to reusable basic materials is only believable if it can maintain the considerable quality level and increase it. The addition of a polyol item with reusable components is needed before the polyurethane process becomes complex. It is primarily used for urethanized polymer modifications. A series of diol and dicarboxylic acid is also available as polyester modules by the correct alteration of reusable basic materials like castor oil or fatty acid. It suggests that the need for reusable parts will gradually grow in the future. The absolute limit for which it can be attracted will be determined by the demand for land crops or any other natural resources that are needed. The first step in dealing with polyisocyanates is to use recycled basic materials. The C5 components, for example, are available in the form of reusable diamines for isocyanate processing. The recent statistics deal with the economic as well as industrial evaluation of these product types. To meet potential concerns, the spectrum of WPU adhesive coating-specific materials is now being exposed to further productive evolution.

19.3 CONCLUDING REMARKS

WPUs are widely used, versatile, flexible, and well-analyzed substances available in many parts of the world. The hardness and strength of metals are combined with the flexibility of rubber and are perfect replenishments in various engineered goods for metals, plastics, and rubber. WPUs have turned out to be an ideal element for research purposes. They have been used in many industrial applications and are designed as an integral skin for multiple items, including elastic fibers, rigid insulations, elastomers, liquid coatings, paints, flexible soft foam, and so on. A broad range of polyols, diisocyanates, and many other chain extenders and associate agents can be used to manufacture WPUs. This allows a broad spectrum of custom materials and can be used for several different applications. The majority of polyols used in the preparation of WPUs originally came from the petroleum sector, but high costs and demands for energy and environmental considerations have enhanced the need for a more suitable replacement for the environment. This recently attracted a great deal of interest in trade and academia on green sources, including vegetable oils. A clear majority of research on vegetable oil usage as a substitute to petroleum-related products for polyurethane processing has been published in literature over the previous decade. There are, however, some weaknesses related to these materials, particularly in terms of efficiency. The use of nanomaterials has been proposed for the desired results. Consequently, there has also been an extensive study of the integration of nano pieces that can restore the rough segments of isocyanate forerunners. Some of the basic materials like carbon nanotubes (CNT), carbon nanofibers (CNF), and clays act as important additives to polyurethane products and are also of great importance. Recyclability of the substance is critical for all the research on WPU. The WPU recycling processes were fortunately identified as cost-effective and realistic. Thus, on its excellent recycling and recovery properties, WPU may be regarded as environmentally friendly and an economical relative to the other traditional polymers. Because of the abovementioned characteristics, WPUs are widely used all over the world.

REFERENCES

1. Z hao, Zhuo, De-Ming Guo, Teng Fu, Xiu-Li Wang, and Yu-Zhong Wang. "A highly-effective ionic liquid flame retardant towards fire-safety waterborne polyurethane (WPU) with excellent comprehensive performance." *Polymer* 205 (2020): 122780.
2. Zhang, Fengyuan, Weiqu Liu, Liyan Liang, Shuo Wang, Hongyi Shi, Yankun Xie, Maiping Yang, and Ke Pi. "The effect of functional graphene oxide nanoparticles on corrosion resistance of waterborne polyurethane." *Colloids and Surfaces A: Physicochemical and Engineering Aspects* 591 (2020): 124565.
3. Cui, Jincan, Jingcheng Xu, Jing Li, Hanxun Qiu, Shiyou Zheng, and Junhe Yang. "A crosslinkable graphene oxide in waterborne polyurethane anticorrosive coatings: Experiments and simulation." *Composites Part B: Engineering* 188 (2020): 107889.
4. Cao, Xianli, Xia Ge, Huanhuan Chen, and Wenbo Li. "Effects of trimethylol propane and AAS salt on properties of waterborne polyurethane with low gloss." *Progress in Organic Coatings* 107 (2017): 5–13.
5. Aïssa, Brahim, Mohamed Adnane Habib, Esam H. Abdul-Hafidh, Mohamed Bououdina, and Mourad Nedil. "Carbon nanotubes materials and their related polymer nanocomposites: Frontiers, challenges and strategic priorities." *International Journal of Materials Engineering Innovation* 6, no. 2–3 (2015): 185–213.
6. Zhang, Chaoqun, Haiyan Liang, Dunsheng Liang, Zirun Lin, Qian Chen, Pengju Feng, and Qingwen Wang. "Renewable castor-oil-based waterborne polyurethane

networks: Simultaneously showing high strength, self-healing, processability and tunable multishape memory." *Angewandte Chemie International Edition* 60, no. 8 (2021): 4289–4299.

7. Ahmadi, Younes, and Sharif Ahmad. "Recent progress in the synthesis and property enhancement of waterborne polyurethane nanocomposites: Promising and versatile macromolecules for advanced applications." *Polymer Reviews* 60, no. 2 (2020): 226–266.
8. Alonso-Lerma, Borja, Izaskun Larraza, Leire Barandiaran, Lorena Ugarte, Ainara Saralegi, Maria Angeles Corcuera, Raul Perez-Jimenez, and Arantxa Eceiza. "Enzymatically produced cellulose nanocrystals as reinforcement for waterborne polyurethane and its applications." *Carbohydrate Polymers* 254 (2021): 117478.
9. Li, Hui, Du Yuan, Pengcheng Li, and Chaobin He. "High conductive and mechanical robust carbon nanotubes/waterborne polyurethane composite films for efficient electromagnetic interference shielding." *Composites Part A: Applied Science and Manufacturing* 121 (2019): 411–417.
10. Jia, Yunpeng, Ruizhou Sun, Yamin Pan, Xin Wang, Zhanyu Zhai, Zhiyu Min, Guoqiang Zheng, Chuntai Liu, Changyu Shen, and Xianhu Liu. "Flexible and thin multifunctional waterborne polyurethane/Ag film for high-efficiency electromagnetic interference shielding, electro-thermal and strain sensing performances." *Composites Part B: Engineering* 210 (2021): 108668.
11. Bernard, C., D. G. Goodwin, X. Gu, M. Celina, M. Nyden, D. Jacobs, L. Sung, and T. Nguyen. "Graphene oxide/waterborne polyurethane nanocoatings: Effects of graphene oxide content on performance properties." *Journal of Coatings Technology and Research* 17, no. 1 (2020): 255–269.
12. Zhu, Huixin, Yaqi Yang, An Sheng, Hongji Duan, Guizhe Zhao, and Yaqing Liu. "Layered structural design of flexible waterborne polyurethane conductive film for excellent electromagnetic interference shielding and low microwave reflectivity." *Applied Surface Science* 469 (2019): 1–9.
13. Sriseubsai, Wipoo, Arsarin Tippayakraisorn, and Jun Wei Lim. "Robust design of PC/ABS filled with nano carbon black for electromagnetic shielding effectiveness and surface resistivity." *Processes* 8, no. 5 (2020): 616.
14. Kemona, Aleksandra, and Małgorzata Piotrowska. "Polyurethane recycling and disposal: Methods and prospects." *Polymers* 12, no. 8 (2020): 1752.
15. Magnin, Audrey, Eric Pollet, Rémi Perrin, Christophe Ullmann, Cécile Persillon, Vincent Phalip, and Luc Avérous. "Enzymatic recycling of thermoplastic polyurethanes: Synergistic effect of an esterase and an amidase and recovery of building blocks." *Waste Management* 85 (2019): 141–150.

20 Waterborne Polyurethanes for Sensors

Charles Oluwaseun Adetunji
Applied Microbiology, Biotechnology and Nanotechnology Laboratory, Department of Microbiology, Edo State University Uzairue, PMB 04, Auchi, Edo State, Nigeria

Abel Inobeme
Department of Chemistry, Edo State University Uzairue, PMB 04, Auchi, Edo State, Nigeria

Kshitij RB Singh
Department of Chemistry, Govt. V. Y. T. PG. Autonomous College, Durg, Chhattisgarh, India

Olugbemi T. Olaniyan
Laboratory for Reproductive Biology and Developmental Programming, Department of Physiology, Edo State University Uzairue, PMB 04, Auchi, Edo State, Nigeria

John Tsado Mathew
Department of Chemistry, Ibrahim Badamasi Babangida University Lapai, Niger State, Nigeria

Jay Singh
Department of Chemistry, Institute of Science, Banaras Hindu University, Varanasi, U.P., India

Vanya Nayak and Ravindra Pratap Singh
Department of Biotechnology, Indira Gandhi National Tribal University, Amarkantak, M.P., India

DOI: 10.1201/9781003173526-20

CONTENTS

20.1 Introduction....334
20.2 Fabrication and Component of Waterborne Polyurethanes for Sensors....335
20.3 Specific Researches on Applications of Waterborne Polyurethanes for Sensors....336
20.4 Medical Application of Waterborne Polyurethanes for Sensors....341
20.5 Environment and Agriculture Application of Waterborne Polyurethanes for Sensors....345
20.6 Conclusion and Prospects....349
Acknowledgment....350
References....350

20.1 INTRODUCTION

Recently, waterborne polyurethanes (WPUs) were used for diverse industrial applications and have gained tremendous attention among biomedical and environmental scientists due to their physiochemical eco-friendly nature. Today, WPUs are being utilized as sensors through their renewable reactants and additives like chitosan, starch, and cellulose, thus enhancing green sustainable strategies in the environment, health, tissue engineering applications, biomedical, finishing agents, textile coatings, purification approaches, and membranes for pollutants [1]. Jiang et al. [2] revealed that WPU as an adhesive-based foam combined with carbon nanotubes could fabricate smart fabric material for strain sensing to detect electrical signals through the acidified carbon nanotube conductive layer. Their results demonstrated that carboxyl and hydroxyl groups were attracted to the surface of the WPU as adhesive-based foam and a carbon nanotube conductive layer by the acidification method, resulting in the dispersion. They showed that the combination of these materials revealed quality conducting tendency, high tensile strength, and low resistance level. Cakic et al. [3] demonstrated the synthesis of WPU dispersions utilizing isophorone diisocyanate, polycarbonate diol, dimethylolpropionic acid, ethylenediamine, and trie-thylamine ionomer structures. The authors evaluated the physicomechanical characteristics, like adhesion test, hardness, and gloss, of the dried films. They showed that the adhesive nature of WPU dispersions to diverse substrates and solvents; resistance to chemicals and water; flexibility; and abrasion resistance makes it a good chemical and mechanical agent. Zhao et al. [4] analyzed the properties of the gas sensing characteristics of waterborne/carbon black polyurethane composites in very low organic vapors. Their study revealed that the conducting composites of the polyurethanes were made using latex blending to increase the sensitivity to any polar vapor or nonpolar solvents due to the high microphase separating the water-based structure polyurethane. After a thorough analysis of the composites, they proposed the Flory–Huggins principle or model to explain their observations. In their study, Prabhakar et al. [5] showed the carcinogenic nature of hexamethylphosphoramide and thus the need to develop a highly sensitive polyaniline-coated polyurethane membrane to detect this compound rapidly. They adopted the resistance

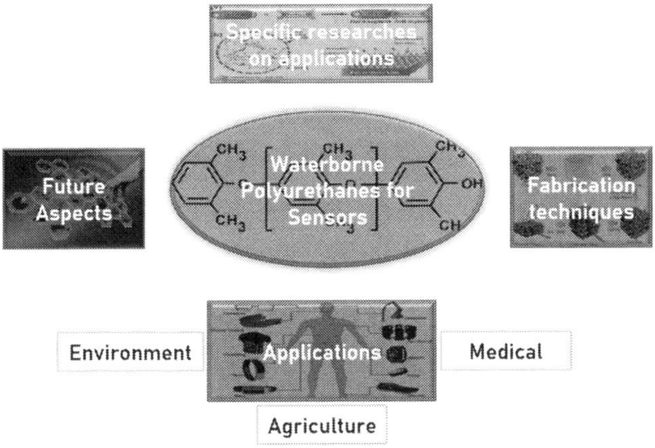

FIGURE 20.1 Systematic illustration of the details discussed in this chapter.

assessment technique in developing this device utilizing polyurethane membrane coated polyaniline, an electrically conductive coating, and sensitivity to carcinogenic compounds due to the big resistance change in coated polyaniline gas sensor to hexamethylphosphoramide. Wang et al. [6] revealed that WPUs could be utilized to monitor large and small-scale body movements, such as joints and muscle movements, through the production of high-quality electrical signals on the skin by the wearable, stretchable, dry and self-adhesive strain sensors. Various high-performance WPU coatings have been developed for use in fabricating humidity sensors [7]. Therefore, this chapter will give a general overview of several components (Figure 20.1) utilized in the fabrication of WPUs for sensor technology and the diverse applications of WPU sensors in medicine, the environment, and agriculture.

20.2 FABRICATION AND COMPONENT OF WATERBORNE POLYURETHANES FOR SENSORS

In the past, polyurethane was used in various fields because of its outstanding structures, including good abrasion resistance, excellent mechanical strength, low-temperature flexibility, and toughness. More precisely, polyurethane might certainly be "tailor-made" to meet exact demands. Based on the development of standard of living, normal polyurethane products may not meet the rising desires for higher quality, novelty, and comfort. In the middle of the main uses, polyurethane is part of the noticeable coating materials and retaining agents in the leather industry [8].

The environmentally friendly and inexpensive fabrication approach for microfluidic paper based on analytical strategies could restrict the diffusion of organic solvents and surfactant solutions through application-resistant surfactant solutions (10 wt.%, SDS, Triton X-100, and CTAB) as well as organic solvents (isopropanol, methanol, DMSO, and DMF). These are useful in analyzing complex biological samples on microfluid, and the PUA originator is environmentally friendly water [9]. WPU coatings have some advantages: low cost, environmental protection, and toughness. On the other hand, the poor wear, lower water contact angle, and unique resistance to decomposition make them incompatible for use with super-hydrophobic coatings, including self-cleaning coatings used for outdoor facilities, antipollution flashover coatings used in transmission lines, as well as waterproof textiles. As the hydrophobic scheme constituent rose, the complex's water contact angles initially improved and then remained stationary, though the corrosion and adhesion resistance initially improved and then reduced [10].

Like piezoresistive ones, to be specific, some of the polymer-based keen constituents are improving and are being applied in a variety of uses, ranging from motorized mechanisms to therapeutic strategies. Water-derived piezoresistive polymer mixtures established by thermoplastic poly (vinyl alcohol) occupied through carbon nanotubes are used for fabricating advanced sensors and performance piezoresistive materials. Poly (vinyl alcohol) is a synthetic polymer that is soluble in water by way of appropriate mechanical characteristics to expand great strain sensors. Poly(vinyl alcohol)/carbon nanotube nanocomposites have good linearity among electrical and deformation resistance deviation as well as with gauge factors. Consequently, polymer-based piezoresistive sensors were established by employing a green method and were applied through accessible printing skills [11]. Further, innovative waterborne acrylic resin improved by utilizing glycidyl methacrylate. It was effectively produced through homogeneous solution polymerization using isopropyl alcohol and solvent exchange through water. A silane coupling agent, aminopropyl triethoxysilane, which can crosslink through iron and resin base material, was applied as a curing agent for coagulating this glycidyl methacrylate improved resin. The study opens an innovative way to produce high-performance waterborne acrylic resin [12].

20.3 SPECIFIC RESEARCHES ON APPLICATIONS OF WATERBORNE POLYURETHANES FOR SENSORS

Wang et al. [6] used dopamine for surface modification to enhance the deposition efficiency of the coatings on WPU. The test carried out to ascertain the electromechanical properties revealed a wide working range of 37% in parallel to the fiber. The sensing efficiency was reported to be remarkable in the detection of movements of human fingers and elbows.

Lee et al. [13], in a related study, fabricated a polyurethane-based piezoresistive sensor for pressure with remarkable repeatability and speedy response. The sensor was based on the characteristic coating of silicone rubber, which possesses a wide range for measuring the pressure. It was developed by using polyurethane foam for coating silicon rubber. The sensor was found to be suitable for the measurement of low pressure as well as high pressure like the underneath of the heel.

Petr et al. [14] produced a thermoplastic elastomer polyurethane (PU/MWCNT-N) network and assessed it as a strain-sensing device by measuring macroscopic resistance. At the initial stage of preparation, multiwalled carbon nanotube (MWCNT) dispersion in aqueous form was filtered through polyurethane membrane through electrospinning technique. The filtering membrane was a vital component of the final composite when the filter supported an MWCNT network fixed on the surface of the polyurethane tensile specimens through compression molding, and the filter was then converted into polyurethane adhesive film. The composite can sustain large deformation, and the strain could be detected by measuring resistance in real-time deformation.

Babar et al. [15], in their work, focused on the detection of hexamethylphosphoramide through the use of polyaniline-coated polyurethane membrane. The device was constructed to reveal its practical relevance through the resistivity detection technique. The electrospinning technique was employed in the preparation of the polyurethane nanofiber. A modified synthetic approach was employed for the coating of polyaniline into a polyurethane membrane, and it was observed to be efficient in preparing a uniform coating that is electrically conductive and has the potential for sensing hexamethylphosphoramide, a carcinogenic agent. The sensing performance can be credited to the high resistance variation in polyaniline due to proton removal when exposed to hexamethylphosphoramide in the environment. The study concluded that the polyaniline-modified polyurethane membrane is an effective gas-sensing device to detect carcinogenic compounds like hexamethylphosphoramide.

He et al. [16], in their work, formulated an MWCNT WPU sensor using a macroporous structure that is ultra regular in design through a directional drying process. This new polymeric material can widen the usage of polymer-based gas sensing devices from static to moving organic vapor like acetone, toluene, and hexane, which was reported to be most effective for detecting tetrahydrofuran. Due to the specific micro-phase separation structure of WPU, the conducting sensor poses a wide use in the area of nonpolar and polar organic vapors.

Moreover, by using complex hydrophilic groups, dimethylol propionic acid (DMPA) and ethoxylated capped polymeric diol, WPUs were synthesized that exhibited high solid content (45%). Further, to enhance the performance coating cast from dispersions of WPU, the polymer matrix was integrated with alkyl-grafted silica via in-situ polymerization, which resulted in enlargement of the particle size distribution and increased emulsion viscosity, as shown in Figure 20.2. This study showed that the silica functionalized surface enhances compatibility with the WPU matrix, improves thermal stability and solvent resistance, and leads to the homogeneous dispersion of silica particles. This facile technique to synthesize WPU can be a promising approach for preparing WPU hybrids with enhanced performance [17].

In a related study, Chen et al. [18] synthesized a conductive composite made up of WPU and carbon black (CB). Aside from the low threshold of percolation (0.7–0.95 wt.%), the composites were observed to be more sensitive to organic solvent vapors irrespective of their varying polarities as detected by the rapid variation in conductivity. For polar solvents, it was observed that positive and negative vapor coefficient occurrence of the composites was remarkably noticed to increase in CB. It was also observed that different reaction routes were accountable for the wider applications of the composites used in gas sensing devices due to their interaction between the solvent molecules, matrix polymers, and filler particles.

Chen et al. [19] studied the gas sensitivity of carbon black-containing composites of WPU and its effects on vapor adsorption behavior. They reported that the conventional swelling model, which was used in reversible changes in the electrical resistance of the composite initiated by the solvent uptake, was affected.

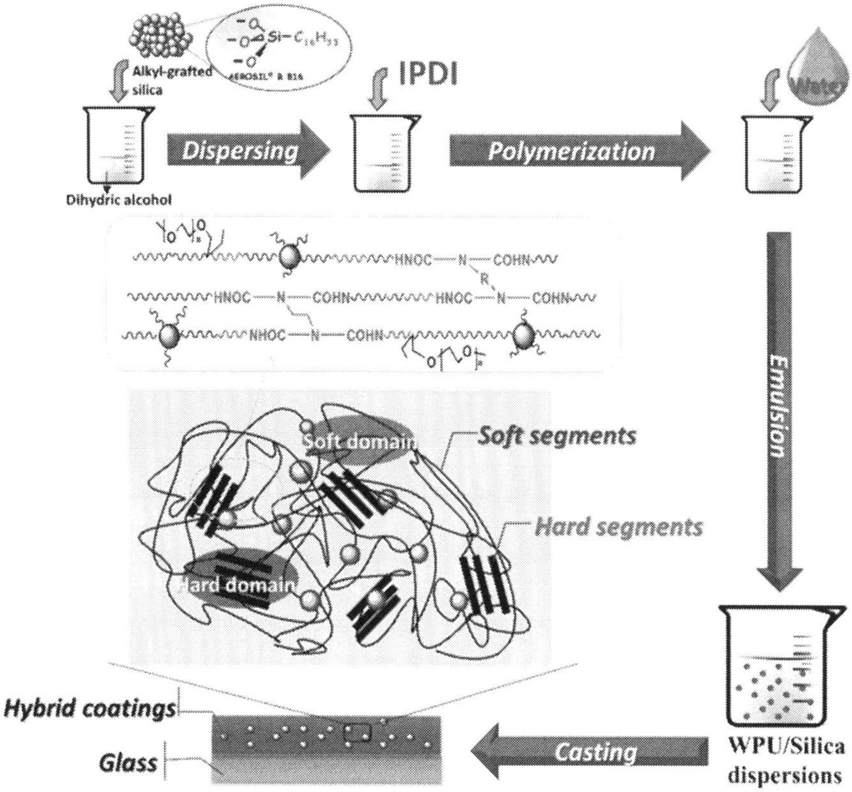

FIGURE 20.2 Preparation of dispersed waterborne polyurethane/silica (WPU/Silica), which exhibits high solid content and their coatings (reproduced with permission from Han et al., [17] [CC BY 4.0]).

In other words, a higher adsorption quantity did not directly imply greater vapor sensitivity. The matrix polymer–solvent association and the chemical properties of the solvents played a vital role in determining the nature of the relationship existing between the electrical sensitivity of the composite and the vapor adsorption tendency. Moreover, Aksoy et al. [20] documented that the application of polymeric membranes as gas detectors and sensors is a rapidly improving area. They emphasized that polyurethane membranes are promising in producing gas sensors for the next generation, and this formulation is based on membrane diffusion. Their review work focused on evaluating, comparing, and discussing recent advancements in the utilization of polyurethane membrane in detection technologies based on gas with electrical, optical, and chemical techniques. They further explained that in all the measurement methods, the polyurethane structures usually act as a semi-permeable membrane, making it a good matrix for conducting additives in the coating of polymeric films.

In a different study, Chen et al. [18] produced a conductive composite made of WPU and carbon black through latex blending. Different from the traditional polymer composites, which are only capable of responding to a specific class of solvent vapors, the composite produced had high electrical resistance responses to a category of vapors of polar and nonpolar solvents due to the molecular structure of polyurethane, which is blocked. The rate of response was observed to increase with temperature, while there was a decrease in the maximum response magnitude due to desorption of vapor. Further, Chen et al. [21], in another study, reported that the exposure of CB that is filled with WPU composite to vapors of organic solvent brought about a rapid rise in the electrical resistance of the materials. The addition of a crosslinking agent resulted in the formation of an intramolecular crosslinked network structure within the matrix polymer of composites. They also observed an enhancement of performance and gas sensitivity.

Zhao et al. [4] produced a conductive polymeric composite made of WPU and carbon through latex blending. This was observed to exhibit high detection of nonpolar and polar vapor of solvent due to polyurethane structure that shows microphase separation. Through a thorough investigation of the composite's absorption and electrical response behavior under a low vapor pressure, a model was put forward based on the Flory-Huggins mechanism aimed at discussing the specificity in the composite performance, including the impact of test temperature and vapor pressure.

Jia et al. [22] addressed a simple technique for the fabrication of conductive polymer composite films that are aided on silver flakes WPU through the process of spraying. This was fabricated for possible use as a sensor in monitoring human movement and some wearable electronic devices. The WPU/silver composite possessed an outstanding electromagnetic property and flexibility with very high shielding efficiency and sustained a high standard of mechanical deformation. The films produced also had a high electro-thermal impact and a high gauge factor and durability. Moreover, a novel method was utilized to synthesize green WPU using 2,4-diaminobenzenesulfonate (SDBS) through an organic solvent-free process. The SDBS was used as a chain extender to eliminate catalysts and decrease the temperature and reaction time of the chain extension step to conserve energy. Similarly, water was utilized as the solvent and to prepare a homogenized solution. Further, a step-by-step process to synthesize the green WPU is described in Figure 20.3, where prepolymer is prepared using PEG-1000 and PTMEG-2000 at 80°C, which is then followed by dissolving of SDBS in the deionized water for 10 min at 25°C. The mechanical properties of green WPU are similar to conventional WPU synthesized using organic solvents. Further, the study of their applications is still under process [23].

In a study by Lawal [24], the piezoresistive coating was fabricated for use as a sensor using WPU and graphene nanoparticles. The morphology of the nanomaterials produced in the investigation was analyzed using a scanning electron microscope (Field Emission). The rheological properties of the polymeric mixture loaded with 3.5 wt.% of the graphene nanoplates were assessed at varying concentrations of the water component (up to about 20 wt.%) to obtain the most appropriate viscosity necessary for air spraying. Further, covering agents were also employed in assessing the impact of humidity on the electrical property of the films of the nanocomposites. The piezoresistive response of the sensor was determined

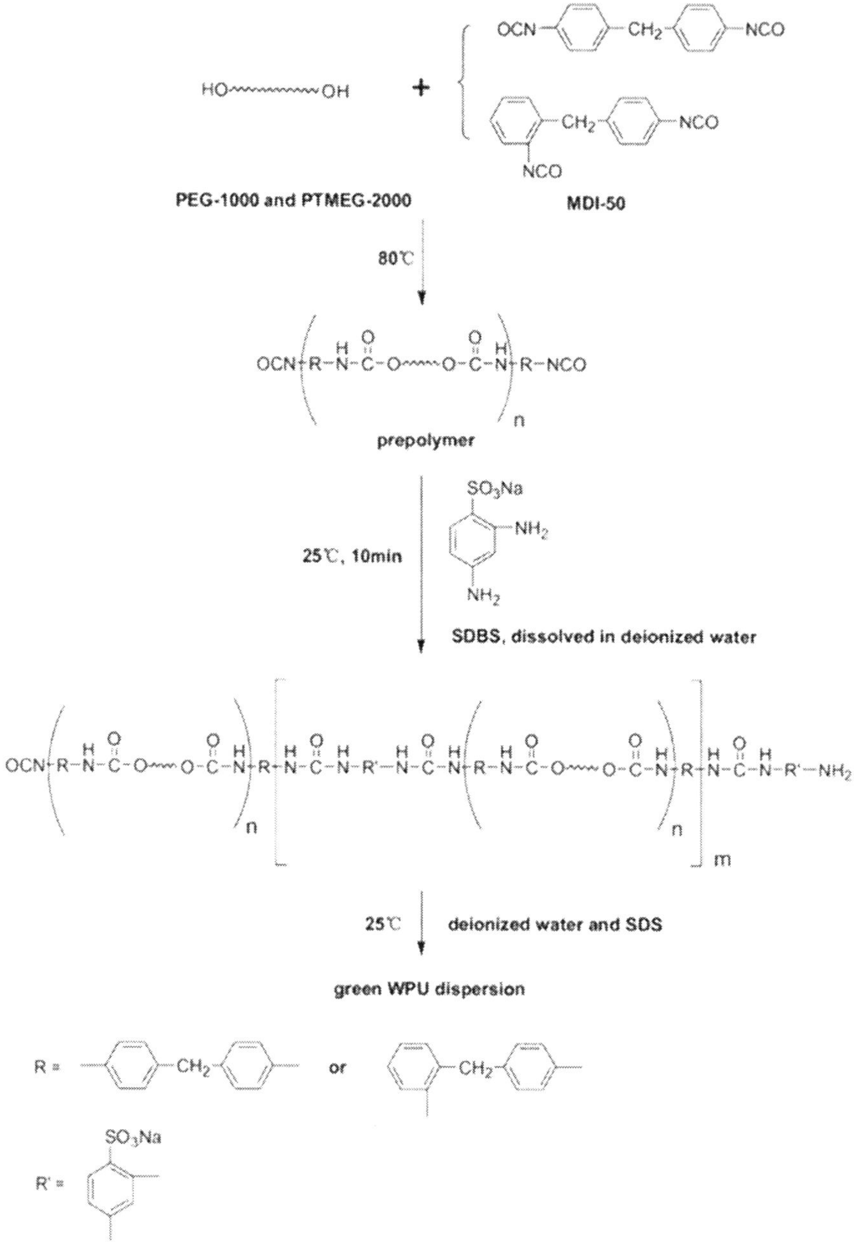

FIGURE 20.3 Stepwise synthetic route for the synthesis of green WPU (reproduced with permission from Xiao et al., 2016).

using flexural tests of three-point and dc amperometric measurements. Electromechanical test results revealed a rise in the sensor's sensitivity with the deformation applied to show the feasibility of the paint in structural health monitoring sensors.

Wang et al. [25] demonstrated the fabrication of wearable strain sensors that are also stretchable with vital applications in various areas. They identified a high level of noise as one of the major challenges in using this sensor in monitoring the body's movement. They attributed the noise to motion artifacts associated with the poor connection between the skin and the sensor. The sensor was fabricated by biocompatible WPU, which is elastomeric alongside a sensing layer composed of carbon nanotubes and graphene oxide that are nonadhesive. The role of the adhesive layer is to ensure the conformation of the sensor to the skin. They concluded that the sensitivity observed in the sensor fabricated from the elastomeric WPU is remarkably high; hence, it is suitable in monitoring movement along with two directions that are perpendicular. Similarly, a novel and promising method to develop flexible ultrasensitive polymer-based proximity wearable sensors was developed using thermoplastic polyurethane (TPU)/carbon nanotubes (CNTs) composites that can detect the presence of an external object from a wide range of distance. Figure 20.4 demonstrates the experimental setup and the method used to analyze the change in capacitance. To eliminate the noise, the sensor probes were mechanically co-planned, forming a 45°C angle between them. This proximity sensor showed an unprecedented detection distance of 120 mm, with a resolution of 0.3% mm with minimum noise. Therefore, it can be used in the biomedical field, which can monitor motion analysis and be used to develop artificial electronic skin [26].

Sousa et al. [27] carried out a study emphasizing the possibility of using a composite of CB and polyurethane in piezoresistive sensors. They documented in their study that CB composite dispersed into the matrix of multicomponent polyurethane showed the desired combined properties of the polymers, such as electrical parameters of the conducting particles, mechanical properties, and elasticity. The electrical voltage-current assessment found that the percolation threshold through the sample ranged from 0.6 to 0.8 vol.%. The extent of repeatability in the piezoresistive property was noticed under the treatment of numerous loading cycles. They concluded that the material could be used in the production of the sensor.

20.4 MEDICAL APPLICATION OF WATERBORNE POLYURETHANES FOR SENSORS

Polyurethanes have made meaningful input in the medical sector. Due to their inherent unique properties, they have promising potentials in futuristic medicine and technology. Polyurethanes have been employed in various areas of medical relevance, such as hospital bedding, dressing of wounds, general tubing, and injection of molded devices. Although the traditional polyurethane, which is solvent-based, has been used in the medical field for tubing, catheters, dressing, and antibacterial membrane, among others, the development of the WPU has aided in solving the challenges and inherent setbacks associated with the conventional polyurethanes.

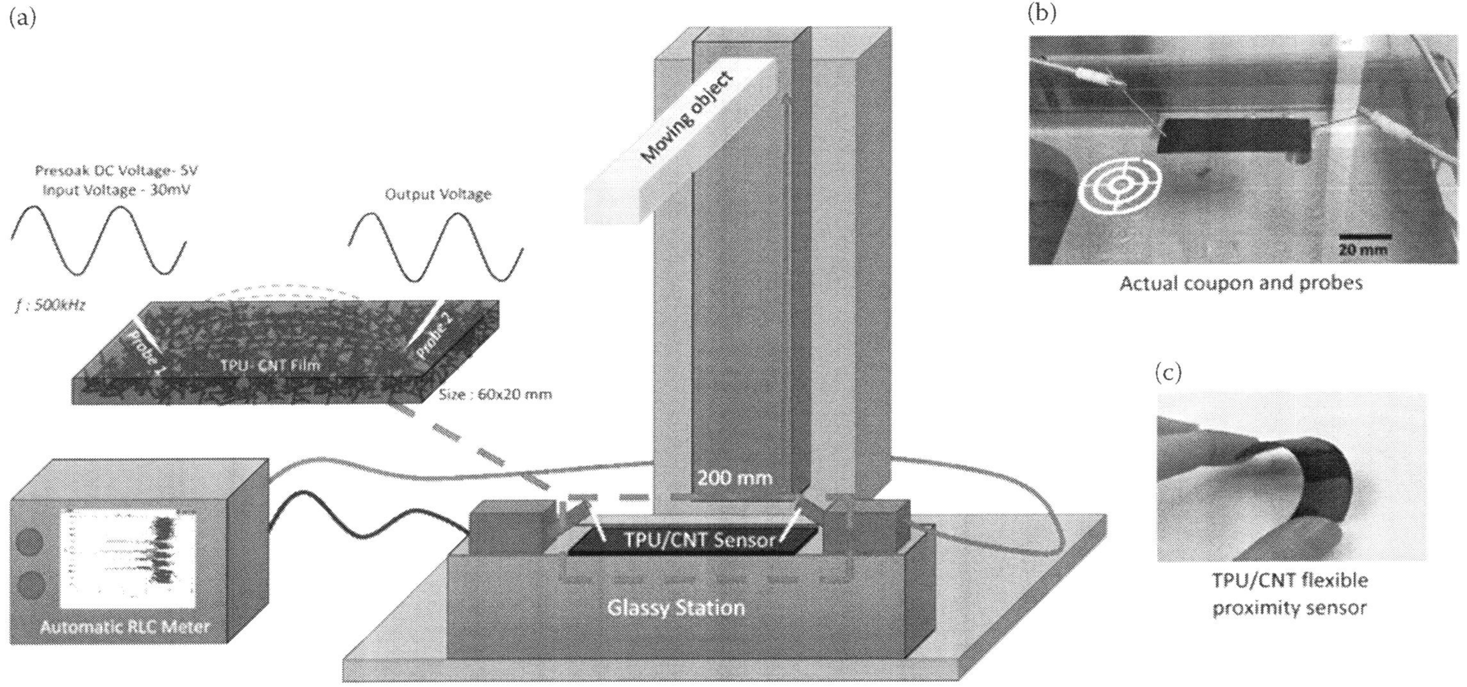

FIGURE 20.4 Systematic illustration of a TPU/CNT proximity sensor experimental setup. (a) Using probing station distance of about 220–20 mm range was applied (Keithley 4200-SCS, Tektronix, USA). The film was fixed over a glassy substrate to eliminate the noise, and the sensing object made up of a brass bar (10 × 20 × 200 mm) approached the sample with a speed of 6.6 mm/s. (b) The sensor probes were mechanically semi-planned with an angle of 45° to eliminate the noise. (c) Sensor being bent to show its thickness and flexibility (reproduced with permission from Reza et al., [26]).

This has further enhanced the process of dissolving polyurethane using water as a dispersant against the use of toxic organic solvents. WPU has several advantages in medical applications, including a low level of utilization of organic solvents, making them environmentally friendly; the absence of traces of isocyanates, making them less toxic; and unique structures and diversity of properties, giving them wider applicability. Hence, they have become the spotlight in the field of medicine [3]. The advent of WPU has allowed provision of a remarkable level of biocompatibility, high resistance to abrasion, remarkable tensile strength, and flexural endurance. These properties are vital in supporting novel applications, including feeding tubes, devices in dialysis, medical garments, and balloon pumps for intraaortic applications. WPUs are efficient in outperforming several other known materials in terms of resistance to tear and abrasion. This is because several devices used in this area are capable of rubbing against others repeatedly. The waterborne process brought about a decrease in the environmentally related issues associated with the poisonous organic solvents used in the preparation of conventional polyurethane. More recently, WPUs have gained significant attention due to their versatility and the fact that they are environmentally friendly with no toxicity [28].

With the introduction of WPU, there is a significant breakthrough in the medical field, leading to eliminating various health-based challenges. They have therefore found relevance in cardiovascular diseases due to their biocompatibility and mechanical strength. Most of the WPUs have unique elastomeric properties, including tear resistance, toughness, and abrasion resistance. They have been used in the fabrication of lead insulators and artificial hearts. Patients using WPU medical devices would prefer them over other materials due to better comfort. Soft polyurethanes are more comfortable and stronger than other soft materials; hence, they are better materials for this application. Further, WPU is also relevant in smart actuators, textiles, and other useful medical devices due to the fact its soft and hard segments are adjustable; hence, it is possible to tailor some of their applications depending on the desired properties. This observation is based on the molecular exciton theory of Kasha, which has unique relevance in polymeric phenomena, where the luminophores polymerization brings about excitonic coupling and improves forward and backward crossing of the intersystem [4].

Yang et al. [29] formulated a new polyurethane elastomer with healable properties using a double network (DN) system of structures that suggest a promising usage in flexible sensors for future generation medical field and technology. In this system, a chemical network loosely crosslinked was acquired using photocurable double bonds of acrylic, which acts as a molecular backbone and ensures the polymer's elasticity. Due to the efficient design, the elastomer prepared showed remarkable properties, such as high stretchability, tensile strength, unique resilience, and outstanding self-healing potential. The healable elastomeric material enhances the easy production of composite conducting material, suggesting a promising potential as flexible sensors in the coming generation of stretchable electronic components, strain sensors, and robots.

In their work, Feng et al. [30] aimed at finding a solution to the challenge of inflammatory responses and adhesion in tissue. They produced bacterial cellulose in plasticized form (BCG) and composite WPU using antibacterial function as the new nasal stent. They observed that the gelation property of BCG enhanced the protection of the paranasal sinus mucosa, while the WPUs having enhanced mechanical properties were targeted for acting as a support to the narrow nasal cavity. There was also the possibility of adjusting the supporting force, thickness, and size of the nasal stent depending on particular conditions of the nose. The properties of the composite were investigated using various instrumental techniques, including thermogravimetry, water absorption, and contact angle analysis. It was concluded from the study that the prepared composite materials could ensure antibacterial activity for 12 days.

Hsieha et al. [31], in their study, produced the complex structure of tracheal scaffold using WPU composite and 3DP ink to copy the morphology of the trachea of a native rabbit. Small molecular drugs alongside bioactive factors were easily introduced in ink to enhance chondrogenesis of the stem cells. The 3DP scaffolds were in strong resemblance to the native trachea with similar mechanical properties. An airflow test was also done, which showed the tightness of the gas at both the negative and positive air pressure. In related work, Dai and Dai [32] produced several films of WPUs using varying proportions of ionic groups to select an efficient anti-adhesion film. The WPU was placed, and characterization was done for the assessment of their physicochemical parameters. Their findings show that the various films produced were nontoxic and have suitable physical and chemical properties, including mechanical strength based on zeta potential, thickness change, contact angle, and mechanical features. The anti-adhesion features were also evaluated using rabbit tendons that were seriously injured and then sutured using the adapted Kessler core suture method, arranged around the injured tendon. The results reveal that the WPU films showed outstanding anti-adhesion properties when compared to the commercially available products.

One of the unique revolutions in the medical sector is 3D printing technology (3DP). It has led to high demand for printers that are of 3D relevance, whose materials and mechanisms of functioning are in line with the medical demands. WPU has been relevant as promising medical-grade filaments can be used in modeling modern 3D printers for medical applications [33]. Further, WPU elastomers are known to have molecular structures closely related to human proteins. It was observed that the absorption of protein that marks the starting of the cascade of blood accumulation was relatively slower or less compared to other materials. This is accountable for their ideal usage in various areas in medicines that demand biomimetic and adhesive strength or antithrombotic features. For example, WPU is used as a unique sealant for the binding of bundles of hollow fibers in artificial dialysis.

Hendessi et al. [34] fabricated a nanocomposite film of WPU, which incorporates carvacrol loaded as an antibacterial nanofiller, bringing about a continuous discharge of antimicrobial agent. The nanocomposite showed significant inhibition of the growth of the bacteria. Moreover, in a study, Wattanodorn [35] fabricated anionic WPU incorporated with silver, showed consistent silver discharge for 21 days, bringing about unique potencies against bacteria and enhanced Young modulus and tensile strength. Further, in their work, Wu et al. [36] documented that cationic WPU with nanocomposite shows unique potency against *E. Coli* and *S. aureus*.

There was a reduction in the number of colonies formed when compared with the original values. Within a contact time of 3 h, it got to 100%. In a related study, Zhang et al. [37] produced highly crosslinked WPUs containing ammonium salts with high stability, biocompatibility, and antibacterial potential. The fabrication was achieved using polyoxytetramethylene glycol, poly (ethylene glycol), and diisocyanate through a simple method. Further, Ajorlou et al. [38] documented that the WPU nano-micelles produced had inhibitory properties on the growth of cancerous cells and the weight and sizes of the tumors in mice, which decreased during the treatment using nano-micelles decorated with folates. Also, the overall performance of the nano-micelles fabricated was relatively higher than the commercially available drug.

Wang et al. [39] had prepared a silver nanowire (AgNW) based strain motion sensor on thermoplastic polyurethane (TPU) flexible film via a transfer-printing technique. The AgNW/TPU-based composite strain sensor was used to sense the huge human motions, such as finger, elbow, and knee bending, subtle muscle contractions, etc. Because of its high flexibility and optical transmittance, this AgNWs/TPU film strain sensor can be applied well on irregular wearable devices. Further, Figure 20.5a shows the change in resistance detected by the sensor when fixed on the bending finger. The relative change in the resistance determines the bending degree of the finger. Figure 20.5b shows the relation between the rate of resistance and the bending degree of the elbow joint when the sensor is placed on the elbow joint. As the bending degree of the elbow joint increased, the change in the rate of the resistance also increased. In Figure 20.5c, the monitoring of the motions of the knee joint is shown. Further, Figure 20.5d shows the capability of the sensor to differentiate various gait behavior like slow walk, leg lift, and run, at different curvatures, like 30°, 45°, 60°, and 90°, of the knee, which was observed in the change in the rate of resistance change value (Rc) at each curvature. Therefore, this precise measurement of different gait behaviors can be used to monitor joint movements of rehabilitation training for patients with limb dyskinesia, and simultaneously, large motions of joint and subtle motions can also be monitored. Figure 20.5e shows the use of a strain sensor to monitor different stages of saliva swallowing motion by attaching it to the neck. The use of a strain sensor can detect muscular relaxation by monitoring muscle movement on the forearm when making a fist [39].

20.5 ENVIRONMENT AND AGRICULTURE APPLICATION OF WATERBORNE POLYURETHANES FOR SENSORS

Xie et al. [40], in their work, reported that the utilization of WPU and other polymeric materials as sensors in various areas of agriculture lies on their advantages, which include low cost and simplicity of fabrication. Also, the physicochemical properties of such sensing polymers change in interaction with external conditions, such as moisture, nutrients, temperature, pH, and humidity. The change in the observable and measurable physical parameters is then converted to stress and detected by the sensing device. For example, in the case of WPU material that is capable of swelling on reaction with some outer molecules in the environment,

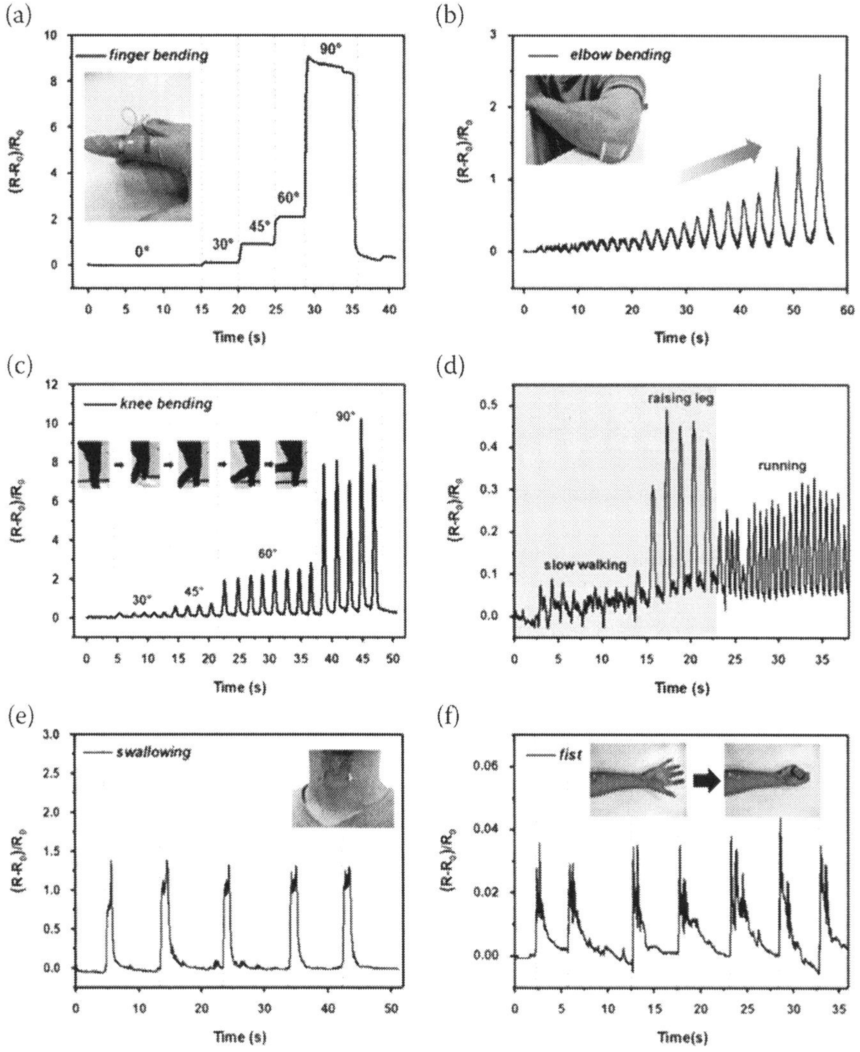

FIGURE 20.5 Detection of real-time human activities by the AgNW/TPU film adhered directly on the skin surface. (a) adhered onto a finger; (b) adhered on the elbow; (c) adhered on the knee; (d) monitoring of different states of motion with the film adhered on knee; (e) adhered onto throat; (f) adhered onto forearm muscle for monitoring of muscle relaxation (reproduced with permission from Wang et al., [39]).

the swelling effect can be sensed due to the stress created causing a coated cantilever beam to deflect. The deflection is measurable when a standard technique is employed, such as variation in reflectance or resistance.

PH sensors used in agriculture are also fabricated through the use of conducting polymers and WPUs. Such devices can detect the concentration of hydrogen ions in the soil environment. The variation in the concentration of the ions between the electrode selective to ions and the electrolytic solution brings about the variation in the gate potential, which is responsible for the modulation of current flowing between the drain and source [41].

Various WPU materials have been employed for the measurement of soil pH. A typical example is a photocurable polymer, which is made by mixing pre-polymer compositions like oligomer aliphatic urethane diacrylate and a suitable crosslinking agent, which provides a remarkable adhesion for the surface. Also, tetrapolymers have been reported for use in pH measurement as a result of their high composition of acrylic acid, which makes a provision for suitable stability and loading [42].

WPU sensors have also found relevance in the detection of soil moisture contents, though there are standard electrical approaches for the measurement of soil moisture content, such as the frequency dormain reflectrometry. Most related techniques are expensive and require much time. The use of polymeric sensors, therefore, presents numerous advantages. WPU could be deployed in the selective coating of micro-electro-mechanical systems (MEMs) for the measurement of moisture content. The composition of this sensing device includes a MEMS cantilever beam, which has a nano polymer that is sensitive to water as coating a temperature sensing component. The absorption of the moisture is accomplished by the polymeric material coated on the beam. As a result of weak forces, there is the production of stress on the MEMS beam, bringing about the deflection of the beam [40].

WPU sensors have also been employed in the measurement of atmospheric humidity. Humidity at a particular pressure and temperature refers to the ratio of water vapor content to saturated water vapor in the air. This significantly affects various environmental and agricultural parameters. For efficient growth of plants, the relative humidity must be within the range of 60%–70%. MEMS technology alongside polyurethane materials has been useful in the fabrication of humidity sensors. The component of the sensor includes a humidity generator component, where the humidity can be varied from 10% to 95% while the capacitance is measured using an LCR meter [32].

Temperature is a vital variable in environmental monitoring and assessment. Soil temperature also affect the growth of various crops and other environmental balances and processes. The temperature of soil helps in determining the proper seeding time, which is vital since germination is paramount immediately after seeds are cultivated. The fabrication of temperature-sensing devices using polymeric materials permits the development of cheap and easy-to-design temperature-sensing MEM devices base on WPU. Various polymers have been documented for their usefulness in the fabrication of temperature-sensing devices [43].

Cho et al. [44], in their work, fabricated enzyme sensors using an enzyme immobilization approach based on WPU ion-selective membranes. The fabrication of the enzyme membrane was through the coating of a tiny hydrophilic polyurethane sheet directly combined with the enzyme over the underlying polyurethane base membrane, which is ion-selective. The approach was found to be single and reliable for the development of a potentiometric biosensor suitable for the quantification and detection of organophosphorus pesticides in the environment. Optimization of the enzyme electrode was achieved through a constant variation of the compositions of the enzyme membrane for evaluating the impact of changes in the sensor responses. The fabricated sensor based on polyurethane was applied successfully in the detection and quantification of organophosphorus pesticides and paraoxon.

Ecologically friendly WPU coatings have pH stability, low-temperature flexibility, water resistance, outstanding weathering resistance, desirable mechanical and chemical properties, and superior solvent resistance. This makes them highly suitable for use in the fabrication of various kinds of sensing devices for environmental monitoring and measurements. UV-cured WPU varnishes make available a significant class of eco-friendly and green coatings through outstanding rapid curing and mechanical properties. Hyperbranched polyurethanes reveal some outstanding properties: reactivity, good rheological and high solubility behavior due to several end groups, diminishing chain entanglement, and compact molecular structure. Essentially, WPU coatings have decreased mechanical strength and stiffness, which could be improved by adding nanoparticles, such as Cu, Ag, SiO_2, and TiO_2 [45]. The addition of various kinds of nanoparticles and development of various composites make them useful in the fabrication of sensors, which are economical and durable.

WPUs could be formulated into adhesives and coatings, with no co-solvent or containing little, which form films by ambient temperature. These coating materials are commonly employed during the development of various kinds of sensors. They display outstanding linkage to numerous surfaces, such as polymeric fibers and glass. Some of these environmentally friendly polymers are nonflammable and nontoxic, and they cannot contaminate the produce wastewater or air. Generally, polyurethanes are insoluble in water as well as hydrophobic in nature. Consequently, to scatter them in water, they have to be improved through, for instance, nonionic hydrophilic segments into the polymer structure and incorporating ionic groups [46].

WPUs have also been employed in the fabrication and design of smart textiles with strain-sensing properties that are also environmentally friendly. Tensile strain-sensing devices have been fabricated through the integration of conductive polymers made from WPUs, which solve the inherent limitations associated with encapsulation of other elastic polymeric materials [2].

Babar et al. [47] developed hexamethylphosphoramide sensors by using polyaniline (PANI) coated polyurethane membrane structures to detect hexamethylphosphoramide (HMPA), which is highly soluble in both organic and aqueous matter and is considered a very important solvent to produce organic and organometallic compounds. Moreover, it also exhibits special features as UV inhibitors for plastics and polymerization catalysts. However, it shows certain negative effects on humans through inhalation, skin absorption, or contamination of groundwater. Therefore, HMPA vapor detection in the air is very crucial. The hexamethylphosphoramide sensors showed good reproducibility and sensitive and reversible responses.

Polyurethane ureas and polyurethanes, mainly their water-based dispersals, have increased significantly as a highly useful area established on ecologically friendly methods.

The development of their production approaches and the characteristics of the compounds involved in developing progressively maintainable routes have situated these dispersals as an essential and relevant product for various uses frameworks. Therefore, the study consists of the basic perceptions of polyurethane chemistry to develop the dispersal formulation approaches. This study similarly revealed the dispersal strengths with different renewable additives, including cellulose, chitosan, and starch. In the same way, dispersion's prospective to be treated utilizing diverse techniques is revealed, showing, through diverse examples, their practical relevances in a diversity of circumstances due to their adaptability for high constraint uses [1].

Xiao et al. [48], prepared green synthesized nitrogen-doped carbon dots (N–C-dots) that were grafted on the backbone of WPU to produce a film. This film was found to be self-healing and specifically identified the Fe^{3+} ion. Due to the presence of the hydrogen bonds between the WPU and N–C dots, the film showed the self-healing capacity, whereas WPU–N–C-dots film exhibited fluorescent characteristics, which helped them to specifically recognize the Fe^{3+} ion. Moreover, the film showed a wide detection range, high selectivity and sensitivity, excellent tensile strength, and increased thermal stability, which showed that the WPU–N–C dots can be utilized in the environmental domain to detect heavy metal ions.

The steadiness of polyurethane is of serious significance for uses consisting of biomaterials and the coating industry. To remove the ecological crisis on polyurethane production that comprises the application of aqueous-based, organic solvents or WPUs has been established. However, WPU could be unsteady in an electrolyte-rich environment. The steadiness of biodegradable WPU in the protected saline solutions assessed utilizing atomic force microscopy. Numerous WPU films that are biodegradable were organized through the spin coating on cover-slip glass using a viscosity of ~300 nm [49].

Remote sensing is regarded through spatial resolution, spectral resolution, as well as temporal resolution. Spatial resolution is said to be the last area that could be famous in appearance. Improvements in hyperspectral imaging have controlled enhancements in spectral resolution for some time. Nowadays, hyperspectral imaging schemes could amount to several very constricted connecting spectral bands through the near-infrared, visible, thermal infrared, and mid-infrared portions of the electromagnetic range [50]. Furthermore, WPU-based sensors for agricultural utilities have not been demonstrated and need to be further investigated by researchers in this domain.

20.6 CONCLUSION AND PROSPECTS

Novel approaches to the development and design of different WPUs have been highlighted in line with several applications in medicine, the environment, and agriculture. WPU utilizes polymer-based composites with very high-quality physiochemical properties to detect, monitor, and sense various biomolecules. These properties give WPU advantages over conventional polyurethane, but further research and studies should be carried out to develop different materials, screens, and spray-printed sensors that are stable, strong, nonbiodegradable, nonflammable, nontoxic WPU in the electrolyte-rich environment through the utilization of various nanoparticles.

ACKNOWLEDGMENT

COA, AI, OTO, and JTM are thankful to their respective universities for helping them to prepare this chapter. KRBS would like to thank Professor A.K. Singh for providing constant support and guidance throughout this work. JS expresses their gratitude for the DST-INSPIRE faculty Fellowship, BHU (IoE grant), and UGC New Delhi for providing financial support. RPS and VN are thankful to Vice-Chancellor, IGNTU, Amarkantak, India, for providing constant support financially and for motivating them to do good science.

REFERENCES

1. Santamaria-Echart, A., Fernandes, I., Barreiro, F., Corcuera, M.A., & Eceiza, A. (2021). Advances in waterborne polyurethane and polyurethane-urea dispersions and their eco-friendly derivatives: A review. *Polymers*, 13:409. doi:10.3390/polym1303 0409
2. Jiang, W. (2021). Functional fabric with strain sensing based on foam finishing. IOP Conference Series: Earth and Environmental ScienceIOP Conf. Ser. *Earth & Environmental Science,* 697:012020#
3. Cakic, S.M., Spirkoya, M., Ristic, I.S., & Poreba, R. (2013). The waterborne polyurethane dispersions based on polycarbonate diol: Effect of ionic content. *Material Chemistry and Physics*, 138(1):277–285. doi:10.1016/j.matchemphys. 2012.11.057
4. Zhao, B., Fu, R.W., Zhang, M.Q. Zhang, B., Zeng, W., Rong, M.Z., & Zheng, Q. (2007). Analysis of gas sensing behaviors of carbon black/waterborne polyurethane composites in low concentration organic vapors. *Journal of Material Science*, 42:4575–4580. doi:10.1007/s10853-006-0517-6
5. Prabhakar, P.K., Pallipurtath, A., & Doble, M. (2011). Biocompatibility studies on polyaniline and polyaniline-silver nanoparticle coated polyurethane composite. *Colloids and Surfaces B: Biointerfaces,* 86(1):146–153 doi:10.1016/j.colsurfb. 2011.03.033
6. Wang, S., Fang, Y., He, H., Zhang, L., Li, C., & Ouyang, J. (2020). Wearable stretchable dry and self-adhesive strain sensors with conformal contact to skin for high-quality motion monitoring. *Advanced Functional Materials*, 2007495. doi:10.1002/adfm.202007495.
7. Liu X., Hong W., & Chen, L. (2020). Continuous production of water-borne polyurethanes: A review. *Polymers,* 12(12):2875. doi:10.3390/polym12122875
8. Tian, S. (2020). Recent advances in functional polyurethane and its application in leather manufacture: A review. *Polymers*, 12(9). doi:10.3390/polym12091996
9. Lin, D., Li, B., Qi, J., Ji, X., Yang, S., Wang, W., & Chen, L. (2019). Low cost fabrication of microluidic paper-based analytical devices with water-based polyurethane acrylate and their application for bacterial detection. *Sensors and Actuators B: Chemical*, 127213. doi:10.1016/j.snb.2019.127213
10. Wang, R., Xu, W., Shen, W., Shi, X., Huang, J., & Song, W. (2019). A highly stretchable and transparent silver nanowire/thermoplastic polyurethane film strain sensor for human motion monitoring. *Inorganic Chemistry Frontiers*, 6(11): 3119–3124.

11. Gonçalves, B. F., Oliveira, J., Costa, P., Correia, V., Martins, P., Botelho, G., & Lanceros-Mendez, S. (2017). Development of water-based printable piezoresistive sensors for large strain applications. *Composites Part B: Engineering*, 112:344–352. doi:10.1016/j.compositesb.2016.12.047
12. Guo, X., Ge, S., Wang, J., Zhang, X., Zhang, T., Lin, J., & Guo, Z. (2018). Waterborne acrylic resin modified with glycidyl methacrylate (GMA): Formula optimization and property analysis. *Polymer*, 143:155–163. doi:10.1016/j.polymer.2018.04.020
13. Lee J., Kim J., Shin Y., & Jung, I. (2019). Ultra-robust wide-range pressure sensor with fast response based on polyurethane foam doubly coated with conformal silicone rubber and CNT/TPU nanocomposites islands. *Composites Part B: Engineering,* 177(15):107364
14. Petr, S., & Robert, O. (2011). Effect of functionalized nanotubes with HNO_3 on electrical sensory properties of carbon nanotubes/polyurethane composite under elongation. *Materials Science*, Corpus ID: 138318759
15. Babar, D.G., Olejnik, R., Slobodian, P., & Matyas, J. (2016). High sensitivity sensor development for hexamethylphosphoramide by polyaniline coated polyurethane membrane using resistivity assessment technique. *Measurement,* 89:72–77.
16. He, H., Xu, X., & Zhang, D. (2013). An aligned macro-porous carbon nanotube/waterborne polyurethane sensor for the detection of flowing organic vapors. *Sensors and Actuators B Chemical*, 176:940–944. doi:10.1016/j.snb.2012.09.095
17. Han, Y., Hu, J., & Xin, Z. (2018). In-Situ incorporation of alkyl-grafted silica into waterborne polyurethane with high solid content for enhanced physical properties of coatings. *Polymers*, 10(5):514.
18. Chen, S.G., Hu, J.W., Zhang, M.Q., Wei, M., & Rong, M.Z. (2004). Gas sensitivity of carbon black/waterborne polyurethane composites. *Carbon*, 42(3):645–651.
19. Chen, S.G., Hu,Z.L., Hu, J., Zhang, M.Q., Rong, M.Z., & Zheng, Q. (2006). Relationships between organic vapor adsorption behaviors and gas sensitivity of carbon black filled waterborne polyurethane composites. *Sensors and Actuators B: Chemical*, 119:110–117.
20. Aksoy, B., Sel, E., Savan, E.I., Ates, B., & Koytepe, S. (2020). Recent progress and perspectives on polyurethane membranes in the development of gas sensors. doi:10.1080/10408347.2020.1755823
21. Chen, S.G., Hu, J.W., Zhang, M.Q., Rong, M.Z., & Zheng, Q. (2006). Improvement of gas sensing performance of carbon black/waterborne polyurethane composites: Effect of crosslinking treatment. *Sensors and Actuators B: Chemical*, 113(1):361–369.
22. Jia, Y., Sun, Y., Pan, Y., Wang, X., Zhai, Z., Min, Z., Zheng, G., Liu, C., Shen, C., & Liu, X. (2021). Flexible and thin multifunctional waterborne polyurethane/Ag film for high-efficiency electromagnetic interference shielding, electro-thermal and strain sensing performances. *Composites Part B: Engineering*, 210(1):108668.
23. Xiao, Y., Fu, X., Zhang, Y., Liu, Z., Jiang, L., & Lei, J. (2016). Preparation of waterborne polyurethanes based on the organic solvent-free process. *Green Chemistry*, 18(2):412–416.
24. Lawal, A. L. (2020). Recent progress in graphene based polymer nanocomposites. *Cogent Chemistry*, 6(1):231–250. doi:10.1080/23312009.2020.1833476
25. Wang, Y., Li, W., Zhou, Y., *et al.* (2020). Fabrication of high-performance wearable strain sensors by using CNTs-coated electrospun polyurethane nanofibers. *Journal of Material Science,* 55:12592–12606. doi:10.1007/s10853-020-04852-8
26. Moheimani, R., Aliahmad, N., Aliheidari, N., Agarwal, M., & Dalin, H. (2021). Thermoplastic polyurethane flexible capacitive proximity sensor reinforced by CNTs for applications in the creative industries. *Scientific Reports*, 11:1104. doi:10.1038/s41598-020-80071-0

27. Sousa, E.A., Lima, T.H., Arlindo, E.P., & Sanches, A. O. (2020). Multicomponent polyurethane–carbon black composite as piezoresistive sensor, *Polymer Bulletin*, 77:3017–3031 doi:10.1007/s00289-019-02888-8
28. Xiano, Y., Fu, X., Zhang, Y., Liu, Z., & Lei, J. (2016). Preparation of waterborne polyurethanes based on the organic solvent-free process. *Green Chemistry*. doi:10.1039/C5GC01197C
29. Yang, Y., Ye, Z., Liu, X., & Su, J. (2020). A healable waterborne polyurethane synergistically crosslinked by hydrogen bonds and covalent bonds for composite conductors. *Journal of Materials Chemistry C*, 15(2):233.
30. Feng, Z., Li, M., Jin, X., Zheng, Y., Liu,J., Zhao, L., Wang, Y., Li, H., & Zuo, D. (2020). Design and characterization of plasticized bacterial cellulose/waterborne polyurethane composite with antibacterial function for nasal stenting. *Regenerative Biomaterials*, 7(6):597–608. doi:10.1093/rb/rbaa029
31. Hsieha, C., Liaob, C., Daic, N., Tseng, C., & Hsuad, L. (2018). Printing of tubular scaffolds with elasticity and complex structure from multiple waterborne polyurethanes for tracheal tissue engineering. *Applied Materials Today*, 12: 330–341.
32. Dai, L., Hung, Y.M., & Dai, N. (2018). Evaluation and characterization of waterborne biodegradable polyurethane films for the prevention of tendon postoperative adhesion, 5485–5497. *International Journal of Nanomedicine*, 13. doi:10.2147/IJN.S169825
33. Przybytek, A., Gubańska, I., Kucińska-Lipka, J., & Janik, H. (2018). Polyurethanes as a potential medical-grade filament for use in fused deposition modeling 3D printers—a brief review. *Fibres & Textiles in Eastern Europe*, 26(6):132–140.
34. Hendessi, S., Sevinis, E.B., Unal, S. & Unal, H. (2016) Antibacterial sustained-release coating from halloysite nanotubes/waterborne polyurethanes. *Progress in Organic Coatings*, 101:253–261. doi:10.1016/j.porgcoat.2016.09.005
35. Wattanodorn, Y. (2014). Material performance: antibacterial anionic waterborne polyurethanes/Ag nanocomposites with enhanced mechanical rpoperties. *Polymer Testing*, 40:163–169. doi:10.1016/j.polymertesting.2014.09.004
36. Wu. Y., Lin, W., Hao, H., Li, J., Luo, F., & Tan, H. (2017). Nanofibrous scaffold from electrospinning biodegradable waterborne olyurethane/poly(vinyl alcohol) for tissue engineering application. *Journal of Biomaterials Science Polymer*, 28(7): 648–663. doi:10.1080/09205063.2017.1294041
37. Zhang, L., Du, W., Nautiyal, A., Liu, Z., & Zhang, X. (2018). Recent progress on nanostructured conducting polymers and composites: Synthesis, application and future aspects. *Science China Materials*, 61(3), 303–352. doi:10.1007/s40843-017-9206-4
38. Ajorlou, E., Khosroushahi, A.Y., & Yeganeh, H. (2016). Novel waterborne polyurethane nanomicelles for cancer chemotherapy: Higher efficiency of folate receptors than TRAIL receptors in a cancerous Balb/C mouse model. *Pharmaceutical Research*, 33(6):1426–1439. doi:10.1007/s11095-016-1884-6
39. Wang, F., Feng, L., Li, G., Zhai, Z., Ma, H., Deng, B., & Zhang, S. (2019). Fabrication and properties of superhydrophobic waterborne polyurethane composites with micro-rough surface structure using electrostatic spraying. *Polymers*, 11(11), 1748. doi:10.3390/polym11111748
40. Xie, D., Jiang, Y., Pan, W., Dan, Li., Wu, Z., & Li, Y. (2002). Fabrication and characterization of polyaniline-based gas sensor by ultra-thin film technology. *Sensors and Actuators B: Chemical*, 81:158–164.
41. Chan, Y. H., Jin, Y., Wu, C., & Chiu, D. T. (2011). Copper (II) and iron (II) ion sensing with semiconducting polymer dots. *Chemical Communications*, 47:2820–2822. Crossref, Google Scholar.

42. Tajau, R., Rohani, R., Alias, M.S., Mudri, N.H., Halim, K. (2021). Emergence of polymeric material utilising sustainable radiation curable palm oil-based products for advanced technology applications. *Polymers*, 13:11. doi:10.3390/polym13111865
43. Jorquera, C. J., Orozco, J., Baldi, A. (2010) ISFET based microsensors for environmental monitoring. *Sensors*, 10:61–83.
44. Cho Y.A., Lee H.S., Cha G.S., & Lee, Y.T. (1999). Fabrication of butyrylcholinesterase sensor using polyurethane-based ion-selective membranes. *Biosens Bioelectron*, 14(4):435–438. doi:10.1016/s0956-5663(99)00016-0.
45. Noreen, A., Zia, K. M., Zuber, M., Tabasum, S., & Saif, M. J. (2015). Recent trends in environmentally friendly waterborne polyurethane coatings: A review. *Korean Journal of Chemical Engineering*, 33(2):388–400. doi:10.1007/s11814-015-0241-5
46. Honarkar, H. (2017). Waterborne polyurethanes: A review. *Journal of Dispersion Science and Technology*, 39(4):507–516. doi:10.1080/01932691.2017.1327818.
47. Babar, D. G., Olejnik, R., Slobodian, P., & Matyas, J. (2016). High sensitivity sensor development for hexamethylphosphoramide by polyaniline coated polyurethane membrane using resistivity assessment technique. *Measurement*, 89:72–77. doi:10.1016/j.measurement.2016.03.078.
48. Xiao, L., Shi, J., Chen, W., Zhang, F., & Lu, M. (2020). Highly sensitive detection of fe^{3+} ions using waterborne polyurethane-carbon dots self-healable fluorescence film. *Macromolecular Materials and Engineering*, 305(3):1900810.
49. Lin, Y. Y., Hung, K.-C., & Hsu, S. (2015). Stability of biodegradable waterborne polyurethane films in buffered saline solutions. *Biointerphases*, 10(3):031006. doi:10.1116/1.4929357
50. Li, S., Simonian, A., & Chin, B. A. (2010). Sensors for agriculture and the food industry. *The Electrochemical Society Interface*, 19(4):41–46. doi:10.1149/2.f05104if

21 Waterborne Polyurethanes for Sealants

Mehrdad Fallah and Amir Ershad Langroudi
Color and Surface Coating Group, Polymer Processing Department, Tehran, Iran
Polymer & Petrochemical Institute (IPPI), Tehran, Iran

Aida Alavi
Color and Surface Coating Group, Polymer Processing Department, Tehran, Iran
Polymer & Petrochemical Institute (IPPI), Tehran, Iran
Department of Chemistry, University of Isfahan, Isfahan, Iran

CONTENTS

21.1 Introduction 355
21.2 Classification of Various Types of Sealants 356
21.3 Overview of Types of Sealants and Families 358
21.4 Polyurethane Sealant 364
 21.4.1 Chemistry of Polyurethane Sealants 364
 21.4.2 Classification of Polyurethane Sealants 365
 21.4.3 Polyurethane Sealant Applications 366
21.5 Waterborne Polyurethane Sealants 366
 21.5.1 The Classification of Isocyanate Involved in Waterborne Polyurethane Sealant 368
  21.5.1.1 Aromatic Isocyanate 368
  21.5.1.2 Aliphatic Isocyanate 369
21.6 Conclusion 372
References 372

21.1 INTRODUCTION

"Sealing" is described as "the art and science of preventing leaks," in which a liquid, paste, or foam material is applied to a joint and forms a seal against liquids or gases while still allowing the joint to move. Joints and holes between two or

DOI: 10.1201/9781003173526-21

more substrates are sealed with sealants. Sealing is an old problem that ancient people noted for having a better quality of life.

Ancient people used natural "sealants" to protect their homes against the wind, water, etc. Some natural sealants include earth, loam, grass, and reeds. The difference between sealants and adhesives is not always very obvious, and the classification of a product with both adhesion and sealing capabilities is vague. That's why the classification does not have an accurate demarcation line.

Sealants' primary function is to seal a joint while also providing important adhesion and movement properties. Adhesives, on the other hand, are planned to hold materials together by surface attachment, often as alternatives to mechanical fastening systems. Adhesives' primary role is to pass load between adjoining surfaces through adhesion and structural strength.

Some products, referred to as structural sealants, are adhesive-sealants that serve a dual purpose of bonding and sealing. All sealants must fulfill three critical functions:

1. Fill the joint or void properly to achieve a sealed environment.
2. Put together a gas or liquid flow barrier.
3. Maintain the sealing in operation while enabling a specific amount of movement in the sealed parts [1].

By sealing a joint between different components or materials, a sealant prevents air, water, and other substances from entering or leaving a structure. Sealing is used to create a barrier against specific natural stimuli, such as moisture, driving rain, draughts, sand, dust, and so on, depending on the intended use. Their primary function is to protect sealed surfaces from environmental conditions, such as temperature, dampness, and daylight exposure, as well as to absorb shear, compression, and extension stresses [2].

21.2 CLASSIFICATION OF VARIOUS TYPES OF SEALANTS

A wide-ranging sealing material and the variety of sealant-forming methods, types, formulations, and chemical crosslinking systems make troublesome sorting and obscure the grouping of the material, mainly as the diversity stays propagation by new product development [3]. It makes it difficult to understand the composition of a material (adhesive or sealant). Then, there was not an exact ordering of adhesives [4,5].

As described, because of the diversity of sealant formulation, they have no distinct classification. Accordingly, many kinds of research have been carried out to classify sealants based on application, hardening ability, cure type, type of polymer, and so on. In this regard, Petrie et al. [3] reported the classification of sealants according to the crosslinking sort, hardening ability, application, and functioning.

Figure 21.1 shows the first classification, which is based on hardening and non-hardening types. Hardening sealants, when are applied to a joint, make a rigid or flexible solid seal. Hardening sealants consist of two categories, rigid and flexible sealants.

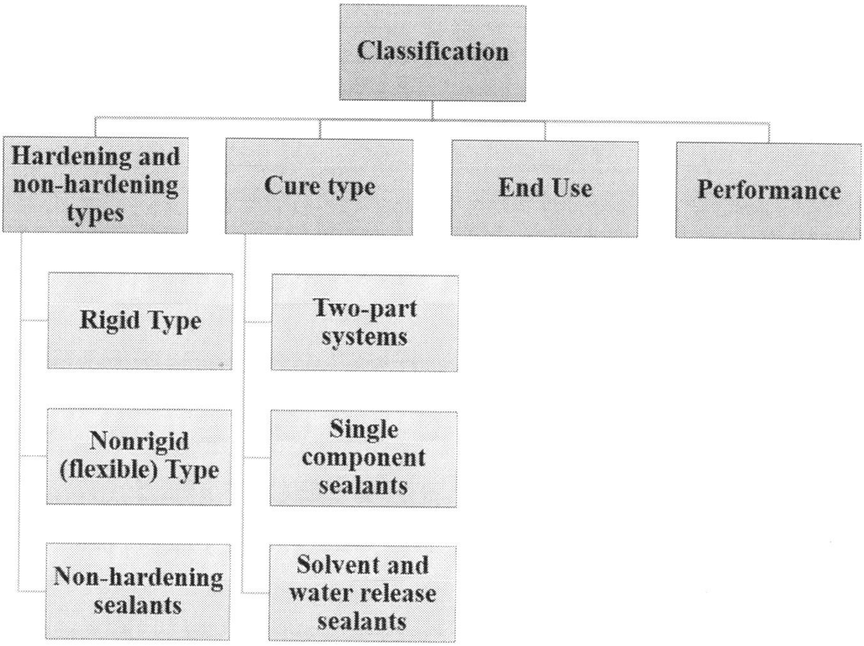

FIGURE 21.1 Classification of sealants in different bases.

Rigid sealants have great resistance to flexibility. These kinds of sealants are usually used for both sealing and joining purposes. The polymers that are used in this category include epoxy, modified epoxy, and acrylic. Flexible sealants have flexibility and resiliency properties and generally are based on elastomer polymers, such as silicones, polyurethane (PU), and polysulfide. The physical properties of non-hardening sealants stay changeless, and their flowability after applying is evident, which shows that non-hardening sealants persist in being soft, even after application, solvent evaporation, chemical interactions, and some other crosslinking mechanism. Natural resins and oil-based sealants, which are obtainable in drying and non-drying types, are classified as non-hardening sealants.

Another characterization is based on the curing type. The parameters that influence curing are temperature, moisture, and catalyst type. According to their curing type, the sealants are divided into three general groups: two-part systems, single component sealants, and solvent or water release sealants. Also, the sealant is sometimes classified by its function and end use. The sealants are used in many conditions and places, such as automobiles, insulated glass, concrete, construction, etc., and the last characterization in this study is about the performance of sealants [3].

In another study, Galimzyanova and coworkers reported different types of classification of sealant, including application, the nature of the polymer base, and the nature of the transition to working condition.

One of the essential groups of sealants that have a vital role in most industry areas is polymer-based sealing materials. They have many consumptions to solve

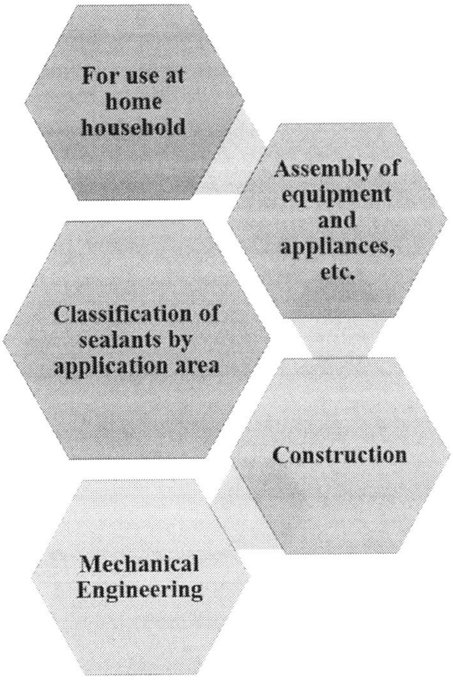

FIGURE 21.2 Classification of sealants by application [4].

everyday problems and have worthy application in different parts, including the aerospace field, construction and automotive, and sealing plumbing. In Figure 21.2, polymer-based sealants applications, which are divided into four main groups, are shown.

One of the most important features of sealants is that they are polymer based. Different polymer bases are used for the manufacture of sealants depending on the application and performance of the sealant in various conditions. In Figure 21.3, these polymers are shown.

The last classification discussed in this study is a classification of sealant by the nature of the transition to working conditions (hardening and non-hardening).

As explained in the former section, hardening sealants include plastic, rigid, and elastic sealants that work in both chemical and physical ways. Non-hardening sealants consist of rigid and plastic sealants (Figure 21.4).

21.3 OVERVIEW OF TYPES OF SEALANTS AND FAMILIES

During the sealing process, the sealing materials must adhere to fill the holes and cavities between two or more substrates, taking into account the full range of temperature-induced movement, mechanically induced movement, and environmental stress. The main component of a sealant, a binder (usually a polymer), holds the individual components together and is one of the most critical factors in product efficiency. Plasticizers, antioxidants, catalysts, tackifiers, diluents, preservatives,

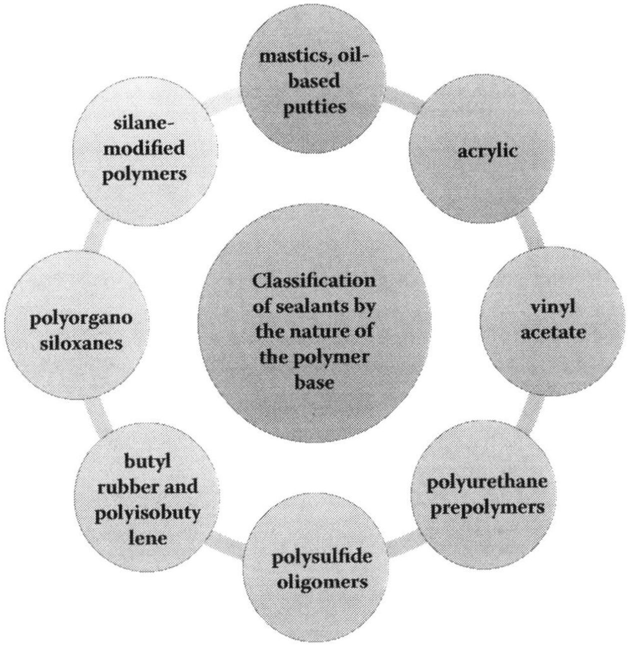

FIGURE 21.3 Classification of sealants by the nature of the polymer base [4].

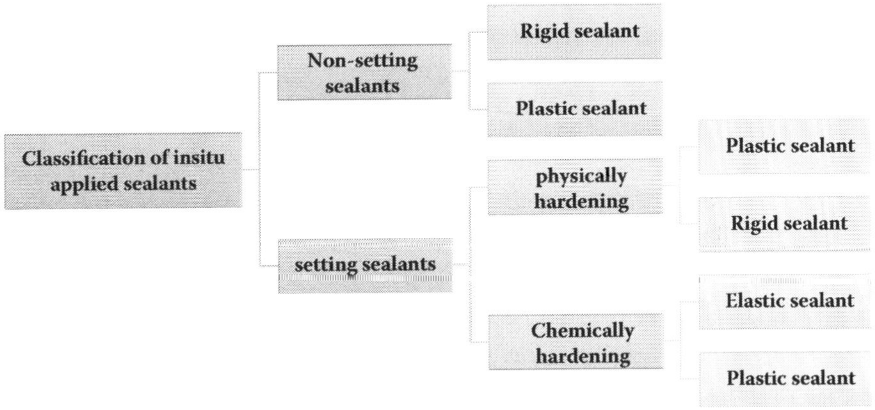

FIGURE 21.4 Classification of sealants by the nature of the transition to the working condition [4].

and solvents are among the other materials used in the formulation to achieve the properties required to meet the performance and cost requirements of a particular application [5].

In almost all areas of industry, polymer-specific sealing materials (binders) are particularly utilized. A significant amount of sealants used is in buildings and automobiles. The use of sealants will improve enterprises' energy production and reduce greenhouse gas emissions to the atmosphere. For instance, when assembling a windshield, adhesive sealants may increase vehicle protection in mechanical engineering.

Sealants are also implemented in advanced fields, such as fuel tanks sealing airplanes, and solar panel projects. Simultaneously, sealants are utilized to work out a broad spectrum of household issues, such as closing plumbing. Generally, sealants can overcome many complex problems in many specialist applications.

Different fields for the use of sealants and adhesive materials contribute to developing a broad range of stuff. Several thousand adhesive and sealant formulations are challenging to recognize and also challenging to differentiate from each other on the market. The manufacturers present sealing and adhesive products for a wide range of applications and are prepared to produce customer orders directly.

In this section, two kinds of classifications of sealant families are discussed by two different studies. First, the Petrie research explored the different standardized families of sealant products. More than 15 families of polymers are used in individual or mixture models for each application to achieve the storage properties, application features, physical efficiency, and longevity. Also, the different sealant groups were principally categorized by their movement ability. As demonstrated in Figure 21.5, Petrie considers three main categories for sealant families. In this classification, each polymer is located in one category. In the second study, Klosowski et al. [5] offered a high-level description of the different general types of sealants depending on the efficiency category. They used four main categories to identify sealant families, as demonstrated in Figure 21.6.

A significant proportion of rigid sealants are single-component materials applied with a putty knife or a sealant gun in the case of thixotropic putties. To achieve a satisfactory service life, putties generally need painting. In addition, putties have a limited range of motion, so they can't be used in joints that move.

Silicones, acrylics, PUs, and polysulfides are high-performance, elastic sealants that withstand joint movements of up to 25%. Each sealant type's base polymer has inherent properties that make it useful in specific applications. For example, silicones are weather tolerant for use in applications that are mainly concerned with durability. PUs are tough, abrasion immune, with good adhesion to plenty of substrates, which helps in many weather-resistant applications, particularly in areas of high aggression. Polysulfides have excellent resistance to chemicals and low gas permeation; therefore, they are commonly used for IG production and civil engineering applications. Acrylic latex sealants are highly weatherable, cost-effective, and convenient to use, and they are popular with customers and specialists. A new class of sealants, called organic or hybrid silane-modified sealants, has recently been marketed. These sealants provide increased adhesion and weather resistance and properties traditionally correlated with organic sealants like paintability.

Cold-applied plastic sealants is a vast range of different forms and enhancements of composition with very different characteristics. A common character is their distinct plasticity (their elasticity is zero or none). These single-component, non-crosslinking sealants are sprayed cold with a gun, either because they are too viscous and gluey (acrylics and butyls solvent-resistant) or because their composition

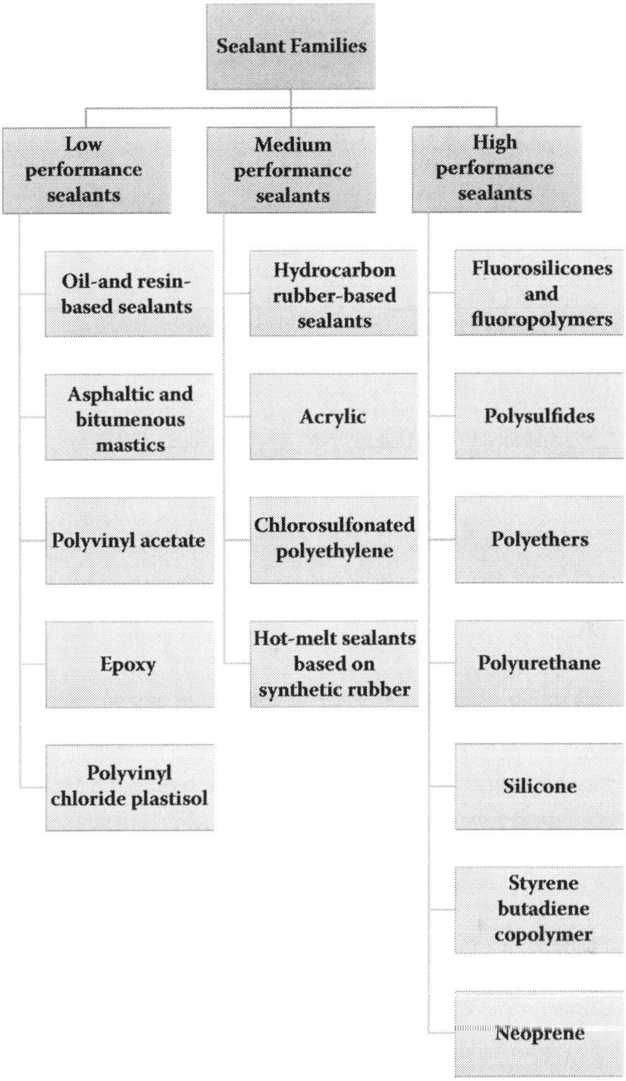

FIGURE 21.5 Classification of sealant families based on their performance.

is too fluidic (latex acrylic). Butyl rubber materials are also accessible in the form of ropes and tapes that are not curable.

Solvent-borne acrylic-based sealant is employed to seal low motion joints (although they remain limited to expert applicators and decline as substitute sealing agents of comparable value in the longer term). Therefore, solvent-borne acrylic-based sealant has some characteristics, such as sticky and hard to apply, shrinking through curing, outstanding adhesion without undercoat to most surfaces, minimal movement, superb weather ability, and proper paintability. Butyl sealants are available as tape sealants, and ropes have the same properties as acrylic-based sealants.

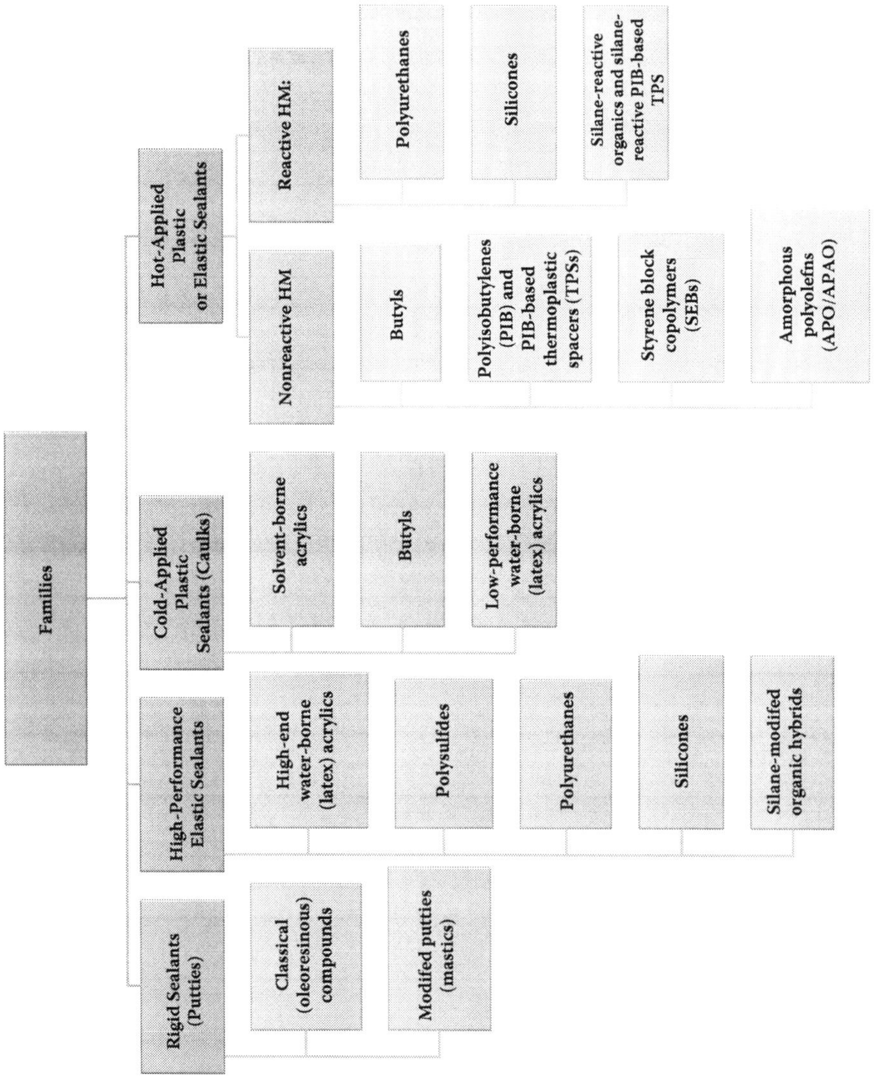

FIGURE 21.6 Classification of sealant families based on binder performance and movement capability.

One conventional sealant is a low-end acrylic latex sealant that cures via unhanding water (for pigmented and transparent products, it has 15%–25% and 50%–55% shrinkage, respectively). The most easy-to-use sealant has a low string, is easy to clean with water, is made from solvent-free material with low volatile organic compound (VOC) content, has strong adhesion to widely used construction materials, and is suitable for consumer usage.

High-temperature thermoplastic sealants, or hot-melts, are typically 100% solid without a solvent. These materials are solid at room temperature, but when heated above their softening point, they become viscous sticky liquids that flow and wet the substrates. Through the cooling process, hot-melt materials endure a reverse-phase change. This thermal conversion helps improve their inner strength easily and finally becomes a solid of ample structural integrity to act as a sealant. Compared to the usual bonds, hot-melted sealants deliver various advantages.

For example, they are 100% solid and free of solvent materials, resulting in low shrinkage, low to no VOC, and no need for drying or curing, all of which contribute to improved assembly line performance. They are also ideal for high-speed automated production due to quick initial bond strength (green resistance) and tremendous adhesion on various substrates.

But then, facing its drawbacks, when the substrate is subjected to high heat stress, the application requires advanced equipment. Most hot-melt sealants develop strong bonds and are generally easy to manipulate, but they differ significantly in functioning features due to their polymer structures. HM sealants are also distinguished by polymer form, e.g. butyls, silicones, and PUs. There is no chemical cross-linking with the great majority of hot-melt sealants in building and construction. The setup mechanism is physical, ensuring that the liquid–solid transformation remains reversible during the service life of the sealing materials. This function restricts the temperature tolerance of non-reactive hot-melt sealants and induces them to creep, particularly at high temperatures, when undergoing continuous stress. The melting temperature in the solid–liquid–solid phase transition must be significantly higher than the phase set for functioning, usually within a range between 100–180°C to reach an acceptable higher service temperature limit. Following solidification, reactive thermal hot-melts (RHM) are further cured (usually atmosphere-induced moisture), thereby integrating with traditional hot-melts via the advantages of chemically crosslinking sealants. RHM sealants get their initial bond strength as they are cooled from the solidified nonreactive polymers or oligomer components, whereas the final strength is obtained through chemical reactions. The last chemical crosslinking step enhances the resistance of the final RHM to the external threats; thus, compared to existing hot-melt sealants, RHM sealant products give the following advantages:

- Improved resistance to heat and creep
- Enhanced solvent and hydrolysis resilience
- Capacity for structural connection

Hot-melt sealants function in the working life of the sealants (set temperature mode), dependent on the wet out obtained when applied (melt temperature phase).

When hot-melt is applied to a heated surface and then allowed to cool at a slower rate, its adhesion time is significantly longer than when applied to a cold surface. However, the primary sealing material on the spacer's surface may be marginal, particularly in the winter, when shop floor temperatures are much lower than in the summer. As a result, consumers of hot-melt sealants should track not only the sealant's application temperature but also the temperature of the surface it is applied to.

21.4 POLYURETHANE SEALANT

One of the most predominantly used innovative sealants for new building and reconstruction plans is PU sealants accessible in single and multi-component formulations and primarily used in diverse applications like horizontal or vertical joints. Excellent abrasion and tear resistance, appropriate chemical and weather resistance, applicable in a broad type of joints, and the diversity of substrates adherence (metal, wood, and stone) are some of PU sealants' advantages. Sealants are practical to bond nonporous substrates like glass. But, it should be noted, they have some limitations; the most important one is that PUs are recommended for dry substrates and are not efficient in wet conditions [6]. PU sealants give great push recuperation to hold shape after being bowed or pulled quick curing rates and indeed adherence to nonprimed concrete. PU sealants can even be painted to coordinate the vicinal surfaces. Whereas many special holders of surfaces next to each other cannot be painted, PU sealants have this property [7]. Since PUs can be made from various raw materials with different properties, they can have a broad range of properties that can be used in a variety of industries. Therefore, the chemistry of PU sealants is one of the most prominent issues that should be further evaluated.

21.4.1 Chemistry of Polyurethane Sealants

Bayer et al. pioneered the polyaddition polymerization reaction of polyisocyanates with di- or polyfunctional hydroxyl or amine materials, developing PU sealants and adhesives in 1937. Since then, researchers have discovered that altering the structure of polyols and isocyanate components can result in a wide range of properties [8]. PUs, notwithstanding the rest of the molecule, are a group of polymers comprising a critical number of urethane groups. A distinctive PU besides the urethane bonds can contain aliphatic/aromatic hydrocarbon, ether, ester, amide, and urea groups. [8,9] The primary and introductory chemistry of PU compounds can be defined as the reaction of isocyanates with polyols, water, and amines. Isocyanates react with all active hydrogen-containing materials. The main reaction, which leads to the formation of urethane bonds, is the reaction of isocyanates with hydroxyl groups (see Figure 21.7) [10,11].

Polyfunctional active hydrogen-containing compounds react with isocyanate groups to create high molecular weight PUs [11]. The reaction of isocyanate groups with di- or polyfunctional hydroxyl materials, such as polyesters or hydroxyl-ended polyethers, is one of the most important reactions. The number of functional groups

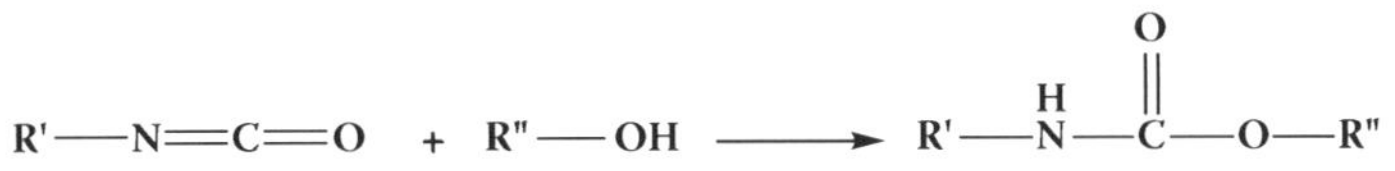

FIGURE 21.7 The polyurethane formation reaction between isocyanate and hydroxyl groups.

of the hydroxyl-containing compound and the isocyanates are often raised to three or more to compose branched or crosslinked polymers. The reactivity of alcohol type is primary (1°)> secondary (2°)> tertiary (3°).

PU sealant formulations usually include three components: (1) an NCO-terminated pre-polymer with at least 2.0 isocyanate functionality, (2) a polyol and chain extender, and (3) an infinitesimal amount of catalyst to catalyze the isocyanate and active hydrogen-contained material reaction [12].

Two types of isocyanates can be used to make a urethane pre-polymer: (1) aromatic isocyanates like toluene diisocyanate (TDI) and diphenylmethane diisocyanate (MDI) and (2) aliphatic polyisocyanates like isophorone diisocyanate (IPDI) and hexamethylene diisocyanate (HDI), with the aromatic group being more reactive than the aliphatic one. In addition, a range of various polyols can be used to manufacture the different types of PU adhesives and sealants, including polyesters and polyethers [13–15].

Ziyan et al [12] introduced that the main component of PU sealants has three main parts:

1. A PU pre-polymer owes at least functionality of 2.0 and molecular weight of 2,000.
2. A tin catalyst to catalyze the reaction of isocyanate and H-contained materials.
3. A small quantity of an organophosphite that has an alkaryl substance or each an acyclic ligand and an aromatic ligand.

In another research study, Li et al. [16] reported the PU sealant preparation for glass insulating. For this purpose, the two-part sealants were synthesized by the reaction of a *p*-chlorophenol-blocked PU pre-polymer and a poly-functional polyamine polymer. Also, organobentonite and dibutyltin dilaurate (DBTDL) were added to the mixture as the catalyst material to advance the curing process at low temperatures.

21.4.2 Classification of Polyurethane Sealants

It is a complicated issue to classify the sealant materials. Still, in PU sealants, the more accepted one is that they are classified into one- or two/component PU sealants based on divers using methods. The curing agent of a one-part system is the present moisture in the air. Therefore, the operating system is easy, and the curing process is slow. Still, on the other hand, as stability is affected by the exterior environment, its storage may be a complex criterion.

Conversely, the fast curing, simple storage, and arduous application system are the characteristics of a two-component PU sealant. The curing agents (alcohol or amine) are added to the system. In one-part PU sealants, a high molecular weight is desired. However, the high cohesive strength, heat resistance, and wide range of elasticity and hardness are properly designed in two-component materials [8,16–18]

Zhao et al [19] claimed the evaluation of track free time of one-part moisture curing PU sealant based on the content of free NCO, plasticizer, catalyst, and dehydrating agent. Recently, Zhukova et al. [20] informed that light-resistant one-part urethane sealant preparation consumes HMDI, TDI, crosslinking agents of aldimines (Jeffamine), and activating agent of methyl iodide. Shen et al. [21] reported the successful preparation of two-component shape memory polyurethane-based (SMP) sealant fortified by rutile TiO_2 particles (1, 3, and 5 wt.%) to seal expansion joint of concrete pavements. In another research, Patri et al. [22] developed a two-component PU sealant with high mechanical and environmental resistance, with the first part (pre-polymer) synthesized by end-capping hydroxy-terminated polybutadiene (HTPB) with TDI and the second part polyol (polyoxypropylene triol), 4,4-diamino-3,3-dichlorodiphenylmethane (DADCDPM) as curing agent (crosslinker) and chain extender, respectively.

21.4.3 Polyurethane Sealant Applications

PU sealants have outstanding properties due to the variety and breadth of reactants used in their production, including excellent tensile strength, low-temperature stability, and excellent adhesion to different materials and layers. As a result, PU sealants are used in buildings, highways, aircraft, and marine equipment [5,22]. Because of simples operation, low weight, and appropriate adhesion to glass substrates, PU sealants are a well-known candidate in sound sealing and heat isolation in construction glasses in apartments to avoid the waste of energy [16,23,24]. PU sealants may be used to seal expansion joints and prevent their failure in dynamic moving joints like pavement concrete and expansion joints [21]. Transportation and automobile are some of the most consuming PU sealants that utilize it in glazing, adhesion of glass into the metal (to neutralize the effect of movement and vibration), energy insulation, and so on [8,25]. The abrasion and corrosion resistance of PU sealants, as well as their high adhesion to jointing surfaces, make them ideal for sealing pipes and tanks [26,27].

21.5 WATERBORNE POLYURETHANE SEALANTS

PUs based on solvents have been used as high-performance materials for a long time. On the other hand, chemical solvents evaporate due to the use of solvent-based PUs and their processing. While the amount of VOCs in the atmosphere increases, several environmental protection agencies have taken basic measures to minimize VOCs in the industrial sector. In addition, environmental bodies have taken essential measures to reduce VOC emissions from the manufacturing sector. As a result, researchers have altered their research in the production of environmentally friendly polymeric materials significantly. Waterborne polyurethane

(WPU) is an essential eco-friendly polymeric material. It is a fabrication when a PU pre-polymer comprising an isocyanate functional group is dispersed in water. Further, after dispersion, these PUs are chain extended with a diamine in the water phase.

WPU is essential due to its excellent characteristics, including lower levels of VOC. In addition, water-based PUs are versatile and environmentally friendly coating materials used in a wide range of film hardness and solid content. They include no free isocyanate residuals, are weather resistant, have long-term flexibility, are UV resistant, and have high abrasion resistance.

Water is used in WPU as the primary dispersion solvent. The resultant WPU materials have many advantages:

1. Zero or deficient levels of VOCs (environmentally friendly)
2. Absence of isocyanate residues (nontoxic)
3. Good applicability, versatility, and a wide range of superior properties, such as abrasion resistance, impact strength, and low-temperature flexibility

WPU products are nonflammable, nontoxic, low in odor, and highly flexible. WPU is used in a wide range of PU products, consisting of coatings, adhesives, sealants, etc. Further, it is used as a binder in hygiene coatings, insulating coatings, concrete sealers, industrial coatings, architectural coatings, paper coatings, leather and textile finishing, printing ink, and various other applications. [28,29]

The companies in the WPU market invest extensively in developing new and innovative products to expand the applicability areas and target new industries.

Due to the importance of complying with environmental restrictions in marketing, WPU has been replaced with solvent-based PU. Also, the increased usage of WPU in fast-rising end-use industries affected the market growth.

The sealant usage section is considered to be the fastest-rising usage of the WPU market. The fast-rising of WPU-based sealant in the construction industry is regarded as the reason for the high growth of the section, particularly in the Asia-Pacific region. In 2014, after Asia-Pacific, Europe was predicted as the leading consumer of WPU.

The leading manufacturers of WPU in the United States are Dow, Axalta, and PPG, and in Germany, they are Bayer and Henkel Companies. These companies employ various strategies to increase their market share, such as introducing new products or designing WPU materials [30].

Due to the requirement for innovative and environmentally friendly PU products in the construction industry of emerging countries, demand for WPU is significantly increased. Furthermore, the rising knowledge of the harmful effects of solvent-based PUs, as well as the industrial growth in Third World and developing countries, has provided a growing opportunity for the WPU industry.

WPU sealants provide more robust seals and create a flexible and elastomeric joint on curing. In addition, the UV-resistant adhesive offers excellent adhesion to most industrial materials. High-strength, water-resistant WPU sealants are used in road repair, plumbing, and construction. The demand for WPU sealants is growing because of these applications [31].

21.5.1 The Classification of Isocyanate Involved in Waterborne Polyurethane Sealant

The reaction of OH groups in a diol, NCO groups in a diisocyanate, and two or more alcohol groups result in the formation of urethane repeating groups in a WPU sealant. To formulate a WPU pre-polymer for sealant, the requirements are a hydrophilic chain extender, diol or amine, catalyst, and other additives. Water is then applied to the system, emulsifying and dispersing the pre-polymer to produce WPU dispersion. Until now, the effects of polyols and hydrophilic chain extenders on the properties of WPU have mainly been studied [32]. Aliphatic diisocyanate compounds, such as IPDI, HDI, and bis(4-so-cyanato cyclohexyl)methane (H12MDI), are popular isocyanates for WPUs. Generally, aliphatic IPDI is chosen for improved hardness. HDI is selected to provide flexibility and H12MDI for toughness, but a mixture of these isocyanates may also be utilized in WPU composition. Owing to the high reactivity of aromatic isocyanates with water like toluene TDI and MDI, they are less commonly performed in WPUs. Therefore, the use of blocked isocyanate chemistry could allow aromatic isocyanates to be incorporated into WPUs [29].

21.5.1.1 Aromatic Isocyanate

Aromatic diisocyanates are performed to manufacture WPU pre-polymers of sealants, which supply lower cost, better mechanical properties, lower curing time, and upper melt viscosity than the aliphatic type. However, due to their high reactivity with water, they are rarely used in WPU formulation. As a result, manufacturers use more aromatic isocyanates instead of aliphatic isocyanates. However, aromatic rings can quickly produce chromophores when exposed to UV light. Therefore, UV stabilization agents may be applied to the aromatic isocyanate sealant system to reduce or postpone color alteration and mechanical strength depletion [1].

21.5.1.1.1 TDI-Based Waterborne Polyurethane Sealant

TDI is an aromatic diisocyanate compound with many economic advantages, including the lowest cost of any polyisocyanate and its two isocyanates having different reactivity. The most common TDI combination is the 80:20 blend of 2,4-TDI and 2,6-TDI (80–20.TDI). However, it is both volatile and poisonous. Active, unreacted TDI monomer can be highly harmful. It is worthy of mentioning after the reaction of the first NCO, the reaction rate of the second isocyanate is significantly reduced. As we can see in the case of automobile sealants, these delays cure and impairs adhesion. The structure of 2,4 TDI and 2,6 TDI are shown in Figure 21.8.

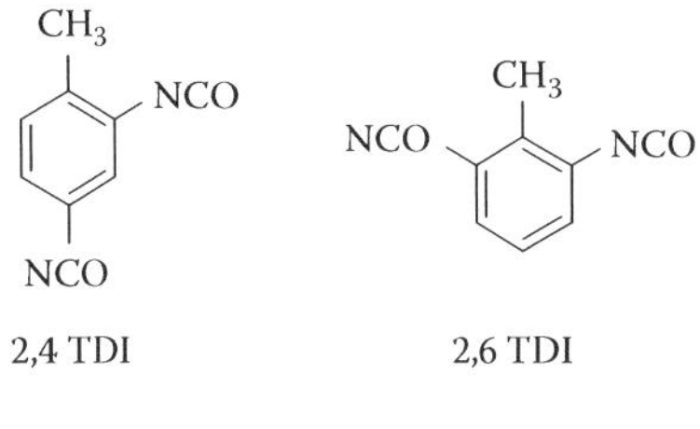

FIGURE 21.8 Chemical structures of 2,4 TDI and 2,6 TDI.

21.5.1.2 Aliphatic Isocyanate

The most well-known aliphatic polyisocyanates are IPDI, HDI, and H12MDI. In addition, aliphatic isocyanates will be selected if an anti-yellowing sealant is needed [33].

21.5.1.2.1 IPDI-Based Waterborne Polyurethane Sealant

IPDI, or 3-isocyanatomethyl-3,5,5-trimethyl cyclohexyl isocyanate, is a popular isocyanate with consistent color, a long pot life, and a fast curing rate. IPDI is widely used in the production of PU-based, light-stable coatings. Interestingly, it is also used in military applications, where it is a critical component in the production of binders in many cast-cured solid propellants and plastic-bonded explosives. Isocyanates are generally dangerous, but the lower instability and volatility of IPDI make it a good selection than other options. Furthermore, IPDI is an aliphatic that is commonly used in sealants. It has the benefits of being nonyellowing as well as having two NCO groups with somewhat different activities. The slower NCO is relatively unreactive to have excellent life durability and nonreactivity with latent hardeners such as oxazolidines and ketimines (before cure) [34,35]. The physical properties and chemical structure of IPDI isocyanate are shown in Table 21.1 [36] and Figure 21.9.

TABLE 21.1
Physical properties of isocyanates

Physical Properties	IPDI
Empirical formula	$C_{12}H_{18}N_2O_2$
Molecular weight	222.28
NCO contents (wt.%)	37.5
Physical form	Colorless to slightly yellow liquid
Melting point (°C)	–60
Vapor pressure (mmHg at 20°C)	0.0003
Boiling point (°C)	450
Flash point (closed cup)	155°C
Density (20°C)	1.062 g/cm^3

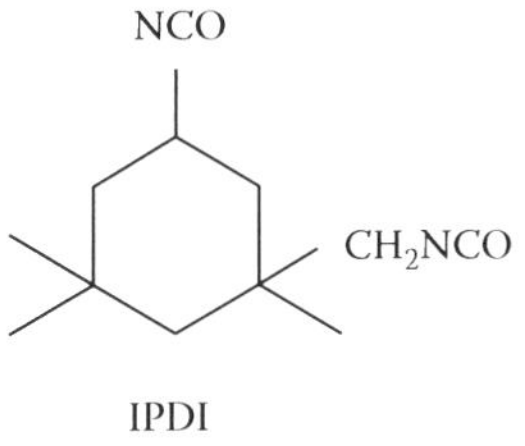

FIGURE 21.9 Chemical structures of IPDI.

Licht et al. [37] synthesized aqueous dispersions containing a lipophilic PU by a mini-emulsion process and the reaction between a polyisocyanate (mix of IPDI and TDI) and a compound containing isocyanate-reactive groups (polyether polyol and polyester polyol).

The innovative dispersions were exceptionally well suited to produce the aqueous coating materials, adhesives, and sealants with excellent efficiency properties. Preparation of sheets and films, as well as impregnate textiles are some other applications of this dispersion.

Berezkin and coworkers [38] introduced an isocyanate-terminated pre-polymer with 2–7 wt.% isocyanate content, one polyether diol, and one active-hydrogen material containing two or three hydroxyl groups, one carboxy group and hydroxy alkane carboxylic acid, a polyether monol, a polyether triol, and IPDI and HMDI. The output is a stable compound with proper visco-elastic film properties with very little to no co-solvent [38,39].

Joen et al. [40] used a pre-polymer mixing process to make aqueous PU dispersions from IPDI, poly (tetramethylene adipate) glycol (PTAd), and dimethylolproprionic acid (DMPA). It was discovered that raising the molecular weight of PTAd increased the thermal, mechanical, and adhesion properties of PUs due to the well-ordered structures of soft segments with a high phase separation between soft and hard segments. Furthermore, in terms of hybridization effects, the highest tensile strength and elongation at break with quick-drying rate are attributed to the complete IPNs. At the same time, semi-IPNs had the most increased initial and final adhesion, meaning that a definite amount of chain mobility may enhance the penetrations of adhesive molecules into soft PU foam surfaces.

Bedri et al. [41] developed a sealant formulation containing an ultra-high solid PU dispersion. This dispersion has about 65 wt.% of solid content and a viscosity of fewer than 5,000 cps at 20 rpm at 21°C. The ultra-high solid PU dispersion contains a first PU pre-polymer emulsion (i.e. synthesized by an IPDI and polyether polyol), a second component having a media phase selected from the group including a second PU pre-polymer emulsion, a low solid content PU dispersion, a seed latex, and combinations and a chain extender.

21.5.1.2.2 TMXD-Based Waterborne Polyurethane Sealant

One important aliphatic isocyanate is TMXDI (META) (tetramethyl xylene diisocyanate), which has a diversity of uses in polymer applications.

Performance coatings, roofing compositions, flooring, sealants, adhesives, advanced dispersions, and elastomeric compositions for casting and injection molding may benefit from these materials.

TMXDI is a monomer comprising a reactive double bond and a reactive isocyanate group. Therefore, specialized polymers with specific properties can be made by selectively reacting to these groups. For example, polymers with isocyanate functionality can be made by copolymerizing TMXDI with acrylic, styrene, and other monomers. Further, It can be reacted with diols, triols, or polyols to produce polymers with vinyl functionality. The physical properties of these isocyanates are shown in Table 21.2.

TABLE 21.2
Physical properties of Isocyanates

Physical Properties	*m*-TMXDI	*p*-TMXDI
Empirical formula	$C_{14}H_{16}N_2O_2$	$C_{14}H_{16}N_2O_2$
Molecular weight	244.3	244.3
NCO contents (wt.%)	34.4	34.4
Vapor pressure (mmHg at 100°C)	0.5	0.4
Physical form	Colorless liquid	White crystals
Melting point (°C)	−10	+72
Boiling point (°C)	150°C/3 mm	150°C/3 mm
Flash point (closed cup)	450°C	450°C

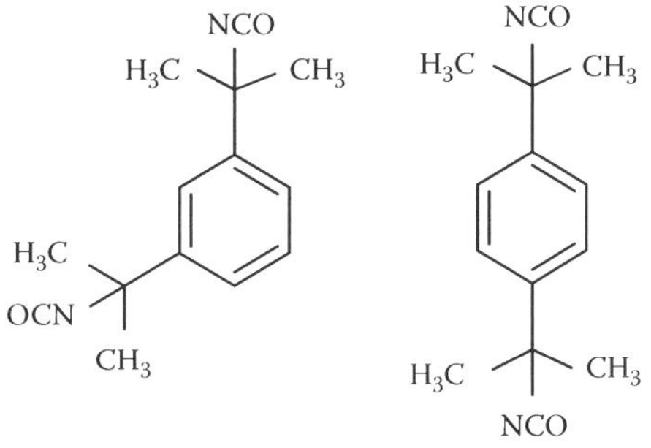

m-TMXDI *p*-TMXDI

FIGURE 21.10 Chemical structures of *m*-TMXDI and *p*-TMXDI.

The chemical structures of *m*- and *p*-TMXDI are shown in Figure 21.10. They are structural isomers of one another, one of the main differences in physical properties being that the *meta*-isomer is liquid at room temperature and the *para*-isomer is a solid, crystalline material melting at 77°C [35].

Tomko et al. [42] prepared a low VOC penetrating constituent for staining and protecting porous substrates, such as wood, concrete, brick, and so on, containing aqueous dispersions of PU-ureas. They improve a water-based PU dispersion by adding active hydrogen such as polyester polyols and having a long aliphatic chain, such as 1,4 cyclohexane dimethanol, part of the pre-polymer reaction mixture. This composition reacts with the diisocyanate groups, *m*-TMXD diisocyanate, during the pre-polymer formation. As a consequence, the pre-polymer is isocyanate-functional as well as having a hydrophobic end group. As a result of the inclusion of the hydrophobic aliphatic end party, the pre-polymer was neutralized, dispersed in water that penetrates porous substrates, and exhibits excellent water repellency.

As explained in this section, suitable isocyanates for the preparation of WPU used in sealant are included any linear or branched aliphatic or cycloaliphatic organic compound that has at least two free isocyanate groups per molecule, such as diisocyanates $X(NCO)_2$, that X can be an aliphatic or a cycloaliphatic hydrocarbon. Among many diisocyanates, some of them, such as TDI, MDI, IPDI, and TMXD, are more used in sealants. Due to the high reactivity of aromatic diisocyanates in water, aliphatic diisocyanates are mostly used in the synthesis of WPU in the industry.

21.6 CONCLUSION

WPU is essential due to its excellent characteristics, including lower levels of VOCs. In addition, WPUs are versatile and environmentally friendly coating materials used in a wide range of film hardness and solid content. They have low abrasion resistance, weather resistance, long-term flexibility, and UV resistance, and they have no free isocyanate residuals. Due to the importance of complying with environmental restrictions in marketing, WPU has been replaced with solvent-based PU. Also, the increased usage of WPU in fast-rising end-use industries has affected market growth. The sealant usage section is considered to be the fastest-rising usage of the WPU market. To formulate a WPU pre-polymer for sealant, the requirements are a diisocyanate, diol, hydrophilic chain extender, catalyst, and some other additives. The reaction of OH groups of a diol with NCO groups of diisocyanate results in urethane repeating groups in WPU sealants. Therefore, the type of diisocyanate plays an essential role in the final physical and chemical properties of WPU. Diisocyanates have two main categories, aliphatic and aromatic. Aliphatic diisocyanate compounds, such as IPDI, HDI, and H12MDI, are popular isocyanates for WPUs. Generally, aliphatic IPDI is chosen for improved hardness. HDI is selected to provide flexibility, and H12MDI for toughness, but a mixture of these isocyanates may also be utilized in the WPU composition. Owing to the high reactivity of aromatic isocyanates with water like TDI and MDI, they are less commonly performed in WPUs. Therefore, the use of blocked isocyanate chemistry could allow aromatic isocyanates to be incorporated into WPUs.

REFERENCES

1. Mittal, K.L. and A. Pizzi, *Handbook of sealant technology*. 2009: CRC Press.
2. De Buyl, F., *Silicone sealants and structural adhesives. International Journal of Adhesion and Adhesives*, 2001. **21**(5): p. 411–422.
3. Petrie, E.M., *Handbook of adhesives and sealants*. 2007: McGraw-Hill Education.
4. Galimzyanova, R.Y., M.S. Lisanevich, and Y.N. Khakimullin, *Sealing materials based on polymers. Key Engineering Materials*, 2020. **869**: p. 93–100.
5. Klosowski, J. and A.T. Wolf, *Sealants in construction*. 2016: CRC Press.
6. https://www.bostik-industrial.com/polyurethane-sealants/.
7. https://polyurethane.americanchemistry.com/CASE/.
8. Segura, D.M., et al., *Chapter 3 Chemistry of polyurethane adhesives and sealants. In:* P. Cognard (Ed.), *Handbook of adhesives and sealants*. 2005: Elsevier Science Ltd. p. 101–162.
9. Saunders, J. and K. Frisch, *High polymers, Vol. XVI, part 1, polyurethanes, chemistry and technology*. 1962: Interscience Publishers.

10. Saunders, J. and R. Slocombe, *The chemistry of the organic isocyanates. Chemical Reviews*, 1948. **43**(2): p. 203–218.
11. Akindoyo, J.O., et al., *Polyurethane types, synthesis and applications–a review. Rsc Advances*, 2016. **6**(115): p. 114453–114482.
12. Wu, Z., *Polyurethane sealant compositions*. 2003: Google Patents.
13. McGrath, J., M.A. Hickner, and R. Höfer, Introduction: Polymers for a sustainable environment and green energy. In: *Polymers for a sustainable environment and green energy*. 2012: Elsevier. p. 1–3.
14. Dutta, A.S., Polyurethane foam chemistry. In: *Recycling of polyurethane foams*. 2018: Elsevier. p. 17–27.
15. Dodangeh, F., et al., *Synthesis and characterization of alkoxy silane modified polyurethane wood adhesive based on epoxidized soybean oil polyester polyol. Composites Part B: Engineering*, 2020. **187**: p. 107857.
16. Li, W., et al., *Two-component modified polyurethane sealant for insulating glass: Design, preparation, and application. Journal of Applied Polymer Science*, 2019. **136**(46): p. 48219.
17. Feldman, D., *Polyurethane and polyblend sealants. Polymer Engineering & Science*, 1981. **21**(1): p. 53–56.
18. Delmonte, J., *Liquid urethane elastomers as encapsulating compounds*. 1964: Insulation.
19. Zhang, B.J.Z.G.-Z. and H.-l.Z. Lian-Hong, *Study on the tack free time of one-component moisture-curing polyurethane sealant. Chemistry and Adhesion*, 2007. **6**: p. 1395–1405.
20. Zhukova, I.V., et al. *One-component urethane sealants synthesis*. In: *Key engineering materials*. 2020. Trans Tech Publ.
21. Shen, D., et al., *Development of shape memory polyurethane based sealant for concrete pavement. Construction and Building Materials*, 2018. **174**: p. 474–483.
22. Patri, M., S. Rath, and U. Suryavansi, *A novel polyurethane sealant based on hydroxy-terminated polybutadiene. Journal of Applied Polymer Science*, 2006. **99**(3): p. 884–890.
23. Krishnan, B. and L.A. Grier, *Polyurethane sealant based on poly (butylene oxide) polyols for glass sealing*. 2017: Google Patents.
24. Yao, R., B. Li, and K. Steemers, *Energy policy and standard for built environment in China. Renewable Energy*, 2005. **30**(13): p. 1973–1988.
25. Polyntsev, E. and A. Kvitko. Using foam polyurethane sealers for strengthening of soils of a road bed of transport constructions. In: *IOP conference series: materials science and engineering*. 2020. IOP Publishing.
26. Kelly, P.B. and G.D. Edwards, *Pipe sealant compositions*. 1968: Google Patents.
27. Ligon, E.R. and T.M. Medved, *Pipe joint of polyurethane*. 1969: Google Patents.
28. Shin, E.J. and S.M. Choi, Advances in waterborne polyurethane-based biomaterials for biomedical applications. In: H.J. Chun et al. (Eds.), *Novel biomaterials for regenerative medicine*. 2018: Springer. p. 251–283.
29. Anıl, D., et al., Chapter 11—Recent advances in waterborne polyurethanes and their nanoparticle-containing dispersions. In: P. Zarras, M.D. Soucek, and A. Tiwari (Eds.), *Handbook of waterborne coatings*. 2020, Elsevier. p. 249–302.
30. https://www.marketresearch.com/MarketsandMarkets-v3719/Waterborne-Polyurethane-Application-Coating-Sealant-9068265/.
31. https://www.researchandmarkets.com/reports/5178160/waterborne-polyurethane-market-by-application.
32. Zhou, X., et al., *Recent Advances in synthesis of waterborne polyurethane and their application in water-based ink: a review. Journal of Materials Science & Technology*, 2015. **31**(7): p. 708–722.

33. Cognard, P., *Handbook of adhesives and sealants: basic concepts and high tech bonding*. 2005: Elsevier.
34. Evans, R.M., *Polyurethane sealants: technology & applications*. 2014: CRC Press.
35. Szycher, M., *Szycher's handbook of polyurethanes*. 1999: CRC press.
36. https://en.wikipedia.org/wiki/Isophorone_diisocyanate.
37. Licht, U., et al., *Aqueous polyurethane dispersion*. 2004: Google Patents.
38. Berezkin, Y., et al., *Polyurethane dispersions for sealants*. 2009: Google Patents.
39. Johnston, J., C. Sear, and A. Pasquini, *Sealant compositions with a polyurethane dispersion and a hydroxy-functional compound*. 2015: Google Patents.
40. Jeon, H.T., S.K. Lee, and B.K. Kim, *Improved adhesion of waterborne polyurethanes by hybridizations. The Journal of Adhesion*, 2008. **84**(1): p. 1–14.
41. Bedri, E., et al., *Sealant composition*. 2009: Google Patents.
42. Tomko, R.R., B.J. Varone, and R.A. Martuch, *Penetrating stains and sealants from polyurethane dispersions*. 2001: Google Patents.

22 Waterborne Polyurethanes for Packing Industries

Saima Zulfiqar, Rida Badar, and Muhammad Yar
Interdisciplinary Research Center in Biomedical Materials, COMSATS University Islamabad Lahore Campus, Lahore, 54000, Pakistan

CONTENTS

22.1 Introduction 375
22.2 Chemistry of PUs 376
22.3 Structure of WPUs 377
22.4 Development, Classification, and Synthesis of WPUs 379
22.5 Applications of WPUs in the Packaging Industry 379
22.6 Use of WPUs in Packaging Industries 382
22.6.1 Waterborne Polymer-Based Coatings 382
22.6.2 Waterborne Nanocomposite-Based Coatings for Packing 383
22.7 Latest Innovations in WPUs for Packaging Applications 385
22.7.1 Antibacterial WPU Based on Nanoparticles for Packing 385
22.7.2 UV-Cured WPU Coatings for Packaging 386
22.7.3 Medical Uses of Polyurethanes 387
22.8 Health and Safety Concerns of WPUs 387
22.9 Polyurethanes: A Problem Solver for Today's Packaging Challenge 389
22.10 Conclusion 390
References 390

22.1 INTRODUCTION

Polyurethane (PU) is a polymer manufactured as a substitute to typical rubber in WW-II by Dr. Bayer using polyol and di-isocyanate as precursors. PU polymers find application as coating material against corrosion, as composites for insulating purposes, and as adhesives. Starting from simple diisocyanates, now PU manufacturers have begun to use aliphatic polyisocyanates, which are leading to a rapid revolution in the PU industry. Continuous research in this field has led to more adaptive, chemically stable, and flexible PUs for future applications. One of the strikingly interesting features is they can be either as flexible as typical rubber or

DOI: 10.1201/9781003173526-22

as durable as metal. Owing to integrated properties of flexibility, hardness, and durability, they are being employed in numerous commodity-type applications, replacing fiber, rubber, and metallic material [1].

PU manufacturing is quite different and arduous than the traditional polymers available in markets. For PU, the manufacturer only sells parent materials and does not carry out the polymerization, which in turn is carried out by the converter. However, other plastics are polymerized by chemical industries and sold to the buyer in product form, i.e. powder form. During manufacturing, a range of densities, grades, flexibilities, and hardnesses could be achieved for the desired field of application [2].

Two key factors governing the success of the PU market are:

I. Cheap raw materials and production process
II. Tailoring of product according to preferred applications

This wide range of material properties is achieved by opting the right composition and judicious choice of ingredients. The ester polyols provide hardness while ether-based polyols are ideal for the flexibility of the article. Similarly, the right choice of isocyanate, i.e. aromatic or aliphatic, is essential for imparting desired material properties, giving the PU market endless applications.

22.2 CHEMISTRY OF PUS

PU is a polymerization adduct of urethane linking units, formed by joining a polyol and di-isocyanate. The simple and basic constitutes with their estimated percentage have been summarized in Figure 22.1

A slight variation in the ratio or type of polyol and isocyanate could drastically alter the final article properties.

Isocyanate (30%) could be aromatic/aliphatic and bi- or polyfunctional, with the former being used as coatings and the later as flexible foam, as shown below, in Figure 22.2.

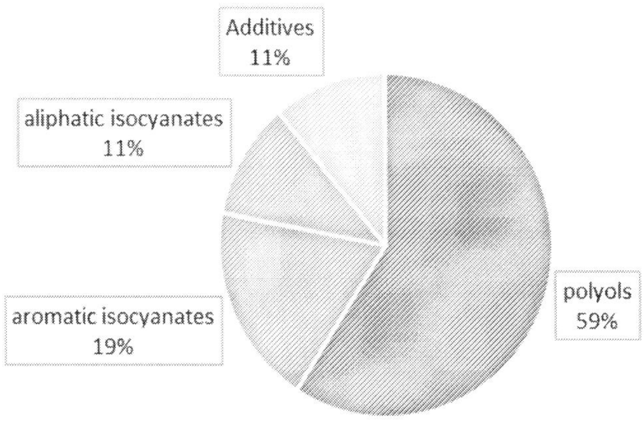

FIGURE 22.1 Composition of polyurethanes.

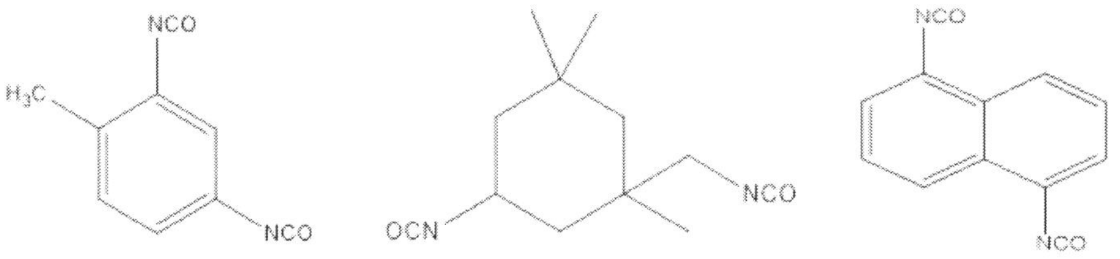

FIGURE 22.2 Commonly used bi-isocyanates.

As far as transparency in coatings is concerned, the aliphatic isocyanate is a better choice than the aromatic analog, in which UV damage yellows the coating. TDI type di-NCO is more reactive than its aliphatic analog due to aromatic stabilization. Moreover, the role of the substituent is also critical as the EW groups increase the reactivity while ED groups decrease the reactivity of the isocyanate functional group. Isocyanates have higher reactivity toward nucleophiles, i.e. –OH [3].

Polyhydroxy alcohols (59%) with amide, ether, or ester functionality are the second basic constituent of PUs. Among these, ether and ester functional groups containing polyols are most important. Ester polyols are expensive and impart greater tensile strength to the article than ether polyols. Polyols with less molecular weights result in increased hardness of the urethane polymer. On the other hand, high molecular weight polyols having fewer urethane groups would build a rather flexible product. This property is important in tailoring the flexibility or hardness of the article. Some other additives include fillers, coloring agents, catalysts, plasticizers, and crosslinkers, which impart mechanical support, tensile strength, colors, reduce hardness, and reaction activator [4].

22.3 STRUCTURE OF WPUS

WPUs consist of two major parts. One is the backbone (main structure), and the other is the emulsifier. Emulsifiers are amphiphilic polymers and hydrophilic in nature, such as grafted polymers. They stabilize the dispersion of PUs in water. Emulsifiers can be classified as internal and external. Nonionic centers, such as polyethylene oxide, are internal emulsifiers. The scheme for the synthesis of nonionic dispersions is given below in Figure 22.3.

WPUs are formed from two main portions: the main structure, or backbone, and the emulsifier. The emulsifiers are hydrophilic materials and amphiphilic polymers, such as grafted polymers. They stabilize the dispersion of PU in water. The rigid segments with ionic groups are hydrophilic, and the soft segments play the role of hydrophobic parts. Emulsifiers can be divided into two classes: internal and external. The internal emulsifiers include nonionic centers, such as polyethylene oxide, and ionic centers, such as anionic, cationic, and zwitterions. Therefore, the nonionic PU dispersions contain a hydrophilic soft segment pendant group, such as polyethylene oxide. These dispersions are colloidal stable in a wide range of pH levels.

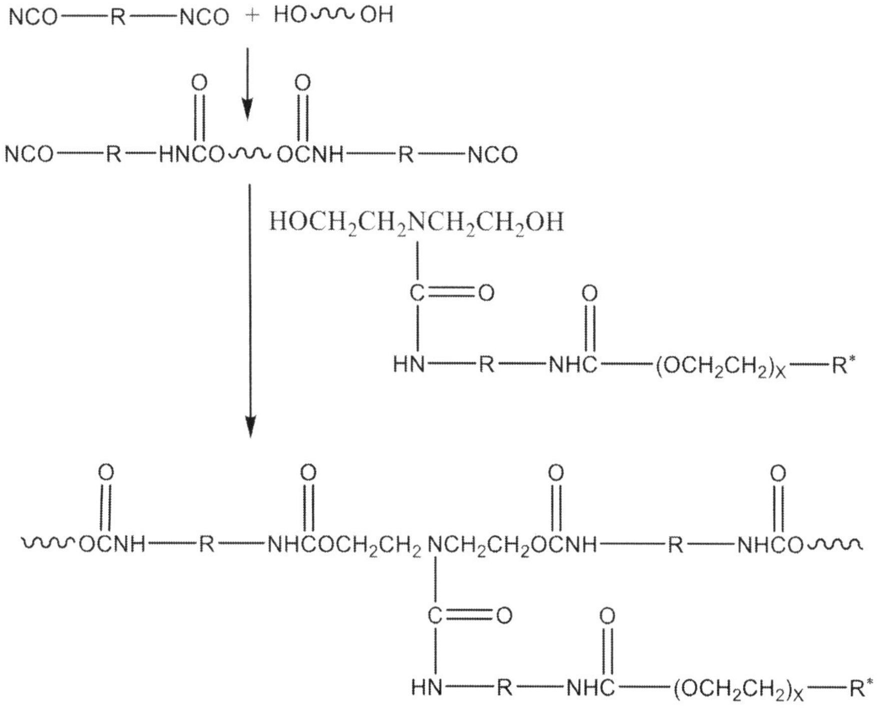

FIGURE 22.3 Preparation of PU dispersions with nonionic hydrophilic components. Adapted with permission from reference [11]. Copyright (2018) Taylor and Francis Ltd.

This type of PU dispersion has environmentally friendly properties. The properties, low viscosity with high solid contents, belong to nonionic PU dispersion even though they show stability toward the cold environment, changes in the pH level, and different types of electrolytes and solvents. As it was referenced earlier, the ionic groups incorporate into the chains via reaction agents (ionic diols or ionic di-isocyanate) or the polymerization of PU with an ionic reagent. The obtained component is named PU ionomer.

Figure 22.3 shows the synthesis scheme of the nonionic dispersion.

A simple, innovative approach to the preparation of a single WPU matting resin portion does not involve the accumulation of inorganic matting mediators. During the reaction route, 2,2-aminoethyl amino ethyl sulfonic acid sodium salt and hydrazine hydrate were employed to prepare WPU. Latex particles were diffused in water along with a little A95, which has high hydrophilicity, in the synthesis of sulfonate aqueous PU. The properties, like exceptional biocompatibility and low surface tension, were shown in the WPU film materials by using A95.

The surface mechanical properties of the WPU film materials are affected by the type of hydrophilic extender. When A95 is used in the synthesis of WPU, the resultant product exhibits excellent biocompatibility and low surface tension; when hydrazine hydrate is used and it plays its role as a chain extender, the product has a rigid structure that cannot be deformed. Thus, the prepared WPU emulsion has a regular distribution of molecular, the latex particle is fixed, and it forms regular microspheres, which can give the WPU film good low gloss, high covering power, and excellent emulsion stability. Besides, the WPU films conceal coating surface defects and enhance coating adhesion resistance, scrub resistance, folding resistance, and scratch resistance, etc. This kind of novel WPU is potentially used in polyvinyl chloride (PVC), PU, and synthetic leather surface [5].

22.4 DEVELOPMENT, CLASSIFICATION, AND SYNTHESIS OF WPUS

Melt dispersion process, acetone process, pre-polymer mixing process, and ketamine process are some common synthetic routes used to obtain WPU dispersion. In all these processes, the first step is preparation of low molecular weight pre-polymer by reaction of diols or polyols with a molar excess of diisocyanates and polyisocyanates. In the reaction mixture, a diol containing either an ionic group or a nonionic group is usually an internal emulsifier that develops the part of the core chain of pre-polymer. The main step in which the different synthetic pathways vary is the dispersion of pre-polymer in aqueous media and the molecular weight buildup. The pre-polymer blending and acetone processes are good approaches for creating WPU dispersion, as shown in Figures 22.4a and 22.4b. The WPU dispersions are primarily investigated by interactions between segments (soft/hard) and also by the ionic group interactions. The latest studies documented that the molecular weight of the polyol, the kind of chain extender, the segmented structure, the ratio of hard and soft segments, and the ionic group content decide the WPU dispersion properties. A large number of polyols have been used in the preparation of WPU dispersion to change its structure as well as accordingly enterprise their properties. Polycaprolactone diols, polycarbonate, polyester, and polyether are the most common macro diols used in the synthesis of PUs.

One component-based WPU dispersions have now obtained a significant market value. These dispersions, which usually depend on aliphatic diisocyanates, remain known for their elasticity and durability. One component-based WPU dispersions may be used to make storage stable thermoplastic coatings [6].

22.5 APPLICATIONS OF WPUS IN THE PACKAGING INDUSTRY

In food and beverage packaging, coatings usually take account of thin layers that can be either in direct interaction with surroundings or sandwiched between layered polymers. The thickness of such layers generally ranges from tenths of a nanometer to a few micrometers. Nowadays, when these are used in packaging of food and beverages, they can afford the material many benefits in terms of barrier performance, optical as well as mechanical properties, characteristics of the external surface, and new properties including antimicrobial and antioxidant properties. These sheets of coating can be functional to a variety of objects. Roll-to-roll manufacturing, physical vapor deposition, dipping chemical vapor deposition, spinning, and spraying and are some of the methods used. Scientists

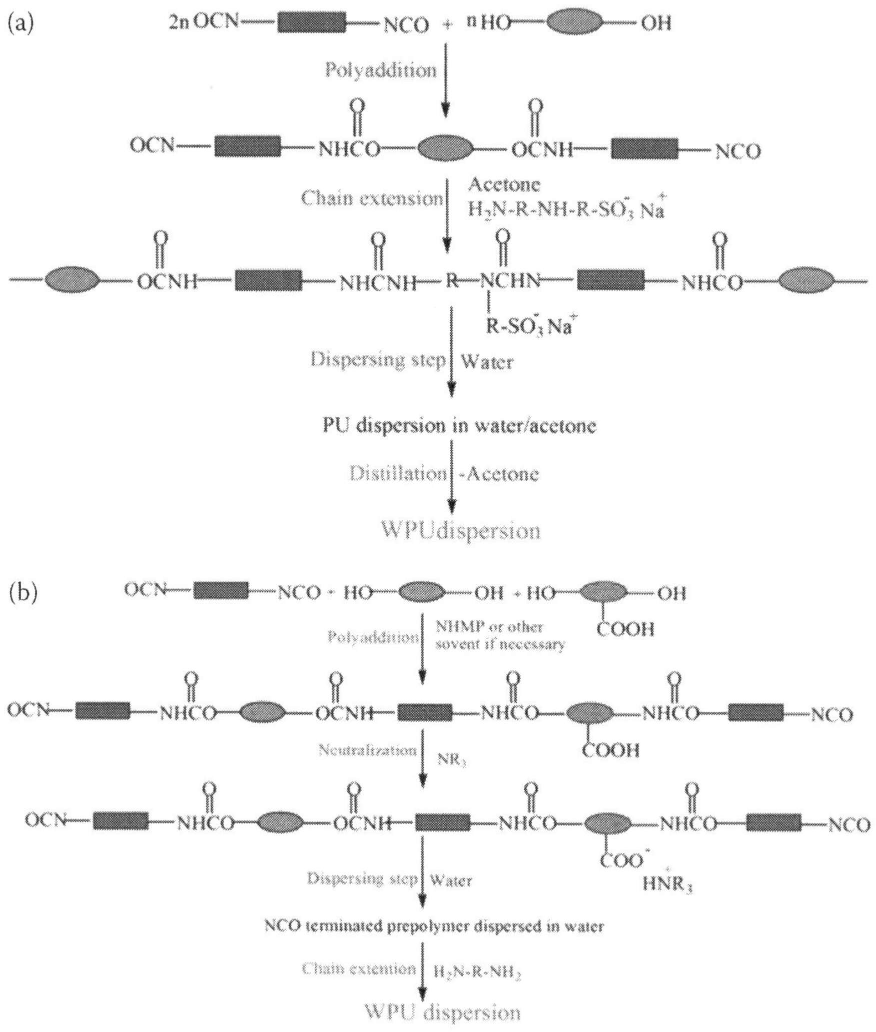

FIGURE 22.4 A simplified representation to synthesize WPU dispersion via acetone process (a) and mixing of prepolymer (b). Adapted with permission from reference [11]. Copyright (2018) Taylor and Francis Ltd.

working on food and beverage packaging products are still concerned with barrier properties. In general, high barrier packaging is becoming more common, especially in terms of vapor transmission of gas as well as water throughout the package. Glass was traditionally applied as a barrier medium, replaced by aluminum in the production of cans. Despite the fact that both of these constituents have a complete obstruction toward gas and water vapor, tinplate and aluminum may also act as a light barrier. Glass coatings, on the other hand, enable customers to have transparent containers, which is a desirable publicizing feature. Although these coating materials have served the packaging industry well, some disadvantages related to their designs have led to the development of specific new useful solutions. The weight of glass, for example, reflects a logistic constraint on both an economic and environmental level. At present, the options for the market are restricted, particularly because most applications require transparent packaging, which rules out metalized films. In packing food/beverages, coatings manufacturers prepare polymeric coatings that excel in terms of mechanical resistance, barrier properties, sealing, and optical properties. Ethylene-vinyl alcohol polyvinylidene chloride, polyvinyl alcohol, and other polymers are commonly used in these coatings. Synthetic phenolic resins have been produced ever since the early 1900s, in addition to the aforementioned systems. These resins may be used to make food cans that prevent the food from coming into direct contact with metal surfaces. These coatings must have a high degree of stability to prevent harm from external impacts, which can be achieved by mixing phenol and cresols in different ratios. Bisphenol A is still used as a precursor for the production of epoxy resins as coatings on the inner side of food and beverage containers, despite numerous health issues since 2008. The application of waterborne coatings has increased dramatically in recent years. As a result, water-based coatings in the forms described above are given special attention [7].

Solvent-based inks release volatile organic compounds (VOCs) during the printing process, which is harmful to the environment. Ink binders must be developed using easy and environmentally friendly methods to minimize dependency on crude oil. The world has been facing a petroleum crisis since 1970, and everyone has been looking for alternate sources to minimize its consumption. Solvent-based ink binders are also obtained by petroleum resin production. Solvent-based inks tried to replace water-based and UV-curing ink, which were more reliable. A material with good physical properties is considered ideal for the adhesive industry and can be used to replace solvent-based binders. WPU possesses extremely good abrasion resistance, high tear and tensile strength, a range of hardness, high elasticity, low-temperature stability, and good resistance. These properties provide an alternative to the extensively used solvent-based PU binder.

The use of WPU binder in ink films reduces printing speed due to its low curing speed and water resistance. On the other hand, water-based binders are preferred due to their better curing speed and water resistance. To increase the potential of WPU binders, Fang and its co-researchers introduced dispersions with low crystallinity and high solid content as adhesive material. It is observed by using a transmission electron microscope and laser diffraction instrument that drying speed is enhanced by reducing solvent content. Solvent content is reduced by introducing nanosized solid content, which is preferred when substrate essentially requires deep penetration. Rapid drying in printing is achieved by using highly solid and nanosized particles.

Different ways are introduced by Lei and co-workers to prepare several WPU dispersions. These dispersions showed useful potential application for inks. Diethylenetriamine, tri-ethylene tetramine, and ethylenediamine were used as post chain extenders and perform various functions. The synthesis of WPU dispersions was modified to increase the water resistance by crosslinking of prepolymers. The resulting product was made compatible with waterbone ink binders and organic solvents without being removed. The printing process using the slow drying speed of water-based inks showed limited application with WPUs due to its use in small-scale printing industries. The thickness of WPU can be enhanced by the addition of solid content, which is the ultimate solution to minimize this problem. Vegetable oil-based hyperbranched polymers are preferred over linear polymers to get improved material properties and the advantage of green.

Linear polymers possess higher molecular weight distribution compared to counter hyperbranched polymers, which have less molecular weight and viscosity. Higher molecular weight polymers with water-based inks fulfill the need for rapid-speed drying with more strength. This is why hyperbranched PU is preferably employed as a water-based ink binder due to its maximum ability for rapid printing. Application of sunflower oil is reported in the synthesis of hyperbranched and linear PU. The following synthetic materials are reported: toluene diisocyanate, poly(ε-caprolactone), butanediol with pentaerythritol, and monoglyceride of oil without pentaerythritol as the multifunctional unit used for the synthesis of WPU dispersion. Monoglyceride is obtained from sunflower oil by standard method. The desired properties, such as physio-mechanical, thermal, and biodegradability, are obtained by biosynthesizing HBPU. It shows modern multifaceted applications as a thin film material. Castor oil and monoglycerides of castor oil (by A2 and B3 standard approach) are also used to synthesize hyperbranched PU. PUs as advanced surface coating material behaved as dielectric material as shown by its dielectric constant and loss factor experiments. It was concluded that molecular weight and thickness impart an important role in the solid content of hyperbranched WPU. Therefore, it is recommended to focus on the synthesis of hyperbranched WPU to use as an ink binder [8].

22.6 USE OF WPUS IN PACKAGING INDUSTRIES

In packaging industries, WPUs are used in the form of polymer- and nanocomposite-based coatings, as described below.

22.6.1 Waterborne Polymer-Based Coatings

Several efforts have been made by scientists in recent years to lessen the impacts caused by human activities. As a result, a lot of effort has been made to find renewable polymer-based coatings to replace petroleum-based coatings. In the fields of packaging, food industry, and biomedical sciences, many natural polymer-based coatings have been comprehensively premeditated. As an abundant plant protein, soy protein isolates (SPI) are biodegradable, biocompatible, and made from films. SPI-based compounds have good oxygen, fragrance, and lipid barrier properties in general, which can be used to coat edible or nonedible packaging. SPI coatings are very breakable, and resistance toward water is quite poor. Thus, different modification methods have been performed to improve the properties of SPI-based coatings. Wang and co-workers developed polybutylene-based WPU, which could be fused with SPI to manufacture blend films [9].

22.6.2 Waterborne Nanocomposite-Based Coatings for Packing

In packing industries, waterborne nanocomposite-based coatings are being used due to their water-repellent property. For this purpose, several inorganic compounds are added during the synthesis of PUs to enhance their use in the packing industry. Solvent-borne PUs are preferred in humid media as compared to WPUs due to their strong mechanical properties, thermal stability, and poor hydrolytic strength. These properties can be improved in WPUs by the addition of nanoparticles. Several experiments were performed by researchers to improve the properties of WPUs. Silica nanocomposites were introduced with WPU by blending both dispersions, which result in improved properties. Commercially available WPU nanocomposites include colloidal silica- Ludox AS and Ludox TMA. Both of these commercial products have different-size shapes and counter ions. An increase in water resistance, functional properties, and thermal stability is observed to be 5%, 32%, and 50% by weight, respectively, by the addition of silica. Silica possessed a better crosslinking effect, which is shown by its better performance compared to TMA. Cellulose nanocrystal-based WPU nanocomposites are chemically made using different CNC contents. Different contents showed different results as low CNC content in the crystallization of soft fragments. It could result in more CNCs and lower crystallization. This decrease in crystallization by increasing the content of CNC is due to the interfacing of cellulose nanocrystals and PU chains. The addition of CNC also helped to improve thermal stability, hydrophilic character, and mechanical properties. Silica is dispersed in poly tetrahydrofuran glycol, preformed polymer, and S2 to synthesis WPU/nanosilica composites. It is observed that nanosilica is more compatible with the S1 series compared to the S2 series due to its better dispersion. The S1 series is preferred due to its better thermal stability, physical properties, and water resistance compared to S2.

A diol and polyhedral oligomeric silsesquioxane (POSS) was used in 0.3–4.6% by weight to synthesize a series of aqueous PU nanocomposites by pre-polymer mixing method.

The particle size, viscosity, modulus, Tg, tensile strength, and thermal stability increased by increasing POSS content. The incorporation of nanoparticles into hard segments gives a homogeneous structure, as indicated by morphological results. The anionic internal emulsifier (2-hydroxyethyl-2-amino ethane sulfonic acid sodium salt, BES sodium salt) used as a functional group encouraged the formation of PU-POSS nanocomposites.

Scratch-resistant, waterborne, UV-curable, transparent coating reported being synthesized by silica nanoparticles. Acis and alkaline silica showed a less compact structure compared to modified nanosilica. The modified nanosilica was homogeneously dispersed throughout the PU matrix, which results in transparent film [10].

The addition of inorganic and organic fillers into WPU dispersion was discovered to enhance the resultant characteristics of the nanofilms. The most contemplated are silica, silver, nanocellulose, and nano-clays to boost hardness, microbial resistance, thermal stability, and mechanical and barrier properties, respectively, without lowering the concentration of bio constituents. Furthermore, by dispersing particles and polymers in the same medium, the construction of nanocomposites is highly facilitated. Xia and the team constructed a WPU from castor oil by adding 3-APTES that co-reacted the leftover isocyanate [11]. Powerful stirring and water resulted in forming aqueous dispersion produced with various quantities of silica. As anticipated, this enhanced the mechanical (Young's modulus, tensile strength and toughness, and elongation at break) as well as thermal resistance properties of the product. Similarly, Fu et al. constructed a bio WPU inorganic-organic dual substance by chemically adding nanosilica in its structure and using castor oil as a precursor of thiol-ene coupling reaction to getting alkoxysilane castor oil and carboxyl castor oil to ultimately construct SiWPU nanocomposites. The suspension showed superficial central shell geometry due to encapsulation of nanosilica by WPU. Increased amounts of silica also increased the thermal stability roughness and hydrophobicity of films.

The effects of nano-clay addition via in situ polymerization to jatropha oil-based WPU were studied. The oil-derived poly alcohol was synthesized through ring-opening and epoxidation, getting an OH ratio of 184 mg KOH/g polyol. The researchers synthesized WPU utilizing DMBA as an intrinsic emulsifier. They employed MT, AT, and FT at a ratio of 2 wt.% in PUs and in nano-clays incorporated with APTES. The layered structure and better dispersion in WPU rendered the modified montmorillonite (MT) as the excellent clay for enhancing the thermal stability of substances, hence behaving as a heat insulator. The Tg of PUs is also enhanced due to extra hydrogen bonding. These alterations served as a link with the increased mechanical characteristics found in composites [12]. Panda et al. synthesized nanocomposites using nano-clays. For this research, they got a castor oil-based WPU and utilized organically altered nano-clay cloisite as filler. The nano-clay was mixed after the construction of WPU, and it was noted that these fillers helped in hydrogen bond formation with PUs, affecting films characteristics. Three wt.% clay enhanced tensile strength along with Young's modulus.

The incorporation of silver nanoparticles to the construction of WPUs was regarded to incorporate antibacterial characteristics into the substance. For instance, Fu and his team fabricated a chain of antibacterial castor oil-based PUs by alterations of silver nanocomposites. These fillers have been Ag-HNT, which could be well distributed in WPU. Gao and colleagues demonstrated the outcomes on WPU derived on polyethylene glycol and castor oil filled with eucalyptus CNC. Due to the good compatibility between cellulose and PUs, the dispersion of CNC was regular, whereas the size poly dispersion of CNC had been very vast in this study, despite the L/D being in the order of 23. The good compatibility, larger aspect ratio, and uniform dispersion resulted in better improvement on mechanical characteristics of elastomeric PUs (e.g. 4 wt.% incorporations of CNC enhanced the value of Young's modulus four times than that of PU).

The increase in the mechanical characteristics was variable and demonstrated by the fabrication of agglomerates on increasing concentrations above a threshold value linked to the percolation threshold. In this specific study, the matrix was partitioned PU, and there was a very minor decline in the tensile elongation at a breakpoint, which was linked to the development of hydrogen bonding largely between the hard part of nanocomposites films and nanoparticles, hence affecting the elongation of soft parts of PUs less. Due to the smaller size of the WPU particles in aqueous distributions and the very small size of the inorganic and organic particles incorporated while formulating, the films commonly keep transparency or little translucency. Other nanocomposites have also resulted from a WPU synthesized from castor oil-based macro diol with 70% C obtained from energy-efficient resources, and OH number of 80 mg KOH/g.

Hexamethylene diisocyanate (HDI) was an isocyanate monomer, and CNCs from sisal were the nanoparticles utilized to synthesize nano-reinforcement to PUs. These substances exhibited the typical consequences noted when hard particles are incorporated into a polymeric composite, i.e. enhanced Young's modulus and reduced strain. The force at break relies on the greater stress supported by the composites and the reduced elongation at break, and in this scenario, it was affected particularly by the extension at the break. Overall stress reduced from 11.6 MPa to 7.21 MPa.

Similar results can be derived while including microfibrillar cellulose nanocomposites comparing a WPU based on PCL and bio-based. The writers exhibited that interaction filler matrix had been more severe in PCL-based substances, and hence, the characteristics relied more on MFC incorporation in that chain of nanocomposites [11].

22.7 LATEST INNOVATIONS IN WPUS FOR PACKAGING APPLICATIONS

Despite PUs' complicated synthetic processes, they are still being used in almost all industries, such as paint, inks, foam, etc. A simple scheme of their synthesis and use is summarized in Figure 22.5.

22.7.1 Antibacterial WPU Based on Nanoparticles for Packing

Owing to their medical use (medical implants/thermal insulation/packaging), microbes can attack them, leading to drastic effects. So, it is requisite to enhance their

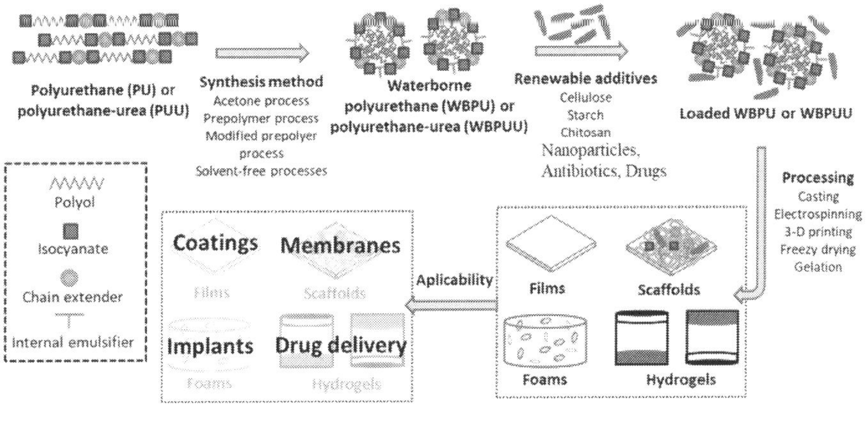

FIGURE 22.5 A general schematic illustration of synthesis and uses of WPU [13].

characteristics (physical/chemical), via either coupling or embedding active agents that include Ag, Au, ZnO, and TiO2 nanoparticles. These govern positive societal effects in the compass of both scientific as well as regulatory communities. The following section describes the effects of nanoparticles on the applications of WPUs.

Silver nanoparticles (Ag NPs) embedded WPU reduce the number of bacteria and augment mechanical attributes, such as tensile strength, Young's modulus, etc. A little quantity of silver nanoparticles is effective, but a higher amount causes toxicity to humans and animals. However, Ag-WPUs are widely used in the pharmaceutical and packing industries. On the other hand, gold (Au) nanoparticles also have antibacterial potential, but these require peculiar conditions.

Titania (TiO_2) and ZnO nanoparticles are also widely used due to their fascinating characteristics of large surface area, extraordinary reaction capability, non-toxic behavior, and economical cost. Generally, these play an important role in catalysis, self-cleaning process, UV-shield, and environmental protection. They kill bacteria, fungi, and cancerous cells by producing either hydroxyl radicals or reactive oxygen species (ROS) on their surface that react with phospholipid (polyunsaturated) constituents of the cell membrane. When encapsulated onto WPU, they not only affect the thermal and mechanical behavior of WPU but also inactivate microbes (bacteria/fungi). Still, some environmental problems are associated with TiO2 nanoparticles. They increase mechanical resistance, corrosion, and catalytic activity; that is why they are used in pharmaceutical, medical, and packing industries. ZnO-loaded WPUs manifest bactericidal applications, being a source of water via releasing zinc cations and reactive oxygen species [14].

22.7.2 UV-Cured WPU Coatings for Packaging

Waterborne UV-curable coatings have application in packaging materials, magazines, flexographic printing, and industries (synthetic leather, metal, fabrics, paper, and automobile parts), where these play a role as substrate protectors. UV-WPUs are restricted to flat surface substances because UV light does not approach shadow points, resulting in insufficient curing. On the other hand, thick surfaces have insubstantial UV infiltration. So, curing efficiency is lowered due to suppressing the level of molecular oxygen radicals. Adverse effects of limited oxygen radicals are prevented by not only using inert gases and physical gauze but also enhancing the intensity of UV light and the quantity of photo-initiators. Alternatively, click reaction is used for consistent UV curing to reinforce low stress. When UV-WPUs are coated onto fabric (cotton), there is a bit of color change only in case of exposure for a long period. On the other hand, sharp color on fabric can be developed through higher intensity of UV light along with the percentage of C=C. Therefore, fabric coated with dye, UV-cured castor oil-based cationic WPU, and acryloyl chloride modified lysozyme exhibits no bacterial growth. UV-cured engineering is eco-friendly; such compositions not only discharge a small amount of VOCs, but also insufficient energy [15–18].

Hence, these coatings protect the internal or external surfaces from environmental stress as humidity, irradiations, bio-decomposition, and chemical or mechanical impairment. These also inhibit corrosion due to their gelatinous cascade. In this case, linear frameworks of UV-WPU coatings have more efficacy than that of cyclic structures [19].

22.7.3 Medical Uses of Polyurethanes

Because of their accessibility, better biocompatibility, and mechanical properties, PUs have been employed in medical and paramedical practice, such as surgical drapes, medical unit bedding, tubing, catheters, wound dressings, and many more applications. Still, the most common use of PUs in medicine is a shot-period transplant. PUs are famous for medical applications due to their toughness, cost-effectiveness, and stability. These characteristics have helped surpass other substances like metals, ceramics, and alloys. The market for bio PUs was about 1,554 tons in 2012. Currently, the revenue for bio polyol is approximately USD 5.03 billion according to a GVR, Inc. study. One study used pre-polymers to enhance the bio decomposability of PUs utilizing water as a chain extender. Moreover, such characteristics have been found highly suitable in applications like joint endoprostheses. Because of alteration in pH during sexual activity, specific drug delivery systems such as microbicide vaginal pessaries that could assist in resisting STDs like AIDS are recommended.

For such applications, extremely smart and sensitive PUs for vaginal drug transport have also been fabricated. PUs have established applications in drug delivery systems for intra-vaginal as well as colon rings. Some reports have suggested the suitability of carbohydrates as constituents of ecofriendly PUs. Such ecofriendly PUs have been constructed utilizing propylene glycol and numerous carbohydrates as crosslinkers. According to a study, several physical properties of PUs change due to incorporating carbohydrates, which subsequently render them useful for biomedical purposes. This study also supports similar studies related to the incorporation of polysaccharides in PUs structures. Recently, cost-effective eco-friendly aliphatic PUs have been developed that have high basic stability up to body temperature without losing mechanical aspects. Furthermore, recent studies also employed chitosan-based PUs for eco-friendly antibacterial electroactive PUs for heart tissue engineering. Recently, Bahrami et al. conducted a study on electroactive PUs with graphene composites, which showed they have better outcomes with increased electro conductivity after 2% graphene ratio and bio evaluation exhibited that these nontoxic substrates promote cell adhesion and expansion. Similarly, Wang et al. researched enhanced antibacterial and corrosion resistance characteristics by coating Mg with functionalized PUs and gathered hopeful results regarding both properties. The most significant result was enhanced compatibility with human blood [2,20].

22.8 HEALTH AND SAFETY CONCERNS OF WPUS

During the synthesis of WPUs, raw material disturbs the environment as well as human health due to adverse side effects. One of the common components is isocyanate that is produced by phosgene, harming the atmosphere and human health. Isocyanate itself, with atmospheric water, leads to atmospheric pollution, serious diseases, as well as problems in manufacture, storage, and transport. Diisocyanates modified from isocyanates cause complications in respiration, dermatological, and chronic disease depending upon exposure time, while chain extenders and co-solvents have cancerous effects. To reduce the destructive effects of the chemicals, the European Commission has limited their use up to a certain point, as mentioned in Table 22.1.

It is mandatory to save the environment and human health from the adverse effects of chemicals. So, when WPUs are used as coatings and paints, Directive 2004/42/CE restricts the emission of VOCs from them up to 140–840 g/L and 30–750 g/L, respectively. In the case of their medical applications, Regulation 745/2017 forced them to either eliminate chemicals that cause endocrine disruption, cancer, and mutations, and have toxic effects on reproduction or use a safe alternative [21–23].

TABLE 22.1
The permissible limits of raw material for the synthesis of typical WPU and WPUU according to actual European legislation. Adapted from reference [13] (An open access source)[1,2]

Raw Materials		Level of Restriction	Regulation/Directive
Diisocyanate	4-4′-Dicyclohexylmethane diisocyanateIsophorone diisocyanate4,4′-Diphenylmethane diisocyanate	1 mg/kg (in end product as isocyanate moiety)	Commission Regulation 1149/2020
Catalyst	Dibutyltin dilaurate Stannous 2-ethylhexanoate	1 mg/kg (in end product as dibutyl moiety)	Regulation (EC) No 1907/2006–
Internal Emulsifier	Dimethylol propionic acid *N*-Methyldiethanolamine	SML (1) = 0.05 mg/kg SML = 0.05 mg/kg	Commission Regulation10/2011
Neutralizing Agent	Triethylamine	3	–
Chain Extenders	Hydrazine monohydrate Diethylenetriamine Ethylenediamine	Not allowed SML = 5 mg/kg SML = 12 mg/kg (2)	Commission Regulation 1272/2008 Commission Regulation 10/2011 Commission Regulation 10/2011
Co-Solvent	Acetone *N*-methyl-2-pyrrolidone	Without limitations 3 mg/Kg in final product	Commission Regulation 588/2018

[1]Specific migration limit defined according to food contact Regulation (EU) No 10/2011.
[2]Pre-registered on SVHC List, under evaluation.

22.9 POLYURETHANES: A PROBLEM SOLVER FOR TODAY'S PACKAGING CHALLENGE

The secret to the success of PU chemistry is the reactivity and high variability of polyols and polyisocyanates groups toward all types of nucleophiles. It results in the synthesis of required polymer blocks, therefore making it feasible that obtained materials' properties and applications are very diverse and fine-tuned. At present, climate change is a key challenge and demands a significant reduction in the climate-damaging greenhouse gasses, in particular CO_2, caused by the combustion of fossil fuels. Modern PU systems still mainly contain petrochemicals based on petroleum. The CO_2 footprint of PU raw materials has been gradually lowered in recent years, but PUs are the most important contributors to energy and resource efficiency through their intended use. A lot of energy is utilized for heating or cooling buildings. During its durability, PU thermal insulation saves the energy needed to produce the insulating materials more than 100 times. The quick spoiling of food is prevented by PU thermal insulation in coolers and the transport chain. They are crucial for nourishing a world population that is constantly increasing. Weight savings in mobility directly equate to reduce fuel use. PU foams are also essential for taping/cushioning, sound absorption, and as a matrix material in addition to contributing to sustainability. The low viscosity of the PU components allows their relatively easy processing into polymer materials, thus making them cost-competitive, especially in production runs. The excellent resistance to wear and durability of PUs in coatings extends the product life and increases its value retention. PU foams and PU adhesives enable effective composite products to be made from a varied range and in particular in lightweight structures, which are therefore crucial to modern material concepts.

PUs have proved their versatility and adaptability to new problems for over 75 years now, and the potential for future breakthroughs is far from exhausted. Formerly solvent-borne coatings could not be matched with typical aqueous solutions for industrial coatings with superior performance. Solvent-borne PUs are known as they perform on a higher level. Therefore, it was a logical step in trends of reducing solvents to convert this technology to the aqueous phase.

The use of WPUs in binding organic materials and long-strand lumbers, laminated veneer lumber, oriented strand boards, particle boards, medium density fiberboards, and straw boards are being addressed. Industries such as wood panels, foundry, rubber, etc. are using them due to their exceptional thermal stability, chemical resistance, and binding capability. In the home, these are employed to bind carpets with PU foams, to replace the conventional toxic organic solvents. Their coatings can be used for various applications due to distinctive formulations. These have been used as a fixture in automotive finishes, as these have a clean surface. Even though, these reduce VOCs, their practical use is lacking. In the case of metal covering (industrial installations, ships, offshore facilities, steel bridges, etc.) with paints, WPU could be applied via a two-component system mixed in a solid solvent in one pot to get satisfactory thickness, but here, bot components react, resulting in unstable paint. So, there is a need to develop such WPU-based coating that may have stability and weather resistance. Aside from that, these might be used in rocket propellants as well as polymer-bonded explosives (PBX) due to the excellent mechanical properties of the raw material [13].

22.10 CONCLUSION

PUs are miracle materials since their discovery due to the relationship between their structure and characteristics. These have adjustable particular mechanical, physical, biological, and chemical features depending upon their raw materials, either polyols or isocyanates. When the raw materials are altered, a synthetic procedure is changed, or sophisticated characterizations are used, application-specific WPU (flexible foams, rigid foams, chemical-resistant coatings, specialty adhesives and sealants, and elastomers) could be attained. In this chapter, we have enlightened the chemistry, types, synthesis, and applications of different kinds of PUs. In the last decade, researchers are paying more attention to the synthesis and application of antimicrobial WPUs and their cast films. The antimicrobial and degradable PU dispersions and their films can be widely used in industrial applications, such as self-sanitizing medical devices, environmental surfaces, food packaging, etc. The degradation behavior of PU dispersions and their films are strongly dependent on the chemical structure of the PU backbone, particularly the type of polyols or the soft segments and the polarity and concentration of the internal surfactant. Generally, structure modification of PUs, such as the incorporation of cleavable groups into the PU structure, confers biodegradability. For the successful chemical modification of PU, it is crucial to improve the level of biodegradability to maintain a good balance between the desired mechanical properties and the biodegradation rate. PU scaffolds can be manufactured via 3D printing techniques, and the chemical structure can be tailored for excellent biocompatibility, good flexibility, as well as excellent mechanical properties. Thus, it can be concluded that with all these properties and advancements PU will become one of the best materials we have ever seen in terms of performance and versatility.

REFERENCES

1. E. Sharmin and F. Zafar, Polyurethane: an introduction. In: *Polyurethane*, InTech, pp. 3–16, 2012.
2. A. Das and P. Mahanwar, "A brief discussion on advances in polyurethane applications," *Adv. Ind. Eng. Polym. Res.*, vol. 3, pp. 93–101, 2020.
3. G. Oertel and L. Abele, *Polyurethane handbook: chemistry, raw materials, processing, application, properties*, Gunter, 1994.
4. D. K. Chattopadhyay and K. Raju, "Structural engineering of polyurethane coatings for high performance applications," *Prog. Polym. Sci.*, vol. 32, no. 3, pp. 352–418, 2007.
5. G. T. Howard, "Biodegradation of polyurethane: a review," *Int. Biodeterior. Biodegradation*, vol. 49, no. 4, pp. 245–252, 2002.
6. T. Kochanė, S. Budrienė, K. Pielichowski, and J. Pielichowski, "Application of polyurethane-based materials for immobilization of enzymes and cells: a review," *Chemija*, vol. 17, no. 4, pp. 74–89, 2006.
7. D. K. Chattopadhyay and D. C. Webster, "Thermal stability and flame retardancy of polyurethanes," *Prog. Polym. Sci.*, vol. 34, no. 10, pp. 1068–1133, 2009.

8. J. Li, W. Zheng, W. Zeng, D. Zhang, and X. Peng, "Structure, properties and application of a novel low-glossed waterborne polyurethane," *Appl. Surf. Sci.*, vol. 307, pp. 255–262, 2014.
9. A. Noreen, K. M. Zia, M. Zuber, S. Tabasum, and M. J. Saif, "Recent trends in environmentally friendly water-borne polyurethane coatings: a review," *Korean J. Chem. Eng.*, vol. 33, no. 2, pp. 388–400, 2016.
10. A. Javadi, A. Cobaj, and M. D. Soucek, "Commercial waterborne coatings," in *Handbook of waterborne coatings*, Elsevier, 2020, pp. 303–344.
11. H. Honarkar, "Waterborne polyurethanes: a review," *J. Dispers. Sci. Technol.*, vol. 39, no. 4, pp. 507–516, 2018.
12. F. E. Golling *et al.*, "Polyurethanes for coatings and adhesives–chemistry and applications," *Polym. Int.*, vol. 68, no. 5, pp. 848–855, 2019.
13. A. Santamaria-Echart, I. Fernandes, F. Barreiro, M. A. Corcuera, and A. Eceiza, "Advances in waterborne polyurethane and polyurethane-urea dispersions and their eco-friendly derivatives: a review," *Polymers (Basel).*, vol. 13, no. 3, p. 409, 2021.
14. Z. Farrokhi, A. Ayati, M. Kanvisi, and M. Sillanpää, "Recent advance in antibacterial activity of nanoparticles contained polyurethane," *J. Appl. Polym. Sci.*, vol. 136, no. 4, p. 46997, 2019.
15. S. Chen, W. Wu, G. Zhao, T. Jin, and T. Zhao, "Fabrication and properties of superparamagnetic UV-curable nanocomposites based on covalently linked waterborne polyurethane/functionalized hollow Ni 0.3 Zn 0.5 Fe 2 O 4 microspheres," *RSC Adv.*, vol. 5, no. 59, pp. 47788–47797, 2015.
16. J. Xu *et al.*, "Preparation, characterization of UV-curable waterborne polyurethane-acrylate and the application in metal iron surface protection," *J. Appl. Polym. Sci.*, vol. 130, no. 5, pp. 3142–3152, 2013.
17. Y. Dai *et al.*, "UV-curable electromagnetic shielding composite films produced through waterborne polyurethane-acrylate bonded graphene oxide: preparation and effect of different diluents on the properties," *e-Polymers*, vol. 14, no. 6, pp. 427–440, 2014.
18. Y. T. Dai *et al.*, "Preparation and properties of UV-curable waterborne graphene oxide/polyurethane-acrylate composites," *Plast. Rubber Compos.*, vol. 43, no. 2, pp. 54–62, 2014.
19. T. Zhang, W. Wu, and Y. Mu, A novel nanosilica-reinforced waterborne UV-curable material. In: *Second International Conference on Smart Materials and Nanotechnology in Engineering*, 2009, vol. 7493, p. 74936I.
20. L. D. Agnol, F. T. G. Dias, H. L. Ornaghi Jr, M. Sangermano, and O. Bianchi, "UV-curable waterborne polyurethane coatings: a state-of-the-art and recent advances review," *Prog. Org. Coatings*, vol. 154, p. 106156, 2021.
21. P. Vermette, H. J. Griesser, G. Laroche, and R. Guidoin, *Biomedical applications of polyurethanes*, Landes Bioscience, 2001.
22. K. Taylor, "Ten Years of REACH—An Animal Protection Perspective," *Altern. to Lab. Anim.*, vol. 46, no. 6, pp. 347–373, 2018.
23. U. G. Sauer *et al.*, "Local tolerance testing under REACH: accepted non-animal methods are not on equal footing with animal tests," *Altern. to Lab. Anim.*, vol. 44, no. 3, pp. 281–299, 2016.

23 Waterborne Polyurethanes for Automobile Industries

Ayesha Kausar
Nanosciences Division, National Center For Physics,
Quaid-i-Azam University Campus, Islamabad, Pakistan

CONTENTS

23.1 Introduction....................393
23.2 Eco-Friendly Waterborne Polyurethane....................394
23.3 Eco-Friendly Waterborne Polyurethane for Automobiles....................394
    23.3.1 Eco-Friendly Waterborne Polyurethane for Automobile Interiors....................394
    23.3.2 Eco-Friendly Waterborne Polyurethane Adhesives in Automobiles....................397
    23.3.3 Waterborne Polyurethane Coatings With Anticorrosion and Nonflammability for Vehicle Parts....................399
    23.3.4 Shape Memory Eco-Friendly Waterborne Polyurethane for Automobiles....................400
23.4 Outlook and Summary....................402
References....................403

23.1 INTRODUCTION

Polyurethane is an imperative category of polymers [1]. It has significant chemical, thermal, and mechanical stability properties [2,3]. Crosslinking is an important property of polyurethanes [4–6]. Polyurethanes have been applied in a range of automotive and automobile solicitations [7,8]. Waterborne polyurethane is a type of eco-friendly polyurethane [9]. It has structural, solubility, and physical property advantages relative to pristine polyurethane. Waterborne polyurethanes are prepared using appropriate polyurethane precursors and fabrication techniques [10]. They have been deliberated as lightweight, low density, and high-performance materials for the automotive, motorized, packaging, and construction fields [11]. Applications of eco-friendly waterborne polyurethanes are wide-ranging in the automotive field. Waterborne polyurethane-based nanocomposites have also been used to form eco-friendly automotive components. Eco-friendly waterborne polyurethanes have been applied as adhesives in automotive constituents [12]. Various coatings based on eco-friendly waterborne

DOI: 10.1201/9781003173526-23

polyurethanes have been designed for corrosion resistance and flame resistance for vehicle parts [13,14]. Another important use of eco-friendly waterborne polyurethanes is shape memory components [15]. This is a state-of-the-art chapter regarding the design and development of waterborne polyurethanes for automobile applications. Primarily, the fundamentals of eco-friendly waterborne polyurethanes are established. Subsequently, use of eco-friendly waterborne polyurethanes in the automobile interior, as adhesives, coatings, and shape memory components, is discussed.

23.2 ECO-FRIENDLY WATERBORNE POLYURETHANE

Polyurethane is a significant high-performance polymer [16]. Polyurethanes are generally developed through the reaction of di- or polyisocyanate (R–(N=C=O)$_n$ and di- or polyhydroxy (R–(OH)$_n$ compounds [17]. Consequently, polyurethanes possess repeating urethane/carbamate units in their backbone. Di- or polyisocyanates and polyols are polymerized under the appropriate catalytic/temperature conditions. Polyurethanes have essential features of hydrogen bonding interactions between their chains [18]. Usually, polyurethanes are insoluble in water. Waterborne polyurethane is also known as water-based polyurethane [19]. Waterborne polyurethane has hydrophilic groups in the backbone to attain high water solubility. It has been modified using various functionalities, such as carboxylate, sulfonate, ammonium groups, etc. An example is shown in Figure 23.1 [20]. Eco-friendly waterborne polyurethanes possess properties of heat constancy, chemical stability, nonflammability, mechanical features, and physical properties [21]. They have potential for coatings [22], electronic devices [23], radiation shielding [24], textiles [25], etc.

23.3 ECO-FRIENDLY WATERBORNE POLYURETHANE FOR AUTOMOBILES

23.3.1 Eco-Friendly Waterborne Polyurethane for Automobile Interiors

Polyurethane is always an attractive polymer for the automotive industry, intended for vehicle engineering. Typically, polyurethanes are thermoset polymers;

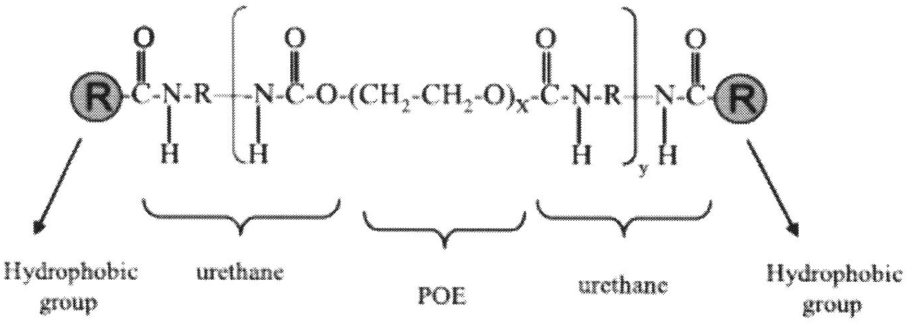

FIGURE 23.1 Chemical structure of a waterborne polyurethane. POE = polyoxyethylene. Adapted with permission from reference [20], Copyright 2009, Elsevier.

however, thermoplastic polyurethanes are also known. Polyurethanes are thermoset polymers because of extensive crosslinking between the chains, however, some thermoplastic polyurethanes have also been researched. Polyurethanes can be formable as thin films, sheets, foams, and other forms. The foamed polyurethanes are an important component of the automotive industry. Like common polymers, polyurethane foams do not show melting behavior. Polyurethane foams are extremely lightweight, durable, supple, as well as strengthened materials for the vehicle parts. Therefore, the automotive industries are continuously in the production of polyurethane foam materials. Moreover, polyurethane foams are rigid and have heat insulation properties. Polyurethane foams have been employed in seats, armrests, headrests, B-pillars, bumpers, wadding panels, headliners, and other interior parts in vehicles. The purpose of polyurethane foams in automobiles is comfort, to reduce exhaustion and anxiety while driving. Another important purpose for the use of polyurethane foams is protection from any of hazards and shocks in automobiles. Then, the benefits in terms of energy conservation can be attained by reducing the overall weight of the vehicle. The reduction in the inclusive weight of the vehicles may result in high fuel efficiency and better ecological enactment. Moreover, the use of eco-friendly polyurethanes has been considered due to the environmental concerns of the global automobile market. Traditionally, metals or steel were probable primary materials for structuring automobile structures. The metals generally have high strength grades. The metal components were also preferred for the design of the electric vehicles. The eco-friendly polyurethane foams have been used to replace the roofs, panels, hoods, cladding, etc. in routine automobiles as well as in electric vehicles. A major issue of automobiles is the greater fuel consumption. The use of eco-friendly polyurethane foam has led to a lightweight solution to reduce the vehicle weight, favorably decrease car emissions, and decrease the overall fuel consumption by the vehicles. Thus, eco-friendly polyurethane foams are always a safe, ecological, and lightweight solution to the metal components and metal automobile body parts. When the insulation properties of the eco-friendly polyurethane foams materials are discussed, these include the insulation from the heat, the insulation from the engine or internal noise, and the insulation from any external shockwave. In other words, the eco-friendly polyurethane foams are sanctification for the auto designers and manufacturers to meet the high-performance stipulations deprived of adding weight to the automobiles. Even after a long time of use of the eco-friendly polyurethane foams based car interior parts maintain the innovative steadfastness, original shape, and pristine suppleness. For the automobile components and parts, a major technique used for the formation of the seats, panels, pillars, bumpers, liners, other interiors, and even the outer car body parts using eco-friendly polyurethane foam, is the injection molding method. The car components formed by the polyurethane foam through the injection molding technique are not only lightweight and durable but also have shock absorption features. The greater employment of the injection molding technique is in the automotive industry. The injection molding technique and the polyurethane foams together offer freedom to the researchers and engineers to generate several design innovations. This in turn provides low fuel, gas, or petrol use and high car mileage.

Up till now, metal or metal alloys have been applied for automobile external portions. However, the use of metal not only increases the weight of the vehicle but also is vulnerable to damage, dents, and corrosion relative to eco-friendly polymeric materials. Further research in this regard can replace the use of the metal components in the interiors and exterior parts with polyurethanes and eco-friendly polyurethane foams.

Eco-friendly waterborne polyurethanes may find imperious use in the automotive area [26]. The main use of eco-friendly waterborne polyurethanes is in frothy sandwich panels for the automobile interior [27]. The car body's internal components have high strength, toughness, and energy absorption characteristics. Eco-friendly waterborne polyurethane possesses thermal conductivity properties for high-performance interior components [28]. Wong et al. [29] produced a blend of polyurethane (PU) and polycarbonate (PC) in different compositions. They produced sandwich panels for car interior seats and structure. The panels, prepared with different compositions, are given in Figure 23.2. The neat polyurethane has a thermal conductivity of 0.1867 W/mK. The thermal conductivity of the sandwich panels was measured and found to decrease with the increasing PU content [30,31]. The 50:50% PU:PC was found to be 0.1776 W/mK. A further decrease was not observed with the increasing PU contents. The decrease in the sample porosity and density increase was responsible for the change in the thermal conductivity of the panels. Subsequently, low absorption of the heat was observed. Future research

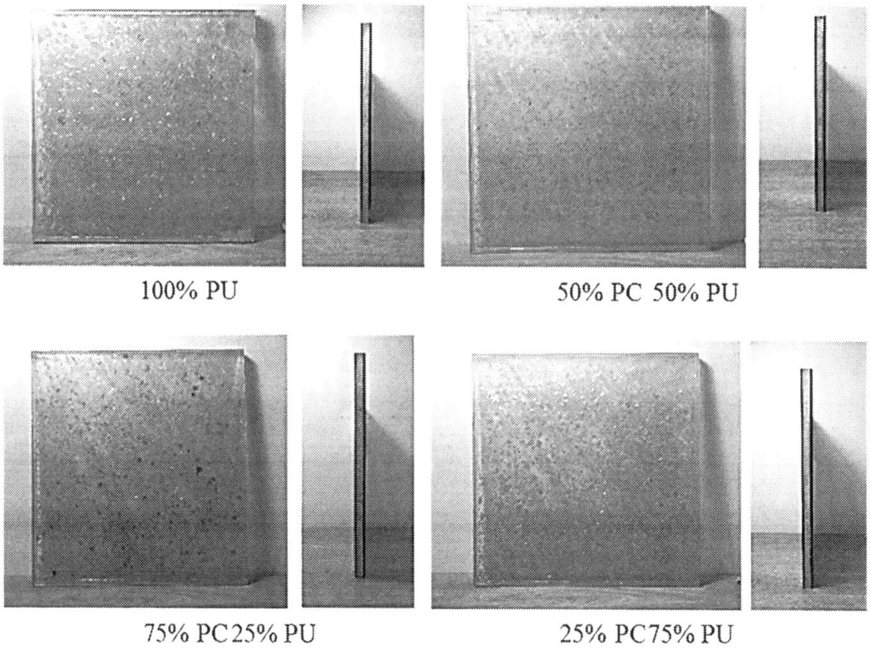

FIGURE 23.2 Samples of the investigated sandwich panels (length = 300 mm; width = 300 mm; thickness = 18 mm) in different compositions. Adapted with permission from reference [29], Copyright 2020, Elsevier.

must turn toward the manufacturing of eco-friendly waterborne polyurethanes using natural or eco-friendly materials and resources and recyclable polymers [32,33].

23.3.2 Eco-Friendly Waterborne Polyurethane Adhesives in Automobiles

Waterborne polyurethane-based automotive adhesives have been employed for automotive interior parts, including on the molding board, in the panels, in the seat fixing, in the armrest fixing, in the headrest fitting, in the pillars, on the bumpers, in the wadding panels, and on several other components. Eco-friendly waterborne polyurethane-based automotive adhesives can be employed at various temperatures. The thickness or viscosity of eco-friendly waterborne polyurethane adhesives can be used at room temperature or high temperature. Automotive interiors also include the use of eco-friendly waterborne polyurethane-based adhesives on woven or nonwoven fabrics on the carpets, ceiling, side door, and other interiors. For applying eco-friendly waterborne polyurethane-based adhesives to automobile parts and components, mostly spraying techniques have been applied. Eco-friendly waterborne polyurethane-based adhesives are usually applied in one or two layers on the automobile components. The thickness of the adhesive layer should not increase to a certain level. Mostly, eco-friendly waterborne polyurethane-based adhesives can be applied in the amount of 100 grams per square meter. In other words, thin layers are preferred. After the application of the adhesive on the automobile part surface, it needs to be dried for a certain time to develop the bond between the surfaces. Eco-friendly waterborne polyurethane-based adhesives are known to have virtuous bond strength.

Eco-friendly waterborne polyurethanes possess intrinsic ionic or hydrophilic groups in structure; therefore, they have been employed as adhesives for automobiles [34,35]. Various studies have explored the structure–properties connection in eco-friendly waterborne polyurethane-based adhesives [36,37]. Different eco-friendly waterborne polyurethanes have been analyzed for their thermal stability, strength, and adhesive properties [38,39]. Orgiles-Calpena et al. [20] proposed eco-friendly waterborne polyurethane adhesives and poly(vinyl chloride) (PVC) based adhesive systems. Eco-friendly polyurethane adhesive was used as a thickener. The adhesion properties of the PVC/polyurethane adhesive/PVC were measured using T-peel tests (Figure 23.3). Both the adhesive failure and the cohesive failure were observed for the adhesive. The T-peel strength of the PVC/polyurethane adhesive/PVC was measured with the increasing thickener contents [40]. After 72 h, the polyurethane thickener showed higher adhesion to the PVC matrix. The 3 wt.% thickener content was found efficient for the joint formation.

Li et al. [41] prepared eco-friendly waterborne polyurethanes using toluene diisocyanate, polypropylene glycol, and ethylene glycol monomers. Eco-friendly waterborne polyurethanes were designed using different soft and hard segment contents. Toluene diisocyanate formed hard segments, whereas polypropylene glycol constituted the soft segments in the eco-friendly polyurethane. Figure 23.4 shows the differential scanning calorimetry (DSC) curves of waterborne polyurethanes. The mixing degree of the soft and hard segments was analyzed. The glass transition temperature (Tg) and enthalpy (ΔH) were premeditated from the DSC

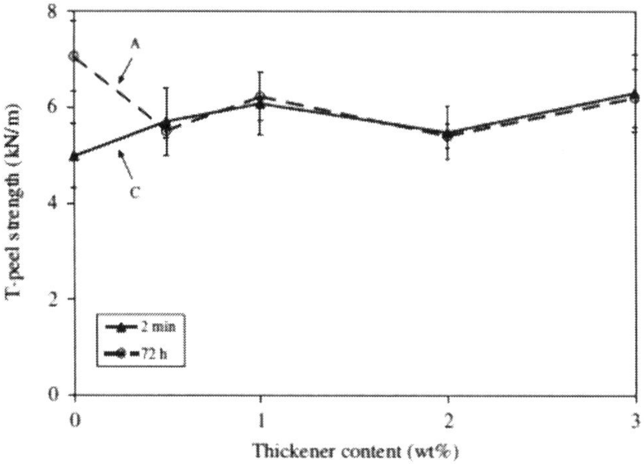

FIGURE 23.3 Variation of T-peel strength as a function of the amount of thickener in the PVC/polyurethane adhesive/PVC adhesive [20]. A = adhesive failure; C = cohesive failure in the adhesive. PVC = poly(vinyl chloride). Adapted with permission from reference [20], Copyright 2009, Elsevier.

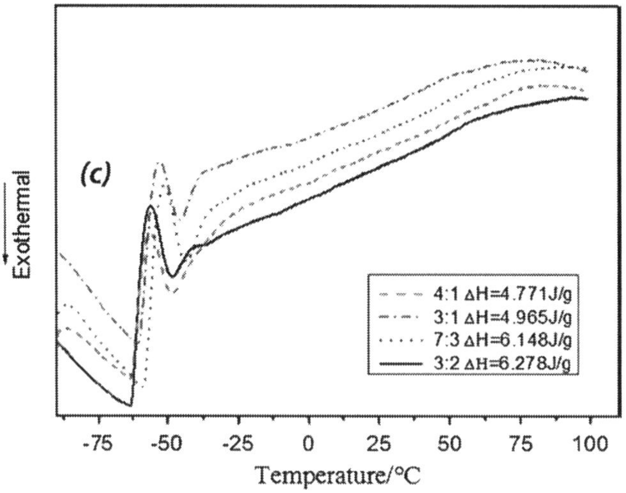

FIGURE 23.4 DSC curve of waterborne polyurethane films. Adapted with permission from reference [41], Copyright 2019, Elsevier.

thermograms. The 3:2 ratio of soft and hard segments revealed phase separation, as confirmed by the higher ΔH value of ~6.278 J/g among all the compositions. Phase separation caused the lower Tg value for this composition. Endothermic peaks indicated the crystallinity of the adhesives. In the case of the T-peel strength of the adhesives, the 4:1 ratio of the segments caused the higher value of 136.14 N/25 mm (Figure 23.5). Adhesive strength was found to be lower for the 3:2 sample, at 21.55 N/25 mm. The lower adhesive strength of the 3:2 soft and hard segments also

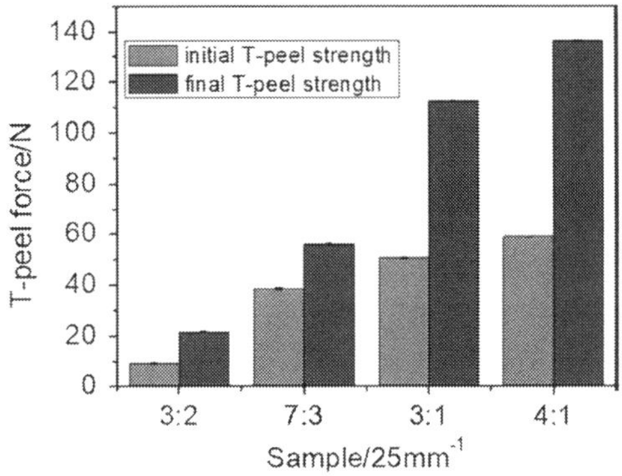

FIGURE 23.5 T-peel strength of waterborne polyurethanes films. Adapted with permission from reference [41], Copyright 2019, Elsevier.

indicated phase separation in waterborne polyurethane. Lesser phase separation properties produced lesser porosity in adhesives, thus proceeding the adhesive properties. Therefore, eco-friendly waterborne polyurethane adhesives have been found to be efficient at forming strong bonds in automobile applications.

23.3.3 Waterborne Polyurethane Coatings With Anticorrosion and Nonflammability for Vehicle Parts

Around the globe, governments are conscious of environmental protection when manufacturing automobile bodies and internal components. Waterborne polyurethane-based coating technology is a safe and ecological method for protecting automobile parts from any external damage. Various waterborne polyurethane-based coatings have been developed for automobile components. These coatings have been applied through processes such as coating procedures, painting, spraying, etc. Different types of waterborne polyurethane-based coatings include primers, base coatings, top coatings, finishes, etc. Exterior automobile coatings may reduce toxic gas emissions from cars to the environment; therefore, they are essential from the ecological point of view. Waterborne polyurethane-based coatings protect automobile parts from environmental heat impact. Moreover, they are important for preventing rust or heat damage of automobile parts and components. Waterborne polyurethane-based coatings have been found to be beneficial for the car's exterior body as well as internal car components. The most frequently used technology for the outer car coating is the spraying method. This method may form eruption-free thin and fine layers. The spraying method can be appropriate for a wide range of operational situations. For corrosion prevention of automobile components, thermoplastic waterborne polyurethane nanocomposites have been employed. However, for heat and flammability protection of auto gears, thermoset waterborne polyurethane nanocomposites have been frequently used. Both thermoset and thermoplastic waterborne polyurethane nanocomposites have offered all-inclusive, lightweight protection resolutions for automobile engine part coverings, batteries, charging units, fuel cells, and also for electric vehicle builds.

Waterborne polyurethane coatings are now also moving to other transportation industries, such as the motorcycle industry, passenger plane industry, military or defense airplane market, aerospace industry, etc.

Automotive coatings have been employed to coat automobile components and outer surfaces [42,43]. In this regard, eco-friendly polyurethanes have been used. In the automotive industry, eco-friendly waterborne polyurethanes have been useful for heat resistance, flame resistance, and radiation shielding properties [44,45]. The heat and flame resistance performance of automotives can be enhanced using eco-friendly waterborne polyurethanes. Moreover, waterborne polyurethanes have shown anticorrosion properties [46,47]. These properties led to the formation of eco-friendly waterborne polyurethane-based coating materials for automobile components. Two types of coatings have been developed, i.e. flame retardant coatings and anticorrosion coatings. Pathan et al. [48] designed eco-friendly waterborne polyurethane-based thermally stable and anticorrosion coatings. Melamine formaldehyde (MF) and 3-isocynatopropyl triethoxy silane (IPTES) were used to crosslink eco-friendly waterborne polyurethane MF/IPTES. Dibutyltin dilaurate (DBTDL) was used as a catalyst to form waterborne polyurethane. The formation scheme is illustrated in Figure 23.6.

Crosslinked eco-friendly waterborne polyurethane possesses improved mechanical, thermal, adhesion, and anticorrosive properties. It revealed good adhesion with metal substrate. The anticorrosion properties were attributed to permeation properties of the coating against the corrosive species (Figure 23.7). The existence of polar groups (urethane, ester,–OH, Si–O, etc.) has consequences for anticorrosive performance [49,50]. For high-performance automotive applications, locomotive industries need to integrate waterborne polyurethane/nanocarbon-based eco-friendly systems. Consequently, eco-friendly waterborne polyurethane nanocomposite-based eco-automotives must be designed for future success in this industry [51].

23.3.4 Shape Memory Eco-Friendly Waterborne Polyurethane for Automobiles

Shape memory polymers are an emergent section of advanced smart materials [52]. Shape memory polymers possess the ability to demonstrate shape memory upshot [53]. Characteristically, shape memory polymers may display shape alteration upon suitable stimulus, such as heat, light, electric current, voltage, stress, pH, water, and other environmental effects. An imperative feature is use of an appropriate type of polymer. Generally, segmented polymers are preferred, including polyurethanes, polyesters, epoxies, etc., and other segmented polymers. Such segmented polymers generally can mimic shapes under the effect of some external force, as mentioned earlier [54]. The automobile industry is shifting from the use of solvent-borne polyurethane to waterborne polyurethane technology. Shape memory waterborne polyurethanes have been found to be beneficial in various car parts and internal components. The use of shape memory waterborne polyurethanes not only reduces the overall weight of the car compared to the metal body but also prevents problems, such as external metal damage by corrosion or denting.

Accordingly, shape memory waterborne polyurethanes can recover their shape upon exposure to heat, light, voltage, pH, etc. They have gained immense

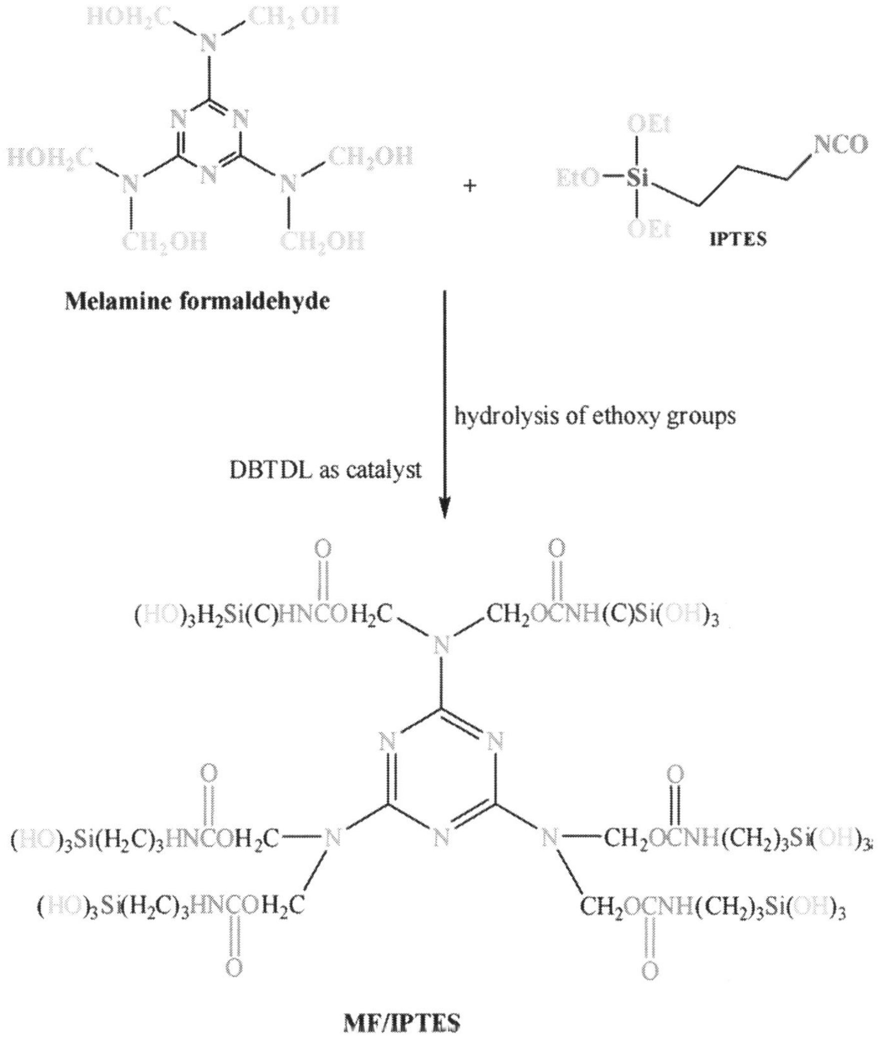

FIGURE 23.6 Functionalization of MF with IPTES. Adapted with permission from reference [48], Copyright 2018, Elsevier.

importance in several technical industries [55]. In addition to eco-friendly shape recovery, polyurethanes have advantages of mechanical strength, thermal stability, nonflammability, electrical conductivity, and other physical properties. Shape memory eco-friendly waterborne polyurethanes have been prepared and employed in the automotive industry [56]. They mostly possess thermo-responsive and electro-activate shape recovery effects. In automobile parts, external heating has been applied to simulate the shape memory effect. Waterborne polyurethanes possess easy thermo-actuation in the materials. Waterborne polyurethanes usually have thermo-elastic phase transformation in the materials at the relevant T_g. The electro-actuation of these polyurethanes has also been employed in vehicles [57].

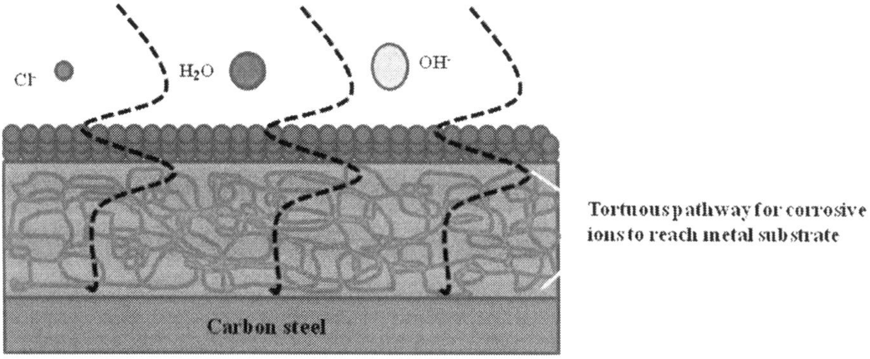

FIGURE 23.7 Schematic illustration of anticorrosive coatings cured with MF/IPTES. Adapted with permission from reference [48], Copyright 2018, Elsevier.

In eco-friendly shape memory waterborne polyurethanes, nanoparticles have been dispersed to improve shape memory performance [58]. The homogeneously dispersed nanoparticles may form the interconnected networks in eco-friendly waterborne polyurethanes. The interlinked structure in eco-friendly waterborne polyurethanes may show better shape memory performance by enhancing modulus properties [59]. Thus, enhanced shape fixing and shape recovery properties have been observed in automotive applications. New research is needed to develop ways to form eco-friendly waterborne polyurethane functional high-performance nanocomposites using modified polymers, functional nanofillers, and new actuation approaches. The upcoming research on eco-friendly shape memory waterborne polyurethanes may lead to several advancements in the automotive industry.

23.4 OUTLOOK AND SUMMARY

The ecological waterborne polyurethanes in the form of sheets, coatings, and adhesives have been reported for various eco-automotive-related applications. Consequently, research has focused on design, characterization, and use of ecological waterborne polyurethanes. Nanostructured ecological waterborne polyurethane materials can be developed to boost the properties of the ecological automotive. The anticorrosion effects, heat stability, flame retardancy, shape recovery, and peel strength can be augmented using the higher nanofiller contents. Functional waterborne polyurethane and functionalized nanoparticles may cause recovering effects on the properties. Accordingly, eco-polyurethanes have been synthesized and altered for the development of next-generation eco-friendly automobiles. Nonetheless, challenges in numerous functional waterborne polyurethane designs, processing methods, and applications must be overcome to achieve complete usage of eco-friendly automotive technology. In brief, eco-friendly waterborne polyurethane is a multipurpose polymer. Eco-friendly waterborne polyurethane-based materials possess better processability, formability, chemical constancy, heat stability, flame retardancy, mechanical strength, peel strength, thermal conductivity, corrosion resistance, and other physical characteristics. Thus, eco-friendly waterborne polyurethanes have high recital properties and wide-ranging automotive applications. In the case of nanomaterial samples, the nanofiller's dispersion may influence the overall nanocomposite performance. The important application areas found for eco-friendly waterborne polyurethane-based materials include ecological anticorrosion waterborne polyurethane coatings, ecological nonflammability waterborne polyurethane coatings for automobile parts, eco-friendly waterborne polyurethane factual for automobile interior parts, and ecological waterborne polyurethane-based shape memory automotive materials. Ecological waterborne polyurethane and nanoparticle-based nanocomposites need to be developed. Eco-friendly waterborne polyurethane-based nanomaterials with various nanofiller contents may have advantageous thermal, mechanical, and shape memory properties. Hydrogen-bonding interactions may occur between polyurethanes and nanoparticles. The inclusion of nanoparticles may increase the T_g and storage moduli, and shape the recovery performance of these nanocomposites. Henceforth, numerous ground-breaking combinations of eco-friendly waterborne polyurethane using glass fiber, carbon fiber, carbon nanotube, graphene, graphite, carbon black, and metal nanoparticles can be prepared for shape memory, nonflammability, anticorrosion, and other beneficial properties of automotive materials.

In summary, eco-friendly waterborne polyurethane-based materials have been established for eco-automotives. Eco-friendly waterborne polyurethane-based nanomaterials have also been developed using eco-polyurethanes and nanoparticles. Eco-friendly waterborne polyurethane-based materials have been produced and applied in various aspects of eco-friendly automobiles. Developments in the field of eco-friendly waterborne polyurethane-based materials have led to numerous technical applications in automotive components and parts. Eco-friendly waterborne polyurethane-based materials comprise the coatings, adhesives, interior components, and shape memory components of eco-automotives. Eco-friendly automotives have been explored for their flame- and heat-resistance properties. Challenges and advancements in the application areas of eco-friendly automotives can be overcome by using new designs of eco-friendly waterborne polyurethane-based materials, modified polymers, advanced fabrication methods, and processing conditions.

REFERENCES

1. Kausar A (2018) Polyurethane composite foams in high-performance applications: A review. *Polym Plast Technol Eng* 57: 346–369.
2. Wang S, Yang Y, Ying H, Jing X, Wang B, Zhang Y, Cheng J (2020) Recyclable, self-healable, and highly malleable poly (urethane-urea) s with improved thermal and mechanical performances. *ACS Appl Mater Interfac* 12: 35403–35414.
3. Gurunathan T, Chung JS (2016) Physicochemical properties of amino–silane-terminated vegetable oil-based waterborne polyurethane nanocomposites. *ACS Sustain Chem Eng* 4: 4645–4653.
4. Oliveira ML, Orlando D, Mulinari DR (2020) Aluminum powder reinforced polyurethane foams derived from castor oil. *J Inorg Organomet Polym Mater* 30: 5157–5167.

5. Chen H, Deng C, Zhao Z-Y, Wan L, Yang A-H, Wang Y-Z (2020) Novel piperazine-containing oligomer as flame retardant and crystallization induction additive for thermoplastics polyurethane. *Chem Eng J* 400: 125941.
6. Yu F, Luo M, Xu R, Huang L, Zhou W, Li J, Tay FR, Niu LN, Chen JH (2020) Evaluation of a collagen-reactive monomer with advanced bonding durability. *J Dental Res* 99: 813–819.
7. Engels HW, Pirkl HG, Albers R, Albach RW, Krause J, Hoffmann A, Casselmann H, Dormish J (2013) Polyurethanes: Versatile materials and sustainable problem solvers for today's challenges. *Angew Chem Int Ed* 52: 9422–9441.
8. Shikha, Meena M, Jacob J (2020) Pentaerythritol derived phosphorous based bicyclic compounds as promising flame retardants for thermoplastic polyurethane films. *J Appl Polym Sci* 50375.
9. Noble K-L (1997) Waterborne polyurethanes. *Prog Org Coat* 32: 131–136.
10. Mou J, Wang X, Yu D, Wang S, Miao Y, Liu Q, Chen B (2021) Practical two-step chain extension method to prepare sulfonated waterborne polyurethanes based on aliphatic diamine sulphonate. *J Appl Polym Sci* 138: 50353.
11. Honarkar H (2018) Waterborne polyurethanes: A review. *J Dispers Sci Technol* 39: 507–516.
12. Kim BS, Kim BK (2005) Enhancement of hydrolytic stability and adhesion of waterborne polyurethanes. *J Appl Polym Sci* 97: 1961–1969.
13. Zhang P, Zhang Z, Fan H, Tian S, Chen Y, Yan J (2016) Waterborne polyurethane conjugated with novel diol chain-extender bearing cyclic phosphoramidate lateral group: synthesis, flammability and thermal degradation mechanism. *RSC Adv* 6: 56610–56622.
14. Li J, Gan L, Liu Y, Mateti S, Lei W, Chen Y, Yang J (2018) Boron nitride nanosheets reinforced waterborne polyurethane coatings for improving corrosion resistance and antifriction properties. *Eur Polym J* 104: 57–63.
15. Lee SK, Yoon SH, Chung I, Hartwig A, Kim BK (2011) Waterborne polyurethane nanocomposites having shape memory effects. *J Polym Scit A Polym Chem* 49: 634–641.
16. Kemona A, Piotrowska M (2020) Polyurethane recycling and disposal: Methods and prospects. *Polymers* 12: 1752.
17. Sekkar V, Gopalakrishnan S, Devi KA (2003) Studies on allophanate–urethane networks based on hydroxyl terminated polybutadiene: Effect of isocyanate type on the network characteristics. *Eur Polym J* 39: 1281–1290.
18. Sung CSP, Schneider N (1975) Infrared studies of hydrogen bonding in toluene diisocyanate based polyurethanes. *Macromolecules* 8: 68–73.
19. Ren L, Lin C, Lei P (2021) Preparation and properties of catechol-based waterborne polyurethane based on thiol-ene click chemistry reaction. *Prog Org Coat* 157: 106303.
20. Orgiles-Calpena E, Aran-Ais F, Torró-Palau AM, Orgilés-Barceló C, Martín-Martínez JM (2009) Addition of different amounts of a urethane-based thickener to waterborne polyurethane adhesive. *Int J Adhes Adhes* 29: 309–318.
21. Agnol LD, Dias FTG, Ornaghi Jr HL, Sangermano M, Bianchi O (2021) UV-curable waterborne polyurethane coatings: A state-of-the-art and recent advances review. *Prog Org Coat* 154: 106156.
22. Ni L, Li S, Liu Y, Jiang X, Cai P, Feng L, Zhang S, Gao X (2021) Fabrication of active corrosion protection waterborne polyurethane coatings using cerium modified palygorskite nanocontainers. *J Appl Polym Sci:* 50899. doi:10.1002/app.50899.
23. Wang S, Cheng H, Yao B, He H, Zhang L, Yue S, Wang Z, Ouyang J (2021) Self-adhesive, stretchable, biocompatible, and conductive nonvolatile eutectogels as wearable conformal strain and pressure sensors and biopotential electrodes for precise health monitoring. *ACS Appl Mater Interfac* 13: 20735–20745.

24. Yousefi K, Parvin N, Banaei N, Mazraedoost S (2021) Water-borne polyurethanes for high-performance electromagnetic interference shielding. *Adv Appl NanoBio-Tech* 2: 35–45.
25. Wang X, Li Y, Zhao Q, Liu G, Chai L, Zhou L, Fan Q, Shao J (2021) High structural stability of photonic crystals on textile substrates, prepared via a surface-supported curing strategy. *ACS Appl Mater Interfac* 13: 19221–19229.
26. Santamaria-Echart A, Fernandes I, Barreiro F, Corcuera MA, Eceiza A (2021) Advances in waterborne polyurethane and polyurethane-urea dispersions and their eco-friendly derivatives: a review. *Polymers* 13: 409.
27. Huo X, Liu H, Luo Q, Sun G, Li Q (2020) On low-velocity impact response of foam-core sandwich panels. *Int J Mech Sci* 181: 105681.
28. Reddy MM, Vivekanandhan S, Misra M, Bhatia SK, Mohanty AK (2013) Biobased plastics and bionanocomposites: Current status and future opportunities. *Prog Polym Sci* 38: 1653–1689.
29. Wong YC, Mahyuddin N, Aminuddin AMR (2020) Development of thermal insulation sandwich panels containing end-of-life vehicle (ELV) headlamp and seat waste. *Waste Manag* 118: 402–415.
30. Li R, Shan Z (2017) Research on structural features and thermal conductivity of waterborne polyurethane. *Prog Org Coat* 104: 271–279.
31. Li R, Shan Z (2018) Asynchronous synthesis method of waterborne polyurethane with the differences of structural features and thermal conductivity. *J Polym Res* 25: 1–11.
32. Du Y, Yan N, Kortschot MT (2013) An experimental study of creep behavior of lightweight natural fiber-reinforced polymer composite/honeycomb core sandwich panels. *Compos Struct* 106: 160–106.
33. Din AT, Hafiz NM, Rus AZM (2018) Exploration to find green building materials from recycled solid wastes. *J Adv Res Fluid Mech Therm Sci* 47: 35–44.
34. Rahman MM, Kim HD (2006) Synthesis and characterization of waterborne polyurethane adhesives containing different amount of ionic groups (I). *J Appl Polym Sci* 102: 5684–5691.
35. Rahman MM, Kim H-D (2007) Characterization of waterborne polyurethane adhesives containing different soft segments. *J Adhes Sci Technol* 21: 81–96.
36. Du H, Zhao Y, Li Q, Wang J, Kang M, Wang X, Xiang H (2008) Synthesis and characterization of waterborne polyurethane adhesive from MDI and HDI. *J Appl Polym Sci* 110: 1396–1402.
37. Rahman MM, Kim H-D, Lee W-K (2009) Properties of waterborne polyurethane adhesives: Effect of chain extender and polyol content. *J Adhes Sci Technol* 23:177–193.
38. Gadhave RV, Mahanwar PA, Gadekar PT (2017) Bio-renewable sources for synthesis of eco-friendly polyurethane adhesives. A review. *Open J Polym Chem* 7: 57–75.
39. Kim S-M, Lee J, Lee S, Park CY, Lee W-K (2019) Effect of nonionomer emulsifier on particle stability of waterborne polyurethane adhesives. *Molecul Crys Liq Crys* 688: 22–28.
40. Chang PP, Hansen NA, Phoenix RD, Schneid TR (2009) The effects of primers and surface bonding characteristics on the adhesion of polyurethane to two commonly used silicone elastomers. *J Prosthodont* 18: 23–31.
41. Li R, Loontjens JT, Shan Z (2019) The varying mass ratios of soft and hard segments in waterborne polyurethane films: Performances of thermal conductivity and adhesive properties. *Eur Polym J* 112: 423–432.
42. Akafuah NK, Poozesh S, Salaimeh A, Patrick G, Lawler K, Saito K (2016) Evolution of the automotive body coating process—A review. *Coatings* 6: 24.

43. Ahmadi Y, Ahmad S (2020) Recent progress in the synthesis and property enhancement of waterborne polyurethane nanocomposites: Promising and versatile macromolecules for advanced applications. *Polym Rev* 60: 226–266.
44. Tabatabaee F, Khorasani M, Ebrahimi M, González A, Irusta L, Sardon H (2019) Synthesis and comprehensive study on industrially relevant flame retardant waterborne polyurethanes based on phosphorus chemistry. *Prog Org Coat* 131: 397–406.
45. Gu L, Luo Y (2015) Flame retardancy and thermal decomposition of phosphorus-containing waterborne polyurethanes modified by halogen-free flame retardants. *Indus Engineer Chem Res* 54: 2431–2438.
46. Liang H, Lu Q, Liu M, Ou R, Wang Q, Quirino RL, Luo Y, Zhang C (2020) UV absorption, anticorrosion, and long-term antibacterial performance of vegetable oil based cationic waterborne polyurethanes enabled by amino acids. *Chem Eng J:* 127774.
47. Zhang F, Liu W, Liang L, Wang S, Shi H, Xie Y, Yang M, Pi K (2020) The effect of functional graphene oxide nanoparticles on corrosion resistance of waterborne polyurethane. *Colloid Surf A Physicochem Eng ASP* 591: 124565.
48. Pathan S, Ahmad S (2018) Green and sustainable anticorrosive coating derived from waterborne linseed alkyd using organic-inorganic hybrid cross linker. *Prog Org Coat* 122: 189–198.
49. Ghosal A, Rahman OU, Ahmad S (2015) High-performance soya polyurethane networked silica hybrid nanocomposite coatings. *Indus Engineer Chem Res* 54: 12770–12787.
50. Ghosh T, Karak N (2020) Cashew nut shell liquid terminated self-healable polyurethane as an effective anticorrosive coating with biodegradable attribute. *Prog Org Coat* 139: 105472.
51. Wainwright H (2016) Design, evaluation, and applications of electronic textiles. In *Performance Testing of Textiles*, Elsevier, pp. 193–213.
52. Leng J, Lan X, Liu Y, Du S (2011) Shape-memory polymers and their composites: Stimulus methods and applications. *Prog Mater Sci* 56: 1077–1135.
53. Del Nero D, Joshi-Imre A, Voit W. (2019) Measuring the electric properties of thin film shape memory polymers in simulated physiological conditions. Proc. 2019 IEEE 69th Electronic Components and Technology Conference (ECTC), 2019: 1848–1852.
54. Lendlein A (2010) *Shape-Memory Polymers*, Springer. doi:10.1007/s11095-010-0062-5.
55. Meng Q, Hu J (2009) A review of shape memory polymer composites and blends. *Compos A Appl Sci Manufact* 40: 1661–1672.
56. Wan T, Chen D (2017) Synthesis and properties of self-healing waterborne polyurethanes containing disulfide bonds in the main chain. *J Mater Sci* 52:197–207.
57. Paik IH, Goo NS, Jung YC, Cho JW (2006) Development and application of conducting shape memory polyurethane actuators. *Smart Mater Struct* 15: 1476.
58. Leng J, Lv H, Liu Y, Du S (2007) Electroactivate shape-memory polymer filled with nanocarbon particles and short carbon fibers. *Appl Phys Lett* 91: 144105.
59. Kausar A, Ur Rahman A (2016) Effect of graphene nanoplatelet addition on properties of thermo-responsive shape memory polyurethane-based nanocomposite. *Fuller Nanotub Carb Nanostruct* 24:235–242.

24 Non-Isocyanate-Based Waterborne Polyurethanes

Marcin Włoch and Iga Carayon
Gdańsk University of Technology, Faculty of Chemistry,
Department of Polymers Technology,
Gdańsk, Poland

CONTENTS

24.1 Introduction 407
24.2 Waterborne Non-Isocyanate Polyurethanes 408
24.3 Examples of Waterborne Non-Isocyanate Polyurethanes 409
24.3.1 Solutions and Dispersions 409
24.3.2 Coatings 417
24.3.3 Nanoparticles 424
24.3.4 Hydrogels 425
24.4 Summary 428
References 429

24.1 INTRODUCTION

Non-isocyanate polyurethanes (NIPUs), also called isocyanate-free polyurethanes, are an alternative to conventional polyurethanes synthesized using di- or polyisocyanates, polyols, low-molecular-weight chain extenders, and others. The development of NIPUs resulted from the disadvantages of isocyanates like, for example, toxicity, phosgene-based synthesis, and moisture sensitivity. Per the literature, there are a lot of works that present reviews related to the synthesis and characterization of NIPUs [1–4].

In comparison to conventional polyurethanes, NIPUs can be synthesized in several different ways, for example:

- Polyaddition of five-, six-, seven-, or eight-membered cyclic carbonates (5CCs, 6CCs, 7CCs, or 8CCs) and di- or polyamines [5–8],
- Polycondensation of bischloroformates and diamines [9],
- Polycondensation of dimethyl dicarbamates and diols [10],
- Self-polycondensation of dihydroxyurethanes [11],
- Polycondensation of polyaldehydes and polycarbamates [12].

DOI: 10.1201/9781003173526-24

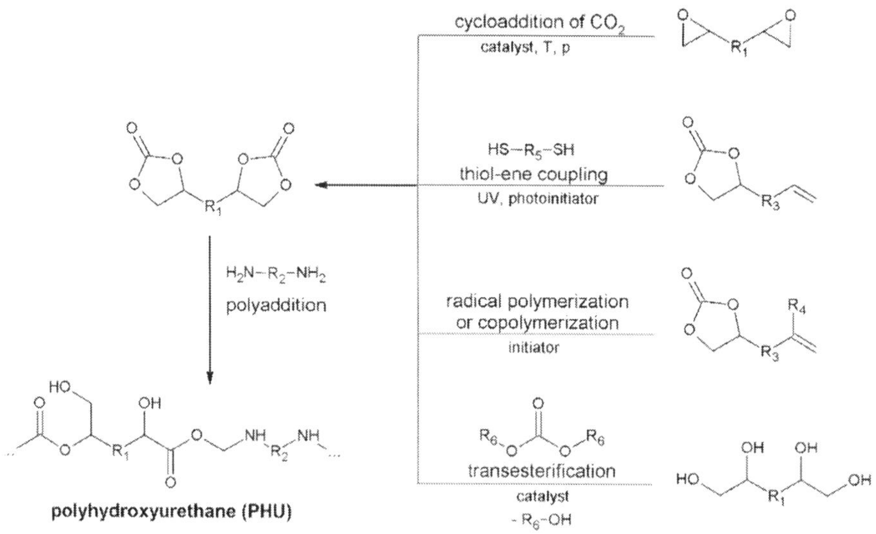

FIGURE 24.1 Synthesis of five-membered cyclic carbonates and non-isocyanate polyurethanes (polyhydroxyurethanes) by their reaction with diamines.

The most investigated NIPUs are obtained by the reaction of five-membered cyclic carbonates and diamines (DAs) or polyamines (Figure 24.1). 5CCs are mostly obtained by the cycloaddition of carbon dioxide into epoxides at elevated temperature and pressure of CO_2 in the presence of a catalyst (e.g. tetrabutylammonium bromide, or TBAB). Other methods involve transesterification of 1,2-diols using dimethyl or diethyl carbonate in the presence of the catalyst (e.g. potassium carbonate or sodium methoxide). 6CCs, 7CCs and 8CCs can be obtained by the cyclization of 1,3-diols, 1,4-diols, and *N*-substituted diethanolamines (1,5 diols) using ethyl chloroformate or triphosgene.

NIPUs, similar to isocyanate-based polyurethanes, can be prepared in the form of elastomers [13,14], foams [15,16], or coatings [17–19]. Also, waterborne non-isocyanate polyurethanes (WNIPUs) can be prepared using the mentioned-above reactive intermediates. The chapter review presents the synthesis and characterization of WNIPUs.

24.2 WATERBORNE NON-ISOCYANATE POLYURETHANES

In accordance with the literature, WNIPUs are mostly obtained by the reaction of five- or six-membered cyclic carbonates and amines or by using hydroxyurethane intermediates synthesized this way. Some works are related to polycondensation methods involving the reaction between carbamates and diols/diamines.

When the molar excess of amine is used for the reaction of CCs and DAs, amine-terminated polyhydroxyurethanes are obtained. The resulting macromolecules can be further functionalized by the reaction of unsaturated or saturated monomers (e.g. succinic acid, methacrylic acid anhydride, or methacrylates bearing CC moieties), hydroxylamine groups presented in the chemical structure of oligomers/polymers subjected to chemical modification. The introduction of carboxylic functionalities is connected with the neutralization of formed groups by using sodium bicarbonate or triethylamine. This kind of modification permits obtaining stable water dispersions of NIPU.

WNIPUs can be obtained by different strategies, including the following:

- Reaction of cyclic carbonates with amines in water and optional further chemical modification of obtained polyhydroxyurethane,
- Reaction of cyclic carbonates with amines in bulk (with or without further functionalization) and dissolving, immersing, or dispersing of resulted polymers in water,
- Dissolving of unsaturated hydroxyurethane oligomers or polymers in unsaturated monomers (e.g. methacrylates or acrylates) followed by their further emulsion polymerization (in the presence of surfactants).

WNIPUs are prepared as water solutions, dispersions (latexes), or hydrogels. Latexes can be further processed into thin films or coatings. Exemplary works will be discussed.

24.3 EXAMPLES OF WATERBORNE NON-ISOCYANATE POLYURETHANES

24.3.1 Solutions and Dispersions

Matsukizono and Endo [20] synthesized water-soluble poly(carbonate-hydroxyurethane)s (PCHUs) composed of hydroxyurethane-carbonate-hydroxyurethane repeating units with well-controlled molecular weights (M_n around 1,500–2,000 g/mol) and polydispersities (2.08–2.76). PCHUs were synthesized with the use of trimethylolpropane (TMP) as a starting molecule and investigated for their hydrolytic behavior in aqueous media varying pH values (from 7.0 to 10.6). The urethane bonds of the PCHUs were detected to be stable, while carbonate and ester bonds may gradually undergo hydrolysis in alkaline aqueous media at room temperature.

The whole process of water-soluble PCHU preparation involves the following:

- Synthesis of bifunctional six-membered cyclic carbonate bridged (TMPC3) by an acyclic carbonate bond by the reaction of TMP with diphenyl carbonate (yield of the reaction is around 37%)—general reaction scheme is presented in Figure 24.2,
- Polyaddition of bifunctional 6CC and 1,3-diaminopropane at −10°C to form amino-terminated polyhydroxyurethane,
- Reaction of terminal amine groups of PHU with 1,4-bis(aminomethyl) benzene to form thio-urea linkages (–NH–C(=S)–NH–) in the structure, which extend the chain length of the PCHUs without crosslinking by hydroxyl side chains,
- Reaction of hydroxyl side chains of the PCHUs with succinic anhydride

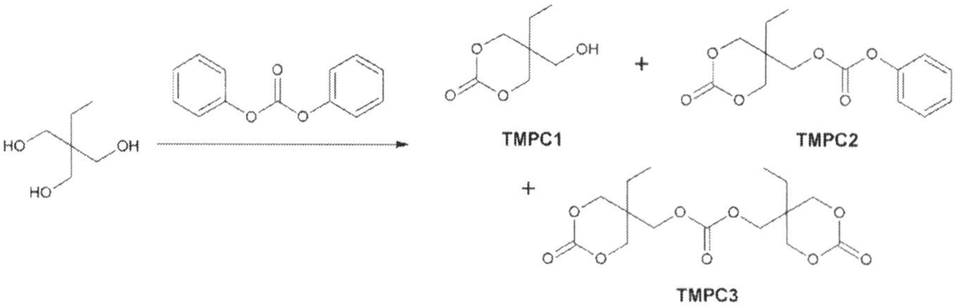

FIGURE 24.2 Chemical structures of mono- and difunctional trimethylolpropane-based cyclic carbonates obtained by the reaction of trimethylolpropane with diphenyl carbonate.

- Reaction of mixture neutralization with sodium bicarbonate—final product is presented in Figure 24.3(a).

In other work, Matsukizono and Endo [21] described the phosgene-free syntheses of water-soluble PHUs bearing ester, carbonate, or ether moieties (Figure 24.3(b)). The process involves the following steps:

- Reaction of bifunctional six-membered cyclic carbonates (coded as M-0, M-1, M-2 in Figure 24.4) using trimethylolpropane or ditrimethylolpropane as substrates,
- Polyaddition of formed 6CCs and 1,3-propanediamine to obtain amino-terminated polyhydroxyurethane,
- Reaction of primary amine and hydroxyl groups in PHUs with succinic anhydride,
- Neutralization of obtained product with $NaHCO_3$.

Prepared water-soluble PHUs were hydrolyzed under basic conditions and their hydrolytic rates depending on the type of linkages in main chains. Carbonate linkages were degraded faster than ester ones, while ether bonds were the slowest hydrolyzed. The authors proved that the hydrolytic stability of prepared water-soluble PHUs can be modulated by using the mixture of bifunctional CCs during the synthesis of NIPUs.

Sardon et al. [22] synthesized polyurethanes in an aqueous solution via an isocyanate- and catalyst-free polymerization process. This efficient and environmentally friendly method led to five PUR systems in a two-step process:

- The first step involves activating 1,6-hexanediol and poly(ethylene glycol) with bis(pentafluorophenyl)carbonate.
- The next step comprises the polycondensation reaction of the 1,6-hexanediol/poly-(ethylene glycol)-derived activated carbonates (at various molar ratios) with JEFFAMINE ED-2003 (M_n = 2000 g/mol).

Obtained polymers were characterized by the following:

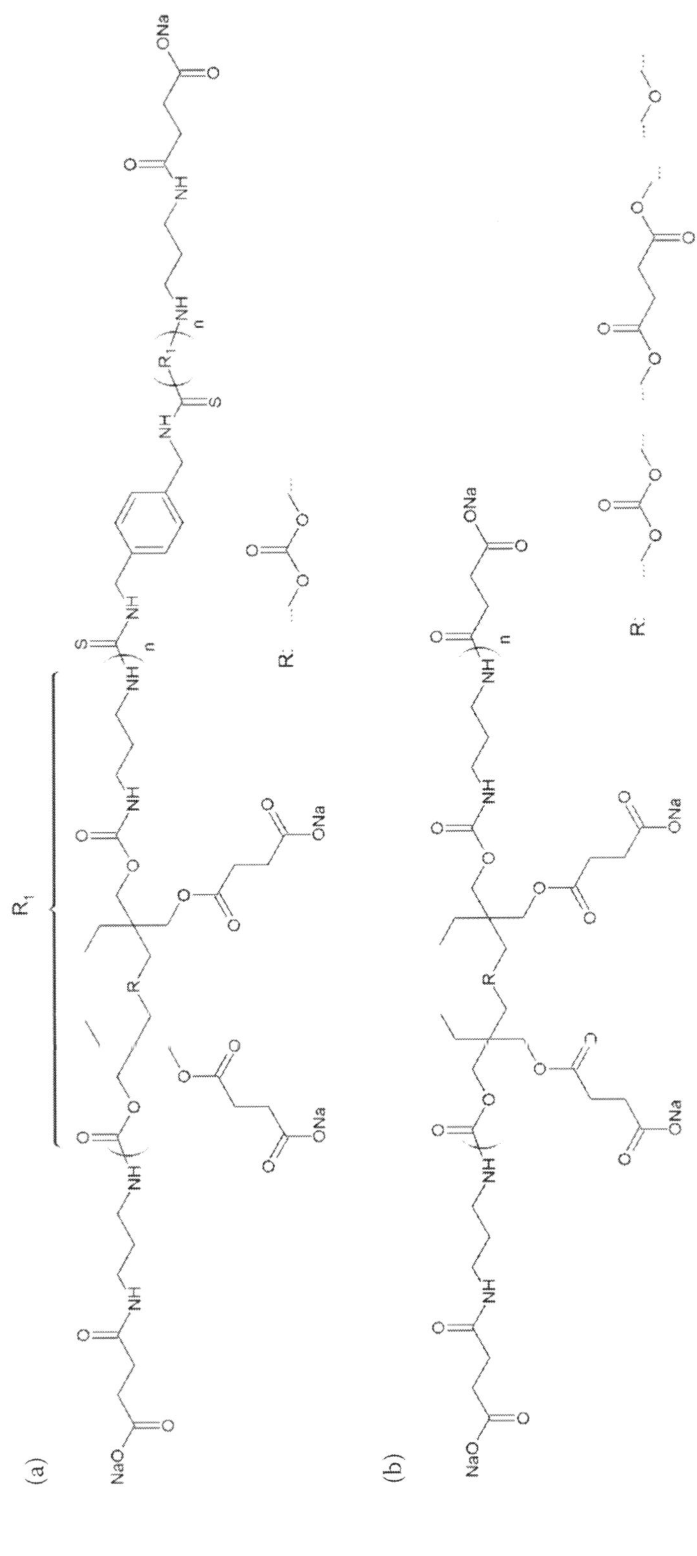

FIGURE 24.3 Chemical structure of water-soluble polyurethanes containing (a) carbonate and thiourea linkages and (b) ester, carbonate, or ether linkages depending on synthesis conditions, prepared by Matsukizono and Endo [20,21].

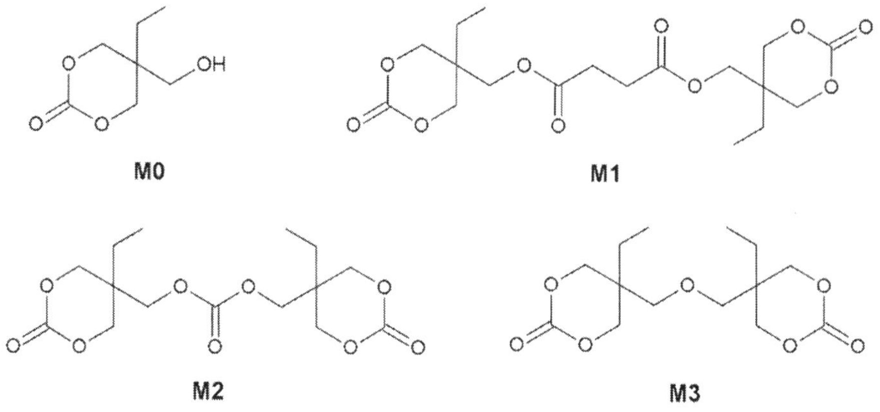

FIGURE 24.4 Mono-(M0) and bifunctional (M1, M2, and M3) six-membered cyclic carbonates prepared by Matsukizono and Endo [21]. In bifunctional intermediates, cyclic carbonate moieties are linked by ester, carbonate, or ether bonds, respectively.

- Melting temperature in the range from 27 to 33°C (determined by DSC), which was linearly dependent on the 1,6-hexanediol/poly(ethylene glycol) ratio, and an increasing trend was observed with greater poly(ethylene glycol) content,
- Degree of crystallinity (similarly to melting temperature) proportional to the poly(ethylene glycol) content,
- Glass transition temperature in the range of −56 to −53°C (determined by DSC),
- High molecular weights (M_n = 15.0–16.4 kDa, PDI = 1.92–1.98) of obtained PURs was easily achieved due to the high reactivity of activated carbonates. It should also be pointed out that molecular weights were not significantly affected by the chain length of activated carbonate.

Bizet et al. [23] prepared waterborne poly(hydroxyurethane)s)/methacrylic hybrids by the following procedure:

- Synthesis of NIPUs by bulk reaction of cyclic carbonates (i.e. sebacic acid-derived biscyclic carbonate) and amine derivative of fatty acids dimers (Priamine 1075, Croda),
- Dissolving of obtained product in butyl methacrylate and miniemulsion polymerization of resulting mixture.

The performed polymerization of butyl methacrylate resulted in the formation of hybrid particles (with the size ranged from 115 to 142 nm) with a core–shell morphology in which the shell is formed by NIPU. The increasing amount of polyhydroxurethane results in decreases in tensile strength and modulus. Microscopic investigations of the films showed that phase segregation occurred, and large domains of polyhydroxyurethanes were formed, which reduced the interaction between the NIPU and the methacrylate.

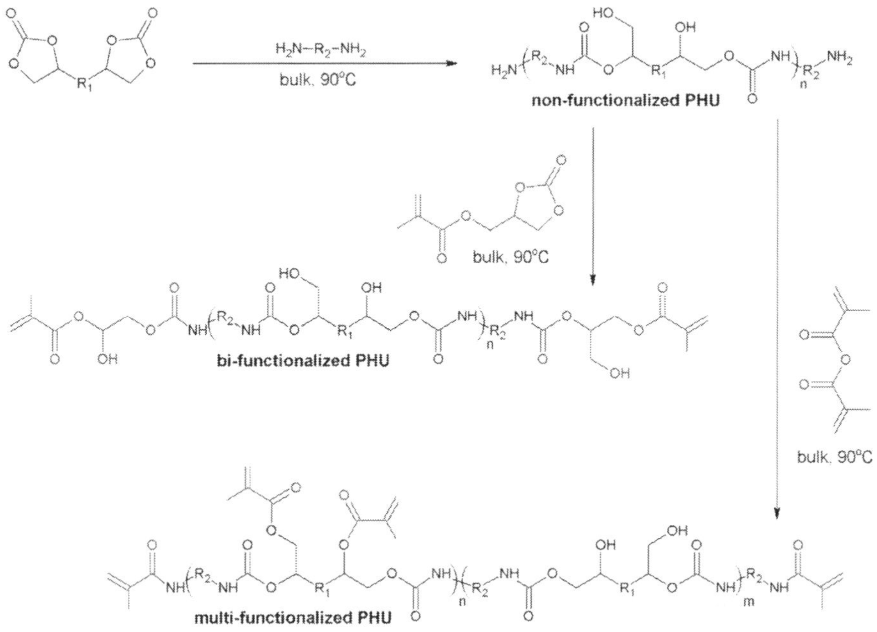

FIGURE 24.5 Non-, bi- and multifunctionalized polyhydroxyurethanes prepared by Bizet et al. [23,24] and used for the preparation of hybrid poly(hydroxyl urethane)*s*-poly(butyl methacrylate) latexes.

In another work, Bizet et al. [24] prepared crosslinked non-isocyanate poly(hydroxyl urethane)s-poly(butyl methacrylate) hybrid latexes. In comparison to previous work, two additional polyhydroxyurethanes (Figure 24.5) were prepared:

- Bifunctionalized PHUs by reacting amine-terminated PHUs with monocarbonated glycidyl methacrylate,
- Multifunctionalized PHU, in which methacrylate groups were grafted along the PHU chain by reacting the hydroxyl groups of the PHUs (formed during the reaction of CCs with amine) with methacrylic anhydride.

The average particle size of non-, bi-, and multifunctionalized PHUs in the prepared latexes were 138, 152, and 106 nm, respectively. Prepared dispersions were used for the preparation of thin films by the casting method, whose properties are presented in Table 24.1. All obtained polymer hybrids were characterized by two glass transition temperatures—the lower one (in the range from −20 (nonfunctionalized PHU) to −12°C (multifunctionalized PHU)) corresponds to PHUs, while the higher one (26–27°C) is related to the methacrylic polymer. The increase of T_g after functionalization of PHU indicates that degree of grafting increased. An increasing amount of PHU resulted in the formation of crosslinks, which were connected with the increase of tensile strength and modulus, while elongation at break decreased.

TABLE 24.1
Tensile properties of cast films formed from non-isocyanate poly(hydroxyl urethane)s-poly(butyl methacrylate) hybrid latexes (temperature of casting 30°C)

Type of PHU	PHU [wt.%]	E [MPa]	σ_b [MPa]	ε_b [%]	References
–	0	336 ± 49	9.9 ± 3.7	5.7 ± 2.5	[23]
Nonfunctionalized PHU (without wetting agent)	10	161 ± 13	4.8 ± 1.7	6.7 ± 5.1	
Nonfunctionalized PHU (without wetting agent)	20	111 ± 11	3.1 ± 0.8	20 ± 12	
Nonfunctionalized PHU (without wetting agent)	30	63 ± 21	2.7 ± 0.7	39 ± 10	
Nonfunctionalized PHU (with wetting agent)	20	52 ± 11	1.5 ± 0.1	41 ± 6	[24]
Bifunctionalized PHU (without wetting agent)	20	73 ± 25	3.3 ± 0.5	48 ± 19	
Bifunctionalized PHU (with wetting agent)	20	46 ± 11	1.3 ± 0.4	76 ± 9	
Multifunctionalized PHU (with wetting agent)	20	86 ± 20	2.2 ± 0.7	5 ± 4	

Rix et al. [25] obtained fatty acid-based poly(hydroxy urethane)s (PHUs) by bulk polymerization of fatty acid-based bis-carbonates (the first one is obtained from fatty acid dimer and glycerol carbonate, BGC-C_{36}, while the second one is obtained from oligo(ricinoleic acid) diacid and glycerol carbonate, BGC-E_{1450}), with fatty acids dimer diamine (Priamine® 1075), which lead to prepare NIPUs. The authors realized the synthesis of polyhydroxyurethanes by bulk and mini-emulsion polymerization using the same monomers. Water-borne NIPU latexes were characterized by particle size in the range from 53 to 116 nm, and during their synthesis different surfactants and hydrophobic additives (hexadecane or olive oil) were tested. It was found that mini-emulsion NIPUs represented lower molar masses (5.8–7.1 kg/mol) in comparison to bulk ones (9.3–17.7 kg/mol). This probably resulted from the hydrolysis of cyclic carbonates. Synthesized PHUs exhibited low glass transition temperature values (in the range from −55°C to −15°C), so application as adhesives or coatings will be possible.

Cobaj et al. [26] prepared a homogeneous and core–shell latexes using urethane methacrylate monomer, i.e. 2-((methylcarbamoyl)oxy)-ethyl methacrylate (MEM), synthesized via non-isocyanate pathway. Synthesis of MEM was a two-step reaction, wherein the first step, hydroxyalkyl carbamate (MA-EC), was obtained by the reaction ethylene carbonate and methylamine, while in the second step, MA-EC, was reacted with methacrylic anhydride in the presence of the catalyst and inhibitor.

Latexes were synthesized by seeded semi-batch monomer-starved emulsion polymerization processes, and seeds were synthesized from a masterbatch seed latex prepared via semi-continuous emulsion polymerization, by the following procedure:

- Preparation of the solution of sodium bicarbonate and anionic surfactant (i.e. sodium dodecyl sulfate (SDS)) in deionized water and placing this solution in the reactor (followed by stirring and heating up to 75°C),
- Preparation of the pre-emulsion by mixing deionized water (80 g), $NaHCO_3$, SDS, methyl methacrylate (MMA), and butyl acrylate (BA).
- Preparation initiator solution containing ammonium persulfate (APS) and deionized water,
- Simultaneous feeding of pre-emulsion and initiator solution for 4 h and polymerization of seed for another 2 h at 75°C.

The same master-batch seed (characterized by particle size 79 nm with a polydispersity of 0.03) was used for all the homogeneous and core–shell latexes, whose preparation procedures are presented in Table 24.2. After that, the synthetic procedure latexes were filtered and neutralized to a pH of 8.0 with ammonium hydroxide aqueous solution [26].

Core–shell architecture was proved by the presence of two glass transition temperatures (one related to shell material and one related to core) and TEM observations. The performance of materials (including thermal properties, viscoelastic, mechanical properties, and morphology) is affected by the concentration of MEM. It was stated that:

- Increasing concentration of urethane methacrylate monomer into homogeneous and core–shell latexes results in increasing viscoelastic and mechanical properties,
- Core–shell latexes exhibited higher mechanical properties (higher storage modulus, Young's modulus, tensile strength, and hardness) in comparison to homogeneous latexes due to the urethane monomer present in the continuous phase, but higher MEM concentration decreased flexibility (connected with decrease of elongation at break and impact resistance) of the final coatings,
- Increasing the concentration of the urethane methacrylate monomer in homogeneous and core–shell latexes resulted in the formation of smoother films and lowered the minimum film formation temperature [26].

A similar investigation was previously reported by Meng et al. [27], who obtained urethane methacrylate monomer (i.e. 2-[(butylcarbamoyl)oxy]ethyl methacrylate, BEM), which was further used with methyl methacrylate (MMA) and butyl acrylate (BA) to prepare NIPU latexes. Synthesis of BEM was a two-step reaction, wherein the first step, hydroxyalkylcarbamate (BA-EC), was obtained by the reaction ethylene carbonate and butylamine, while in the second step MA-EC was reacted with methacrylic anhydride in the presence of the catalyst and inhibitor.

TABLE 24.2
Preparation procedures and properties for homogeneous and core–shell latexes prepared by Cobaj et al. [26]

Latex Type	Homogeneous Latexes	Core–shell Latexes
Process type	One-step semi-batch monomer-starved process	Two-step semi-continuous emulsion polymerization process
Procedure of latex preparation	• Placing of seed in the reactor and heating to 75°C • Separate, simultaneous, and continuous adding of pre-emulsion solution ($NaHCO_3$, surfactants and methyl methacrylate (MMA), butyl acrylate (BA) and MEM in deionized water), and the initiator solution (APS dissolved in deionized water) for 240 min into the reactor.	• Placing of seed in the reactor and heating to 75°C • Separate, simultaneous, and continuous adding of core pre-emulsion solution of (MMA and BA) and initiator solution for 4 h into the reactor • Separate, simultaneous, and continuous adding of shell pre-emulsion solution of (MMA, BA, and MEM) and the initiator solution (APS dissolved in deionized water) for 4 h into the reactor
Characterization of particles containing MEM	• Particle size: 207–217 nm • M_n: 100–198.1 kg/mol • T_g: 19.6–22°C	• Particle size, DLS: 209–215 nm • M_n: 106.2–152 kg/mol • Shell T_g: 7.8–10.9°C • Core T_g: 33.7–36.2°C

Latexes were synthesized by seeded emulsion polymerization processes, and seeds (characterized by particle size 134 nm with a polydispersity of 0.02) were prepared from single seed latex prepared batchwise, by the following procedure:

- Preparation of the solution of sodium bicarbonate and surfactant in water and placing this solution in the reactor (followed by stirring and heating up to 75°C),
- Preparation of the pre-emulsion by mixing water, $NaHCO_3$, surfactant, MMA and BA, and adding to the reactor
- Preparation initiator solution containing ammonium persulfate (APS) and water, and adding to the reactor,
- Polymerization of seed for 1.5 h at 75°C.

Authors performed investigation connected with the preparation and characterization of latexes prepared by:

- Bath (BFL) and semi-batch (SBFL) emulsion polymerization with a constant concentration of BEM in pre-emulsion (10 wt.%) (Table 24.3),

TABLE 24.3
Preparation procedures and properties of latexes prepared by bath and semi-batch emulsion polymerization by Meng et al. [27]

Latex Code	BFL	SBFL
Process type	Batchwise polymerization	Semi-batch polymerization
BEM [wt.%]	10	10
Procedure	• Charging of seed to reactor and heating to 75°C • Adding a pre-emulsion (BA, MMA, and BEM) once with initiator (APS) • Completion of the reaction (after feeding) for 4 h at 75°C	• Charging of seed to reactor and heating to 75°C • Adding a pre-emulsion (BA, MMA, and BEM) along with initiator (APS) continuously for 4 h • Completion of the reaction (after feeding) for 4 h at 75°C
Particle size [nm]	Around 230	Around 260
T_g [°C]	4,3	4,6

- Semi-batch emulsion polymerization containing 5, 10, or 20 wt.% of BEM in pre-emulsion (FLH), core pre-emulsion (FLC-S), and shell pre-emulsion (C-FLS) (Table 24.4).

It was found that BFL latex showed a less homogeneous composition of polymer chains (due to compositional drift that occurred during the batch polymerization) connected with higher tensile modulus and strength with lower elongation at break than that of the SBFL. The compositional drift that occurred during the batch polymerization was connected with the formation of the methacrylates-rich polymers at early stages and the later formation of acrylate-rich polymers.

Homogeneous (FLH) and core–shell structures (C-FLS and FLC-S) were obtained, which was confirmed by several techniques (e.g. DSC or DMTA). The incorporation of urethane functionality improved the tensile properties of the place, where the urethane monomer is located. The observed improvement resulted from the hydrogen bonding that occurs between urethane moieties. The C-FLS system is characterized by the best properties (in comparison to homogenous and core–shell systems) due to the higher concentration of urethane present in the continuous phase.

The concentration and location of urethane moieties affected the properties of latexes and final thin films (Figure 24.6). The increase of urethane moieties concentration results in the formation of a continuous urethane phase (like in C-FLS series), which increased coating tensile strength.

24.3.2 Coatings

Zhang, Wang, and Zhou [28] synthesized WNIPU epoxy hybrid coatings using the following:

TABLE 24.4
Preparation procedures and properties of homogenous and core-shell latexes prepared by Meng et al. [27]

Latex Code	FLH	FLC-S	C-FLS
Latex type	Homogenous	Core-shell	Core-shell
Process type	Single stage semi-batch emulsion polymerization	Two-stage semi-batch emulsion polymerization	
BEM [wt.%]	0, 5, 10, or 20	5, 10, or 20 (in core)	5,10, or 20 (in shell)
Procedure	• Charging of seed to reactor and heating to 75°C • Adding a pre-emulsion (BA, MMA, and BEM) along with initiator (APS) continuously for 4 h • completion of the reaction (after feeding) for 4 h at 75°C	*Procedure for core preparation:* • Charging of seed to reactor and heating to 75°C • Adding a pre-emulsion (BA, MMA, and BEM for FLC-S or BA and MMA for C-FLS) along with initiator (APS) continuously for 4 h • Completion of the reaction (after feeding) for 4 h at 75°C *Procedure for shell preparation:* • Adding of the core product to the reactor and heating to 75°C • Adding a shell pre-emulsion (BA and MMA for FLC-S or BA, MMA, and BEM for C-FLS) along with initiator (APS) continuously for 4 h • Completion of the (after feeding) reaction for 4 h at 75°C	
Particle size [nm] *	394, 391, or 390	384, 383, or 374	385, 380, or 390
T_g [°C] *	−1, 2, or 4	–	–
T_g CORE [°C] *	–	21, 21, or 26	24, 24, or 30
T_g SHELL [°C] *	–	−12, −7, or −7	−14, −12, or −12

Notes
* Values presented with respect to the concentration of BEM.

- Waterborne amine-terminated NIPU prepared using diglycerol dicarbonate, 3,3′-diamino-*N*-methyldipropylamine, and amine derivatives of fatty acids,
- Waterborne epoxy chain extender synthesized using diethanolamine and trimethylolpropane triglycidyl ether.

Coatings were prepared using bio-based substrates (i.e. diglycerol dicyclocarbonate and amine derivative of fatty acids dimers), whose amount ranged from 27.4 to

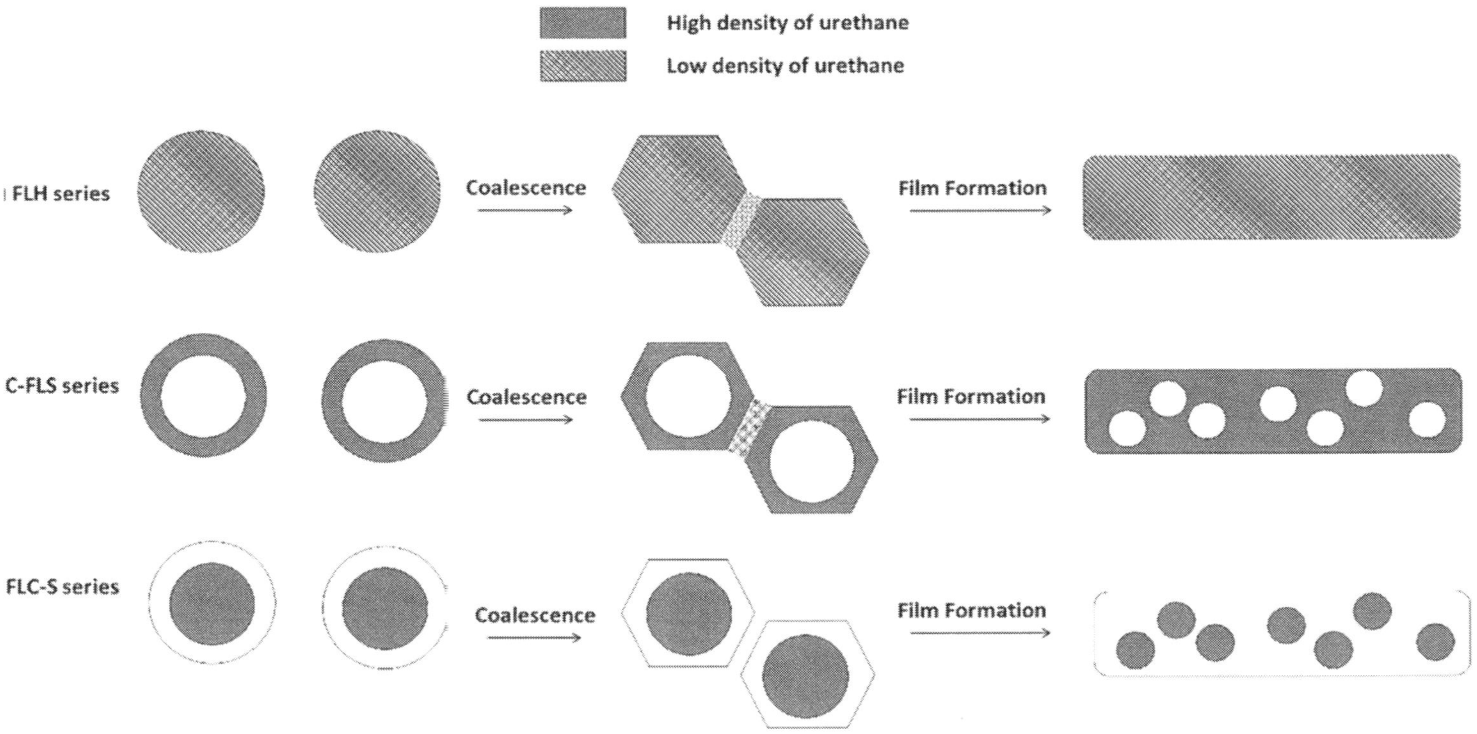

FIGURE 24.6 A possible mechanism during the film formation process for (a) FLH series, (b) C-FLS series, and (c) FLC-S series latexes. Adapted with permission from reference [27]. Copyright (2017) Elsevier.

68.8 wt.%. Mechanical properties and thermal stability are mainly affected by the amount of amine derivative of fatty acids dimers (used for the preparation of final material), which act as soft segments and possess higher thermal stability than urethane moieties.

Wu et al. [29] prepared WNIPU coating (Figure 24.7) by mixing the following:

- Aqueous cyclic carbonate dispersions (with particle sizes ranging from 63 to 174 nm), where cyclic carbonate intermediate containing carboxyl groups were prepared by the reaction of sorbitol-based cyclic carbonate with selected anhydrides (including succinic anhydride, *o*-phthalic anhydride, or maleic anhydride),
- Diamine (1,2-ethanediamine (EDA), 1,4-butanediamine (BDA), 1,6-hexanediamine (HDA), 1,8-octanediamine (ODA) or isophorone diamine (IPDA)) or mixture of IPDA and HDA.

Coatings prepared using a mixture of diamines are characterized by excellent adhesion, gloss, mechanical properties (flexibility and impact resistance), and thermal stability when the molar ratio of IPDA to HMDA is in the range from 1:1 to 1:3.

Ma et al. [30] prepared WNIPU/epoxy coatings (Figure 24.8), characterized by good mechanical and thermal properties (Table 24.5), using the following:

- Carboxyl polyhydroxyurethane synthesized by the reaction diphenolic acid-based biscyclocarbonate and 1,2-ethanediamine, followed by reaction of formed hydroxyl groups with succinic anhydride—prepared NIPU were dispersed in deionized water with different amounts of sodium bicarbonate (the mass ratio of $NaHCO_3$/NIPU was equal 1:1, 1:5, 1:7, or 1:10) under ultrasonic assist,
- Commercial bisphenol A-based epoxy resin emulsion (used as curing agent) under ultrasonic treatment (Figure 24.8).

The degree of ionization of the carboxyl groups, presented in the formed NIPU, through the salification between –COOH and $NaHCO_3$, affects the size of emulsion particles (in the formed products of particles with size range from 78 to 321 nm, Figure 24.9). The increasing amount of $NaHCO_3$ resulted in decreasing thermal stability and glass transition temperature, as a result of reduced crosslinking density.

Bossion et al. [31] obtained self-healable WNIPU-based supramolecular networks by ionic interactions. Water-dispersions of polyhydroxyurethanes (with particle size around 200 nm) were prepared by the reaction of a mixture five-(diglycerol dicarbonate) and eight-membered cyclic carbonates (70 mol% of 5CC and 30 mol% of 8CC), and amine-terminated poly(dimethylsiloxane) by phase inversion polymerization were generated by mixing of carboxylic acids (adipic acid or citric acid) with the NIPU dispersion, in the absence of any organic solvents, permitting supramolecular assemblies based on ionic interactions (due to the presence of tertiary amino groups in the structure of macromolecules). The process of

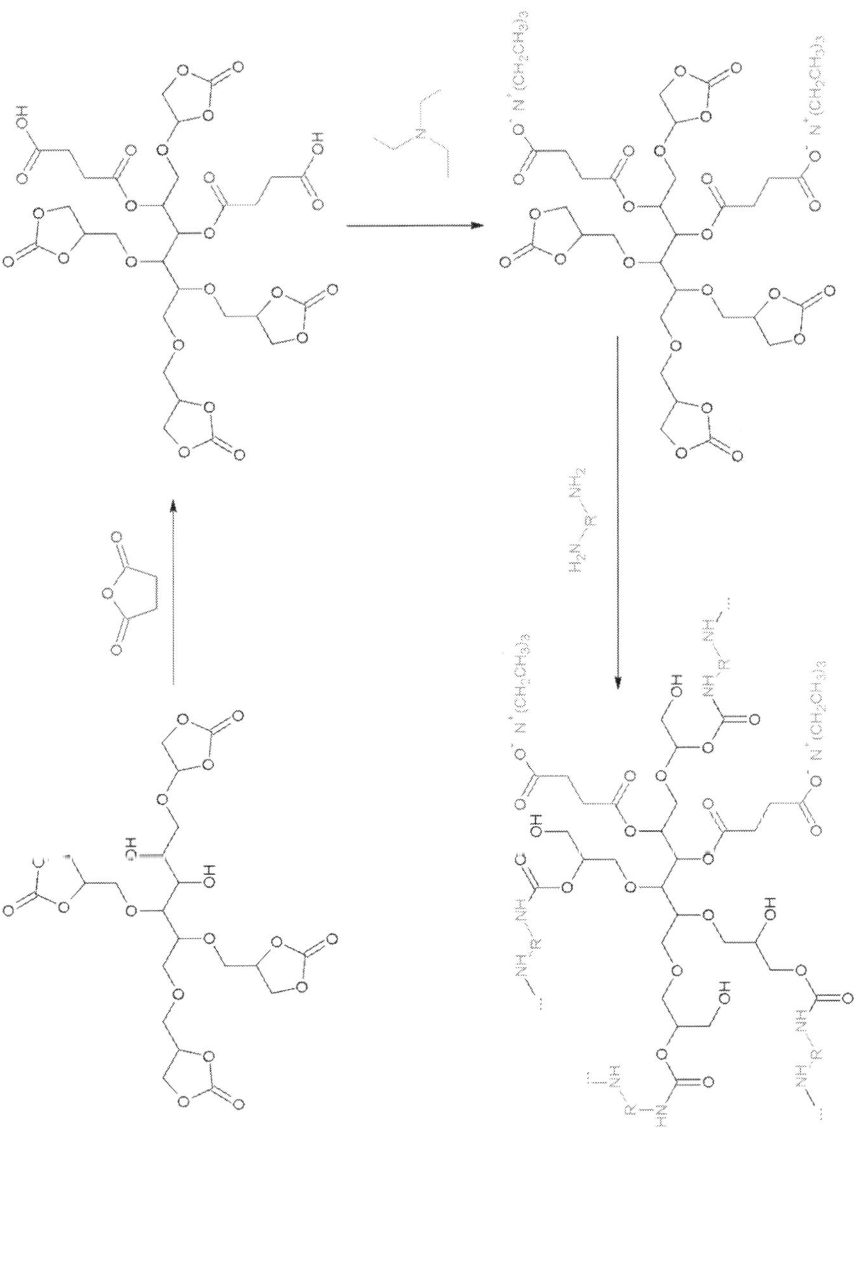

FIGURE 24.7 Preparation of waterborne non-isocyanate polyurethane coating (including chemical modification of polyhydroxurethane intermediate by reaction with carboxylic anhydride) using sorbitol-based cyclic carbonate.

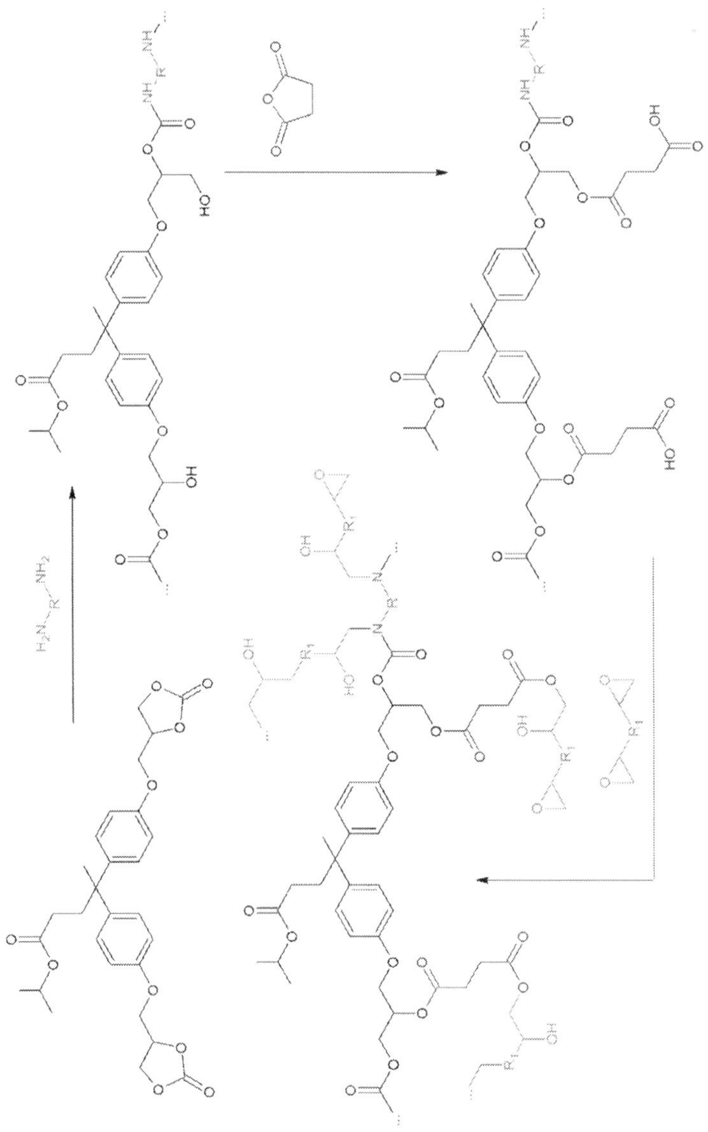

FIGURE 24.8 Synthesis of waterborne non-isocyanate polyurethane/epoxy coatings performed by Ma et al. [30].

TABLE 24.5
Selected works related to waterborne non-isocyanate polyurethane coatings and their properties

Application of Waterborne NIPU/Mechanism of NIPU Formation and CC synthesis	Selected Mechanical and Thermal Properties	References
Waterborne non-isocyanate polyurethane epoxy hybrid coatings • NIPU formed by polyaddition of CCs and DAs • 5CCs obtained by the reaction of polyol with dimethyl carbonate	• E = 0.4–9.3 MPa • σ_b = 0.5–2.4 MPa • ε_b = 26–134% • T_g = 21–26°C	[28]
Waterborne non-isocyanate polyurethane coatings: • NIPU formed by polyaddition of 5CCs and DAs • 5CCs obtained by the cycloaddition of CO_2 into epoxide	• T_g = 68.9–91.1°C • $T_{d5\%}$ = 259–269°C	[29]
Waterborne non-isocyanate polyurethane/epoxy coatings: • NIPU formed by polyaddition of 5CCs and DAs • 5CCs obtained by the cycloaddition of CO_2 into epoxide	• T_g = 35–63°C • PH = 3B, 2 H or 3 H	[30]

E—Young modulus, σ_b—tensile strength, ε_b—elongation at break, T_g—glass transition temperature, $T_{d5\%}$—temperature at which 5 wt.% mass loss occurs, PH—pencile hardness.

supramolecular assembling/disassembling (in designed by authors materials) is dynamic and reversible, so obtained films are characterized by self-healing properties (Figure 24.10).

Ma et al. [32] described stable WNIPU dispersions obtained by the polycondensation of dicarbamate (i.e. isophorone dimethylcarbamate), selected diamines (i.e. 4,7,10-trioxa-1,13-tridecanediamine, poly(propylene glycol) bis(2-aminopropyl ether) with an average M_n of 400 Da, poly(tetrahydrofuran) bis(3-aminopropyl) with an average M_n of 1100 Da), and 3,3′-diamino-*N*-methyldipropylamine as internal dispersing agent. The resulting coatings were characterized by good impact and solvent resistance. In another work of Ma et al. [33], *t*-butyl-oxycarbonylated diamines were investigated as dicarbamate monomers for the synthesis of non-isocyanate polyureas or polyurethane-ureas by polycondensation method (diamine/dicarbamate polymerizations). The high molecular weight polymers were prepared from stoichiometric polymerizations of diamines or diols with *N*-*N*′-di-*t*-butyloxycarbonyl isophorone diamine using potassium *t*-butoxide as a catalyst. Stable dispersions are obtained from NIPUs with 3,3′-diamino-*N*-methyldipropylamine as internal dispersing agent. The resulting polyurea-based coatings were characterized by higher mechanical properties and solvent resistance compared to the polyurethane-urea coatings synthesized from diols and mentioned earlier carbamate and internal dispersing agent.

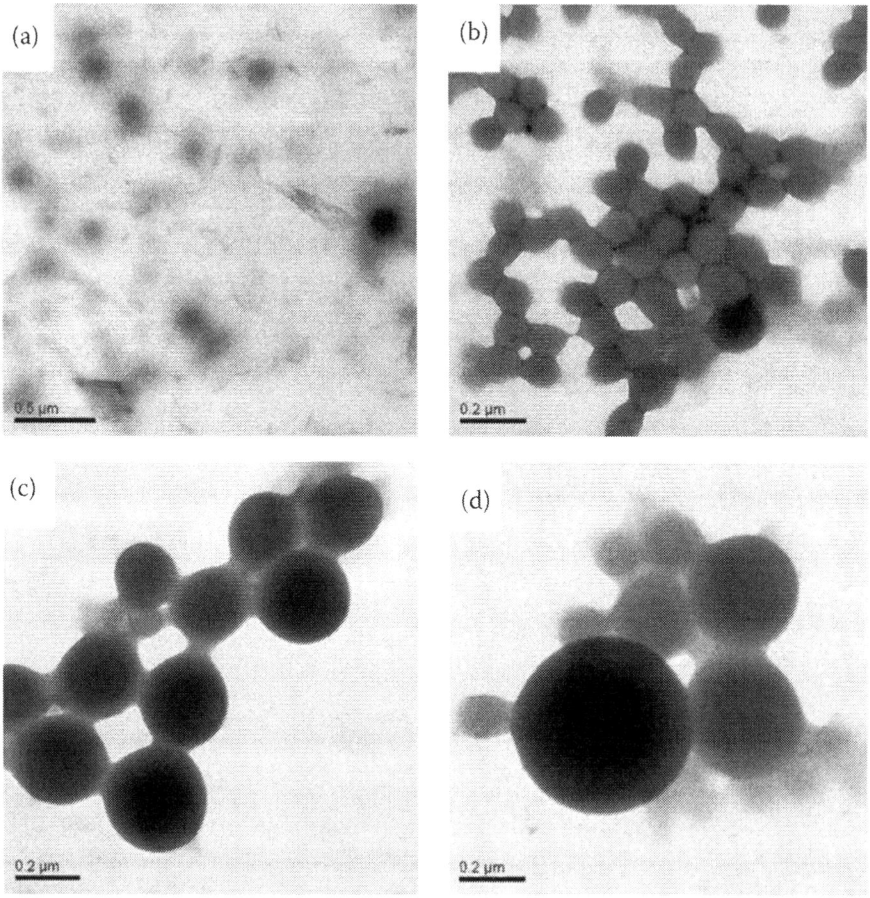

FIGURE 24.9 TEM photographs of NIPU water dispersions. The overall amount of $NaHCO_3$ in solute is in the order of PHU1 (a) > PHU2 (b) > PHU3 (c) > PHU4 (d). Reprinted with permission from reference [30]. Copyright (2017) American Chemical Society.

24.3.3 Nanoparticles

Bossion et al. [34] reported the synthesis of soft polyurethane nanoparticles useful as drug delivery nanocarriers. Applied interfacial polymerization, which employed a non-isocyanate polymerization route (and minimizing side reactions with water), provided stable isocyanate-free nanocarriers useful for doxorubicin encapsulation through ionic interactions. An experiment performed by Bossion et al. is interesting because preparing well-defined PUR nanoparticles required not only miniemulsion polymerization techniques but as well a non-trivial control of the polymerization conditions (the inherent incompatibility of isocyanate-containing monomers and water). Activated pentafluorophenyl dicarbonates were polymerized with diamines and/or triamines by interfacial polymerization in the presence of an anionic emulsifier, which afforded NIPU nanoparticles with sizes in the range of 200–300 nm.

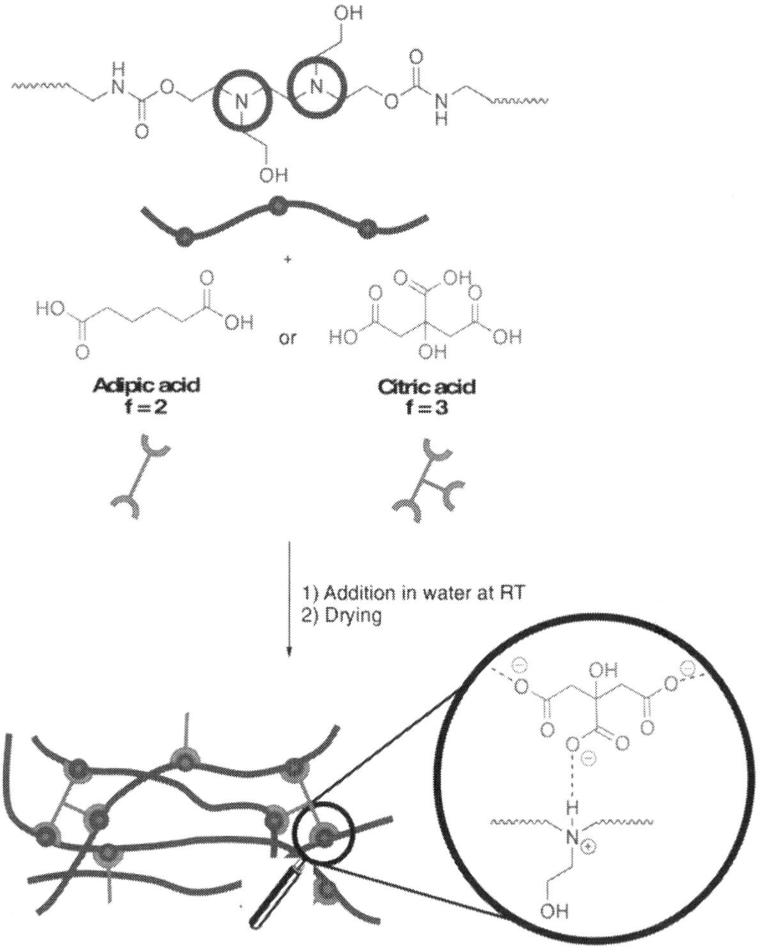

FIGURE 24.10 Preparation of supramolecular ionic structures based on non-isocyanate polyurethane and carboxylic acids. Reprinted with permission from reference [31]. Copyright (2019) Royal Society of Chemistry.

Notably, 5 wt.% of emulsifier was required in combination with a trifunctional amine to achieve stable PU dispersions and avoid particle aggregation. The versatility of this polymerization process allows for the incorporation of functional groups into the PU nanoparticles, such as carboxylic acids, which can encapsulate, for example, a chemotherapeutic doxorubicin drug through ionic interactions. Altogether, this waterborne synthetic method for functionalized NIPU soft nanoparticles holds great promise for the preparation of drug delivery nanoparticles.

24.3.4 Hydrogels

The first NIPU hydrogels were obtained and characterized by Gennen et al. [35]. Polyhydroxyurethanes were obtained in bulk by the reaction of poly(ethylene

TABLE 24.6
Swelling (equilibrium water content, EWC, and equilibrium water absorption, or EWA) and mechanical properties (tensile strength, σ_b, and elongation at break, ε_b) of polyhydroxyurethane hydrogels prepared by Gennen et al. [35]

DA	Monomers Molar Ratio CC:DA:TAEA	EWC [%]	EWA [%]	σ_b [kPa]	ε_b [%]
mXDA	1:0.5:0.33	75.7 ± 0.3	312 ± 5	170.0 ± 44.7	43.6 ± 5.1
mXDA	1:0.6:0.26	75.8 ± 0.9	314 ± 16	217.7 ± 42.5	56.9 ± 3.1
mXDA	1:0.7:0.2	83.6 ± 1.4	505 ± 54	43.4 ± 5.4	61.0 ± 3.4
EDDA	1:0.7:0.2	90.6 ± 0.1	967 ± 9	22.4 ± 8.4	55.6 ± 2.8
ODA	1:0.7:0.2	68.2 ± 0.2	214 ± 2	830 ± 140	74.4 ± 3.3

glycol) dicyclic carbonate with tris(2-aminoethyl)amine (TAEA) and selected diamine (*m*-xylylenediamine (mXDA), 1,8-diaminooctane (ODA) or 2,2′-(ethylenedioxy)diethylamine (EDDA)). The formulation was mixed at 60°C for 10 min and then cured at the same temperature for 24 h. Obtained PHUs (after removing unreacted monomers) were immersed in water for 168 h to prepare hydrogels. Mechanical and swelling properties of prepared PHU hydrogels are affected by the crosslinking ratio, which depends on the amount of TAEA in the monomers formulation (Table 24.6). It was also proved that hydrogels can be successfully reinforced by the incorporation of nanoclay Cloisite 30B (in the amount from 1 to 15 wt.%) into the monomer formulation before polymerization. An increasing amount of nanofiller resulted in increasing tensile properties and decreasing equilibrium water content and absorption.

Bourguignon et al. [36] described WNIPU hydrogels prepared in water at room temperature from five-membered cyclic carbonates (i.e. poly(ethylene glycol) dicyclic carbonate formed by cycloaddition of carbon dioxide into polyethylene glycol diglycidylether) and polyamine (i.e. polyethyleneimine). At the proper pH of the solution (in the range from 10.5 to 11.5), short gel times can be obtained (15–20 min). The properties of final hydrogels are affected by the molar ratio of primary amino groups to five-membered cyclic carbonate moieties, pH (Figure 24.11). The highest crosslink degrees (connected with the highest mechanical properties) were obtained for $[NH_2]/[5CC]$ molar ratios between 0.65 and 0.8 and the initial pH of the formulation at 10.5. The properties of obtained WNIPU hydrogels can be improved by the addition of inorganic filler (clays) or a natural water-soluble polymer (gelatin).

In other work, Bourguignon et al. [37] proposed a different way to synthesize WNIPU hydrogels (Figure 24.12), characterized by adaptable functionality and behavior, including the reaction of substrates in water at room temperature (but at different pH, i.e. 10.5, 11, or 12) without any catalyst. Formulations consisted of the following:

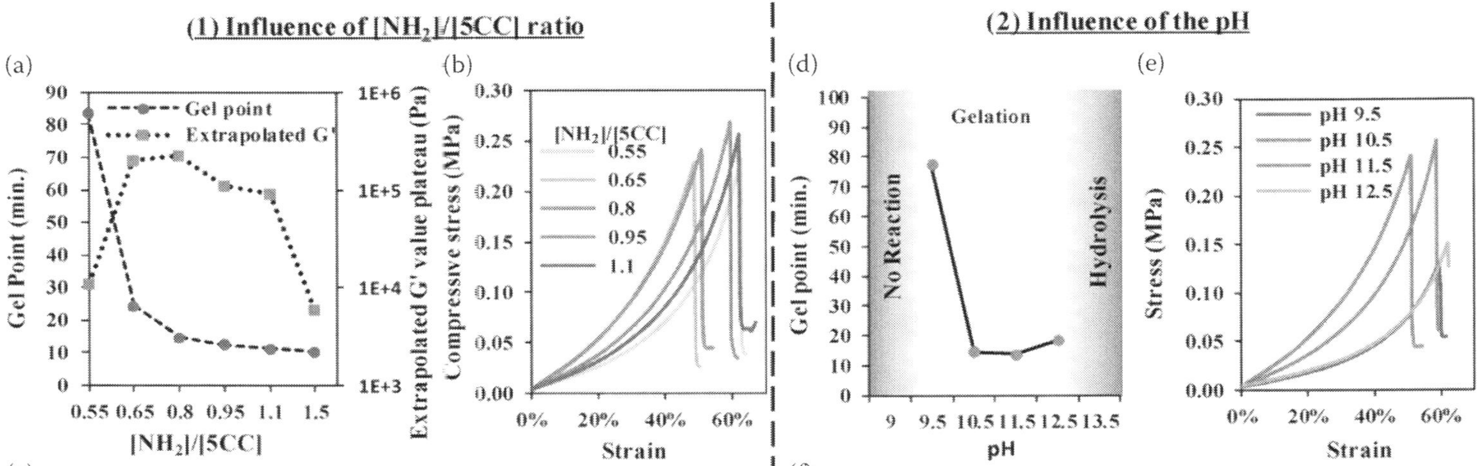

(c)

$[NH_2]/[5CC]$	Young modulus (kPa)	Stress at break (kPa)	Strain at break (%)
0.55	101 ± 14	240 ± 17	62 ± 2
0.65	233 ± 11	176 ± 5	46 ± 5
0.8	233 ± 6	221 ± 16	51 ± 1
0.95	173 ± 16	275 ± 32	58 ± 3
1.1	136 ± 5	228 ± 42	61 ± 3
1.5	-	-	-

(f)

pH	Young modulus (kPa)	Stress at break (kPa)	Strain at break (%)
9.5	68 ± 6	139 ± 24	60 ± 2
10.5	233 ± 6	222 ± 16	51 ± 1
11.5	149 ± 11	233 ± 47	57 ± 2
12.5	73 ± 3	170 ± 20	62 ± 2

FIGURE 24.11 Influence of (1) $[NH_2]/[5CC]$ ratio and (2) pH on the non-isocyanate polyurethane hydrogel formation, and on its mechanical properties in wet state with 30 wt.% of solid content: (A) gel time and storage modulus values; (B and E) compressive curves; (C and F) mechanical properties and (D) gel time. Reprinted with permission from reference [36]. Copyright (2019) American Chemical Society.

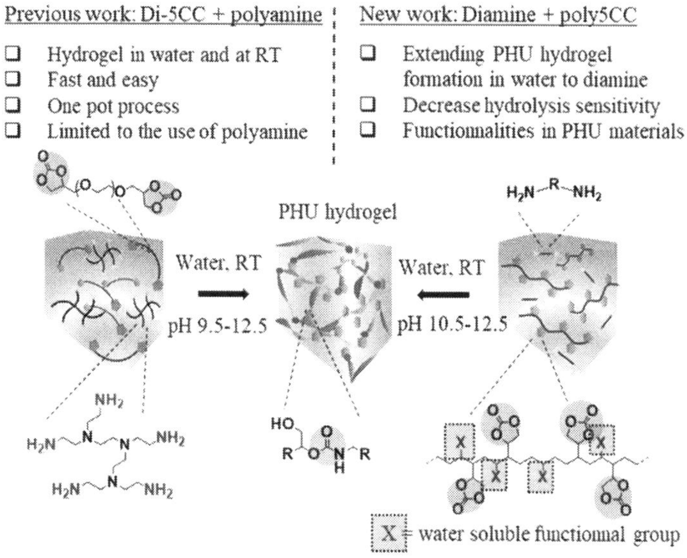

FIGURE 24.12 Scheme of the strategies leading to the formation of PHU hydrogels in water: hydrosoluble dicyclic carbonate crosslinked by polyamine (noted as "previous work") and diamine crosslinked by water-soluble polycyclic carbonate (noted as "new work"). Reprinted with permission from reference [37]. Copyright (2021) John Wiley and Sons.

- Water-soluble polymers bearing cyclic carbonates (synthesized by radical polymerization of glycerol carbonate methacrylate with water-soluble neutral (i.e. poly(ethyleneglycol) methyl ether methacrylate, di(ethyleneglycol) methyl ether methacrylate, methacrylamide or acrylamide), cationic (i.e. [2-(methacryloyloxy)ethyl]trimethylammonium chloride) or anionic (i.e. sodium styrene sulfonate and sodium 2-acrylamido 2-propane sulfonate) comonomers using azobisisobutyronitrile as initiator)
- Diamines (1,6-hexanediamine, *m*-xylylenediamine, 2,2′-ethylenedioxy bis (ethylamine)) or polyamines (triethylenetetramine).

It was also observed that gelation of anionically charged polymer during the reaction with amine, in the presence of calcium chloride, resulted in the in-situ formation of calcium carbonate particles (as a result of a reaction of $CaCl_2$ with carbonate anions formed by solubilized in water CO_2 released during partial hydrolysis of the cyclic carbonate at pH equal 12). The formation of $CaCO_3$ decreased the gel time and strongly improved the storage modulus of the hydrogel (so in-situ formed particles act as reinforcing agents).

24.4 SUMMARY

In this chapter, we focused on the synthesis and characterization of WNIPUs. They are obtained by NIPU transformation into solids, solutions, or dispersions (latexes). To reach sustainable NIPUs, their syntheses are carried out with the use of five- or six-membered cyclic carbonates and di- or polyamines. To follow rules of green chemistry NIPUs were synthesized by phosgene-free routes and sometimes in aqueous media (especially when they are prepared as WNIPUs). Application of bio-based substrates (i.e. glycerol, sorbitol, or fatty acids) in the NIPU's synthesis cannot be omitted. The percentage of bio-based raw materials in NIPU systems can be higher than 50 wt.%. Careful adjustment of raw materials, like in the case of isocyanate-based polyurethanes, can lead to valuable WNIPUs' final products being useful for fabrication of coatings, hydrogels, nanoparticles, etc. Different functional additives were used to improve the stability of WNIPU dispersions and obtained particles, which is a crucial requirement in WNIPU transfer in valuable products. Dispersions with great success may find an application for thin films or coatings formation. Coatings are the largest group of proposed WNIPU applications. They may be characterized by excellent adhesion, gloss, mechanical properties (flexibility and impact resistance), and thermal stability. Another field of WNIPU application is hydrogels, whose mechanical and swelling properties are affected by the cross-linking ratio, which depends on the type and amount of monomers used in the formulation as well as the pH of the medium.

To conclude, WNIPUs are a group of highly specialized materials, whose design will be proposed regarding established needs. WNIPUs can be obtained from bio-based materials through nontoxic synthesis paths, by eliminating toxic compounds like isocyanates, which cause that they will be more eco-friendly than currently known solutions. Thus, that contributes to the sustainable synthesis of polymer materials, which will be targeted in the science and industry of materials due to global requirements in this field.

REFERENCES

1. Rokicki, G.; Parzuchowski, P. G.; Mazurek, M. Non-Isocyanate Polyurethanes: Synthesis, Properties, and Applications. *Polym. Adv. Technol.*, 2015, *26* (7), 707–761. doi:10.1002/pat.3522.
2. Kreye, O.; Mutlu, H.; Meier, M. A. R. Sustainable Routes to Polyurethane Precursors. *Green Chem.*, 2013, *15* (6), 1431–1455. doi:10.1039/c3gc40440d.
3. Ghasemlou, M.; Daver, F.; Ivanova, E. P.; Adhikari, B. Bio-Based Routes to Synthesize Cyclic Carbonates and Polyamines Precursors of Non-Isocyanate Polyurethanes: A Review. *Eur. Polym. J.*, 2019, *118* (June), 668–684. doi:10.1016/j.eurpolymj.2019.06.032.
4. Włoch, M.; Błażek, K. Isocyanate-Free Polyurethanes. In*Polyurethane Chemistry: Renewable Polyols and Isocyanates*; ACS Symposium Series; American Chemical Society, 2021; Vol. 1380, pp 107–166 SE – 5. doi:10.1021/bk-2021-1380.ch005.
5. Lambeth, R. H.; Mathew, S. M.; Baranoski, M. H.; Housman, K. J.; Tran, B.; Oyler, J. M. Nonisocyanate Polyurethanes from Six-Membered Cyclic Carbonates: Catalysis and Side Reactions. *J. Appl. Polym. Sci.*, 2017, *134* (45), 44941. doi:10.1002/app.44941.
6. Yuen, A.; Bossion, A.; Gómez-Bengoa, E.; Ruipérez, F.; Isik, M.; Hedrick, J. L.; Mecerreyes, D.; Yang, Y. Y.; Sardon, H. Room Temperature Synthesis of Non-Isocyanate Polyurethanes (NIPUs) Using Highly Reactive N-Substituted 8-Membered Cyclic Carbonates. *Polym. Chem.*, 2016, *7* (11), 2105–2111. doi:10.1039/C6PY00264A.

7. Tomita, H.; Sanda, F.; Endo, T. Polyaddition of Bis(Seven-Membered Cyclic Carbonate) with Diamines: A Novel and Efficient Synthetic Method for Polyhydroxyurethanes. *J. Polym. Sci. Part A Polym. Chem.*, 2001, *39* (23), 4091–4100. doi:10.1002/pola.10058.
8. Bähr, M.; Bitto, A.; Mülhaupt, R. Cyclic Limonene Dicarbonate as a New Monomer for Non-Isocyanate Oligo- and Polyurethanes (NIPU) Based upon Terpenes. *Green Chem.*, 2012, *14* (5), 1447–1454. doi:10.1039/c2gc35099h.
9. Begines, B.; Zamora, F.; de Paz, M. V.; Hakkou, K.; Galbis, J. A. Polyurethanes Derived from Carbohydrates and Cystine-Based Monomers. *J. Appl. Polym. Sci.*, 2015, *132* (3). doi:10.1002/app.41304.
10. Rokicki, G.; Piotrowska, A. A New Route to Polyurethanes from Ethylene Carbonate, Diamines and Diols. *Polymer (Guildf).*, 2002, *43* (10), 2927–2935. doi:10.1016/S0032-3861(02)00071-X.
11. Ochiai, B.; Utsuno, T. Non-Isocyanate Synthesis and Application of Telechelic Polyurethanes via Polycondensation of Diurethanes Obtained from Ethylene Carbonate and Diamines. *J. Polym. Sci. Part A Polym. Chem.*, 2013, *51* (3), 525–533. doi:10.1002/pola.26418.
12. Silbert, S. D.; Serum, E. M.; LaScala, J.; Sibi, M. P.; Webster, D. C. Biobased, Nonisocyanate, 2K Polyurethane Coatings Produced from Polycarbamate and Dialdehyde Cross-Linking. *ACS Sustain. Chem. Eng.*, 2019, *7* (24), 19621–19630. doi:10.1021/acssuschemeng.9b04713.
13. Beniah, G.; Fortman, D. J.; Heath, W. H.; Dichtel, W. R.; Torkelson, J. M. Non-Isocyanate Polyurethane Thermoplastic Elastomer: Amide-Based Chain Extender Yields Enhanced Nanophase Separation and Properties in Polyhydroxyurethane. *Macromolecules*, 2017, *50* (11), 4425–4434. doi:10.1021/acs.macromol.7b00765.
14. Yuan, X.; Sang, Z.; Zhao, J.; Zhang, Z.; Zhang, J.; Cheng, J. Synthesis and Properties of Non-Isocyanate Aliphatic Thermoplastic Polyurethane Elastomers with Polycaprolactone Soft Segments. *J. Polym. Res.*, 2017, *24* (6), 88. doi:10.1007/s10965-017-1249-9.
15. Cornille, A.; Dworakowska, S.; Bogdal, D.; Boutevin, B.; Caillol, S. A New Way of Creating Cellular Polyurethane Materials: NIPU Foams. *Eur. Polym. J.*, 2015, *66*, 129–138. doi:10.1016/j.eurpolymj.2015.01.034.
16. Clark, J. H.; Farmer, T. J.; Ingram, I. D. V.; Lie, Y.; North, M. Renewable Self-Blowing Non-Isocyanate Polyurethane Foams from Lysine and Sorbitol. *European J. Org. Chem.*, 2018, *2018* (31), 4265–4271. doi:10.1002/ejoc.201800665.
17. Zareanshahraki, F.; Asemani, H. R.; Skuza, J.; Mannari, V. Synthesis of Non-Isocyanate Polyurethanes and Their Application in Radiation-Curable Aerospace Coatings. *Prog. Org. Coatings*, 2020, *138*, 105394. doi:10.1016/j.porgcoat.2019.105394.
18. Wu, Z.; Tang, L.; Dai, J.; Qu, J. Synthesis and Properties of Fluorinated Non-Isocyanate Polyurethanes Coatings with Good Hydrophobic and Oleophobic Properties. *J. Coatings Technol. Res.*, 2019, *16* (5), 1233–1241. doi:10.1007/s11998-019-00195-5.
19. Yu, A. Z.; Setien, R. A.; Sahouani, J. M.; Docken, J.; Webster, D. C. Catalyzed Non-Isocyanate Polyurethane (NIPU) Coatings from Bio-Based Poly(Cyclic Carbonates). *J. Coatings Technol. Res.*, 2019, *16* (1), 41–57. doi:10.1007/s11998-018-0135-7.
20. Matsukizono, H.; Endo, T. Synthesis and Hydrolytic Properties of Water-Soluble Poly(Carbonate–Hydroxyurethane)s from Trimethylolpropane. *Polym. Chem.*, 2016, *7* (4), 958–969. doi:10.1039/C5PY01733E.
21. Matsukizono, H.; Endo, T. Phosgene-Free Syntheses and Hydrolytic Properties of Water-Soluble Polyhydroxyurethanes with Ester–Carbonate–Ether Structures in Their Main Chains. *Macromol. Chem. Phys.*, 2017, *218* (18), 1700043. doi:10.1002/macp.201700043.

22. Sardon, H.; Engler, A. C.; Chan, J. M. W.; Coady, D. J.; O'Brien, J. M.; Mecerreyes, D.; Yang, Y. Y.; Hedrick, J. L. Homogeneous Isocyanate- and Catalyst-Free Synthesis of Polyurethanes in Aqueous Media. *Green Chem.*, 2013, *15* (5), 1121–1126. doi:10.1039/C3GC40319J.
23. Bizet, B.; Grau, E.; Cramail, H.; Asua, J. M. Volatile Organic Compound-Free Synthesis of Waterborne Poly(Hydroxy Urethane)–(Meth)Acrylic Hybrids by Miniemulsion Polymerization. *ACS Appl. Polym. Mater.*, 2020, *2* (9), 4016–4025. doi:10.1021/acsapm.0c00657.
24. Bizet, B.; Grau, E.; Cramail, H.; Asua, J. M. Crosslinked Isocyanate-Free Poly (Hydroxy Urethane)s – Poly(Butyl Methacrylate) Hybrid Latexes. *Eur. Polym. J.*, 2021, *146*, 110254. doi:10.1016/j.eurpolymj.2020.110254.
25. Rix, E.; Grau, E.; Chollet, G.; Cramail, H. Synthesis of Fatty Acid-Based Non-Isocyanate Polyurethanes, NIPUs, in Bulk and Mini-Emulsion. *Eur. Polym. J.*, 2016, *84*, 863–872. doi:10.1016/j.eurpolymj.2016.07.006.
26. Cobaj, A.; Mehr, H. S.; Hu, Y.; Soucek, M. D. The Influence of a Non-Isocyanate Urethane Monomer in the Film Formation and Mechanical Properties of Homogeneous and Core-Shell Latexes. *Polymer (Guildf).*, 2021, *214*, 123253. doi:10.1016/j.polymer.2020.123253.
27. Meng, L.; Soucek, M. D.; Li, Z.; Miyoshi, T. Investigation of a Non-Isocyanate Urethane Functional Monomer in Latexes by Emulsion Polymerization. *Polymer (Guildf).*, 2017, *119*, 83–97. doi:10.1016/j.polymer.2017.05.006.
28. Zhang, C.; Wang, H.; Zhou, Q. Waterborne Isocyanate-Free Polyurethane Epoxy Hybrid Coatings Synthesized from Sustainable Fatty Acid Diamine. *Green Chem.*, 2020, *22* (4), 1329–1337. doi:10.1039/C9GC03335A.
29. Wu, Z.; Dai, J.; Tang, L.; Qu, J. Sorbitol-Based Aqueous Cyclic Carbonate Dispersion for Waterborne Nonisocyanate Polyurethane Coatings via an Environment-Friendly Route. *J. Coatings Technol. Res.*, 2019, *16* (3), 721–732. doi:10.1007/s11998-018-0150-8.
30. Ma, Z.; Li, C.; Fan, H.; Wan, J.; Luo, Y.; Li, B.-G. Polyhydroxyurethanes (PHUs) Derived from Diphenolic Acid and Carbon Dioxide and Their Application in Solvent- and Water-Borne PHU Coatings. *Ind. Eng. Chem. Res.*, 2017, *56* (47), 14089–14100. doi:10.1021/acs.iecr.7b04029.
31. Bossion, A.; Olazabal, I.; Aguirresarobe, R. H.; Marina, S.; Martín, J.; Irusta, L.; Taton, D.; Sardon, H. Synthesis of Self-Healable Waterborne Isocyanate-Free Poly (Hydroxyurethane)-Based Supramolecular Networks by Ionic Interactions. *Polym. Chem.*, 2019, *10* (21), 2723–2733. doi:10.1039/C9PY00439D.
32. Ma, S.; van Heeswijk, E. P. A.; Noordover, B. A. J.; Sablong, R. J.; van Benthem, R. A. T. M.; Koning, C. E. Isocyanate-Free Approach to Water-Borne Polyurea Dispersions and Coatings. *ChemSusChem*, 2018, *11* (1), 149–158. doi:10.1002/cssc.201701930.
33. Ma, S.; Zhang, H.; Sablong, R. J.; Koning, C. E.; van Benthem, R. A. T. M. T-Butyl-Oxycarbonylated Diamines as Building Blocks for Isocyanate-Free Polyurethane/Urea Dispersions and Coatings. *Macromol. Rapid Commun.*, 2018, *39* (9), 1800004. doi:10.1002/marc.201800004.
34. Bossion, A.; Jones, G. O.; Taton, D.; Mecerreyes, D.; Hedrick, J. L.; Ong, Z. Y.; Yang, Y. Y.; Sardon, H. Non-Isocyanate Polyurethane Soft Nanoparticles Obtained by Surfactant-Assisted Interfacial Polymerization. *Langmuir*, 2017, *33* (8),1959–1968. doi:10.1021/acs.langmuir.6b04242.
35. Gennen, S.; Grignard, B.; Thomassin, J.-M.; Gilbert, B.; Vertruyen, B.; Jerome, C.; Detrembleur, C. Polyhydroxyurethane Hydrogels: Synthesis and Characterizations. *Eur. Polym. J.*, 2016, *84*, 849–862. doi:10.1016/j.eurpolymj.2016.07.013.

36. Bourguignon, M.; Thomassin, J.-M.; Grignard, B.; Jerome, C.; Detrembleur, C. Fast and Facile One-Pot One-Step Preparation of Nonisocyanate Polyurethane Hydrogels in Water at Room Temperature. *ACS Sustain. Chem. Eng.*, 2019, *7* (14), 12601–12610. doi:10.1021/acssuschemeng.9b02624.
37. Bourguignon, M.; Thomassin, J.-M.; Grignard, B.; Vertruyen, B.; Detrembleur, C. Water-Borne Isocyanate-Free Polyurethane Hydrogels with Adaptable Functionality and Behavior. *Macromol. Rapid Commun.*, 2021, *42* (3), 2000482. doi:10.1002/marc.202000482.

25 Waterborne Polyurethanes: Challenges and Future Outlook

Felipe M. de Souza and Ram K. Gupta
Department of Chemistry, Kansas Polymer Research Center,
Pittsburg State University, Pittsburg, KS, USA

CONTENTS

Acknowledgment 438
References 438

The excessive use of petrochemical-based starting materials for polymer synthesis and the presence of volatile organic components (VOCs) caused several environmental and health concerns that motivated scientists and researchers to find alternative materials and processes that are eco-friendly and cheap. Polyurethane is one the commercial polymers used for various applications, ranging from automobiles to construction to biomedical. The global polyurethane market was valued at around $70 billion in 2020 and is expected to grow at a compound annual growth rate of 3.8% from 2021 to 2028. Bio-polyurethanes cover about half of the global polyurethane market. Waterborne polyurethanes (WPUs) provide additional advantages over bio-polyurethanes as they utilize a reduced (or none) amount of VOCs during their synthesis. Polyols, isocyanate, chain extender, emulsifiers, and catalysts are the main constituents used for the synthesis of WPU. Each component is very crucial for synthesizing high-quality WPU and possesses several challenges to make them suitable for applications in a variety of areas, such as automobiles, coatings, adhesives, biomedical materials, etc. (Figure 25.1).

The main aspect of WPU consists of dispersing the polyurethane into an aqueous media rather than in volatile organic solvents, which is challenging as the soft segments originating from the polyols are aliphatic while the rigid segments originating from the isocyanates are aromatic (hydrophobic). Thus, the only polar regions that can form hydrogen bonding with water are the urethane linkages throughout the polymer chain. This leads to one of the main challenges in the

DOI: 10.1201/9781003173526-25

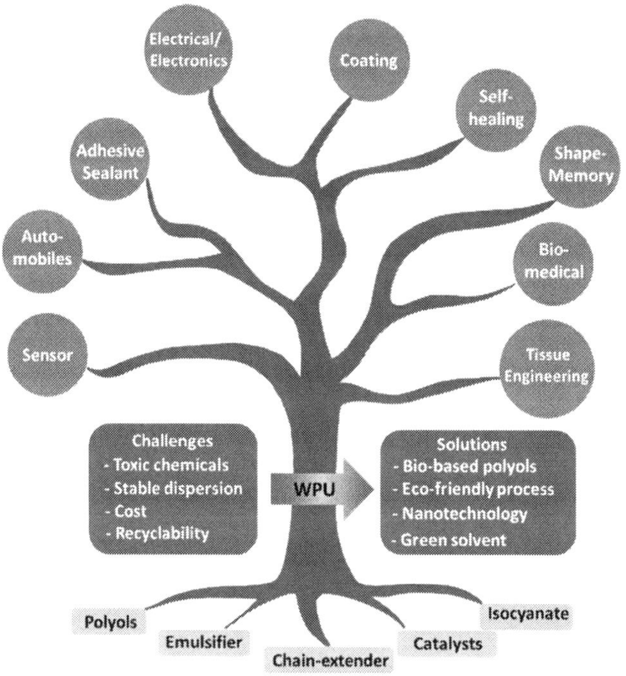

FIGURE 25.1 Materials, challenges, and applications of WPU.

synthesis of WPU, which is obtaining a stable dispersion with a small particle size to reduce the agglomeration or phase separation over time. Also, WPU should possess low viscosity along with high molecular weight to be more processable and efficient for many applications. To tackle these requirements, several methods such as acetone process, use of emulsifiers, hot-melt, ketazine–ketamine, self-emulsifying polymer, atom transfer radical polymerization (ATRP), and reversible addition-fragmentation chain transfer (RAFT) were developed [1]. The first two methods are the most employed methods for the synthesis of WPU due to their lower cost, eco-friendliness, and appropriateness for most polyurethanes. However, there is still room to improve the stability of the dispersion along with decreasing the use of organic solvents, such as acetone or methyl ethyl ketone, which are mostly used in the commercial production of polyurethanes.

The increasing number of polar sites in the polymer chain could be one of the most suitable methods to prepare a stable dispersion. Chemically attaching a compound that contains a dandling charged polar group to the backbone of the polymer chain can improve the dispersibility of the polyurethane in aqueous media. These groups can be a carboxylic acid, sulfonic acid, or tertiary amines, which can be further stabilized by a counter ion to neutralize the charge for a stable dispersion. Emulsifiers also play a crucial role in obtaining a stable dispersion of polyurethane. Some of the challenges lie in balancing the amount of internal emulsifier to stabilize the polymeric dispersion as well as decrease the amount of organic solvent used. Generally, in the acetone process, small particle size along with stable dispersions can be achieved by performing the solvent exchange process slowly accompanied to vigorous stirring.

To make the process eco-friendly, many green approaches are explored for the synthesis of polyurethanes via the conversion of vegetable and essential oils into polyols and chain extenders. Classic examples of such materials are corn oil, soybean oil, castor oil, rapeseed oil, limonene, glycerin, and lignin [2–4]. Polyols obtained from these biosources offer several advantages, such as renewability that provides a stable supply of raw materials, environment credibility, and the ability to harvest in most places around the globe. It will be a wiser approach to use biomaterials, such as castor oil, glycerin, and lignin, that are not used as food stock for the synthesis of WPU for commercial applications. Accompanied with that, several technologies can efficiently convert the unsaturation present in these materials into hydroxyl groups through processes such as epoxidation/ring-opening, thiol-ene, hydroformylation/reduction, transesterification, transamination, ozonolysis, etc. Most of these methods are efficient, facile, and cost-effective. However, there are still some challenges to overcome to reach large-scale production of WPUs for commercial applications. For example, the epoxidation/ring-opening method may require relatively higher quantities of organic solvents in comparison to other methods. On the other hand, the thiol-ene method requires the use of mercaptans, which are toxic and usually petrochemical derived. Ozonolysis requires a large amount of ozone, which is costly and detrimental to the environment. Despite these challenges, recent studies have demonstrated that some polyurethanes derived from renewable materials can surpass the properties of commercial products [5–7].

Isocyanate is another main element in the synthesis of WPU. Isocyanates are toxic, and finding a sustainable alternative for them is also one of the main challenges for the scientific community and industry. Recently, there has been some development in the preparation of isocyanate-free polyurethanes. Transurethanization and polyhydroxy urethanes obtained through the reaction between a diamine and cyclocarbonate are commonly used routes for isocyanate-free polyurethanes. The transurethanization approach is less applied due to the release of byproducts that can be hard to remove. However, the polymers derived using this method have similar properties compared with the commercial isocyanate-based polymers. Polyhydroxy urethanes are advantageous as they utilize CO_2 as a starting material, which makes the process highly sustainable, environmentally friendly, and cost-effective [8]. However, currently, the isocyanate-free routes are mostly used in the coating sectors; therefore, further research is required to expand the applications of isocyanate-free polyurethanes in different areas.

In pursuit of developing eco-friendly and greener processes, the scientific community faces a recurrent challenge in utilizing renewable materials for synthesizing polyols, chain extenders, and neutralizing agents, as well as designing isocyanate-free routes to provide polyurethanes with properties similar to petrochemical-derived polyurethanes. Despite these challenges, polyurethane chemistry shows an important side of its versatility as it not only reaches important and diverse sectors of the industry but also can be synthesized through several routes while using various bio-based starting materials. All these aspects can be combined for the synthesis of WPU for various applications, such as inks, adhesives, coatings with anticorrosion, antibacterial, mechanical impact resistance, biocompatible properties, etc. Based on the wide range of areas where WPU can be used, the applications can be divided into two major areas: engineering and biological. Engineering WPU includes adhesives, self-healing, electromagnetic shielding, sensors, and coatings that can be subdivided into anticorrosive, antibacterial, mechanical protection, flame-retardancy, etc. On the other hand, biological WPU includes drug-delivery systems, scaffolds for growing tissue, and body fluid transport, among others.

Scientists are able to design new synthetic routes that use renewable resources as starting materials to provide polyurethanes with specific properties for specific applications. For example, castor oil-based WPU can be used as adhesives due to the presence of inherent hydroxyl groups in its structure that allow it to be used as a polyol. On top of that, it is a low-cost and nonedible vegetable oil that triggers the interest of scientists for further research. However, castor oil by itself has a low hydroxyl value, which may lead to poor performance in some cases. Hence, performing reactions, such as transesterification, epoxidation/ring-opening, or thiol-ene, are some of the approaches to increase the hydroxyl functionality of castor oil. When applied in adhesives, the increase in functionality is related to the adhesion power due to high crosslinking density [9]. The use of nanotechnology in polymer science has significantly enhanced their properties and thus applications. For example, the addition of nano-silica can significantly enhance the adhesive properties of WPU. Polymer-based corrosion resistive coatings are highly desired in many applications due to their usefulness in protecting the metallic surfaces of many instruments, machinery, vehicles, ships, containers, or any other metallic substrate that can be exposed to corrosive media. Polyurethanes came as a feasible option for anti-corrosive coatings due to their inherent chemical stability and strong adhesion to many substrates. Chromium coating was previously used as a protective layer against corrosion; however, due to its high toxicity and cost, new eco-friendly materials with improved corrosion protection properties are desired. Several eco-friendly WPU nanocomposite coatings were developed by incorporating nanoparticles, such as SiO_2, ZnO, TiO_2, Al_2O_3, $CaCO_3$, graphene and derivatives, and montmorillonite nano clay, to obtain coatings that can provide better performance and present lower cost compared to chromium-based coatings. Each one of these nano additives functions in a different way to implement anti-corrosive properties. For example, SiO_2 is highly hydrophobic, which prevents water from permeating through the coating, therefore providing an effective barrier effect. Nanoparticles of metal oxides provide high stability against UV radiation and most corrosive media due to their chemical stability through the formation of a passive protective layer on the surface. However, the metal oxide coatings can suffer from delamination.

Current approaches on the use of metal oxides rely on functionalizing the surface with hydroxyl groups to promote a covalent bond with isocyanate, leading to stable dispersions to form a strong coating. Graphene and its derivatives are also convenient nanomaterials that can be used in anti-corrosive WPU coatings. Graphene has a good barrier effect against corrosive anions, such as Cl^- or OH^-, due to the high electron density of its π–π interactions. Yet, its main drawback lies in its low dispersibility in water. Despite that, the scientific community found different ways to address this issue by introducing hydroxyl groups into graphene's surface (graphene oxide), in situ polymerization over graphene's surface to increase its adhesion, or chemically bonding graphene oxide to the polyurethane matrix. All these approaches have given effective results; however, there is still room for optimization either by increasing the dispersibility of these nanomaterials into the polyurethane matrix or analyzing other nanomaterials that can be combined to further enhance its properties while aiming for an economically viable process with high performance.

Despite all the properties and tunability that polyurethane possesses, one of its major drawbacks is its inherent poor resistance to fire, which is due to its high surface area and organic composition. Polyurethanes are particularly dangerous when they go through combustion due to the release of toxic fumes, such as aromatic fragments and cyanides, which are extremely toxic. One of the solutions to reduce the flammability of polyurethane is to incorporate flame-retardants. Flame-retardants can be either additive or reactive. Reactive flame-retardants are chemically bonded to the structure of polyurethane, while additives are just physically blended in the polyurethanes. Flame-retardants can also be toxic; for example, halogen-based flame-retardants are toxic for the environment as well as health. Many other flame-retardants are nontoxic compared to halogen-based flame-retardants. Examples of green flame-retardants are compounds of phosphorus and nitrogen, carbon black, expandable graphite, aluminum, magnesium oxides, and hydroxides. These compounds can be added either by physical blending or by covalently attaching them to the polyurethane structure. These flame-retardant compounds work through two mechanisms: gas or solid phase. The gas-phase mechanism relies on the release of relatively stable radical species that can capture more reactive radical fragments originated from the polyurethane's combustion, hence named radical scavengers. The solid-phase mechanism relies on forming or inducing a compact char layer that blocks the entrance of these radical species and oxygen, hence protecting the unburnt polyurethane underneath it. A viable strategy relies upon combining two or more compounds that can provide these two flame-retardancy mechanisms simultaneously.

The constant search for new bio-based components to serve as an alternative to the petrochemical counterparts eventually lead to novel materials that have great potential to be part of the market. In that regard, jatropha oil is a convenient nonedible vegetable oil that has been little explored in comparison to other vegetable oils as starting materials for high-performance coatings. Previous studies have demonstrated that the incorporation of this vegetable oil into polyurethane coatings can surpass the properties of commercial petrochemical ones [10,11]. Vegetable oils such as jatropha oil can vary their functionality based on the synthetic procedure adopted to introduce hydroxyl groups. The change in functionality allows the properties to be tuned for specific requirements. Another abundant bio-based source that can be used to add both mechanical strength as well as UV-resistance properties is lignin and its derivatives. Its rigid structure containing benzene ringed with a high conjugation degree increases mechanical strength and UV absorption. Several approaches can be performed by designing WPU containing lignin along with other bio-based materials, such as jatropha oil or castor oil, for example. Hence, a plethora of WPU with a variety of properties can be produced using these materials.

Their low cost, renewable nature, and abundance make them very attractive for industrial applications. Even though lignin demands pretreatment followed by separation processes, it is a convenient plant-derived starting material due to its large hydroxyl functionality.

The high degree of conjugation can provide high electromagnetic shielding (EMS) characteristics. Graphene, a highly conjugated system, is a suitable nanomaterial that can be incorporated into a WPU to fabricate coatings for EMS applications. Also, despite the inherent insulation properties of polyurethanes, the WPU can be designed in a way that its structure contains ionic segments to allow ionic conduction, branching out its application for electric and electronic industries. Researchers in the biomedical field can also take advantage of ionomeric WPU. Ionomeric WPU can be used to transport blood, serum, plasma, or other body fluids. The recurrent issue is that materials used for this application can develop fouling due to the physical attachment of proteins, which can easily lead to infections. WPU with an ionomeric structure can prevent that because positive and negative charges throughout the polymer prevent proteins from adhering to its surface. Another inherent challenge for biomedical applications resides in the necessity for the materials to be biocompatible and biodegradable. Even though several materials are derived from biosources, that does not necessarily mean that they are also biodegradable. Hence, scientists need to tackle this situation to obtain materials suitable for these applications. Besides transport of body fluids, WPU is also applicable as support to regenerate tissue and transport of medicine in vivo.

WPU can also be chemically modified to acquire self-healing properties. This property can be incorporated in several ways: (1) encapsulation of nanoparticles that release the filler after a crack occurs, (2) presence of disulfide bonds that can reattach after thermal curing, (3) Diels–Alder-based compounds that also present reversible bonds, and (4) complex metal-organic bonds. Such property is highly desirable as it greatly improves the usability, robustness, and shelflife of the product. Thus, it is notable that WPU occupies a plethora of areas and is an extremely versatile polymer. The current challenge for the foreseeable future lies in emphasizing greener and sustainable approaches capable of delivering high-performance materials to compose the market and improve quality of life.

ACKNOWLEDGMENT

The authors wish to thank Ms. Cassia Allison, Ms. Madeline Ellis, and Ms. Anjali Gupta from Pittsburg High School, Pittsburg, Kansas, for their help in drawing Figure 25.1.

REFERENCES

1. Zhou X, Li Y, Fang C, Li S, Cheng Y, Lei W, Meng X (2015) Recent advances in synthesis of waterborne polyurethane and their application in water-based ink: A review. J Mater Sci Technol 31:708–722.
2. Yang Z, Feng Y, Liang H, Yang Z, Yuan T, Luo Y, Li P, Zhang C (2017) A solvent-free and scalable method to prepare soybean-oil-based polyols by thiol-ene

photo-click reaction and biobased polyurethanes therefrom. ACS Sustain Chem Eng 5:7365–7373.
3. Jayavani S, Sunanda S, Varghese TO, Nayak SK (2017) Synthesis and characterizations of sustainable polyester polyols from non-edible vegetable oils: Thermal and structural evaluation. J Clean Prod 162:795–805.
4. Zhang C, Madbouly SA, Kessler MR (2015) Biobased polyurethanes prepared from different vegetable oils. ACS Appl Mater Interfaces 7:1226–1233.
5. Li R, Shan Z (2018) Research for waterborne polyurethane/composites with heat transfer performance: a review. Polym Bull 75:4823–4836.
6. Liang H, Feng Y, Lu J, Liu L, Yang Z, Luo Y, Zhang Y, Zhang C (2018) Bio-based cationic waterborne polyurethanes dispersions prepared from different vegetable oils. Ind Crops Prod 122:448–455.
7. Zhang W, Zhang Y, Liang H, Liang D, Cao H, Liu C, Qian Y, Lu Q, Zhang C (2019) High bio-content castor oil based waterborne polyurethane/sodium lignosulfonate composites for environmental friendly UV absorption application. Ind Crops Prod 142:111836.
8. Ma Z, Li C, Fan H, Wan J, Luo Y, Li B-G (2017) Polyhydroxyurethanes (PHUs) derived from diphenolic acid and carbon dioxide and their application in solvent- and water-borne PHU coatings. Ind Eng Chem Res 56:14089–14100.
9. Cakić SM, Valcic MD, Ristić IS, Radusin T, Cvetinov MJ, Budinski-Simendić J (2019) Waterborne polyurethane-silica nanocomposite adhesives based on castor oil-recycled polyols: Effects of (3-aminopropyl)triethoxysilane (APTES) content on properties. Int J Adhes Adhes 90:22–31.
10. Saalah S, Abdullah LC, Aung MM, Salleh MZ, Awang Biak DR, Basri M, Jusoh ER, Mamat S, Osman Al Edrus SS(2021) Chemical and thermo-mechanical properties of waterborne polyurethane dispersion derived from jatropha oil. Polymers (Basel). 13.
11. Saalah S, Abdullah LC, Aung MM, Salleh MZ, Awang Biak DR, Basri M, Jusoh ER, Mamat S (2018) Colloidal stability and rheology of jatropha oil-based waterborne polyurethane (JPU) dispersion. Prog Org Coatings 125:348–357.

Index

acetone process, 3, 4, 5, 10, 41, 67, 84, 88, 89, 107, 109, 138, 139, 141, 179, 180, 181, 255, 379, 434
acrylic, 5, 39, 72, 74, 75, 144, 163, 164, 199, 261, 273, 274, 275, 278, 279, 290, 336, 343, 347, 357, 360, 361, 363, 370
adhesive, 4, 18, 23, 24, 25, 26, 27, 36, 72, 96, 111, 189, 254, 259, 273, 275, 292, 308, 309, 316, 326, 329, 330, 334, 335, 337, 341, 344, 356, 360, 370, 381, 397, 398, 399, 436
anionic, 2, 3, 8, 9, 10, 18, 21, 33, 36, 37, 49, 57, 58, 67, 68, 78, 106, 107, 109, 110, 214, 216, 254, 255, 271, 308, 326, 344, 377, 383, 415, 424, 428
antibacterial, 6, 7, 11, 37, 58, 73, 76, 79, 87, 95, 96, 112, 163, 193, 194, 195, 196, 197, 198, 199, 201, 217, 222, 260, 276, 277, 341, 343, 344, 345, 384, 386, 387, 436
antifouling, 9, 146, 151, 232
antimicrobial, 36, 76, 87, 193, 194, 199, 262, 344, 379, 390
automotive, 31, 161, 162, 232, 262, 268, 275, 284, 307, 308, 322, 358, 389, 393, 394, 395, 396, 397, 400, 401, 402, 403

bacteria, 3, 6, 7, 10, 11, 27, 37, 76, 87, 88, 163, 193, 194, 195, 197, 198, 199, 241, 244, 344, 386
binders, 23, 27, 101, 111, 275, 307, 322, 359, 369, 381
biocompatibility, 3, 12, 39, 72, 87, 95, 96, 138, 164, 170, 171, 194, 195, 197, 203, 205, 213, 214, 215, 217, 226, 232, 235, 343, 345, 378, 387, 390
biodegradability, 12, 39, 41, 72, 73, 95, 98, 114, 168, 171, 203, 214, 218, 222, 226, 233, 235, 245, 246, 273, 273, 382, 390
biomedical, 1, 3, 7, 10, 12, 14, 39, 41, 43, 48, 95, 96, 170, 171, 205, 215, 225, 322, 334, 341, 382, 387, 433, 438
building, 27, 56, 96, 120, 167, 262, 271, 322, 363, 364

carbon nanotubes, 24, 72, 77, 85, 86, 88, 111, 150, 184, 259, 280, 298, 309, 323, 324, 326, 331, 334, 336, 341
cashew nut, 55, 57
castor oil, 2, 4, 11, 14, 33, 34, 35, 37, 49, 52, 53, 57, 60, 75, 110, 150, 171, 181, 256, 258, 330, 382, 384, 385, 435, 436, 437
catalysts, 21, 32, 34, 36, 53, 66, 79, 86, 101, 102, 114, 269, 274, 307, 339, 348, 358, 377, 433
cationic, 21, 33, 36, 37, 49, 57, 58, 67, 78, 106, 107, 162, 164, 183, 198, 206, 254, 289, 308, 309, 313, 344, 377, 386, 428
chain extenders, 3, 4, 20, 22, 37, 49, 58, 59, 61, 66, 67, 69, 70, 78, 85, 89, 106, 120, 138, 142, 145, 148, 149, 151, 171, 181, 223, 254, 269, 331, 368, 381, 388, 407, 435
char, 95, 113, 126, 127, 129, 130, 131, 132, 437
citric fruits, 11
coating, 9, 10, 11, 12, 14, 18, 19, 23, 24, 25, 26, 27, 36, 37, 53, 58, 72, 74–7, 79, 95, 96, 145, 148, 162, 163, 164, 170, 182, 187, 189, 195, 198, 199, 232, 233, 235, 242, 254, 255, 256, 257, 258, 259, 260, 261, 268, 273, 275, 276, 277, 278, 279, 280, 282, 283, 284, 288, 289, 290–9, 300, 307, 309, 312, 314, 316, 322, 330, 335, 337, 338, 339, 347, 348, 349, 367, 370, 372, 375, 377, 379, 380, 382, 383, 387, 389, 399, 400, 417, 420, 435
composite, 9, 11, 13, 14, 18, 23, 27, 52, 72, 73, 74, 76, 77, 79, 85, 87, 88, 90, 91, 93–8, 110–14, 122, 127, 130, 131, 150, 151, 163, 164, 166–8, 170, 171, 182–5, 187, 189, 197, 201, 215, 217, 218, 222, 223, 226, 232, 255, 256, 258–61, 277, 278, 280, 291–5, 297–9, 306, 310–17, 322–5, 327, 328, 334, 336, 337–9, 341, 343–5, 348, 349, 375, 382–5, 387, 389, 399, 400–3, 436
construction, 48, 161, 162, 164, 213, 268, 276, 284, 308, 309, 322, 325, 330, 357, 358, 363, 366, 367, 383, 384, 393, 433
contact angle, 11, 12, 14, 37, 74, 78, 110, 215, 216, 217, 222, 248, 258, 276, 280, 292, 344
corrosion, 7, 11, 14, 24, 27, 33, 57, 79, 84, 96, 187, 189, 257–61, 268, 275, 277, 278, 280, 282, 284, 287, 288, 289, 290–5, 297–300, 309, 310, 312, 316, 328, 336, 375, 386, 387, 394, 396, 399, 400, 402, 436
cycloaliphatic isocyanates, 3

dibutyltin diacetate, 5
di-methylol propionic acid (DMPA), 2, 57, 60, 72, 149, 274, 334
dispersibility, 2, 5–7, 9, 14, 33, 34, 36, 37, 39, 106, 290–2, 295, 298, 309, 326, 434, 436
dispersion, 2–6, 9, 10, 11, 18–21, 23, 27, 33, 36, 37, 39, 41, 52, 53, 57–61, 67, 68, 70–2, 74–6, 78, 89, 90, 93, 106–12, 122, 138, 141, 142, 150, 162, 168, 179, 180, 182, 183, 186, 195, 201, 217, 219, 225, 233, 235, 239, 240, 254–7, 260, 261, 269, 274, 275, 278–80, 298, 299, 309, 311–13, 317, 321, 322, 325, 334, 336, 337, 367, 368, 370, 371, 377–9, 382–4, 403, 420, 434
drug, 10, 41, 43, 194, 195, 198, 199, 202–6, 215, 218, 219, 345, 387, 424, 425
drug delivery, 10, 39, 41, 43, 96, 194, 199, 202–4, 206, 215, 387, 424, 425, 436

eco-friendly, 1, 11, 12, 14, 23, 24, 36, 39, 48, 49, 61, 70, 178, 187, 232, 268, 275, 284, 308, 314, 315, 334, 348, 367, 386, 387, 393–403, 429, 435, 436
electrical conductivity, 23, 77–9, 88, 95, 259, 260, 309–13, 315, 323, 325, 328, 329
electromagnetic interference, 311, 327, 328
electronic skin, 189, 326, 341
emulsification process, 4, 161
emulsifier, 2, 4, 7–10, 18, 21, 22, 25, 34, 36, 37, 39, 49, 52, 57, 58, 70, 73, 74, 76, 88, 105–7, 162, 178, 179, 186, 214, 233, 235, 254, 257, 274, 377, 379, 384, 424, 425, 434
epoxidation, 33, 34, 43, 52, 54, 108, 109, 162, 234, 384, 435, 436

filler, 52, 114, 127, 166–8, 171, 179, 182, 256, 257, 259, 260, 280, 298, 310, 313, 314, 316, 317, 324, 377, 383–5, 426, 438
flame-retarded, 120–3, 127, 130–3
flexibility, 2, 12, 18, 19, 24, 32, 53, 54, 58, 68, 75, 76, 94, 96, 102, 104, 105, 115, 161, 162, 165, 171, 187, 189, 194, 195, 216, 223, 232, 262, 268, 269, 272, 279, 292, 323, 331, 334, 335, 339, 345, 348, 357, 367, 368, 372, 376, 377, 390, 415, 420, 429
flexible foam, 329, 376
4,4′-diphenylmethane diisocyanate, 19, 306
fourier-transform infrared spectroscopy (FTIR), 120
fungal, 240, 241, 242, 245, 246
furan, 41, 56, 182

gas barrier, 77
glass transition, 52, 141, 144, 151, 160, 166, 272, 397, 412, 413, 414, 415, 420
glycolysis, 52, 330
grafted polymers, 9, 21, 107, 377
grape, 34, 53
graphene, 9, 13, 23, 24, 74, 77, 85, 88, 95, 123, 125, 130, 131, 182, 184, 256, 259, 260, 276, 277, 280, 291, 297, 298, 299, 309–11, 313, 317, 324, 325, 339, 341, 387, 303, 436, 437, 438
graphene oxide, 23, 74, 77, 85, 88, 123, 125, 130, 131, 182, 260, 291, 298, 309, 325, 341, 437
green material, 13, 284

heat Release Rate, 123, 127, 129, 130
hexadecyltrimethoxysilane, 11
hot-melt, 5, 67, 84, 89, 107, 108, 115, 162, 363, 364
hydroformylation, 33, 35, 36, 43, 52, 53, 54, 108, 234, 435
hydrogen bonding, 2, 10, 20, 23, 37, 66, 70, 91, 114, 142, 144, 145, 166, 178, 217, 255, 280, 295, 384, 394, 403, 417, 433
hydrogenation, 34, 35, 52, 53, 54, 162, 330
hydrolysis, 7, 9, 19, 58, 89, 114, 142, 203, 215, 222, 232, 242, 243, 246, 269, 289, 330, 363, 409, 414, 428
hydrophilicity, 2, 3, 13, 39, 57, 72, 78, 106, 162, 178, 199, 216, 218, 219, 221, 235, 248, 257, 258, 295, 378
hydrophobicity, 2, 11, 12, 14, 37, 38, 49, 72, 74, 78, 132, 147, 187, 199, 214, 215, 216, 258, 262, 295, 298, 384

index, 121, 123, 127, 248
ink, 1, 18, 23, 24, 36, 69, 79, 105, 164, 170, 215, 218, 219, 232, 254, 344, 381, 382, 384, 385, 435
isocyanates, 2, 3, 19, 32, 33, 54, 55, 61, 66, 89, 102, 107, 114, 115, 148, 162, 203, 235, 241, 254, 274, 275, 343, 364, 365, 368–72, 377, 388, 390, 407, 429, 433, 435
isophorone diisocyanate, 3, 53, 54, 72, 86, 105, 109, 111, 148, 179, 185, 197, 216, 217, 257, 290, 299, 334

jatropha oil, 49, 52, 109, 257, 384, 437

ketamine–ketazine, 3, 5, 6, 89, 107, 108, 115

lignin, 13, 56, 86, 233, 299, 435, 437, 438
limiting oxygen index (LOI), 121, 123, 127

linseed, 34, 49, 52, 53, 58, 182

macroglycols, 85, 86
mechanical properties, 7, 11–13, 20, 22, 23, 27, 37, 39, 56, 57, 59, 66, 68, 72–4, 76–9, 94, 97, 101, 104, 114, 122, 123, 125, 142, 148, 150, 168, 183, 184, 186, 189, 194, 197, 217, 222, 225, 232, 246, 248, 258, 259, 278, 290, 295, 298, 310, 314, 317, 327, 339, 341, 344, 348, 368, 378, 379, 383, 387, 390, 415, 420, 423, 426
mechanical strength, 36, 49, 76, 77, 84, 87, 96, 104, 114, 122, 138, 148, 163, 171, 189, 197, 216, 217, 232, 241, 259, 275, 290, 307, 314, 323, 335, 343, 344, 348, 368, 401, 402, 437
melt-dispersion, 3, 5, 67
metal oxides, 292, 298, 436
methylene diethanolamine, 2
morphology, 77, 78, 89, 92, 110, 115, 121, 142, 145, 165, 201, 202, 217, 225, 240, 257, 258, 280, 295, 339, 344, 412, 415

nanocomposites, 18, 23, 27, 72, 74, 77, 78, 79, 85, 87, 90, 93–6, 98, 114, 127, 131, 183, 184, 223, 226, 258, 259, 314, 336, 339, 383–5, 393, 399, 402, 403
nanofibers, 85, 87, 111, 198, 331
nanofillers, 14, 74, 317, 402
nanparticles, 87
nature rubber, 76
N-methyl diethanolamine, 3, 10, 21, 37, 58, 106
N-methyl-2-pyrrolidone, 3, 88, 107, 199
NMR, 91, 92, 186, 217, 248
non-isocyanate polyurethanes, 108, 115, 407

Olive, 34, 53, 414
1,4-butanediol, 5, 106, 112, 299
1,4-diazabicyclo octane, 21
1,3 propane sulton, 3
Oxirane, 32, 34, 52
Ozonolysis, 33, 52, 54, 234, 435

Packing, 23, 36, 121, 269, 308, 381, 383, 386
palm oil, 49
particles, 2, 4, 7, 10, 18, 22, 52, 67, 69, 71, 105, 114, 140, 142, 145, 183, 184, 197, 203, 205, 214, 216, 217, 223, 226, 254–61, 274, 278, 293, 295, 321, 324, 337, 341, 350, 378, 381, 383–6, 403, 412, 420, 428, 429
pentamethylene diisocyanate, 54, 56, 235
poly(2-(dimethylamino)ethyl methacrylate), 3
poly(ethylene glycol), 3, 222, 345, 410, 412, 426
poly(oxypropylene glycol), 5
poly(tetramethylene adipate), 19, 370
poly(urethane-co-vinyl pyridine), 10
polyaniline, 85, 88, 259, 293–5, 311, 316, 317, 334, 335, 337, 348
polycaprolactone, 12, 19, 39, 170, 235, 379
polycaprolactone glycol, 19
polycarbonate diol, 7, 76, 105, 216, 257, 334
polydiphenyl amine, 14
polyester, 2, 19, 32, 41, 54, 76, 86, 104, 105, 138, 141, 142, 144, 148, 149, 235, 236, 239, 240–6, 248, 257, 269, 272, 290, 306, 323, 325, 330, 370, 371, 379
polyethyleneglyco, 72
Polymerization, 2, 9, 14, 21, 36, 39, 53, 58, 60, 67, 85, 90, 102, 119, 131, 138, 139, 162, 163, 181, 197, 201, 223, 225, 236, 258, 259, 291, 295, 297, 307, 312, 313, 336, 337, 343, 348, 376, 378, 384, 409, 410, 412, 414–17, 420–6, 428, 437
polyols, 2, 3, 10, 19, 32, 33, 34, 36, 43, 48, 49, 52–4, 58, 61, 66, 70, 75, 76, 86, 88, 101–8, 110, 114, 115, 120, 121, 138, 141, 148, 151, 162, 178, 203, 233–5, 241, 262, 269, 271, 290, 306, 307, 331, 364, 365, 368, 370, 371, 376, 377, 379, 389, 390, 394, 407, 438, 435, 438
polytetramethylene ether glycol, 19, 217
polyurethane, 2–5, 11–14, 18, 23, 32, 33, 34, 36, 37, 39, 41, 48, 52, 68, 70, 72, 88, 90, 102, 108, 109, 119, 120, 122, 127, 129, 130–2, 138, 145, 149, 162, 177, 184, 194, 195, 199, 202, 205, 216, 217, 232, 235, 255, 256, 258, 261, 268, 274, 279, 290, 293, 294, 298, 299, 300, 306, 307, 320–6, 330, 331, 334–9, 341, 343, 345, 347–9, 357, 366, 375, 393–7, 399, 400, 402, 403, 423, 433–5, 437
Prepolymer, 3, 4, 5, 10, 52, 53, 59, 60, 68, 69, 84, 88, 89, 90, 106–8, 110, 115, 119, 142, 217, 258, 294, 339, 381
Pyrolysis, 34, 121, 129, 132, 330

Raman spectroscopy, 91, 92, 186
recycling, 103, 329, 330, 331
renewable, 1, 4, 14, 31, 32, 33, 37, 39, 42, 43, 49, 56, 57, 59, 60, 61, 72, 85, 86, 98, 103, 108, 140, 162, 233, 235, 257, 258, 288, 334, 349, 382, 435, 436, 438
rigid foam, 26
ring-opening, 3, 4, 32, 34, 42, 43, 52, 58, 60, 162, 384, 435, 436

safflower oil, 14
scaffolds, 72, 164, 170, 171, 201, 205, 214, 215, 217–19, 221–3, 225, 390
sealants, 1, 14, 31, 48, 49, 54, 101, 104, 105, 307, 322, 356–61, 363–72, 390

self-healing, 11, 13, 14, 19, 72, 96, 146, 148, 151, 170, 171, 178, 180–7, 189, 259, 293, 295, 298, 299, 327, 326, 327, 343, 349, 423, 436, 438
SEM, 13, 91, 92, 93, 112, 142, 145, 184, 206, 217, 235, 260, 280
sensor, 97, 325–7, 335–7, 339, 341, 345, 347, 348
shape memory, 38, 41, 77, 97, 146, 150, 151, 157–60, 165–8, 170, 221, 222, 226, 366, 394, 400–3
shear strength, 4
smart waterborne polyurethanes, 138
solvent exchange, 4, 5, 41, 336, 435
soybean, 4, 10, 14, 34, 36, 49, 52, 53, 54, 58, 88, 108, 110, 435
storage modulus, 10, 75, 110, 168, 415, 428
superhydrophobic, 11, 97, 146, 258
surfactants, 2, 33, 60, 66, 102, 103, 114, 204, 307, 409, 414
synthesis, 1–5, 7–12, 14, 18–21, 27, 31–4, 36, 37, 41–3, 49, 53, 54, 56, 57, 59, 60, 61, 67, 68, 70, 71, 78, 86, 89, 91, 101, 102–10, 115, 119, 122, 127, 129, 132, 138, 139, 140, 146, 151, 162, 167, 178–81, 189, 197, 198, 203, 217, 221, 223, 225, 233, 235, 243, 254, 257, 269, 295, 299, 307, 308, 310, 326, 327, 334, 377–9, 381–3, 389, 390, 407–10, 412, 414, 415, 423, 424, 428, 429, 433–5

TEM, 91–3, 105, 112, 142, 144, 145, 186, 223, 415
tensile strength, 7, 8, 10, 11, 13, 18, 19, 20, 22–4, 36, 37, 52, 59, 74–6, 78, 94, 105, 110, 114, 115, 125, 127, 131, 148, 162, 171, 189, 195, 197, 216, 222, 248, 279, 312, 316, 334, 343, 349, 366, 370, 377, 381–4, 386, 412–5, 417
thermal stability, 4, 32, 41, 49, 53, 57, 76, 77, 84, 87, 94, 95, 110, 113, 120–3, 133, 150, 162, 186, 197, 215, 216, 222, 259, 272, 273, 295, 298, 337, 349, 383, 384, 397, 401, 420, 429
thermoelectric, 163, 323
thermoplastic, 19, 56, 59, 66, 84, 165, 166, 307, 322, 336, 341, 345, 363, 379, 395, 399
thiol-ene, 53
tissue, 41, 72, 95, 96, 164, 170, 171, 183, 194, 199, 202, 204, 205, 214, 215, 217, 218, 219, 222, 225, 226, 321, 326, 327, 334, 343, 387, 436, 438
tissue Engineering, 72, 95, 96, 164, 170, 171, 214, 215, 217, 218, 225, 334, 387
toluene diisocyanate, 5, 19, 54, 86, 105, 138, 299, 306, 307, 365, 382, 397
total heat release, 123, 127, 129, 130–2
total smoke production, 127
transamidation, 52, 53
transesterification, 33, 36, 52, 108, 162, 235, 408, 435, 436
trimethylamine, 34
2-(dimethylamino)ethyl methacrylate dihydroxy, 3
2-hydroxyethyl acrylate, 75
2,2-bis(hydroxymethyl) butyric acid, 3, 21, 217, 257

UV spectrum, 91

vegetable oils, 32, 33, 48, 49, 52, 53, 54, 57, 58, 59, 75, 86, 108, 109, 290, 299, 331, 437
viscosity, 3, 4, 5, 10, 23, 25, 49, 58, 67, 69, 70, 72, 88–91, 94, 102, 105–10, 115, 139–41, 162, 233, 256–8, 260, 268, 272, 292, 299, 308, 312, 337, 339, 349, 368, 370, 378, 382, 383, 389, 397, 434
volatile organic compounds (VOC), 1, 18, 30, 48, 66, 162, 194, 216, 232, 254, 308, 381

waterborne polyurethanes, 18, 32, 66, 119, 138, 178, 189, 232, 254, 334, 393, 394, 396, 397, 400, 401–3
wearable, 97, 189, 323, 325, 326, 335, 339, 341, 345
wound, 95, 96, 194, 195, 198, 199, 200, 201, 202, 217, 260, 387

Young's modulus, 11, 37, 74, 217, 384–6, 415

zwitterion, 2
zwitterionomers, 9

Printed in Poland
by Amazon Fulfillment
Poland Sp. z o.o., Wrocław
20 August 2023

61045c21-e2cc-4f50-b336-d9b4d6788222R01